MEDICAL MICROBIOLOGY

MEDICAL MICROBIOLOGY

S. RAJAN

Head, Department of Microbiology
Srimad Andavan Arts and Science College
Trichy, Tamil Nadu

MJP PUBLISHERS
Chennai 600 005

ISBN 10: 81-8094-029-2
ISBN 13: 978-81-8094-029-3

Cataloguing-in-Publication Data

Rajan, S (1965 –).
Medical Microbiology / by S. Rajan. –
Chennai : MJP Publishers, 2007
xxx, 652p. ; 23 cm.
Includes Glossary, References and Index.
ISBN 81-8094-029-2 (pbk.)
1. Microbiology, Medical. I. Title.
616.904 1 RAJ MJP 025

ISBN 81-8094-029-2

Printed and bound in India

MJP PUBLISHERS
47, Nallathambi Street
Triplicane
Chennai 600 005

Publisher	:	J.C. Pillai
Managing Editor	:	C. Sajeesh Kumar
Project Editor	:	P. Parvath Radha
Assistant Editors	:	B. Ramalakshmi, S. Revathi
Composition	:	M. Uma, Lissy John, N. Yamuna Devi
Cover Designer	:	M. Uma Maheswari
CIP Data	:	Prof. K. Hariharan

Preface

This book has been written with the aim to provide the basic aspects of medical microbiology in a concise and easily understandable format.

This book has been divided into 6 units which are organized in a hierarchical fashion as followed in the *Bergey's Manual of Systematic Bacteriology* and provides all necessary information about pathogenic microorganisms and the diseases they cause. Description includes causative agent of the disease, characters of pathogens, identifying features, pathogenesis and multiplication of viruses, life cycles of parasites, lab diagnosis, prevention and treatment. The first section provides general information like normal flora, host–microbe interactions, antimicrobial chemotherapy, vaccines, general identifying features and epidemiology of infectious diseases. This book also provides identifying features of bacteria, fungi, protozoa and helminthes based on microscopy, macroscopy, culture on media and biochemical tests.

This book would serve as a complete textbook for undergraduate students of microbiology, biotechnology, zoology, nursing, pharmaceutics and medicine. In addition, the practicals, spotters and model questions included at the end would help students to prepare for their exams. This book would also be a useful source book for preparation for competitive exams like ICMR, CSIR, UGC, NET, SLET and so on.

Constructive criticisms and suggestions for improvement of this book are most welcome.

S. Rajan

Acknowledgement

I wish to acknowledge the help and support offered by my friend Mr. S. Balakumar, Lecturer, Department of Microbiology, SASTRA University, Kumbakonam. I would like to express my sincere gratitude to my family members—amma, appa, brothers, and in-laws—for their constant encouragement. I am indebted to my wife, J. Selvichristy, Lecturer, Shrimathi Indira Gandhi College, without whose support this book would not have been possible. I am also grateful to Shri. V. S. Narasimhan, Secretary, Dr. P. Brindha, Dean, Life Science, Dr. K. Prema, Principal and my colleagues of Srimad Andavan Arts and Science College, for their constant encouragement. I am extremely grateful to my guide, Dr. T. Thirunalasundari, Senior Lecturer, NFMC, Department of Microbiology, Bharathidasan University for her encouragement and support.

I am greatly thankful to Mr. J.C. Pillai, Mr. C. Sajeeshkumar and the staff members of MJP Publishers, Chennai for their encouragement, and the support rendered by them for the successful completion of this project.

S. Rajan

Contents

Part I General Aspects of Medical Microbiology

Part IV Mycology

PART I

GENERAL ASPECTS OF MEDICAL MICROBIOLOGY

1 Introduction and Historical Developments in Microbiology

INTRODUCTION

Microbiology is a branch of life science, which deals with the study of microscopic life including all microorganisms. Microorganisms are not visible to the naked eye and are visible or observed only under a microscope. Microbiology is a new field, which has emerged from the basic biological sciences like botany and zoology. After the discovery of microscopes, developments in microbiology started in the late 19th century but advancements were observed only from the middle of the 20th century.

One of the important health-oriented field in Microbiology is Medical Microbiology, also called Physician-oriented microbiology. It deals with the general nature of disease-causing agents, pathogenesis, epidemiology, lab diagnosis, prevention and treatment. Diagnostic microbiology is the part of medical microbiology, which deals with the isolation and identification of specific microorganisms from the patients as per the instructions given by the physician.

Microbiology is a relatively new discipline and its development is described below.

During the last century, people gained some knowledge about microbiology. Now microbial importance is well known to mankind. Ancient people regarded diseases as having a supernatural origin and sent by God as punishment for the sins of human beings. Varro recorded the diseases in living creatures in the 2nd century BC. Fracastoro (1546) suggested that invisible organisms cause diseases.

Various biologists made a creative contribution on various diseases and its transmission but first clear observations on microorganisms were made by Antoni van Leeuwenhoek (1676). He is considered as the Father of Microbiology due to his pioneering contribution in this field. He was a merchant, Haber Dasher, and owner of a dry goods shop in Delf, Holland. He was a qualified surveyor and the town's official wine taster. In his spare time, he ground pieces of glass into fine lenses. He was not an educationalist but had a keen mind. Magnification of Antoni's

Antoni van Leeuwenhoek

microscope was between 50 and 300 times. He observed minute objects through his lens and named them as animalcules. He observed microbes from rainwater, pepper infusions, saliva, tooth scrapings and excreta. He communicated his findings to the Royal Society of London and the observations were published in 1677 in the *Proceedings of the Royal Society* as a series of letters. In 1680, he was elected as a Fellow of the Society.

After the discovery of microscope by Antoni van Leeuwenhoek, there was no progress in the field of microbiology up to the 19th century because of the lack of developments in microscopy and of the belief in the spontaneous generation theory.

SPONTANEOUS GENERATION

Francesco Redi

The generation of living matter from non-living matter is known as spontaneous generation or abiogenesis.

Aristotle (384–322 BC) said that animals could originate from soil.

Francesco Redi (1668) demonstrated that maggots (fly larvae) do not arise spontaneously from decaying meat by performing the jar experiment. Three sealed jars and three open jars, all containing meat were used. All the meat decayed, but only the open jars produced maggots. The counter-argument to this was that air was necessary for generation. A second experiment in which the open jars were covered with a fine net showed production of maggots on the net, but none on the meat, even though air was present. This was a blow to spontaneous generation, but many believed that the small "animalcules" observed by Antoni van Leeuwenhoek were simple enough to arise by spontaneous generation and others believed that the experimental protocols disturbed the generative forces.

John Needham (1749) proved spontaneous generation. Needham boiled mutton and then tightly stoppered the flasks. Many of the flasks became cloudy. He thought organic matter contained a vital force that could confer the properties of life on non-living matter.

Pouchet (1859) argued that access to air is necessary for spontaneous generation.

Lazzaro Spallanzani (1769) provided evidence that a microbe does not arise spontaneously.

The doubts of Spallanzani were cleared by the experiments conducted by the following scientists:

Franz Schulze (1815–1873) passed air through concentrated sulphuric acid to disprove spontaneous generation.

Theodore Schwann (1810–1882) passed air to enter a flask containing a sterile nutrient solution through red-hot tube to disprove spontaneous generation. The flask remained sterile.

In 1859, a controversy of Pouchet experiment was observed. During this time Louis Pasteur prepared boiled broth in a flask with long, narrow and goosenecked tubes that are open to air. There was no growth because microbes settled in the gooseneck region. At the same time he tilted the flask several times so that sterile broth interacts with air microbial flora. He allowed the flask for 24 hours and observed bacterial growth. Based on this experiment, he disproved this theory (Figure 1.1).

Trapped air escapes from open end of flask

Bacteria and dust settle in the bend

Hours/day

1. Broth sterilized

2. Broth allowed to cool slowly

3. Broth stays sterile indefinitely

4. Flask tilted so that sterile broth interacts with bacteria and dust from the air

5. Bacteria grow and multiply in broth

Figure 1.1 *Pasteur's experiment to disprove spontaneous theory*

Finally John Tyndall (1820–1893) proved that dust carried germs and the spontaneous generation theory was disproved. He also showed evidence for the existence of heat-resistant forms of bacteria.

GERM THEORY OF DISEASE

Causal nature of infectious disease was first established only in the latter half of the 19th century. In the early 1800s, Agostino Bassi (1835–1844), proved that a fungus causes a disease in silkworm called muscardine in France. In the 1840s, Oliver Wendell Holmes wrote about contagiousness of puerperal fever. Ignaz Semmelweis (1847) postulated that blood victims of puerperal fever contained the causative agent of the disease. In 1861, Ignaz Semmelweis introduced the antiseptic technique, and it was not fully realized until the late 1870s, when Joseph Lister demonstrated the value of spraying operating rooms, with phenol. In 1880, Pasteur demonstrated diseases of wine. In 1850, Pollander, Rayer and Davine observed rod-shaped bacteria in the blood of animals (anthrax). In 1857 Bauell found anthrax bacilli in cattle and that they transmitted the disease to sheep. Robert Koch was the first biologist, to demonstrate the role of bacteria in the disease-causing process.

Robert Koch was a German physician and he is called as the Father of Medical Microbiology because of his significant contributions to the field of medical microbiology. He proposed the germ theory of disease in 1876 based on his experiment on transmission of anthrax in animals. He also contributed several findings to the field of medical microbiology which are listed in Table 1.1.

Robert Koch

Table 1.1 Contributions of Robert Koch

Year	Contribution
1876	Demonstrated the causative nature of anthrax
1881	Cultured bacteria on gelatin
1881	Used agar to demonstrate pure culture technique
1882	Discovered tubercle bacilli
1884	Published Koch's postulates
1883	Discovered cholera bacilli

Koch's Postulates

1. The microbes must be present in every case of the disease.
2. The suspected microorganism must be isolated and grown in a pure culture.
3. The same disease must result when the isolated microbe is inoculated into a healthy host.
4. The same organism must be isolated again from the infected host (Figure 1.2).

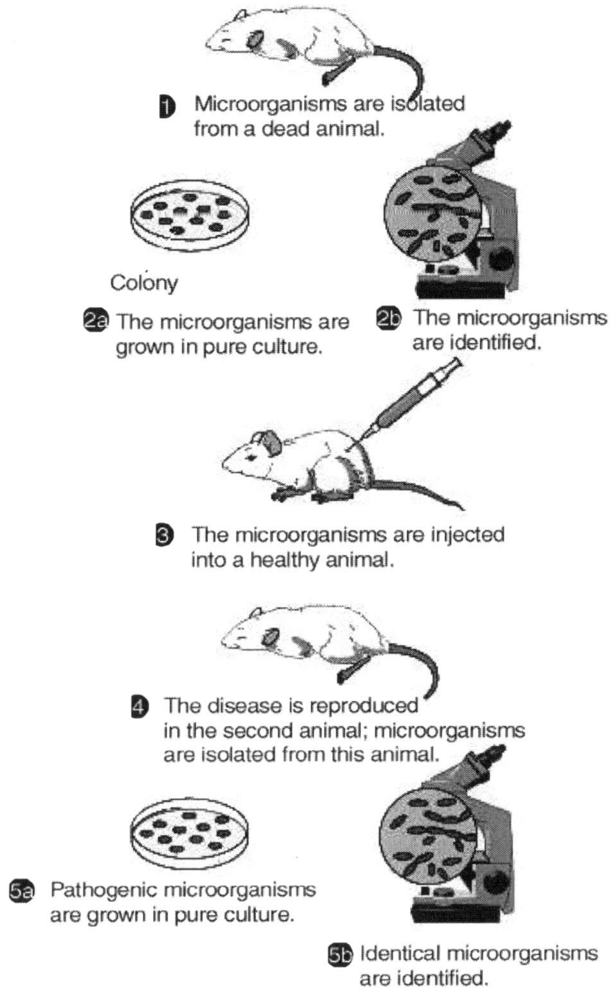

Figure 1.2 Demonstration of Koch's postulates

Due to scientific advancement, development of drugs for treatment and vaccination, some of Koch's postulates cannot be accepted. The exceptions to Koch's postulates are listed in the Table 1.2.

Table 1.2 Exceptions to Koch's postulates

Exceptions	Examples
Some diseases are caused by opportunistic pathogens.	Comedo, acne vulgaris
If the animals are immunized with the pathogen, the disease does not arise.	Chickenpox
Some diseases require cooperation between the pathogens.	Impetigo
Some organisms cannot be obtained in pure culture.	*Treponema pallidum*

LOUIS PASTEUR

Pasteur was a Professor of Chemistry at the University of Lille, France. He was the father of Fermentation Technology. He was the first person to disprove the theory of spontaneous generation by making use of swan-neck flask experiment (1859). He took up the problem of wine industry and introduced the process of pasteurization (1860). He demonstrated that microscopic germs are responsible for silkworm disease (1865). He also introduced hot-air sterilization of glassware.

Louis Pasteur

Other discoveries of Louis Pasteur are as follows:

- Introduced vaccination against anthrax (1881).
- Introduced vaccine for rabies (1885). The term "vaccine" was given by Pasteur to honour Edward Jenner.
- Sugar materials were broken down into simple alcohol by yeasts (1857).
- He showed that lactic acid fermentation is due to microorganisms.

Life without air is called fermentation.
Animals do not survive in the absence of microorganisms.

GENERAL DEVELOPMENTS IN THE FIELD OF MICROBIOLOGY

Biologists develop hypotheses on problems faced by them. They find solutions to their problems by guessing or by scientific experiments. Many microbiologists discovered new findings. Some of them were awarded Nobel Prizes. Fracastoro suggested that invisible organisms cause diseases. Leeuwenhoek microscope magnification was not enough to study all the microorganisms and new microscopes were invented. General developments in microbiology are presented in Table 1.3. The period between 1860 and 1910 is considered as the Golden Age of Microbiology because numerous discoveries were made by many scientists, especially by Louis Pasteur and Robert Koch.

Table 1.3 Developments in microbiology

Year	Scientist	Discovery
1590–1610	Jensen and Hans	Designed the first compound microscope
1786	Mueller	First classified the bacteria
1805	Nicholas Appert	Preserved soups and liquids by heating them in thick bottles (appertization)
	Cohn	Discovered bacterial endospores

(Contd.)

Table 1.3 (Continued)

Year	Scientist	Discovery
1854	John Snow	Showed the reason for cholera outbreak. It is due to contamination of municipal water. He was called the Father of Epidemiology.
1859	Darwin	Postulated about origin of species
1865	Joseph Lister	Introduced aseptic techniques
1870	His	Developed microtome for cutting sections of tissue cells
1878	Ernst Karl Abbe	Introduced oil immersion lens
1884	Fannie Elishemius Hesse, wife of Walter Hesse	Used agar agar as a solidifying agent
1884	Christian Gram	Invented gram-staining technique
1886	Ernst Karl Abbe	Discovered Abbe condenser
1886	Mac Munn	Discovered cytochromes
1887	Richard Petri	Devised petri plate
	Charles Chamberland	Discovered bacterial filter and autoclave
	Ferdin and J.Colin	Discovered the multiplication of bacteria
1890	Sergei Winogradsky	Discovered auxotrophic growth of chemolithotrophs
1894	Ehrlich	Articulated the principle of selective toxicity
1897	Buchner	Discovered cell-free alcohol fermentation
1901	Martinus Beijerinck	Developed enrichment culture methods
1903	Buchner	Discovered enzymes
1906	Tswell	Invented chromatography
1907	Harrison	Developed tissue culture techniques
1912	Carel	Devised the technique for tissue culture
1926	Svedberg	Developed ultracentrifugation
1937	Krebs	Discovered TCA cycle
1932	Knoll and Ruska	Discovered electron microscope
1953	Zernike	Discovered phase contrast microscope
1954	Salk	Introduced inactive polio virus vaccine
1977	Carl Woese	Discovered the Archaeae
1978	Carl Woese	Devised classification of microorganisms
1980		WHO proclaimed smallpox an extinct disease

(Contd.)

Table 1.3 (Continued)

Year	Scientist	Discovery
1986		First hepatitis B vaccine produced by genetic engineering and approved for human use
1995		*Haemophilus influenzae* gene sequenced
1996		Yeast gene sequenced
1997		Discovery of *Thiomargarita namibiensis,* the largest bacterium in the world
1997		*E.coli* gene sequenced
2000		Discovery of the fact that *V.cholerae* has two separate chromosomes

DEVELOPMENTS IN CHEMOTHERAPY

Many diseases are treated with chemotherapeutic agents. They kill pathogens in a specific way and are called magic bullets. Modern chemotherapy begins with the work of Paul Ehrlich, who coined the term chemotherapy and he is considered as the Father of Chemotherapy. Alexander Flemming, Chain and Florey in 1952 were awarded the Nobel Prize for their discovery of penicillin and Waksman for streptomycin discovery in 1952. Other developments are given in Table 1.4.

Table 1.4 Developments in chemotherapy

Year	Scientist	Discovery
1910	Paul Ehrlich	Developed Salvarson, a magic bullet for syphilis
1929	Alexander Flemming	Discovered penicillin
1935	Domag	Discovered sulpha drugs
1944	Waksman	Discovered streptomycin
1962		Discovery of nalidixic acid
1990	Murry and Johnson	Used immunosuppressive agents to perform sucessful transplantation

DEVELOPMENTS IN IMMUNOLOGY

Immunological developments are directly related to medical microbiology. Here scientists develop diagnostic methods and prophylactic measures to cure infections. Edward Jenner laid the foundation by immunizing a boy with cowpox pustules. Followers of Jenner contributed much important findings, which are given in Table 1.5.

Table 1.5 Developments in immunology

Year	Scientist	Discovery
1798	Edward Jenner	Developed smallpox vaccine
1885	Louis Pasteur	Produced rabies vaccine
1884	Metchnikoff	Discovered phagocytosis
	Jules Bordet	Developed complement fixation test (CF)
	Almorth Wright	Described the role of opsonin in phagocytosis
1903	Wright	Discovered antibodies in blood of immunized animals
1913	Richet	Worked on anaphylaxis
1921	Flemming	Discovered lysozyme
1930	Landsteiner	Discovered ABO blood grouping
1931	Lewis	Described pinocytosis
1953	Medavar	Discovered immunotolerance
1955	Jerne and Burnet	Proposed the clonal selection theory
1957	A.Isaacs and J.Lindermann	Discovered interferon
1959	Yalow	Developed the radioimmunoassay technique
1963	Porter	Proposed the structure of IgG
1972	G.Edelman, R.Porter	Researched on the structure of antibodies
1975	Kohler and Milstein	Developed the technique for monoclonal antibody synthesis
1980	Bena Cerraft, G.Snell and J.Daussel	Discovered histocompatibility antigen
1987	Susumu Tonegawa	Discovered the genetic principle for generation of antibody diversity
1996	P.Doherty and R.Zinkerngel	Discovered the mechanism by which T lymphocytes recognize virus-infected cell

DEVELOPMENTS IN MEDICAL BACTERIOLOGY

Hanson was the first person to demonstrate the causative agent of leprosy but he was not able to demonstrate the process of infection because of the lack of culture medium for growth. The First direct demonstration of the role of bacteria in causing disease came from the study of anthrax by Robert Koch. Many new findings were observed between 1884 and 1910. Table 1.6 explains various discoveries in bacteriology up to 1997.

Table 1.6 Developments in bacteriology

Year	Scientist	Discovery
1871	Gerhard Hansen	Discovered the causative agent of leprosy (*Mycobacterium leprae*)
1873	Otto H.F. Obermier (Germany)	Discovered the bacteria responsible for relapsing fever
1884	George Gaffley	Cultivated typhoid bacilli
1884	Loeffler	Cultivated diphtheria bacilli
	Kitasato	Discovered tetanus bacilli
	Albert Calmette	Discovered non-virulent strain of TB bacilli
	Rose	Discovered malaria-causing agent
1892	William Welch	Identified gas gangrene bacilli
1887	Richard Pfeiffer	Identified the cause of meningitis
1894	Kitasato and Yersin	Identified plague-causing agent
1898	Kiyoshi Shiga	Discovered *Shigella dysenteriae*
1876	Koch	Discovered *Bacillus anthracis*
1879	Neissen	Discovered *Neisseria gonorrhoeae*
1885	Escherich	Discovered *E. coli*
1882	Koch	Discovered *Mycobacterium tuberculosis*
1883	Koch	Discovered *Vibrio cholerae*
1886	Fraenkel	Discovered *Streptococcus pneumoniae*
1887	Bruce	Discovered Brucella
	Van Ermengem	Discovered *Clostridium botulinum*
1905	Schandinn and Hoffmann	Discovered *Treponema pallidum*
1906	Bordet and Gengou	Discovered *Bordetella pertussis*
1909	Rickets	Discovered *Rickettsia rickettsi*
1912	McCooy Chapin	Discovered *Francisella tularensis*
1975		Discovery of Lyme disease
1977		Discovery of legionnaire's disease caused by *Legionella pneumophila*
1977		Discovery of enteric disease caused by *Campylobacter jejuni*
1981		Discovery of toxic shock syndrome
1982		Discovery of haemorrhagic colitis caused by *E. coli* O157:H7
1982		Discovery of peptic ulcer by *Helicobacter pylori*
1997		Discovery of paediatric infection by *Kingella kingae*

DEVELOPMENTS IN MEDICAL VIROLOGY

Many believed that bacteria are the etiological agent for diseases until the middle of 19th century. In 1892, Ivanowsky, a Russian scientist, discovered tobacco mosaic virus (TMV) from the leaves of tobacco plant and Beijerinck confirmed this in 1898. During 1932, Ruska discovered the electron microscope, which helped the scientists to study the viruses extensively. Contributions related to viruses are presented in Table 1.7.

Table 1.7 Developments in virology

Year	Scientist	Discovery
1900	W. Reed	Discovered yellow fever causative agent
1903	P. Remlinger	Discovered rabies virus
1903	A. Negri	Demonstrated inclusion bodies of rabies virus
1908	V. Ellermann and O. Bang	Identified avian leukemia causative virus
1911	P. Rous	Identified Avian sarcoma causative virus
1932	Furth	Used mice as a host for virus
1933	R.E. Shope	Identified a virus causing mammalian cancer
1939	G.A. Kausche	Visualized virus under electron microscope
1949	J.F. Enders	Developed human cell cultures for the growth of poliovirus
1953	W.P. Rowe	Discovered adenovirus
1955	F.L. Schaffer	Crystallized poliovirus
1959		Parvovirus discovered
1963		Discovery of occurrence of double-stranded RNA virus
1981	Lun Montagnier	Discovered HIV
1997	S.Prusiner	Discovered prions
2003		Identification of SARS (severe acute respiratory syndrome) being caused by corona virus

Protozoans, nematodes. trematodes, cestodes and fungi cause different types of infections in humans. Developments in the field of parasitology and mycology are presented in Table 1.8.

Table 1.8 Developments in parasitology and mycology

Year	Discovery
Parasitology	
1897	Ross discovered that malarial parasite was transmitted by mosquitoes for which he was awarded the Nobel Prize.
1976	Cryptosporidiosis
1985	Picrosporidiosis
1991	Bebiasis
1998	Myositis-Brachiola
Medical Mycology	
1892	Coccidioidomycosis
1931	Lobamycoses-*Loba lobai*
1956	Phycomycosis
1980	Blastomycosis
1990	Rhinosporidiosis

1. Explain various contributions in the field of bacteriology.
2. Discuss about Nobel Prize winners in the field of microbiology.
3. Discuss the contributions of
   i. Antoni van Leeuwenhoek
   ii. Robert Koch
   iii. Louis Pasteur
   iv. Francesco Redi
   v. Edward Jenner
   vi. Ivanowsky
4. Write short notes on
   i. Koch's postulates
   ii. Spontaneous generation
   iii. Pure culture
   iv. Animalcules

1. What is the importance of pure culture in the field of microbiology?
2. What is the reason behind the study of historical developments?
3. Explain the relationship of microbes with diseases with an example.
4. Do you accept all concepts discussed in this chapter?

Brook,T.D. *Milestones in Microbiology*. Prentice Hall, London. 1961.

Michael, J. Pelczar, Chan, E.C.S. and Noel, R. Crieg. *Microbiology*. Tata McGraw-Hill Publishers Company Ltd. 1995.

Rajesh Bhatia and Rattan Lal Ichpujani. *Essentials of Medical Microbiology,* 3rd edn. Jaypee Publishers. 2004.

www.biology.clc.uc.edu

www.kvo27.com

www.nmm.ac.uk

www.hhmi.princeton.edu

www.krownspellman.com

www.science-art.com

www-micro.msb.le.ac.uk

www.slic2.wsu.edu:82

www.users.stlcc.edu/kkiser

www.infochembio.ethz

www.histmicro.yale.edu

www.angelfire.com

www.micro.cornell.edu

www.bact.wisc.edu

www.austehc.unimelb.edu

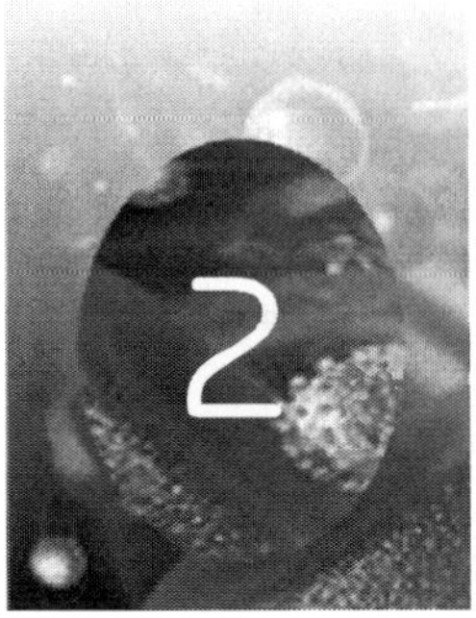

Normal Flora of the Healthy Human Host

INTRODUCTION

There are countless microorganisms in the environment. Human beings encounter these microbes continuously. But our intimate contact is with the large number of microbes that actually live in and on our bodies. It is estimated that the adult human body is a host to at least 100 trillion microbial cells at any time. These indigenous microbes, most of which are bacteria, that inhabit the human body are called the normal flora.

GENERAL FEATURES OF NORMAL FLORA

- Normal flora inhabits the skin and some of the inner parts of the body.
- The species and number of the flora vary according to the particular sites and particular age of the host.
- Sex of the host may also influence the nature of normal flora.
- Normal flora benefits from the host but the host is not adversely affected.
- Many sites of the human body are free from microbes. These include the cerebrospinal fluid, blood, urinary bladder, uterus, fallopian tubes, middle ear, paranasal sinuses and kidneys.
- Normal flora are also called residents of the healthy human host and are permanent occupants. Transients may establish brief contact with the human body, but are excluded by competition with the residents or by the host immune system.
- Some normal flora may become pathogenic when the host defence becomes faulty or when any damage is caused to the human body.
- The normal flora may increase the overall immune status of the body against pathogens having related or shared antigens.

Nature of normal flora depends on the environment, hygienic practices and frequency of washing.

A knowledge of the normal flora of the human body is essential, for several reasons. Four specific reasons include:

1. An understanding of the microbes at specific locations provides greater insight into possible infections.
2. A knowledge of the native organisms in the infected part of body gives the physician-investigator a better perspective concerning the possible source and significance of microorganisms isolated from the site.
3. A knowledge of the indigenous microbiota helps the physician understand the causes and consequences of overgrowth of microorganisms normally absent at specific body sites.
4. An increased awareness of the role that these indigenous microbiota play in stimulating the host immune response can be gained. This awareness is important because the immune system provides protection against potential pathogens.

ORIGIN OF THE NORMAL FLORA

Before birth, a healthy human foetus is free of microbes. Within hours after birth, it begins to acquire a normal microbiota, which stabilizes during the first week or second week of life. From then on, numerous of microorganisms associate with the human body.

GERM-FREE LIFE

Louis Pasteur first suggested that "animals could not live in the absence of microbes". Between 1899 and 1908, scientists raised germ-free chicken. It had limited success because the animals died within a month. Thus it was believed that intestinal bacteria were essential for adequate feeding and health of the chicken. In 1912, 17 germ-free chicken were raised at Pasteur Institute of France. In 1928, James A. Reyniers at the University of Notre Dame in the United States began his work on germ-free animals. He and his colleagues reared several generations of chicken, rats, mice and other animals in the absence of microorganisms.

Germ-free animals or those living in association with one or more known microorganisms are called gnotobiotic. Establishing germ-free colony of rats, mice, guinea pigs or monkeys begins with caesarean sections on pregnant females. The operation is performed under aseptic conditions in a germ-free isolator. The newborn animals are then transferred to other germ-free isolators in which all air, water and food entering are sterile. Once germ-free animals become established under these conditions, normal mating among themselves maintains the germ-free colony.

Characteristics of Germ-free Animals

Germ-free animals are not anatomically and physiologically normal. They possess the following characteristics:

1. Poorly developed immune system
2. Thin intestinal wall
3. An enlarged caecum
4. Low antibody titre
5. Requirement of high amount of vitamin K and B complex
6. Low level of cardiac output and lower metabolic rates
7. High susceptibility to diseases
8. Complete resistance to *Entamoeba histolytica*
9. Resistance to dental caries (if it takes contaminated sucrose-containing food, it will develop dental caries).

Germ-free animals and gnotobiotic techniques provide a good experimental system to investigate the interaction of animals and specific microorganisms. The animals are reared in the presence of one or more known microbial species in order to determine the effect of these organisms on the growth and development of various physiological processes.

NORMAL FLORA OF VARIOUS PARTS OF THE BODY

Skin

Skin is one of the most clearly understood regions in the body. It provides a good barrier system and is considered as the first line defence system in the body. It has epidermis, dermis, subcutaneous connective tissue, sweat glands, sebaceous glands, hair follicles, etc. (Figure 2.1). The skin varies widely in structure and function, depending on its location on the body. These differences determine the types and numbers of microbes that occur on each skin site. Commensal microbes living on or in the skin can be either resident or transient microbiota. Resident microbiota normally grow on or in the skin. Constant exposure to the environment means the skin harbours many transient microbes.

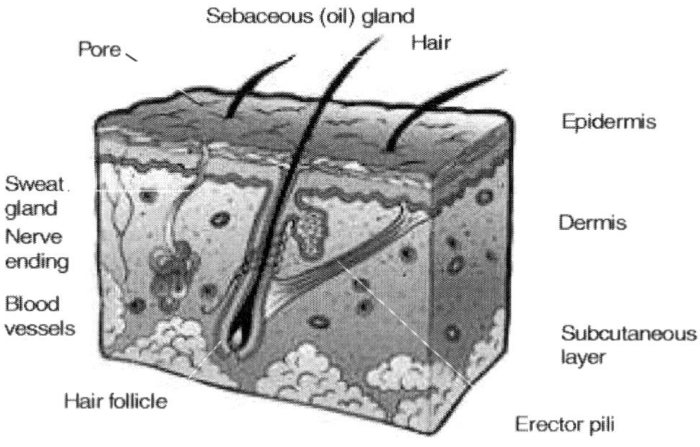

Figure 2.1 Anatomy of human skin

Transient flora of the skin include

- Pathogenic yeast
- *Streptococcus* spp.
- Ringworm fungi
- Coagulase-negative *Staphylococcus*
- Soil organisms
- Gram-negative bacilli
- Gram-positive cocci

Transient microbes are unable to multiply and normally die in a few hours. Several factors are responsible for discouraging microbial growth on the skin.

Dryness The relatively dry surface of the skin is inhibitory to microbial growth. When skin is allowed to dry, many bacteria enter a dormant condition, some species die within an hour. Moist regions of the skin have high number of normal flora.

Low pH The normal pH of the skin is about 3–5. Some organisms produce an organic acid that reduces pH. This low pH can inhibit the growth of many kinds of microorganisms.

Inhibitory substances Sweat glands contain high concentrations of sodium chloride. This makes the skin surface hyperosmotic and it osmotically stresses microorganisms. Sweat glands secrete lysozyme, which lyses gram-positive bacterium. Sebaceous glands secrete complex lipids that may be partially degraded by the enzymes of gram-positive bacteria and forms unsaturated fatty acids such as oleic acids (it has strong antimicrobial activity). Some of the fatty acids are volatile in nature. It contributes to body odour.

In spite of these positive aspects of the skin, there are also some negative aspects that encourage growth of microorganisms. Excretions from sebaceous and sweat glands provide water, amino acids, urea, electrolytes and fatty acids that serve as nutrients primarily for *Staphylococcus* and *Corynebacterium*. The most prevalent bacterium in skin glands is the gram-positive anaerobic lipophilic rod *Propionibacterium acnes*. This bacterium is usually harmless. Due to the over-secretion of sebaceous and sweat glands, *P. acnes* growth is enhanced. This leads to the disease acne vulgaris. Accumulation of large amounts of sebum provides the microenvironment for the growth of normal flora. It also leads to an inflammatory response that causes redness and swelling of the gland duct and produces comedo (a plug of sebum and keratin in the duct). Inflammatory lesions are commonly called "blackheads or pimples".

The normal biota of the skin are listed in Table 2.1 given at the end of the chapter.

Eye

Lining the eyelids and covering the eyeball is a delicate membrane called conjunctiva (Figure 2.2). It is a continuation of the skin. Flow of tears, removes microorganisms from eyes. Lysozyme, a microbicidal substance that inhibits the growth of microbes is present in tears. At birth and throughout human life, bacterial commensals are found in the conjunctiva of the eye. The predominant organisms found are *Staphylococcus epidermidis*, *Staphylococcus aureus*, *Corynebacterium* spp., *Streptococcus pneumoniae*, *Neisseria* spp., *Moraxella*, *Branhamella* spp., *Escherichia* spp., *Klebsiella* spp., *Haemophilus* spp., *Proteus* spp., *Enterobacter* spp., *Bacillus* spp., and few anaerobic bacteria.

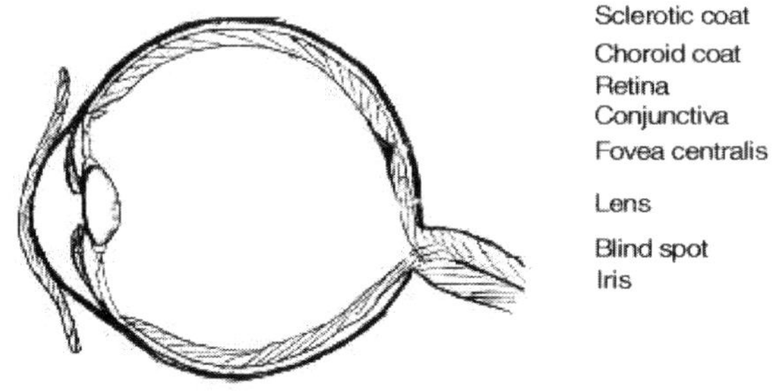

Figure 2.2 Anatomy of human eye

Respiratory Tract

The respiratory tract includes nose, tonsils, nasopharynx, throat, trachea, bronchi and lungs (Figure 2.3). The upper respiratory tract consists of nose, tonsils, nasopharynx and throat. The mucous membrane of the upper respiratory tract is more moist than skin, nevertheless, they can create more problems to microorganisms.

Staphylococcus aureus and *Staphylococcus epidermidis* are the predominant flora of the nose and sometimes it has some microflora similar to the skin of face. Nasopharynx that forms a part of pharynx may contain a few numbers of *Streptococcus pneumoniae*, *Neisseria meningitidis* and *Haemophilus influenzae*. Most of these organisms lack a capsule. Diphtheroids, a large group of gram-positive organisms, are present in the nose and nasopharynx.

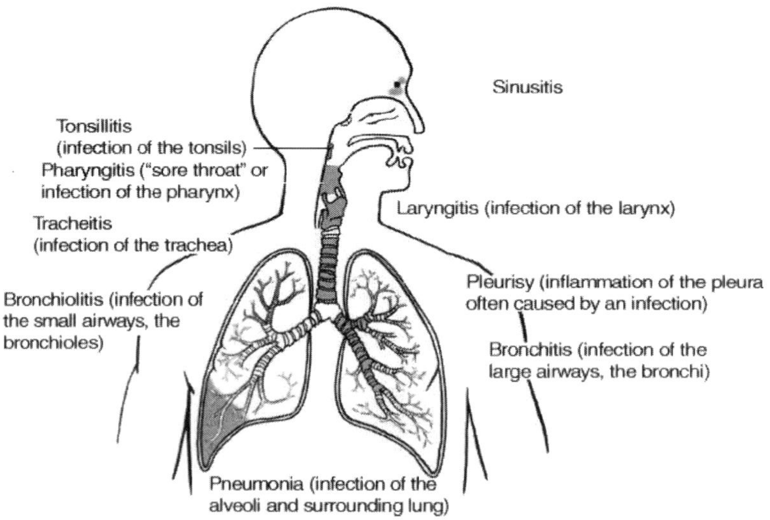

Figure 2.3 Respiratory tract

Oropharynx is a division of pharynx. Important bacteria found in this area are the various alpha haemolytic streptococci (*S. oralis, S. milleri, S. gordonii, S. salivarius*), large number of diphtheroids, and *Branhamella*, small gram-negative cocci. Tonsil harbours a microbiota similar to oropharynx but within the tonsillar crypts, there is an increase in *Micrococcus* and anaerobes (*Porphyromonas* and *Fusobacterium*).

Because of the rhythmic beating of cilia of the nasopharynx, mucous membranes of nasopharynx continuously flow organisms towards the oropharynx. The trapped bacteria are eventually swallowed and may be destroyed by HCl of stomach. In addition to this mechanical removal of bacteria, the enzyme, lysozyme, in the nasal mucosa also kills bacteria.

Normally the lower respiratory tract does not have any normal microbiota, because of the efficient mechanical removal by mucus and cilia. Most of the bacteria usually are engulfed and destroyed by the phagocytic action of alveolar macrophages.

Oral Cavity

Normal microbiota of the oral cavity consists of those organisms that are able to resist mechanical removal by adhering to surfaces like gums and teeth. Oral cavity serves as an ideal environment for microbes because it provides moisture and food material. However, continuous desquamation of epithelial cells, flow of saliva and mechanical flushing action removes microbes from the oral cavity.

At birth the oral cavity is essentially a sterile, warm and moist cavity containing a variety of nutritional substances. The saliva is composed of water, amino acids, proteins, lipids, carbohydrates and inorganic compounds. The newborn establishes its normal flora within few days after birth. The oral cavity is colonized by microorganisms from the external environment. Initially, the microbiota consists mostly of the genera *Streptococcus*, *Lactobacillus*, *Neisseria* and *Actinomyces*. Some yeast is also present. The number and kind of microbes present in the mouth depends on the diet and association with other people, objects such as towels and feeding bottles. The only species isolated from the mouth throughout the life is *S. salivarius*. It has an affinity to epithelial tissues and appears in large numbers on the upper surface of the tongue.

Teeth

The anatomy of teeth is given in Figure 2.4. Until eruption of teeth, most microbes present in the mouth are aerobes and facultative anaerobes. As the first tooth erupts, the anaerobes become dominant due to the anaerobic nature of the gingival groove. As the first tooth grows, *Streptococcus gordonii* and *S. mutans* attach to their enamel surface, *S. salivarius* attaches to the buccal and gingival epithelial surfaces and colonizes the saliva. These streptococci produce various adherence factors. These factors enhance the attachment of oral bacteria. The aggregation of bacteria on organic matter of the teeth surface is termed dental plaque. These organisms allow the secondary invaders and cause dental caries, gingivitis and periodontal diseases. Periodontal disease is caused by several groups of bacteria.

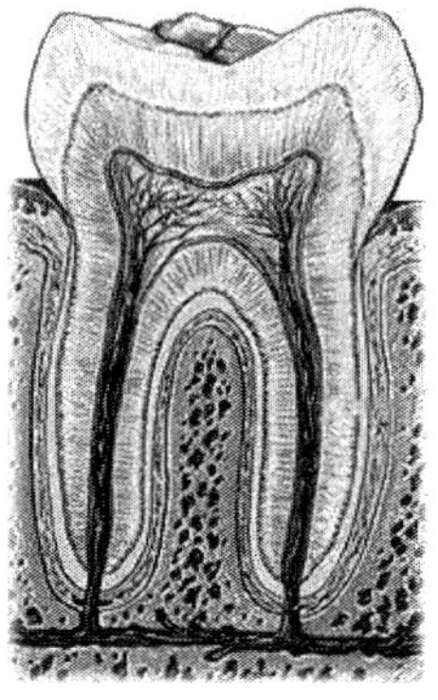

Figure 2.4 Anatomy of teeth

Gastrointestinal Tract

Stomach, small intestine and large intestine are the parts of gastrointestinal tract (Figure 2.5).

Stomach Stomach receives different transient organisms from the oral cavity. The stomach usually contains less than 10 viable bacteria per millilitre of gastric juice. This is due to the bactericidal effect of gastric hydrochloric acid and digestive enzymes. Organisms present in stomach include staphylococci, streptococci, *Lactobacillus, Peptostreptococcus* and yeast. Microorganisms may survive if they pass rapidly through the stomach or if the organisms ingested with food are particularly resistant to gastric pH. Changes in gastric microbiota also occur if there is any increase in gastric pH, which leads to stomach upset. If pH is increased the stomach is filled with both gram-negative anaerobic and aerobic bacteria. Following the ingestion of food, the number of microbes increases but afterwards gradually decreases.

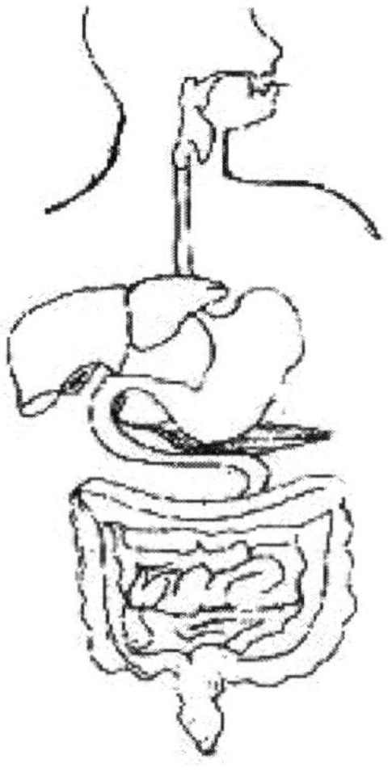

Figure 2.5 Gastrointestinal system

Small intestine The small intestine is divided into three areas—duodenum, jejunum and ileum.

Duodenum contains few microorganisms because the combined action of the strongly acidic environment of the stomach and the inhibitory action of bile and pancreatic secretions restrict their numbers. Of the bacteria present, gram-positive cocci and rods predominate. *Enterococcus faecalis*, lactobacilli, diphtheroids and *Candida albicans* are occasionally found in the jejunum. In the ileum, the microbiota begins to take on the characteristics of the colon microbiota. It is within the ileum that the pH becomes more alkaline. As a result, anaerobic gram-negative bacteria and the members of the family Enterobacteriaceae become established.

Large intestine (Colon) Colon has the largest microbial population in the body. It has been estimated that the number of microorganisms in stool specimen is 10^{12} organisms per gram wet weight, which means that about 25% of faeces is made up of microorganisms. Over 300 different bacterial species have been isolated from human faeces. It has been estimated that an adult excretes 30 million bacterial cells daily through defaecation.

The colon is considered as a large fermentation vessel because it contains anaerobic, gram-negative non-sporing bacteria and gram-positive spore-forming and non-spore-forming bacteria and yeast. Protozoa may occur as harmless commensals. *Trichomonas hominis*, *Entamoeba hartmanni*, *Endolimax nana* and *Entamoeba butschli* are common inhabitants. These organisms do various functions like fermentation.

Various factors tend to remove microorganisms from the large intestine. One factor is the peristaltic movement of the intestinal contents. Desquamation of surface epithelial cells to which bacteria are attached is another factor. The third factor is the mucus, which removes microbes by mechanical process as in the respiratory tract.

The initial residents of the colon of breast-fed infants are members of the gram-positive *Bifidobacterium* because human milk contains a disaccharide amino sugar which is required for the growth of *Bifidobacterium*. In formula-fed infants, *Lactobacillus* species predominate. Composition of the normal flora of the intestine can be influenced by various factors such as strong emotional stress and starvation.

Intestinal bacteria contribute to intestinal odour by producing skatole, amines and gases (CO_2, H, CH_2 and H_2S). The bacteria produce an average of 8.5 litres of gas daily.

Genito-urinary Tract

The upper genito-urinary tract (kidney, ureters and urinary bladder) is usually free of microbiota. The regions of the genito-urinary tract that harbour microflora are the vagina and outer opening of the urethra in female and anterior opening in the males (Figure 2.6a and b). The internal reproductive organs are kept sterile through physical barriers such as the cervical plug and other host defences. The kidney, ureter, bladder and upper urethra are kept sterile by urine flow and regular bladder emptying. Since the urethra in women is so short (about 3.5 cm long) it can form a passage for bacteria to the bladder and lead to frequent urinary tract infections (UTI).

The pH of adult female genital tract (Figure 2.6a) is about 4.4–4.6. This is due to the action of the hormone oestrogen. It stimulates the vaginal mucosa to secrete glycogen, which is metabolized by *Lactobacillus acidophilus*, (often called Doderlein's bacilli) forming lactic acid.

Normal microbiota of the urethra are coagulase-negative staphylococci, streptococci, *Mycobacterium, Bacteroides*, *Fusobacterium,*, *Peptostreptococcus* spp., diphtheroids, etc. Normal microbiota of the vagina *are Lactobacillus,* diphtheroids, *Peptostreptococcus, Streptococcus, Clostridium, Candida* and *Gardnerella vaginalis.*

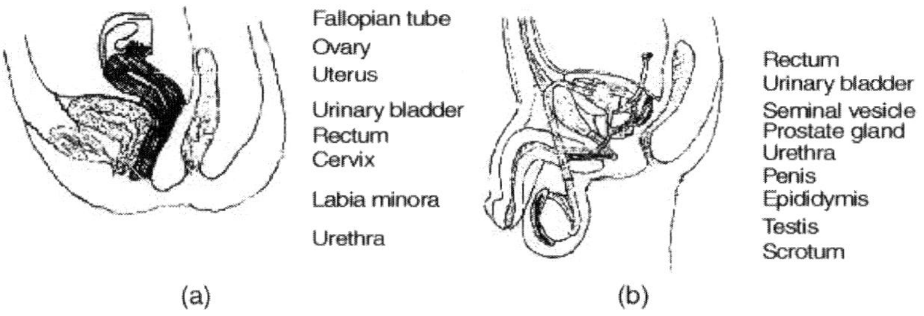

Figure 2.6 Anatomy of human reproductive organs. (a) female (b) male

External Ear

The anatomy of the human ear is given in Figure 2.7. Normal microbiota resemble that of skin with coagulase-negative staphylococci and *Corynebacterium*. Less frequently found are *Bacillus, Micrococcus* and *Neisseria* spp. Gram-negative rods such as *Proteus, E. coli, Pseudomonas* are occasionally present.

Once established, the normal microbiota can be of benefit to the host by preventing the overgrowth of harmful microorganisms. This phenomenon is called microbial antagonism. The relationship between host and normal flora is called symbiosis.

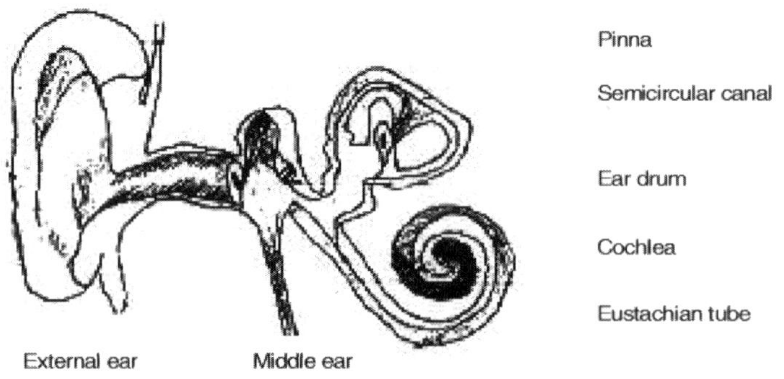

Figure 2.7 Anatomy of ear

Table 2.1 Normal flora of human body

Body site	Normal flora
Skin	Coagulase-negative *Staphylococcus, Staphylococcus aureus, Propionibacterium acnes, Bacillus* spp. streptococci, *Malassezia furfur, Candida* spp., *Pityrosporum*
Conjunctiva	Coagulase-negative *Staphylococcus, Staphylococcus aureus, Haemophilus* spp., *Streptococcus*
Nose	Coagulase-negative *Staphylococcus*, Viridans streptococci, *Staphylococcus aureus*, *Neisseria* spp., *Haemophilus* spp., *Streptococcus pneumoniae*
Mouth and oropharynx	Coagulase-negative *Staphylococcus*, Viridans streptococci, *Staphylococcus aureus, Neisseria* spp., *Haemophilus* spp., *Streptococcus pneumoniae, Veillonella, Fusobacterium, Treponema, Porphyromonas, Prevotella, Branhamella,* diphtheroids, *Candida* spp.
Ear	Coagulase-negative *Staphylococcus,* diphtheroids, *Pseudomonas* spp., Enterobacteriaceae (occasionally)
Stomach	*Streptococcus, Staphylococcus, Lactobacillus, Peptostreptococcus*
Small intestine	*Lactobacillus, Bacteroides, Clostridium, Mycobacterium,* enterococci, Enterobacteriaceae members
Large intestine	*Lactobacillus, Bacteroides, Clostridium, Mycobacterium,* enterococci, Enterobacteriaceae members, *Peptostreptococcus, Actinomyces, Streptococcus, Pseudomonas, Entamoeba coli*
Urethra	Coagulase-negative *Staphylococcus,* diphtheroids, *Streptococcus, Mycobacterium, Bacteroides, Fusobacterium, Peptostreptococcus, Pseudomonas* spp.
Vagina	*Lactobacillus, Clostridium,* diphtheroids, *Streptococcus, Mycobacterium, Bacteroides, Peptostreptococcus* , *Candida* spp. *Gardnerella vaginalis*

1. Define microbiota.
2. Define gnotobiotic.
3. How do you establish germ-free animals?
4. Compare germ-free animals from normal animals.
5. Are animals benefitted from microbes?
6. Explain whether the knowledge of normal flora is necessary or not.
7. Explain the role of skin microbial flora in skin odour.
8. Explain positive response of skin.
9. Define comedo.
10. Define acne vulgaris.

11. Differentiate bottle-fed animals from breast-fed animals.
12. Colon is a large fermentation vessel. Justify.
13. What are Doderlein bacilli?

1. Why is *Entamoeba histolytica* not established in germ-free animals?
2. Nowadays *E. coli* is not considered as a normal flora. Why?
3. Why is there such a huge number of normal flora in the colon?
4. State the reason behind the lungs being free from microorganisms.

Bitton, G. and Marshall, K.C. *Adsorption of Microorganisms to Surfaces*. John Wiley & Sons, New York. 1980.

Brook,T.D. *Milestones in Microbiology*. Prentice Hall, London. 1961.

Draser, B.S. and Hill, M.J. *Human Intestinal Flora*. Academic Press, London. 1974.

Maibach, H. and Aly R. *Skin Microbiology: Relevance to Clinical Infection*. Springer-Verlag, New York. 1981.

Michael, J. Pelczar, Chan, E.C.S. and Noel, R. Crieg. *Microbiology*. Tata McGraw-Hill Publishers Company Ltd. 1995.

Tannock, G.W. *Normal Microflora*. Chapman and Hall, London, UK. 1995.

www.bmb.leeds.ac.uk/mbiology

www.cehs.siu.edu/fix

gsbs.utmb.edu

www.vacadsci.org

www.ncbi.nlm.nih.gov

www.ispub.com

www.mansfield.ohio-state.edu

www.medicine.uiowa.edu

www.probiotictherapy.com

www.bio.net

www.kvo27.com

www.nmm.ac.uk

www.hhmi.princeton.edu

www.micro.msb.le.ac.uk

www.micro.cornell.edu

www.bact.wisc.edu

www.austehc.unimelb.edu

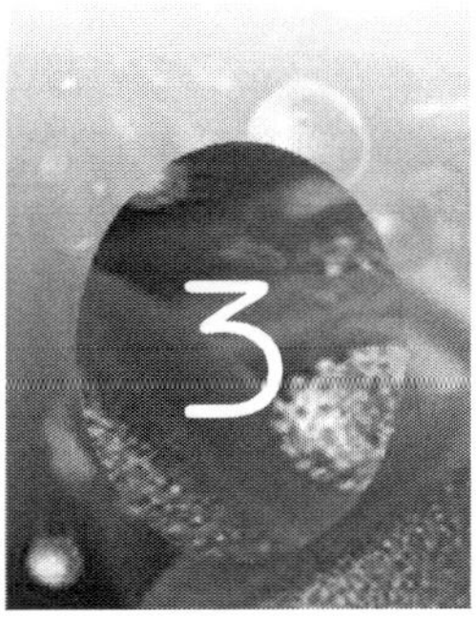

Non-specific Defence Mechanisms

The ability of the host to resist foreign invaders is called immunity. The body of an animal that harbours another organism is the host. The ability to ward off diseases through the defence mechanism is called resistance. Elie Metchnikoff was the first person to describe about the immune responses of animals. Vertebrate animals have two lines of immune response against foreign particles.

- Non-specific immune response and
- Specific immune response.

Immune responses are observed against infectious diseases, cancer, graft rejection, sometimes cells of one's own (autoimmune disease), etc.

NON-SPECIFIC IMMUNE RESPONSE

It refers to defences that protect us from any kind of pathogen. This type of defence is observed in all types of animals. For example Metchnikoff observed phagocytic cells in water flea and starfish. Non-specific defence mechanism of the human body is categorized into general factors, physical factors/mechanical factors/anatomical factors, chemical factors and mobilizing factors/biological factors.

General Factors

The general barriers of human defence mechanisms are categorized into two: direct factors and indirect factors. Nutrition, physiology, fever, age, sex and genetic factors are considered as direct factors. Personal hygiene, socio-economic status and living conditions are the indirect factors.

Nutrition Malnourished individuals are more susceptible to infection than normal healthy individuals.

Age Generally very young and very old individuals are highly susceptible to infection. In children only maternal immunity is observed but in adults, the immune system activity is in suppressed condition.

Genetic trait Temperature of the body, metabolism, physiological and anatomical differences of the host and feeding habits of the animals influences the susceptibility and resistance pattern of the animal or individual. Indians are more resistant to infection than Americans whereas Africans are more resistant than Indians.

Acute-phase reactants It refers to the rapid qualitative and quantitative changes that occur in the host blood plasma during infection. These changes can decrease the virulence of the pathogens and increase the overall defence of the host.

Fever Fever results from the disturbances in the thermostat region of the brain. Hypothalamus is the thermostat region of the human body. It normally sets the body temperature at 37°C (98.6°F). Fever is defined as an oral temperature above 37°C or a rectal temperature above 37.5°C. Lipopolysaccharide, peptidoglycan and *N*-acetyl glucosamine are released when the microbe is phagocytosed, and these are considered as exogenous pyrogens. These pyrogens induce the release of endogenous pyrogens (IL-1). IL-1 causes hypothalamus to release prostaglandin that resets the hypothalamic thermostat at a higher temperature, thereby causing fever. Fever induces our immune system by stimulating leucocytes, by enhancing microbiostasis through decreasing iron available to the pathogens and by enhancing specific activity of immune system.

Physical, Mechanical or Anatomical Factors

Skin Skin and mucus provide the first line defence mechanism of the body. Skin is the largest organ of the body, which comprises over 15% of the body's dry weight. No bacterium by itself is known to be able to penetrate unbroken skin. The normal flora of the skin, which metabolizes substances secreted on to the skin, produces fatty acids that discourage the colonization of skin by potential pathogens. A puncture, cut or scratch, bite of an insect, needle prick, etc. in the skin could introduce the infectious bacteria.

Mucous membrane Mucous membranes consist of an epithelial layer and an underlying connective tissue layer. They line the entire gastrointestinal, respiratory, urinary and reproductive tracts. Mucous membrane lining secretes mucus, which contains antimicrobial compounds including lysozyme and secretory antibodies (IgA). Sometimes phagocytes patrol mucous surfaces (in the lower respiratory tract). Toxins of pathogens may damage the mucous surface and invade the host cell. MALT (mucosal-associated lymphoid tissue) is associated with the mucous membrane.

Respiratory tract Fine hairs and baffles of the nasal membranes entrap bacteria, which are inhaled. Pathogens that pass may stick to mucosal surfaces of the trachea or be swept upward by the ciliated epithelial cells of the lower respiratory tract. Coughing and sneezing also eliminates the pathogens. The lower part of the respiratory tract is well-protected by mucus, lysozyme, IgA and phagocytes.

Gastrointestinal tract Flushing action of saliva, enzymatic activity and normal flora activity remove the pathogens from the oral cavity. Organisms that are swallowed are destroyed by acidity and various secretions of the stomach. Alkaline pH of the lower intestine can discourage other organisms. The peristaltic activity of intestine flushes out organisms, which have not succeeded in colonization. Bile salts and lysozyme inhibit the activities of many microorganisms.

Urinogenital tract The flushing mechanisms of sterile urine and the acidity of the urine, maintain the bladder and most of the urethra free of microorganisms. The vaginal epithelium of females maintains the high population of doderlein's bacilli whose acidic end products (lactic acid) prevent colonization by other types of microorganisms including *Candida albicans.*

Eyes The conjunctiva of the eye is remarkably free of microorganisms. Blinking mechanically removes microbes. The lavaging action of tears washes the surfaces of the eye, and lacrymal secretion contains remarkably large amounts of lysozyme.

Chemical Barriers

Chemical factors play an important role in non-specific defence mechanisms.

Interferons (IFN) Interferons are a family of related low molecular weight, regulatory glycoproteins produced by many eukaryotic cells in response to specific antigens (viral). Five major classes of interferons are recognized.

- IFNα, a group of 20 different molecules derived from virus-infected leucocytes.
- IFNβ derived from virus-infected fibroblasts
- IFNγ produced by antigen-stimulated T cells
- IFNτ produced by placenta
- IFNω produced by placenta

Virus-infected cells induce IFN synthesis. IFNs are released as extracellular materials which bind to the ganglioside of a second cell. It induces the synthesis of oligosynthetase and a special protein kinase.

When interferon-stimulated cells are infected, viral protein synthesis is inhibited by an active endonuclease that degrades viral RNA. Protein kinase inactivates initiation factor 2.

Bacteriocin Many normal flora synthesize a chemical called bacteriocin, which inhibits the growth of other related species, e.g. colicin, staphylococcin.

Fibronectin These are high molecular weight glycoproteins that can interact with certain bacteria and reduce binding of bacteria to the host cell.

Betalysin It is a cationic polypeptide released from blood platelets. It disturbs the cell membrane of gram-positive bacteria.

TNF Tumour necrosis factor alpha and beta are cytotoxic to tumour cells and they enhance phagocytosis and inflammation.

Sebum Sebum covers the skin, which contains unsaturated fatty acids (oleic acid). It prevents the growth and colonization of pathogens.

Lysozyme and sodium chloride Sweat gland secretion contains lysozyme and sodium chloride, which discourage the growth of pathogens on skin. Lysozyme present in tears, mucus and saliva inhibits the activities of peptidoglycan-bearing bacteria.

pH Gastric acidity inhibits the growth of bacteria in the stomach. Vaginal acidity interferes with the colonization of microbes in female genitalia. Alkaline nature of the lower part of intestine inhibits the growth of unwanted flora.

Biological Barriers

Biological barriers are sometimes considered as the second line defence system, which act on the infected cells and other microbial cells. Leucocytes (WBCs) are the most important mobilizing factors that are involved in the clearance of microbial cells from the human body. Reticuloendothelial system (RES) is the collection of WBCs and associated cells that are located in the thymus, liver, spleen, lymph nodes and bone marrow.

Leucocytes (normal value 4000–5000 cells/cu.mm) include granulocytes and agranulocytes. Neutrophils (60–70%), basophils (0.5–1%), eosinophils (2–4%) are granulocytes, which contain large granules in the cytoplasm. Lymphocytes (20–25%) and monocytes (3–8%) are examples of agranulocytes.

Neutrophils are also called polymorphonuclear leucocytes (PMN) that are actively motile cells. Mononuclear phagocytic system (MPS) includes monocytes and free macrophages.

Other names of macrophages are

- Alveolar macrophages in lung
- Kupffer cells in liver
- Histiocytes in skin
- Microglial cells in brain
- Messengial cells in kidney

COMPLEMENT

Complement is a group of 20 different serum proteins, designated by a complex numbering system ranging from C1 through C9. These are highly important for specific and non-specific defence mechanisms.

Complements are sequentially activated in many antigen–antibody reactions resulting in disruption of cell membranes. Complements are involved in phagocytic chemotaxis, opsonization and the inflammatory response.

Complement components are produced in the body including macrophages (C2 and C4), liver cells and epithelial cells of the gastrointestinal mucosa.

Activation of complement takes place by two mechanisms

- Classical pathway
- Alternate pathway or properdin pathway

CLASSICAL PATHWAY

A stepwise illustration of complement cascade mechanism is shown in Figure 3.1.

1. Complement (C1) subunits (C1q, C1r and C1s) bind to Fc portion of two adjacent IgGs and the complement is activated.

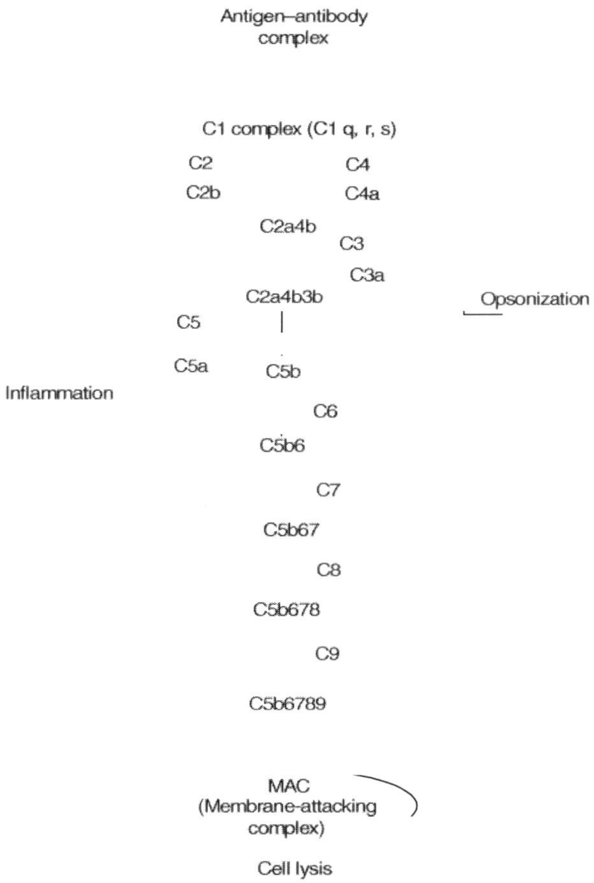

Figure 3.1 Complement cascade mechanism

1. Activated C1 cleaves into C2 (C2a + C2b) and C4 (C4a+C4b).
2. C2a binds C4b forming C2a4b, which attaches to the cell membrane.
3. C2a4b (known as C3 convertase) attracts and enzymatically cleaves C3 into C3a+C3b.
4. C3a is a phagocyte chemotactic attractant, which can bind to a C3a receptor on mast cells inducing degranulation and localized inflammation.
5. C3b can attach to the cell membrane or remain attached to IgG or IgM Fc, which in both cases opsonize cells for phagocytosis.
6. C2a4b+C3b → C2a4b3b.
7. C2a4b3b (C5 convertase) attacks and cleaves C5 into C5a+C5b.
8. C5a induces inflammation.
9. C5b binds C6 and C7 forming C5b67 that attaches to the cell surface.
10. C5b67 attach C8 and C9 forming C5b6789. This complex disturbs the cell membrane (membrane-attacking complex).

PROPERDIN PATHWAY

Properdin is a three-serum protein composed of properdin, factor B and factor D. These proteins comprise 10% of the proteins in serum. Insoluble polysaccharides (LPS, PG and teichoic acid) combine with properdin and activate C3 in the absence of C1, C2 and C4.

Properdin + bacterial component +C3 → C3a + C3b

INFLAMMATION

It refers to a host response to tissue damage characterized by reddening, pain, heat and swelling. The redness is due to increased blood flow. The swelling is due to increased extravascular fluid and phagocytic filtration. The heat is due to increased blood flow and action of pyrogens. The pain is caused by local tissue destruction.

Steps involved in inflammation are

1. Pathogen invasion or tissue injury
2. Release of cytoplasmic constituents
3. Lower pH of the environment
4. Acidity activating kallikrein
5. Activation of bradykinin
6. Bradykinin binding to receptors on the capillary wall opening junctions and allowing leakage of plasma (inflammatory exudates)
7. Increased capillary permeability allowing leucocytes to pass from the vessel tissues

8. Bradykinin also binds to mast cells
9. Influx of Ca^{2+} and intracellular cAMP level drop
10. Release of histamine and heparin
11. Activation of phospholipase A2
12. Arachidonic acid synthesis
13. Prostaglandins, leukotriene synthesis
14. Vasodilation and increased permeability
15. Entry of clotting elements into injured area
16. Blood clot preventing bacterial spread
17. Localized collection of pus
18. Abscess formation
19. Migration of phagocytes into injured area
20. Tissue repair

PHAGOCYTOSIS

Metchnikoff discovered phagocytic cells. The following sequence of events are observed during phagocytosis.

- Delivery of phagocytic cells to the site of infection (chemotaxis)
- Phagocytic adherence
- Ingestion
- Phagolysosome formation
- Intracellular killing
- Intracellular digestion
- Release and presentation

Enzymes involved in phagocytosis are lysozyme, neutral proteases, myeloperoxidases, oxygen radicals, hydrogen peroxidase, hypochlorite, fucosidase, glucosidase, galactosidase, *N*-acetyl glucosamine, hyaluronidase, phospholipase, acid phosphatase, phosphatidic acid, collagenase, elastase, kininase, kininogenase, arylsulphatase, esterase, etc.

1. Describe how nutrition, acute-phase reactants, fever, age and genetics contribute to general defence of the host.
2. Given an account of some non-specific host defence mechanisms that help prevent entry of microbes into host tissues.
3. How does mucus resist parasitic invasion?

4. How does beta lysin act on gram-positive bacteria?
5. How are microorganisms engulfed by phagocytic cells?
6. Write short notes on
   i. Acute-phase reactants
   ii. Fibronectin
   iii. Inflammation
   iv. Interferon
   v. Phagosome
   vi. Respiratory burst

1. How can inflammation be beneficial to a host?
2. Is phagocytosis needed for good health?
3. How can our body react when there is no skin?
4. Is complement useful in excluding a virus?

Baneman, J.B. and Aldereff, J.F. *Microbiology*. ASM Press. 1986.

Baron, E.J., Peterson, L.R. and Finegold, S.M. (eds.). *Bailey and Scott's Diagnostic Microbiology*, 9th edn. C.V. Mosby, St. Louis. 1994.

Baron, S. "Mechanisms of recovery from viral infection". In: Smith, K.M., Lauffer, M.A. (eds.) *Advances in Virus Research*. Academic Press Inc., New York. 10:39–64, 1963.

Keneman, E.W., William, M.J., Stephen, D.A., Schreeken, B. and Washington, C.W. *Introduction to Diagnostic Microbiology*. Lippincott Company. 1994.

Koneman, E.W., Allen, S.D., Schreckenberg, P.C. and Winn, W.C. (eds.): *Atlas and Textbook of Diagnostic Microbiology*, 4th edn. JB Lippincott, Philadelphia. 1992.

Kunin, C.M., *Detection, Prevention and Management of Urinary Tract Infections*, 4th edn. Lea and Febiger, Philadelphia. 1987.

Lwoff, A. *Bacteriol. Rev*. 23:109, 1959.

Murray, P.R., Baron, E.J., Pfaller, M.A. and Tenover, P.C., and Yolken, R.H. (eds.). *Manual of Clinical Microbiology*, 6th edn. American Society for Microbiology, Washington, DC. 1995.

Pennington, J.E., (ed.): *Respiratory Infections: Diagnosis and Management*, 3rd edn. Raven Press, New York. 1994.

Prescott, L.M., Harley, J.P. and Klein, D.A. *Microbiology*, 5 th edn. W.C.B. McGraw-Hill 2001.

Woods, G.L. and Washington, J.A., "The Clinician and the Microbiology Laboratory". Mandell, G.L., Bennett, J.E. and Dolin, R. (eds.). *Principles and Practice of Infectious Diseases*, 4th edn. Churchill Livingstone, New York. 1995.

www.slic2.wsu.edu:82/hurlbert/micro101

www.maexamhelp.com/id106.html

www2.hawaii.edu/~johnb/micro/m130/m130lect12.html

www.jdaross.cwc.net/nonspec.

www.gla.ac.uk/immunology/education/nursing/lectures/immunedefences

www.blackwell-synergy.com/links

www.blackwell-synergy.com

www.gsbs.utmb.edu/microbook/ch049.html

www.slic2.wsu.edu:82/hurlbert/micro101

www.maexamhelp.com/id106.html

www2.hawaii.edu/~johnb/micro/m130/m130lect12.html

www.jdaross.cwc.net/nonspec.

www.gla.ac.uk/immunology/education/nursing/lectures/immunedefences

www.blackwell-synergy.com/links

www.gsbs.utmb.edu/microbook/ch049.html

cat.inist.fr/?aModele=afficheN&cpsidt

vfu-www.vfu.cz/acta-vet/vol67

www.utmb.edu/lsg/labsurvivalguide/sample_procurement.html

www.compostingcouncil.org/pdf/Sample_Collection

www.toolkit.cch.com/tools/letter_m.

www.vgl.ucdavis.edu/forensics/samplecollection.html

www.uga.edu/~sisbl/sampling.

massfatality.dna.gov/Chapter11/SampleCollection

curator.jsc.nasa.gov/lunar/lunar.

www.aphis.usda.gov/ppq/ep/soybean_rust/

www.wada-ama.org/rtecontent/document/urine_testing_

www.dna.gov/lab_services/samplekit

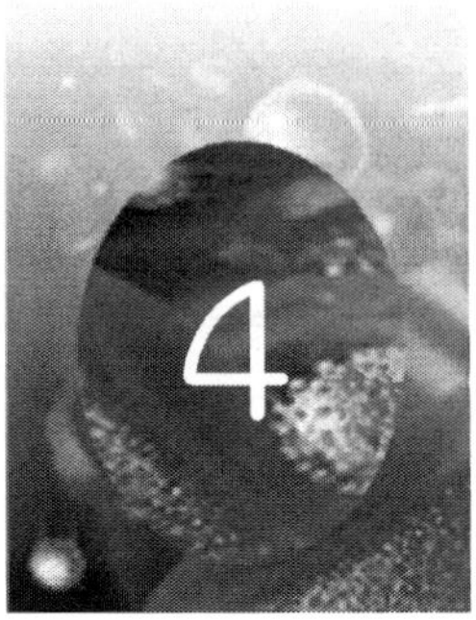

Host–Microbe Interactions

Host refers to the body of an animal that harbours another organism. Humans are the most important hosts for microbes. Humans harbour billions of microbes on their body as commensals. The relationship between normal flora and host is called symbiosis. The interaction between pathogenic microbes and the host is called parasitism. During this type of interaction, microbes utilize nutrients from the host and interfere with its metabolism.

Disease condition occurs as a result of these host–microbe interactions. Host–microbe interactions are very complex. When a parasite tries to establish infection, the host responds by mobilizing an array of defence mechanisms. The ability to ward off disease through the defence mechanism is called resistance. Lack of resistance is called susceptibility. The capability of microbes to cause disease is called pathogenicity.

The outcome of host–microbe interactions depends on three main factors or in other words to induce an infectious disease, a pathogen must be able to perform the following functions:

1. Invasiveness which is the ability of the pathogen to spread to adjacent tissues.
2. Infectivity which is the ability of an organism to establish an infection.
3. Pathogenic potential which refers to the degree to which the pathogens cause morbid symptoms. It mainly depends on toxigenicity, which is the pathogen's ability to produce toxins that are chemical substances that will damage the host and produce disease.

Usually when the number of disease-producing organisms present in the body is more, they cause severe disease. However, even a few number can cause disease if they are extremely virulent or if the host resistance is low. The term virulence refers to the degree or intensity of pathogenicity. Invasiveness, infectivity and pathogenic potential determine it. Virulence is often measured experimentally by determining the Lethal Dose 50 (LD_{50}) or Infective Dose 50. These values refer to the dose or number of pathogens required either to kill or infect the host, respectively.

Transmissibility, attachment, colonization and entry of pathogen into the host depend on its degree of invasiveness. Growth and multiplication depend on its degree of infectivity. Symptom-producing nature depends on its toxigenicity.

Infection may be symptomatic or non-symptomatic.

TRANSMISSIBILITY OF THE PATHOGENS

An essential feature in the development of an infectious disease is the initial transport of the pathogen to the host. The organism is transmitted to the host by the following ways:

1. Airborne transmission
2. Contact transmission
3. Vehicle transmission
4. Vector-borne transmission
5. Carrier transmission
6. Waterborne transmission
7. Food-borne transmission

A source is the location from which the pathogen is immediately transmitted to the host, either directly through the environment or indirectly through an intermediate object. The source is either animate or inanimate. The reservoir is the site or natural environmental location in which the pathogen is normally found living from which infection of the host can occur. Reservoir also can be animate or inanimate.

Airborne Transmission

In this, the pathogen is suspended in the air and travels over a metre or more than the source. The pathogen can be contained within droplet nuclei or dust. Droplet nuclei are small particles, 1–2 μm in diameter. It is derived from evaporated droplets (10 μm or more in diameter). Droplet nuclei remain viable for hours or days and travels long distances. Dust is also an important route of airborne transmission. At times, a pathogen adheres to the dust and is transmitted to the host. Sources of airborne transmission are coughing and sneezing.

Contact Transmission

Contact transmission refers to touching the source or reservoir of the pathogen and host. Contact can be either direct or indirect. Direct contact implies an actual physical interaction with the infectious source. The route is frequently called person-to-person contact, e.g. touching, kissing, sexual contacts, contact with oral secretions, nursing mothers, through placenta and contact with animals.

Indirect contact refers to the transmission of the pathogen from the source to the host via intermediate objects, usually an inanimate object. It is often called vehicle transmission, e.g. thermometers, eating utensils, drinking cups and bedding.

Vector-borne Transmission

Living transmitters of a pathogen are called vectors. Most vectors are arthropods or vertebrates. The transmission can be either external or internal. In external transmission, the pathogen is carried on the surface of the body. In internal transmission, the pathogen is carried within the vector. This type of transmission is of two types—harbourage transmission and biological transmission.

In harbourage transmission, the pathogen does not undergo morphological or physiological changes within the vector, e.g. *Yersinia pestis* causes plague in humans and are transmitted by rats.

In biological transmission, the pathogen goes through physiological or morphological changes within the vector, e.g. *Plasmodium* spp. cause malaria in humans and are transmitted by mosquitoes.

Carrier Transmission

Human hosts are the most important animate source of the pathogen and are called carriers. Carriers play an important role in the epidemiology of disease. Four types of carriers are recognized. They are:

- An acute carrier is an individual who has clinical case of the disease.
- A convalescent carrier is an individual who has recovered from infectious disease but continues to harbour large number of pathogens.
- A healthy carrier is an individual who harbours the pathogen but is not ill.
- An incubatory carrier is an individual who is incubating the pathogen in large numbers but is not ill.
- Convalescent, healthy and incubatory carriers may harbour the pathogens for only a brief period and are called casual, acute or transient carriers. If they harbour the pathogen for a long period, they are called chronic carriers.

Water and Food-borne Transmission

Some of the pathogens are transmitted through food and water. During this type of transmission, organisms are shed in water and enter the food. In food, organisms multiply and produce toxins and cause food intoxication. Faecal material is the main contaminant of water and coliforms act as indicators of contamination.

ATTACHMENT AND COLONIZATION OF THE PATHOGENS

After transmission, pathogens must attach to host cells and tissues to establish infection. Colonization depends on the ability of the pathogen to compete successfully with the host normal flora for essential nutrition. Adherence and colonization depend on the following factors—capsule, fimbriae, lectin, ligand, pili, slime, mucous gel, teichoic acid and cell wall antigens (Figure 4.1).

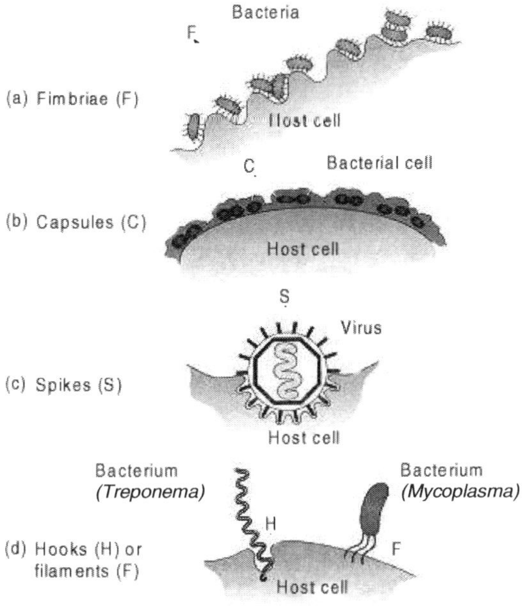

Figure 4.1 Attachment and colonization factors

ENTRY OF THE PATHOGENS

After attachment, a pathogen sometimes enters inside the host cell or it may grow on the surface. This may be accomplished by one of the following lytic substances—coagulase, collagenase, deoxyribonuclease, alkaline protease, exotoxins, haemolysins, hyaluronidase, hydrogen peroxidase, lecithinase, leucocidins, porins, protein A, and streptokinase. These factors alter the host tissue by:

1. Altering the basement membranes of integuments and intestinal linings.
2. Disturbing the cell surface.
3. Disturbing the carbohydrate–protein moiety between cells.

Sometimes pathogens penetrate the cells passively by

1. Small breaks, lesions or ulcers in mucous membranes that permit initial entry.
2. Wounds, abrasions or burns on the skin surface.
3. Wounds caused by mosquito bite.
4. Tissue damage caused by other pathogens.

After entry, the pathogen may penetrate to deeper tissues and continue disseminating throughout the body.

GROWTH AND MULTIPLICATION OF THE PATHOGENS

Successful growth and reproduction of pathogen requires an appropriate environment within the host. During growth, the pathogen produces some metabolic end products, which are often toxic and produce a condition known as septicaemia.

Toxigenicity

Two distinct categories of the disease can be recognized based on the pathogen's role in the disease causing process. They are infection and intoxication. An infectious disease results from the growth and reproduction of the pathogen that often produce tissue alteration.

Intoxication is a disease that results from the entry of a specific toxin into the body of the host. Toxins introduce disease in the absence of microbes. Toxaemia refers to the condition caused by toxins. Microbial toxins are divided into two types—exotoxins and endotoxins.

Exotoxins Exotoxins are soluble, heat-labile proteins that are released by growing microbial cells to the surrounding medium. Exotoxins may be divided into three main categories on the basis of their mode of action.

- Neurotoxins—affect nerve cells, e.g. tetanus toxin
- Enterotoxins—affect intestinal mucosa, e.g. cholera toxin
- Cytotoxins—affect epithelial cells, e.g. *Pseudomonas* exotoxin A

Exotoxins act by:

- Inhibiting protein synthesis
- Inhibiting nerve synapse function
- Disrupting membrane transport
- Damaging plasma membrane

Bacterial exotoxins are the most powerful human poisons. Most protein toxins, notably those that act intracellularly, consist of two components. One component is responsible for attachment (subunit B), the other is responsible for enzymatic activity (subunit A). Generally these toxins are called "A + B toxin". The enzymatic component is not active unless it is released from the native toxin. A and B subunits of the toxin are produced separately in the cytoplasm and assembled only at the periplasmic space (A and B).

There are at least two mechanisms of toxin entry into the target cell. One mechanism is called direct entry; subunit B of the toxin binds to the cell membrane and induces the formation of pore in the membrane through which the subunit A enters into the cell cytoplasm. In the alternate mechanism, the toxin binds to the target cell and receptor-mediated endocytosis takes the toxin into the cell. The toxin is internalized in the endosome. H^+ ions entering into the endosome lower the internal pH, which causes the separation of the subunits. Subunit A enters into the cytoplasm and activates the enzymatic activity. Subunit B remains in the endosome and is recycled back to the cell membrane.

Environmental signals induce the synthesis of protein toxin, and chromosomal, plasmid or phage DNA are responsible for toxin synthesis.

Some of the important exotoxins and their biological activities are as follows:

Cholera toxin ADP ribosylation of G proteins stimulates adenylate cyclase and increases cAMP level. It promotes secretion of fluid and electrolytes in intestinal epithelium leading to watery diarrhoea.

Diphtheria toxin ADP ribosylation of elongation factor 2 leads to the inhibition of protein synthesis resulting in death of cell.

Pertussis toxin ADP ribosylation leads to inhibition of adenylate cyclase and leads to increased cAMP, and reduced phagocytic activity.

Anthrax toxin Lethal factor is a Zn^{++}-dependent protease that induces cytokine release and is cytotoxic to cells. It is also called A1 toxin.

Oedema factor is an adenylate cyclase that causes increased level of cAMP in phagocytes and formation of ion permeability pore on cell membrane. Increased level of cAMP in phagocytes leads to the inhibition of phagocytosis by neutrophils and macrophages, leading to haemolysis and leucolysis.

Tetanus toxin Zn^{++} -dependent protease inhibits neurotransmission at inhibitory synapses and results in spastic paralysis.

Botulinum toxin Zn^{++}-dependent protease inhibits neurotransmission at neuromuscular synapses and results in flaccid paralysis.

Shiga toxin Glycosidase cleavage of rRNA leads to inactivation of rRNA and inhibits protein synthesis and death of susceptible cells.

Leucocidins and Haemolysins

Some pathogens produce extracellular toxins that may kill phagocytic leucocytes and are termed leucocidins. Pneumococci, staphylococci and streptococci produce most of the leucocidins. These toxins cause membrane permeability alteration and degranulation, and release lysozyme in the cytosol. These changes destroy leucocytes and, in turn, decrease host resistance.

Haemolysins form pores on the surface of the plasma membrane of RBC through which haemoglobin and Fe^{2+} ions are released. Streptolysin O produced by *Streptococcus pyogenes*, is inactivated by oxygen. It causes β-haemolysis of erythrocytes on blood agar plates incubated anaerobically. Streptolysin S causes β-haemolysis of erythrocytes on blood agar plates incubated aerobically.

ENDOTOXINS

Gram-negative bacteria have a lipopolysaccharide in the outer membrane of their cell wall. Under certain circumstances, it is toxic to hosts. These lipopolysaccharides are called endotoxins. Endotoxins initially activates Hageman factor, which in turn activates up to four humoral systems—coagulation, complement, fibrinolytic function and kininogen system.

Endotoxins also indirectly induce fever in the host by inducing macrophages, which release endogenous pyrogens that reset hypothalamic thermostat. Recent evidences indicate that important endogenous pyrogens are the lymphokines, IL-1, TNF and IL-6.

Pathophysiological reactions of endotoxins are fever, changes in white blood cell count, disseminated intravascular coagulation, tumour necrosis, hypotension, shock, etc.

Table 4.1 shows some important differences between exotoxins and endotoxins.

Table 4.1 Differences between exotoxins and endotoxins

Exotoxins	Endotoxins
Synthesized by specific pathogens that often have plasmids.	Found in gram-negative pathogens.
Heat labile proteins inactivated at 60°–80°C.	Heat stable.
Very small amount is required to create toxic condition.	Toxic only at high doses.
Highly immunogenic and stimulates the production of neutralizing antibodies.	Weekly immunogenic.
Easily inactivated by formaldehyde, iodine and other chemicals and form toxoids.	Inactivated at temperature 25°C for 30 minutes.
Does not cause fever.	Causes fever.
Often given the name of the disease they cause.	There is no specific name.

HOST–VIRUS INTERACTIONS

Reproduction or replication of viruses is also called as host–virus interactions. Viruses are obligate intracellular parasites. They reproduce only within a host cell. Viruses have the specific efficiency towards the cells. Certain groups of viruses may infect only specific groups of cells. The viruses make use of the metabolic machinery of the host cell to undertake replication, e.g. bacterial viruses infect only bacteria, animal viruses infect only animal cells.

Replication or reproduction of viruses may be divided into several stages. They are

- Adsorption of viruses
- Penetration and uncoating
- Replication of nucleic acid
- Synthesis and assembly of viral capsids
- Release of mature virions

Adsorption of Virion

- It is the first step in the multiplication process. It occurs through random collision on the plasma membrane of the host cell.
- Adherence is an important virulent mechanism of the virus.
- Capsid protein in non-enveloped viruses and spike protein in enveloped viruses play a crucial role in the attachment process, e.g. haemagglutination spike of influenza virus, capsid protein in parvovirus.
- Spike of non-enveloped virus is also involved in attachment process, e.g. adenovirus spike.
- Infection of the virus depends greatly on its ability to bind the cell. Viruses have special structures (sometimes called ligands) binding to the receptors on the host cell surface.
- Receptor is a glycoprotein playing an important role in tissue tropism and host specificity.
- Attachment is in most cases a reversible process—if penetration does not ensue, the virus can elute from the cell surface.

Some examples include the following.

- Influenza A virus binds sialic acid receptor of respiratory tract.
- Rabies virus adsorbs an acetylcholine receptor of neurons.
- Vaccinia virus attaches to epidermal growth factor receptor.
- Rhinovirus binds to intracellular adhesion molecules on the respiratory epithelial cells.
- HIV virus binds the CD4 receptor.
- HAV binds α-2 macroglobin.

Penetration and Uncoating

- Viruses penetrate the plasma membrane and enter a host cell shortly after adsorption. The mechanism of penetration and uncoating must vary with different viruses.
- Penetration of viruses involves 3 processes. They are
    1. Direct penetration
    2. Fusion
    3. Endocytosis

1. *Direct penetration (Translocation)* Naked virus (e.g. poliovirus) undergoes conformational changes after adherence on plasma membrane and released only nucleic acid into the cytoplasm.

2. *Fusion* Enveloped virus fuses with plasma membrane with the help of fusion protein or F protein. It results in entry of capsid protein into the cytoplasm.

3. *Endocytosis* Most of the enveloped viruses enter inside of host cells through receptor-mediated endocytosis and form coated vesicles. Immediately after entry, uncoating takes place by making use of lysozyme or acidity formation within an endocytic vesicle.

Replication of Nucleic Acid in DNA Viruses

Most of the DNA virus genome is replicated within a host cell nucleus with few exceptions, e.g. smallpox virus.

Upon entry, the genome enters into a nucleus and performs transcription by using an enzyme RNA polymerase. In some viruses, two stages of transcription occur. They are early transcription and late transcription. During transcription, mRNA will be formed, which is transferred to the cytoplasm for the translation process.

This process leads to the synthesis of protein responsible for DNA replication. DNA is replicated by using an enzyme DNA polymerase. This leads to the replication of the genome.

DNA (RNA polymerase / Transcription) mRNA (Translation) DNA polymerase

DNA (DNA polymerase / Replication) Multiple copies of virion DNA

Replication and Transcription in RNA Viruses

RNA viruses adopt four strategies for replication. The process of replication depends upon the nature of RNA.

1. Some RNA viruses use their RNA genome as a giant mRNA, e.g. poliovirus. These types of viruses are considered as positive-sense ssRNA viruses.
2. dsRNA viruses like reoviruses carry a virus-associated transcriptase and generate mRNA. RNA polymerase enzyme produces new dsRNA from +mRNA.

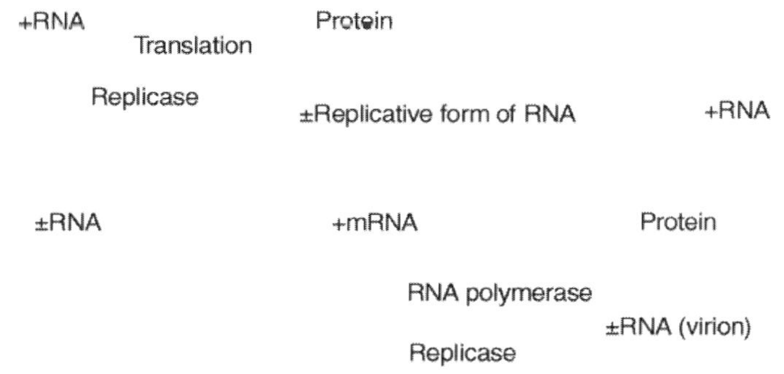

3. In the case of negative-sense ssRNA viruses, viral replicase converts the ssRNA into a double-stranded RNA called the replicative form. This directs the synthesis of specific RNA genome.

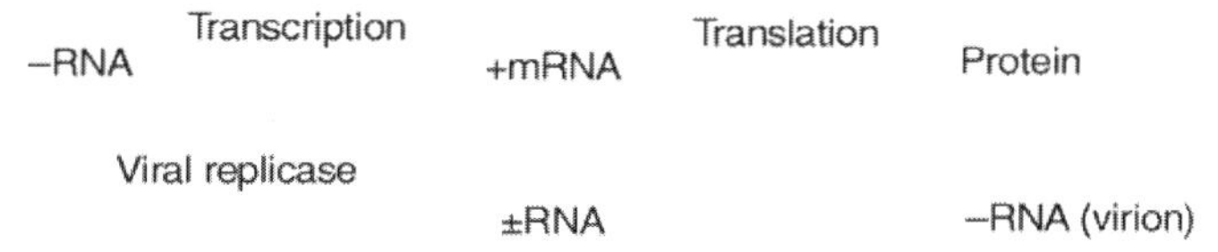

4. Retrovirus performs a different pattern of replication. ssRNA is copied into DNA–RNA hybrid by making use of reverse transcriptase enzyme. Then the ribonuclease H degrades +RNA strand to leave –DNA. After the synthesis of –DNA, the reverse transcriptase copies this strand to produce a double-stranded DNA. It is called proviral DNA, which can direct the synthesis of mRNA and multiple copies of new +RNA virion genome.

   Sites of replication vary and depend upon the type of viruses.

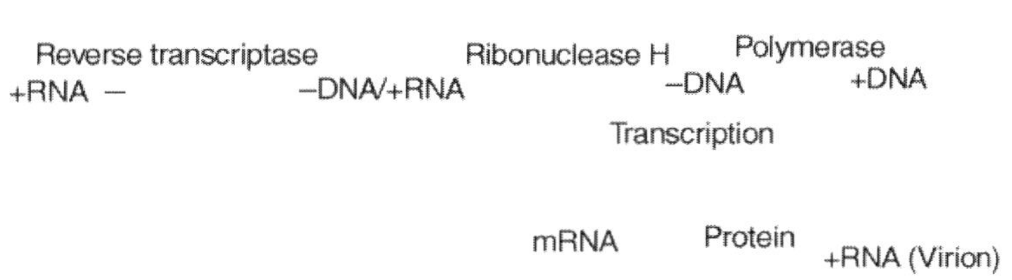

Table 4.2 Sites of replication of different viruses

Virus	Site of replication
Adenovirus	Nucleus
Hepadnavirus	Cytoplasm
Poxvirus	Cytoplasm
Parvovirus	Cytoplasm
Papovavirus	Cytoplasm
Orthomyxovirus	Nucleus
Paramyxovirus	Cytoplasm
Picornavirus	Cytoplasm
Retrovirus	Cytoplasm

Synthesis and Assembly of Capsid

Late genes of virus is responsible for the synthesis of structural proteins (capsid and spike protein). Viruses are assembled after the complete synthesis of structural and functional protein. Assembly takes place in any place of the plasma membrane, nuclear membrane, endoplasmic reticulum or Golgi apparatus (Table 4.3).

Table 4.3 Sites of assembly of capsid of different viruses

Virus	Assembly site
Adenovirus	Nucleus
Hepadnavirus	Cytoplasm
Poxvirus	Cytoplasm
Orthomyxovirus	Nucleus
Rhabdovirus	Cytoplasm

Virion Release

It differs between naked and enveloped viruses. There are three mechanisms available to release viruses from host cells. These are cell lysis or cytopatheic effect, budding and cell degeneration.

Most of the naked viruses are released by lysing the host cells and results in cytopatheic effect.

Enveloped viruses may receive thier envelopes from Golgi apparatus, endoplasmic reticulum or plasma membrane. If the virus receives its envelope from the plasma membrane, it is released by budding. Others may be released by cell lysis.

Viruses such as parvoviruses accumulate within the host cells and are released only after the death of the host cell, which follows the degeneration of host cell. Actin filament of host cell can aid in virion release.

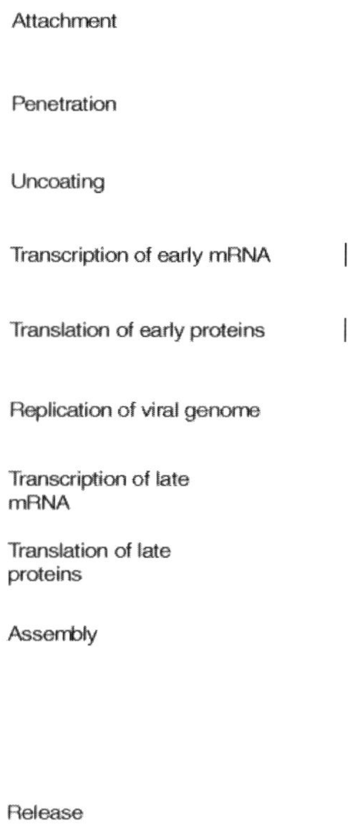

Figure 4.2 Mechanism of host–virus interaction

EFFECT OF VIRUSES ON HOST CELL

1. Viruses inhibit synthesis of host DNA, RNA and protein.
2. Lysosomes may be damaged resulting in cell lysis.
3. Intracellular structures called inclusion bodies are formed.
4. They produce chromosomal disruptions.
5. They may produce transformed cells (cancer cells).

Virus and host interaction (viral replication) may lead to the following types of infection. Some type of interactions lead to cancer.

- Productive infection—complete replication of infectious virion and release of virus.
- Abortive infection—synthesis of viral protein without production of infectious virion.
- Semipermissive infection—complete replication with low yield of infectious virion.

- Malignant transformation—associated with integration of viral DNA and differential viral and cellular gene expression.
- Viral latency—persistence of viral genome in the host cell.

The Figure 4.3 shows the steps in the infectious disease cycle.

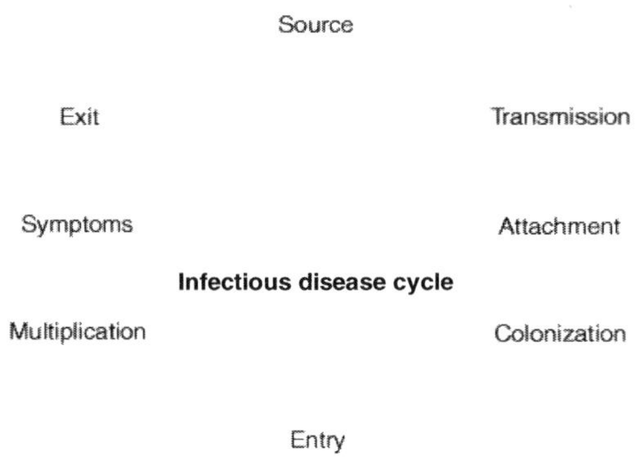

Figure 4.3 Infectious disease cycle

SOME IMPORTANT DEFINITIONS

Abscess A localized infection with a collection of pus surrounded by an inflamed area.

Acute Short but severe course of disease.

Bacteraemia Viable bacteria in the bloodstream.

Chronic Persistence over a long period.

Covert Subclinical, no symptoms.

Cross Transmitted between hosts infected with different organisms.

Focal Existence in circumscribed areas.

Fulminating Infecting agents multiplying with great intensity.

Generalized Affecting many parts of the body.

Iatrogenic Caused as a result of health care.

Latent Infections that persist in tissues for long periods, during which most are symptomless.

Localized Restricted to limited region or to one or more anatomical areas.

Mass Infectious agent occurs in large numbers in systemic circulation.

Mixed More than one organism present simultaneously.

Nosocomial Develops during hospital stay.

Non-cytocidal No change at the cellular level.

Opportunistic Due to an agent that does not harm a healthy host but takes advantage of an unhealthy one.

Overt Symptomatic.

Phytogenic First infection that often allows other organisms to appear on the scene.

Secondary Caused by an organism following an initial or primary infection.

Septic Produced by or due to decomposition by microbes.

Sporadic Occurring only occasionally.

Subclinical No detectable symptoms.

Systemic Spread throughout the body.

Terminal Occurring near the end of a disease and frequently cause death.

Toxaemia Condition arising from toxins in the blood.

Zoonosis Caused by parasitic organisms that are normally found in animals other than human.

1. What is vehicle transmission?
2. Give a brief account of the different types of carriers.
3. Which is virulent factor?
4. Define the term toxaemia.
5. Explain the types of toxins and their mode of action with examples.
6. What is endogenous pyrogen?
7. Differentiate infection and intoxication.
8. Explain the AB model of exotoxin.
9. Give an account on release of mature viruses from the cell.
10. Explain the following terms:
    i. Commensalism
    ii. Symbiosis
    iii. Parasitism
    iv. Resistance
    v. Susceptibility
    vi. Pathogenicity
    vii. Toxigenicity
    viii. LD_{50}
    ix. ID_{50}
    x. Droplet nuclei
    xi. Reverse transcriptase

1. Is there any passive way of microbial entry into a human?
2. What does an organism require to be a parasite?
3. What are characters that the bacteria must possess to be a pathogen?
4. Do environmental factors play any role in transmission of infection?
5. What are the major differences between bacterial and viral infections?

Sathish Gupte. *The Short Textbook of Medical Microbiology*, 8th edn. Jaypee Publishers. 2005.

Michael, J. Pelczar, Chan, E.C.S. and Noel, R. Crieg. *Microbiology*. Tata McGraw-Hill Publishers Company Ltd. 1995.

Bhatia, Rajesh, Ichpujani and Rattan Lal. *Essentials of Medical Microbiology*, 3rd edn., Jaypee Publishers. 2004.

David, M. Knipe and Peter, M. Harley. *Fundamental Virology*, 4th edn. Lippincott Williams & Wilkins. 2001.

John, L. Ingraham and Catherine, A. Ingraham. *Introduction to Microbiology*, 2nd edn. Brooks/Cole. 2000.

Kathleen Talaro and Arthur Talaro. *Foundations in Microbiology*, 2nd edn. W.M.C. Brown Publishers, Chicago. 1996.

Lansing, M. Prescott, John P. Harley and Donald, A. Klein. *Microbiology*, 5th edn. W.C.B. McGraw-Hill. 2001.

www.medvetnet.org
www.onderzoekinformatie.nl
www.intl-pag.org
www.biosecurityboard.gov
www.smbs.buffalo.edu
www.pathogenomics-era.net
www.shopapspress.org
www.mtc.ki.se/pdf/infectionbio
www.wcfs.nl/webdb

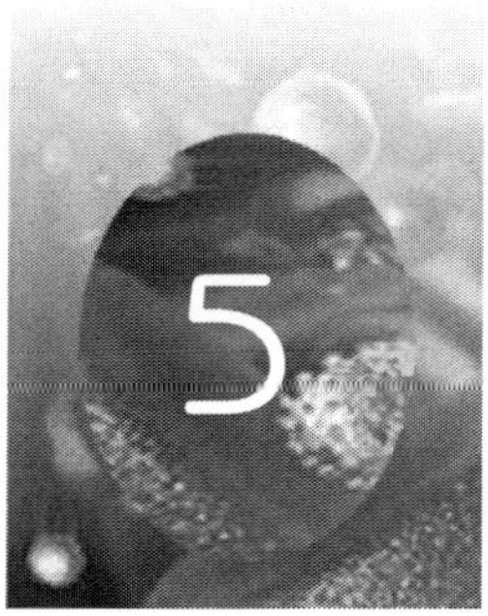

Infective Syndrome and Diagnostic Procedure

INTRODUCTION

The major concern of clinical microbiologists is to isolate and identify microorganisms from clinical specimens rapidly. The purpose of clinical microbiology is to provide the physician with information concerning the presence or absence of microbes in an infectious disease and to perform antibiotic sensitivity when indicated. These tasks will assist the physician in the diagnosis and treatment of infectious disease. Microbiological data are also valuable in monitoring the course of antibiotic therapy and providing epidemiological information for defining the source of infection.

THE DIAGNOSTIC CYCLE

It begins with the patient who consults a physician because of signs and symptoms suggesting infection. On the basis of medical history and physical examination, samples are collected from one or more anatomical sites.

To maintain viability during culture collection and transport, the specimen container must be properly labelled with patient name, location, date, time of collection and type of specimen.

Once the specimen is received in the lab, the information or request form is entered in the log book, and the specimen is examined visually. Depending on the physician's order, microscopic observation and culturing is made.

Each step in the diagnostic cycle must be monitored to assure accuracy and precision of performance.

Symptoms

Patient consults physician

Physician examines patient

Appropriate specimen collected for culture

Written instructions transcribed to a lab

Macroscopic and microscopic examination

Antigen and antibody detection

Cultivation and isolation

Identification

Antimicrobial sensitivity test

Serotyping

Phage typing

Toxin production

Specimen

Specimen represents a portion or quantity of human material that is tested, examined or studied to determine the presence or absence of particular microorganisms. Several guidelines are followed on the safety precautions to be used in selection, collection and handling of the specimen.

- The specimen must be a material from the actual infection site and must be collected with minimum contamination from adjacent tissues, organs or secretions.
- Optimal time and total number of specimens for collection must be established for the best chance of recovery of causative microorganisms.
- A sufficient quantity of specimen must be obtained to perform the culture techniques.
- Appropriate collection devices, specimen containers and culture media must be used to ensure proper isolation.
- Whenever possible, specimen should be obtained before the administration of antibiotics.
- The specimen container must be properly labelled.

The following information should be transcribed along with the specimen.

Name of the patient	:
Hospital name	:
Age of the patient	:
Specimen type	:
Clinical diagnosis	:
Possible isolation	:
Date and time of collection	:

Transporting

Speed in transporting specimen is of prime importance because some laboratories refuse to accept specimens if they have been in transit too long. Certain specimens should be transported in a medium that preserves the microorganisms and helps to maintain the ratio of one organism to another. Special treatment is required for specimens when the organism is anaerobic. Anaerobic specimens are transported to the laboratory within 10 minutes. If any delay is anticipated, the specimen must be injected immediately into an anaerobic vial. Vials should contain a transport medium with an indicator, methylene blue. For anaerobic culturing, aspirates are more useful than swabs.

Commonly available transport media are

- Amies transport medium
- Bile peptone transport medium
- Buffered glycerol saline base
- CVTR medium
- Specimen preservative medium
- Transport charcoal medium
- Transport medium
- Wangs semi-solid medium
- Cary-Blair medium
- Stuart's medium
- Alkaline peptone water

Microscopic Examination

Several authors have emphasized the reason for microscopic examination of clinical material. First, the number and percentage of segmented neutrophils indicates the inflammatory response. Second, the quality of the specimen can be validated and the observation of bacteria, mycelial elements, yeast forms, parasitic structures or viral inclusions may provide sufficient information to render an immediate presumptive diagnosis, leading to specific therapy. Third, direct microscopic examination may also give immediate presumptive evidence that species of anaerobic bacteria are present.

The examination of wet mount of unstained material by phase-contrast microscopy or dark-field microscopy, is useful for demonstrating motility, spirochaetes and endospores. Giemsa, Wright's or acridine orange stains may be helpful in observing bacterial forms that stain poorly. Direct staining of clinical material may also be useful to determine whether a specimen is representative of the site of infection. Quality of sputum specimen is analysed by microscopic examination.

If a specimen has large amount of epithelial cells, it indicates the presence of oral contamination. Positive number of neutrophils indicates the presence of active infection.

Some of the microscopic methods are

- Wet mount method
- Hanging drop method
- Saline mount
- KOH mount
- India ink preparation
- Dark field preparation
- Quellung reaction
- Iodine wet mount
- Observation of inclusion bodies

Culturing

After microscopic examination, the following steps are made to recover and identify microbes.

- Select primary culture media
- Determine the temperature and atmosphere of incubation
- Determine whether the antimicrobial susceptibility tests are required or not
- Streak the sample on specific media for primary cultivation

Identification

The following tests are performed for the identification of bacteria.

- Colony morphology observation (cultural variation)
- Catalase test

- Morphological variation (staining)
- Gram-staining
- Simple staining
- Capsule staining
- Spore staining
- Flagella staining
- Albert's staining
- Spirochaete staining
- Motility
- Indole test
- Methyl red test
- Voges-Proskauer test
- Citrate utilization test
- Carbohydrate fermentation test
- Oxidase test
- Coagulase test
- TSI test
- Urease test
- Bile esculin test
- Optochin susceptibility test
- Deaminase test
- Decarboxylase test
- Starch hydrolysis
- Lipid hydrolysis
- Gelatin hydrolysis
- Nitrate reduction test
- SXT test

SERODIAGNOSIS

Studying antigen–antibody reactions with the help of serum is called serology. Serotyping refers to serological procedures used to differentiate strains (serovars or serotypes) of microbes that differ in the antigenic composition or product. It can differentiate not only microbial species but also strains within a species.

In the early 1930s, Rebecca Lancefield recognized the importance of serological test. She developed a classification system for the streptococci based on the antigenic nature of cell wall carbohydrates. Her system is now known as the Lancefield system (A–O). This system is based on specific antibody agglutination reactions with the cell wall carbohydrate antigen extracted from streptococci. Lancefield also showed that further subdividing of group A into specific serological types was possible, based on the presence of type-specific M protein. More recently *E. coli, Salmonella* and other pathogens are serotyped with the help of antigen–antibody reactions involving H, K and O antigens. Within *E.coli,* 167 different O antigens are there. *E.coli* O55, O111, O27 are responsible for infantile diarrhoea. *Salmonella* has 2000 serotypes.

Antigen and antibody reactions include the following.

Precipitation

- Precipitation ring test
- Radial immunodiffusion
- Immunoelectrophoresis
- Electroimmunoassay

- Double immunodiffusion
- Countercurrent immunoelectrophoresis

Agglutination

- Bacterial agglutination
- Widal test
- Weil-Felix test
- Haemagglutinin test
- Complement fixation test
- Radioimmunoassay
- ELISA
- Immunofluoresence
- Quellung reaction based on capsular antigen (*Streptococcus pneumoniae*)

Gonorrhoea, mycoplasmal pneumonia, syphilis, pertussis and leptospirosis are the important diseases detected with the help of serology.

Gene Probe

Nucleic acid probes are used for the identification of some major microbial pathogens. Gene probes detect specific regions of DNA or RNA that are diagnostic of specific pathogenic microorganisms.

Two advantages of gene probe identification are its speed and accuracy, e.g. hybridization and Southern blot.

Phage Typing

Like the serological test, phage typing looks for similarities among bacteria. Both techniques are useful in tracing the origin and course of a disease outbreak. Phage typing is a test used to determine whether the bacterium is susceptible to phage or not. They usually infect only members of a particular species or even particular strains within a species. One strain might be susceptible to two phages, whereas another strain of the same species may be susceptible to those two phages plus a third phage.

This test is used to detect the source of infection. In the following way phage typing is performed.

1. Make a bacterial lawn on the surface of the agar plate.
2. Then place a drop each of different types of phage.
3. During the incubation period, the phage infects and forms a clearing zone (plaque) if the bacteria are susceptible to phage.

By comparison, the microbiologist assesses the source of infection. For example first some organisms are isolated from surgeons or nurses and their phage typing process is checked. Then organisms isolated from surgical wounds are subjected to phage typing. If both organisms have the same type of phage-infecting capability, microbiologists suggest that the surgeon or nurse is the source of infection.

ISOLATION AND IDENTIFICATION OF MICROBES FROM SKIN AND SUBCUTANEOUS TISSUE

Skin harbours large number of normal flora and has good resistance mechanism. It acts as a first line defence system and is a region of the body that is well-understood. During unfavourable conditions, pathogens and opportunistic pathogens will cause severe infections in the skin.

Infections on the Skin

- Folliculitis
- Furuncles
- Carbuncles
- Scalded skin syndrome (SSS)
- Impetigo
- Echthyma
- Erysipelas
- Cellulitis
- Fasciitis
- Myocitis
- Gasgangrene
- Lyme disease
- Piedra
- Tinea corporis
- Tinea cruris
- Tinea facei
- Ring worm
- Rubeola
- Rubella
- Varicella (chickenpox)
- Wart
- Erythema infectiosum
- Measles
- German measles
- Herpes infection

Aetiological Agents

Bacteria

Gram-negative bacilli

- *Pseudomonas* sp.
- *Aeromonas* sp.
- *Plesiomonas* sp.
- *Pasteurella* sp.
- *Actinobacillus* sp.
- *Streptobacillus* sp.

Gram-positive bacilli

- *Listeria* sp.
- *Nocardia* sp.
- *Clostridium* sp.
- *Bacillus anthracis*
- *Streptomyces* sp.

- *Haemophilus* sp.
- *Francisella* sp.

Gram-negative cocci

- *Neisseria* sp.

Gram-positive cocci

- *Staphylococcus aureus*
- *Micrococcus*
- *Enterococcus* sp.
- *Streptococcus pyogenes*
- *Streptococcus agalactiae*
- *Staphylococcus intermedius*

Fungi

- *Microsporum canis*
- *Trichophyton mentagrophytes*
- *Trichophyton verrucosum*
- *Epidermophyton* sp.
- *Candida albicans*
- *Geotrichum* sp.
- *Cephalosporium* sp.
- *Acremonium* sp.

Viruses

- Rubella virus
- Varicella-zoster virus
- Measles virus
- Herpes simplex virus

Specimen Collection

- Tissue obtained during surgical procedure
- Aspirated material from an abscess or deep wound
- Pus or exudates obtained from an infected site during surgery can be aspirated into the syringe
- Skin swab
- Skin scrapings
- Nail clippings
- Hair

Before sample collection skin surface should be disinfected with 70% alcohol.

Transporting

- Stuart's or Amies transport medium is recommended for swabs.
- Aspirated material should be injected into an anaerobic vial that excludes oxygen.
- Tissue samples for culture should be delivered promptly to the laboratory in a sterile container.

Culture Media used for Cultivation

- Colistin–Nalidixic acid Agar (CNA)
- Phenyl ethyl alcohol agar
- Chocolate agar
- Blood agar
- Bacteroides bile esculin agar
- MacConkey agar
- Thayer-Martin medium
- Baird-Parker agar
- KF Streptococcus agar
- Cetrimide agar
- Thioglycolate medium
- Sabouraud dextrose agar
- DTM (Dermatophyte testing medium)
- Cell culture medium for viruses

Processing of the Specimen

- All specimens should be emulsified in broth to create a homogeneous suspension for the preparation of gram-stained smear.
- Suspected anaerobic tissues should be homogenized in thioglycolate broth to avoid oxygenation, which may affect the recovery of anaerobic bacteria.
- Clear aspirated material should be concentrated by centrifugation.
- Both impression smear and homogenized smear are performed while using tissues.
- Swabs should be vortexed in 0.5 ml–1 ml of broth to extract as much suspension as possible. Each swab should be squeezed against the side of the broth tube to express remaining fluid and discarded.

Microscopy

For bacteria, gram-staining is done.

- Use a drop of homogenized suspension for smearing.
- Air-dry the smear.
- Flood smear with 90–100% alcohol and allow the smear to air-dry for fixing.
- Stain the smear by gram-staining technique.
- Observe the slide under oil immersion lens and report appropriate result.

For fungus, skin scrapings, nail clippings and hair should be examined after preparation with potassium hydroxide (KOH). The KOH preparation is made by emulsifying the specimen in a drop of 10% KOH on a microscopic slide. The purpose of KOH is to clear out any background scales or cell membranes that may be confused with hyphal elements. Clearing can be accelerated by gently heating the mixture over the flame of a Bunsen burner. A coverslip is applied and the specimen is examined for the presence of narrow regular hyphae that characteristically break up into arthroconidia. The visualization was improved by adding calcofluor white to the KOH reagent.

Culturing

- Any two media is selected from the media section as per the recommendation of the clinician or gram-stain observation.
- Both liquid and solid media are used for the cultivation.
- Homogenized tissue materials are inoculated into both bacterial and fungal medium.
- Both aerobic and anaerobic media are used for the cultivation.
- Routine agar should be inoculated in moist atmosphere of 5–10% CO_2 at 35–37°C for a minimum of 48 hours.
- Anaerobic media should be incubated in an anaerobic atmosphere for a minimum of 72 hours.
- All broths, MacConkey agar and fungal media are incubated aerobically.

Examination of the Media

- Examine all broths and plates. Gram-stain and subculture the broth if there is no growth on plates but evident growth in broth.
- Describe the colony type of all isolates.
- Perform standard identification techniques to speciate the isolate.
- Antibiotic sensitivity pattern of the isolate is performed by disc diffusion technique.

ISOLATION AND IDENTIFICATION OF MICROBES FROM UPPER RESPIRATORY TRACT

Infections of the Upper Respiratory Tract

- Sinusitis
- Otitis media
- Pharyngitis
- Laryngitis
- Pertussis
- Epiglottitis
- Tracheitis
- Thyroiditis
- Common cold
- Strep throat
- Diphtheria
- Suppurative parotitis
- Tonsillitis
- Scarlet fever
- Parotitis
- Rheumatoid fever

Aetiological Agents

Bacteria

Streptococcus pyogenes
Other beta haemolytic streptococci
Bordetella pertussis
Streptococcus pneumoniae
Haemophilus influenzae
Corynebacterium diphtheriae
Pseudomonas aeruginosa
Borrelia sp.
Bacteroides melaninogenicus
Neisseria meningitidis
Klebsiella sp.
Staphylococcus aureus

Viruses

Reovirus
Coxsackievirus A
Coxsackievirus B
Echovirus
Rhinovirus
Coronavirus
Mastadenovirus
Herpes simplex virus type 1

Fungus

Candida albicans

Specimen

- Nasopharyngeal swab
- Sodium alginate throat swab for pertussis
- Sinus washings
- Surgical biopsy
- Swab of posterior pharynx
- Swab of tonsil
- Pernasal swab
- Nasopharyngeal aspirates

Nasopharyngeal aspirates Gently pass a sterile catheter through one tonsil as far as the nasopharynx. Attach a sterile syringe to the catheter, and aspirate the specimen of mucopus. Dispense the specimen into a sterile container.

Nasopharyngeal swab or throat swab Bright light should be focused into the oral cavity. The patient is instructed to open the mouth at "aah" position and breathe deeply. The tongue is gently depressed with tongue blade to visualize the tonsillar fossae and posterior pharynx. The swab is extended between the tonsillar pillars and behind the vulva. The tonsillar areas and posterior pharynx should be firmly rubbed with the swab. After collection the cotton swabs are placed into sterile Cary-Blair transport media to prevent desiccation during transit to the laboratory.

Media

- Blood agar plate
- Cystine tellurite blood agar
- Loeffler agar slant
- Chocolate agar
- Charcoal cephalexin blood agar (CCBA)
- Modified tinsdale medium
- Sabouraud dextrose agar
- Thayer-Martin agar
- Cetrimide agar
- New York city agar medium
- Baird-Parker agar
- MacConkey agar

Processing

Swab the affected area with sterile cotton or alginate swab. For 8 hours before swabbing the patient must not be treated with any antibiotics. Within two hours, specimen must be transported to the laboratory.

Microscopy

Albert's staining technique The Albert technique is used to stain the volutin or metachromatic granules of *Corynebacterium diphtheriae*. The granules are most numerous after the organism has been cultured on a protein-rich medium such as Dorset egg or Loeffler's serum. Metachromatic granules can also be found in other *Corynebacterium* species and occasionally in some *Bacillus* species.

Reagents required for the Albert's staining technique are toluidine blue-malachite green stain and Albert's iodine.

Procedure

1. Prepare a thin smear.
2. Fix the dried smear using alcohol.
3. Cover the smear with the toludine blue–malachite green stain for 3–5 minutes.
4. Wash off the stain with clean water.
5. Cover the smear with Albert's iodine for 1 minute. Wash off with water.
6. Wipe the back of the slide clean, and allow the slide to dry.
7. Examine the smear microscopically, first with the 40× objective to check the staining and to see the distribution of the material and then with the oil immersion lens to look for bacteria containing metachromatic granules.

Results

Bacterial cells	Green
Metachromatic granules	Green-black

Culturing

- Inoculate the swab on a plate of blood agar.
- Add a bacitracin disc to the plate. This will help in the identification of *Streptococcus pyogenes*.
- Tinsdale's medium and tellurite blood agar is used for the recovery of *Corynebacterium diphtheriae.*
- Chocolate agar is used for the isolation of *H. influenzae,N. meningitidis*.
- Charcoal cephalexin blood agar and Bordet-Gengou medium is used for the recovery of *Bordetella pertussis*.

- MacConkey agar was used to isolate gram-negative pathogens.
- After inoculation all the plates should be incubated at appropriate temperature for a required time.

ISOLATION OF MICROORGANISMS FROM LOWER RESPIRATORY TRACT

Lower respiratory infections are a major cause of morbidity and mortality. Diagnosis of these infections is often complicated by the contamination of specimens with upper respiratory tract secretions. Because the upper respiratory tract may be colonized with potential pathogens not involved in infection of the lower respiratory tract, the laboratory must ensure that an appropriate specimen is used for processing.

Infections of Lower Respiratory Tract

- Pertussis
- Tuberculosis
- Bronchitis
- Legionellaire's
- Pneumonia
- Cryptococcosis
- Influenza
- Histoplasmosis
- Lung abscess
- Anthrax
- Plague

Possible Pathogens

- *Streptococcus pyogenes*
- *Klebsiella pneumoniae*
- Beta haemolytic group B streptococci
- *Chlamydiae pneumoniae*
- *Pseudomonas aeruginosa*
- *Staphylococcus aureus*
- *Streptococcus pneumoniae*
- *Mycoplasma pneumoniae*
- *Legionella pneumophila*
- *Moraxella catarrhalis*
- *Yersinia pestis*
- *Bacillus anthracis*
- *Histoplasma capsulatum*
- *Cryptococcus neoformans*
- *Cytomegalovirus*
- *Candida albicans*
- *Nocardia asteroides*
- *Blastomyces dermatitidis*
- *Aspergillus fumigatus*
- *Torulopsis* spp.
- *Fusarium* spp.
- *Curvularia lunnatum*
- *Francisella tulariensis*
- *Geotrichum* spp.
- *Toxoplasma gondii*
- *Cryptosporidium*
- Adenovirus
- Influenza virus
- Herpes simplex virus 1

Media Used

- Blood agar
- Chocolate agar
- MacConkey agar
- Baird-Parker agar
- Cetrimide agar
- Charcoal yeast extract medium
- LJ medium
- Buffered charcoal yeast extract agar
- Cell-free medium
- Bordet-Gengou medium
- Thayer-Martin medium

Specimen

For the pathogen isolated from LRT, samples are obtained through needle aspiration method. In this method, a needle is inserted into the throat region and sputum is collected. Needle aspiration includes transtracheal aspiration and lung aspiration. Other methods are bronchial washing and blood collection.

Early morning sputum should be collected because that contains pooled overnight secretions in which pathogenic bacteria are more likely to be concentrated. For collection sterile wide mouthed jar with a tightly firmed screw cap lid can be used.

Transporting

If acid-fast bacilli suspected sputum is transported or stored by adding cetyl pyridinium chloride–sodium chloride (CPC–NaCl). It digests sputum and prevents the overgrowth of other pathogens.

Gram Stain for Lower Respiratory Tract Specimens

Gram stain can aid in rapid diagnosis and appropriate treatment. It is an efficient method for the assessment of the quality of the sputum based on cellular composition of the specimen. The following cells were observed after staining:

PMN (polymorphonuclear leucocytes), squamous epithelial cells, ciliated columnar epithelial cells and bacteria. On the basis of gram-staining, Bartlett grades the sputum sample as follows.

Number of neutrophils per field	Grade
<10	0
10–25	+1
>25	+2
Presence of mucus	+1
Number of epithelial cells per field	Grade
10–25	–1
>25	–2

Smear the purulent portion of the sputum.

Fix the smear by heat and methanol fixing procedure.

Stain the smear by gram-staining technique.

Observe the smear under 40× microscope.

Acid-fast Staining

- Using a piece of stick, transfer a purulent part of the sputum, especially that containing any piece of yellow caseous material to a slide and make a thin smear. Use a circular movement to spread the specimen.
- Allow the smear to air-dry in a safe place. Fix the smear with alcohol by covering it with one or two drops of ethanol or methanol for 2–3 minutes.
- Cover the smear with the filtered carbol fuchsin stain.
- Heat the stain until vapour begins to rise. (Do not overheat.)
- Allow the heated stain to remain on the slide for 5 minutes.
- Wash off the stain with clean water.
- Cover the smear with 3% v/v acid alcohol for 5 minutes or till the smear is sufficiently decolorized, i.e., pale pink.
- Wash well with clean tap water.
- Cover the smear with malachite green for 1–2 minutes.
- Wash the stain with clean tap water.
- Wipe the back of the slide clean and allow the smear to air-dry.
- Examine the smear microscopically.

Reporting

More than10 AFB/field	+++
1–10 AFB/field	++
10–100AFB/100 field	- +
1–9AFB/100 field	- report the exact number

Wayson's Staining Technique

It is also called as bipolar staining technique.

It is helpful to see and differentiate *Yersinia pestis* from others.

Procedure

- Prepare the smear.
- Air-dry.

- Fix the smear with the help of alcohol.
- Cover the smear with Wayson's stain for 10–20 seconds.
- Wash off the stain with clean water.
- Wipe the back of the slide clean, and place in a draining rack for the smear to air-dry.
- Examine the smear microscopically, first with the 40× objective to see the distribution of material and then with the oil immersion objective to look for bipolar-stained bacteria.

Results

- Bacteria stain blue with pink ends.

KOH Mount

Perform potassium hydroxide wet mount for fungal etiology. Transfer a small amount of sputum to a glass slide. Add a drop of 10% potassium hydroxide solution, mix and cover with a cover glass. Examine the preparation under 40× objective.

Culturing

Routine

- Wash the purulent part of the sputum with sterile 5 ml of physiological saline.
- Inoculate washed sputum on plates of blood agar and chocolate agar.
- Add an optochin disc to the chocolate agar. This will help to identify *Streptococcus pneumoniae.*
- Incubate the blood agar plate aerobically and chocolate agar plate in a carbon dioxide-enriched atmosphere at 35–37°C for up to 48 hours.
- Examine the growth and report the result.

For AFB

About 20 minutes before culturing, decontaminate the specimen by mixing equal volumes of sputum and NaOH 40 g/l solution. Shake at intervals to homogenize the sputum.

Using a sterile premarked Pasteur pipette, inoculate 200 microlitre of the sputum on a slope of acid Lowenstein–Jensen medium. Allow the specimen to run down the slope.

Slope turns yellow due to alkalinity of the specimen but it will become green again because the acid in the medium neutralizes the NaOH.

Incubate the tubes at 37°C in rack placed at an angle of about 45° to ensure that the specimen is in contact with the full length of the slope.

After one week, place the slopes in an upright position and continue to incubate the cultures for a further 5–6 weeks, examining twice a week for growth.

For other pathogens Streak the sputum on selective and differential media for isolation and differentiation.

For fungal aetiology Streak the purulent portion of the sputum on the surface of the Sabouraud dextrose agar with cycloheximide and incubate aerobically for 48 hours.

ISOLATION OF MICROORGANISMS FROM GASTROINTESTINAL TRACT

Normal human intestine harbours more than 500 types of microbes. Among these, some of the microbes are considered as pathogens. Pathogens may enter through food and water systems. Most of the intestinal disorders are based on the toxins.

Infection

- Dental caries
- Gum disorders
- Diarrhoea
- Dysentery
- Shigellosis
- Salmonellosis
- Gastritis
- Gastroenteritis
- Enteric fever
- Liver abscess
- Liver granuloma
- Cholera

Possible Pathogens

- *Escherichia coli*
- *Salmonella enteritidis*
- *Shigella* spp.
- *Campylobacter*
- *Vibrio* spp.
- *Plesiomonas* spp.
- *Aeromonas* spp.
- *Yersinia enterocolitica*
- *Clostridium perfringens*
- *Clostridium difficile*
- *Staphylococcus aureus*
- *Streptococcus mutans*
- Coxsackie virus
- Adenovirus
- Echovirus
- Poliovirus
- *Entamoeba histolytica*
- *Giardia lamblia*
- *Giardia intestinalis*
- *Trichomonas hominis*
- *Balantidium coli*
- *Isospora belli*
- *Cryptosporidium*
- *Ascaris lumbricoides*

- *Bacillus cereus*
- *Helicobacter pylori*
- Rotavirus
- *Schistosoma* spp.
- *Taenia* spp.
- *Enterobius vermicularis*

Medium

- MacConkcy agar
- GN broth
- HE agar
- TCBS agar
- Alkaline peptone water
- Campylobacter isolation medium
- Yersinia selective medium
- Deoxycholate citrate agar
- XLD agar
- Wilson and Blair medium
- Selenite broth
- Salmonella-Shigella agar
- MacConkey sorbitol agar
- Cefsulodin Irgasan novobiosin agar
- Cycloserine cefoxin fructose egg yolk agar
- Campy blood agar
- Blood agar
- Blood agar with 10 mg ampicillin
- Inositol brilliant green bile salt agar
- Bismuth sulphite agar
- Vibrio agar
- Entamoeba isolation medium

Specimen

Stool or rectal swab or gastric aspirate or gastric biopsy.

Specimen Collection

The collection of diarrhoeal stool is not difficult. In cases of diarrhoea, stool specimen should be collected in clean, wide-mouthed containers that can be covered with tight fitting lid. In some instances, collection of rectal swab rather than stool specimen may be necessary, particularly in neonates or in severely debilitated adults.

Gastric aspirate is collected with the help of intubation. Sample collection from the hollow tube is called intubation. Long sterile tube is attached with syringe and the tube is either swallowed or passed through a nostril into the patient's stomach. Specimens are withdrawn periodically. The most common intubation tube is Levin tube.

Transport

The specimen should be transported as early as possible. Don't refrigerate the stool specimen if possible because certain species of *Shigella* species are susceptible to cooling and drying.

Processing

Macroscopy

Observe the nature and colour of the specimen.

Unformed stool containing pus, mucus with blood	Shigellosis
Semiformed stool with mucus	Schistosomiasis
Water stool	Rotavirus, ETEC
Rice-water stool	Cholera
Unformed pale coloured stool with unpleasant smell	Giardiasis

Microscopy

By using iodine wet mount technique and gram-staining technique, observe the microscopic nature of the specimen.

Iodine and saline wet mount

- Place a drop of Lugol's iodine and physiological saline in two clean slides.
- Mix a small amount of specimen with each drop.
- Cover each preparation with coverslip.
- Examine the preparation using 10× and 40× objectives.
- Look for protozoan and helminthic parasites.

If cholera is suspected, hanging drop technique is performed to observe the motility.

Culturing

- Gram-negative broth and alkaline peptone water are inoculated with few loopsful of stool specimen and incubated at 37°C for 4–5 hours. Observe turbidity. This step is used to enrich the pathogens.

- Streak Hektoen enteric agar by using gram-negative broth, streak TCBS agar with alkaline peptone water inoculum. Observe the plates for the pathogens after 24 hours of incubation at 37°C.
- At the same time, the remaining media given in the medium section are streaked by direct method.
- Incubation was performed with either aerobic method or anaerobic method based on the medium and causative agent.
- After the completion of primary plating techniques, bacterial pathogens are identified through various biochemical means.

ISOLATION OF MICROORGANISMS FROM URINE

Urine is normally a sterile body fluid. However, unless it is collected properly, it can become contaminated with microbiota from the perineum, prostate, urethra or vagina. The presence of bacteria in urine is called bacteriuria. Significant bacteriuria is usually accompanied by pyuria (pus cells in urine). Infection of the bladder is called cystitis, and infection of kidney is called pyelonephritis. *Escherichia coli* is a commonest cause of urinary tract infection.

Possible Pathogens

Staphylococcus saprophyticus
Pseudomonas aeruginosa
Escherichia coli
Proteus spp.
Klebsiella spp.
Haemolytic streptococcus
Enterococcus spp.

Medium

Blood agar
MacConkey agar
Cetrimide agar
CLED agar
SS agar
KF streptococcus agar
Nutrient agar

Sample Collection

Mid-stream urine is collected in a sterile, dry, wide-necked, leak-proof container. About 20 ml of sample should be collected. Clean catch method is used to collect mid-stream urine, first voided urine is not collected because it is contaminated with microbes from lower portion of the urethra. If immediate delivery to the laboratory is not possible, the urine should be refrigerated at 4°C. If a delay of more than 1 hour is anticipated, boric acid should be added to the urine. Specimen containing boric acid need not be refrigerated.

Cathetral aspiration A catheter is a tubular instrument used for withdrawing fluids from the body cavity. Three types of catheters may be used.

Hard catheter, French catheter and Foely catheter in which multiple samples may be collected at a time.

Suprapubic aspiration Suprapubic aspiration is performed only in neonates, small children and occasionally for adults with clinically suspected UTI but who fail to establish diagnosis. This technique is best performed when the bladder is full. The suprapubic skin overlaying the urinary bladder is disinfected and tap is made. In the immediate site where the tap is to be made, about 1 ml of anaesthetic solution is injected subcutaneously. With the point of a sharply tapered surgical blade, a small lance wound incision through the epidermis is made. Through this wound gently extend an 18-gauge needle into the urinary bladder and aspirate 10 ml of urine into the syringe.

Macroscopy

Note the appearance of the specimen.

- Colour of the specimen
- Nature of the specimen

Normally freshly passed urine is clear and is pale yellow to yellow in colour depending on concentration. When left to stand, cloudiness may develop due to precipitation of urates in acid urine or phosphates and carbonates in alkaline urine. Urates may give the urine a pink-orange colour.

The appearance of urine during various infections is given below.

Colour	Infection
Cloudy	Bacterial
Red and cloudy	Bacterial and urinary schistosomiasis
Brown and cloudy	Black water fever
Yellow-brown and green-brown	Acute viral hepatitis
Yellow-orange	Haemolysis and hepatocellular jaundice
Milky white	Bancroftian filariasis

Microscopic Examination

- Place 3 loopful of well-mixed fresh urine on a slide, and cover with a cover glass. Examine the preparation using the 10× and 40× objectives.
- Using centrifuged urine, perform gram-staining.

Findings of microscopic examination

Bacteria	Casts
White cells	Hyaline
Pus cells	Waxy
Red cells	Cellular
Yeast cells	Granular
Epithelial cells	
Casts	
Crystals	
Parasites	

Culturing

1. Approximate number of bacteria per ml of urine can be estimated by using calibrated loop technique (0.002 ml capacity or 1/500 ml or 20 × 500=10,000).
2. Urine is diluted up to 10^{-5} and perform pour plate technique is performed by using nutrient agar (normal urine had less than 10^4 bacteria).
3. For selective isolation, mix urine properly and inoculate a loopful of urine on blood agar, MacConkey agar, CLED agar, cetrimide agar and SS agar.
4. Incubate all plates under aerobic condition.

ISOLATION OF MICROORGANISMS FROM BLOOD

Infections

- Septic shock
- Endocarditis
- Vascular encephalitis
- Pericarditis
- Myocarditis
- Septicaemia
- Tularemia
- Brucellosis
- Plague
- Infectious mononucleosis
- Dengue
- Malaria
- Anthrax
- Typhoid fever
- Puerperal sepsis
- Rheumatic fever
- Relapsing fever
- Yellow fever

Possible Pathogens

- *Staphylococcus aureus*
- Viridans Streptococci
- *Streptococcus pneumoniae*
- *Streptococcus pyogenes*
- *Salmonella typhi*
- *Escherichia coli*
- *Klebsiella pneumoniae*
- *Corynebacterium diphtheriae*
- *Yersinia pestis*
- *Leptospira* species
- *Brucella* species
- Beta haemolytic group B streptococci
- *Proteus* species
- *Haemophilus influenzae*
- *Neisseria* species

Media

- Thioglycolate broth
- Tryptone soya diphasic broth
- Blood agar
- Chocolate agar
- MacConkey agar
- SS agar
- EDTA
- Materials for WBC and differential count
- Giemsa stain

Sample Collection

Blood collection was performed by needle aspiration procedure. To reduce the contamination during vein puncture, the following method should be followed.

1. Wash with green soap.
2. Rinse with clean water.
3. Apply 1–2% tincture of iodine and allow to dry for 1–2 minutes.
4. Remove iodine with 70% alcohol.
   i. Blood should be collected before antimicrobial treatment has been started and at the time when the patient's temperature begins to rise.
   ii. Blood for culture should be taken by vein puncture.
   iii. 10–20 ml of blood is collected from adults.
   iv. 1–2.4 ml is collected from young infants.
   v. 2.4–5 ml is collected from old infants.
   vi. Two blood cultures should be performed for each patient to confirm the causative agent.

Culturing

- Prepare thioglycolate and tryptone soya diphasic medium and sterilize at appropriate temperature.
- Withdraw appropriate volume of blood from patient.

- Divide the blood into three portions. Inoculate two portions into thioglycolate medium and tryptone soya diphasic medium.
- Inoculate the remaining portion into a bottle containing EDTA. This blood is used to perform total count, differential count, staining and also for streaking (one loopful of blood) into the SS agar.
- Incubate all culture bottles at 37°C.

Examination and Subculturing

For thioglycolate broth

- Examine daily for up to 14 days.
- Look for visible signs of bacterial growth such as turbidity above the red cell layer.
- A strict aseptic technique must be used to avoid contamination.
- Using a sterile needle and small syringe, insert the needle through the rubber liner in the cap, and withdraw 1 ml of broth culture.
- Inoculate the broth on blood agar, chocolate agar and MacConkey agar.
- Incubate blood agar and chocolate agar anaerobically for 48 hours and MacConkey agar plate aerobically overnight.

For tryptone soya diphasic culture

- Examine daily for 7 days and twice a week for up to 4 weeks.
- Look for colonies on the agar slope and for signs of bacterial growth on broth.
- If growth is present, subculture on blood agar, chocolate agar and MacConkey agar.
- Examine gram-stained smear for the colonies.
- If large gram-positive rods resembling *C. perfringens* are seen, subculture on lactose egg yolk milk agar and incubate the plate anaerobically.
- If *Brucella* is suspected, increased attention should be given to it, and marked as HIGH RISK.

ISOLATION OF MICROORGANISMS FROM GENITAL TRACT

The normal human genital tract is lined with a mucosal layer composed of transitional columnar and squamous epithelial cells. Many indigenous microorganisms colonize these surfaces. Female genital tract infections arise from endogenous microorganisms.

Infections

Female

- Gonorrhoea
- Syphilis
- Amnionitis
- Bartholinitis
- Cervicitis
- Endometritis
- Urethritis
- Vulvovaginitis
- Salpingitis

Male

- Epididymitis
- Orchitis
- Prostatis
- Urethritis
- Gonorrhoea
- Syphilis

Aetiological Agents

- *Neisseria gonorrhoeae*
- *Haemophilus influenzae*
- *Chlamydia trachomatis*
- *Bacteroides fragilis*
- *Streptococcus pyogenes*
- *Proteus mirabilis*
- *Gardnerella vaginalis*
- *Pseudomonas aeruginosa*
- *Treponema pallidum*
- *Listeria monocytogenes*
- *Ureaplasma urealyticum*
- *Peptostreptococcus* spp.
- *Streptococcus agalactiae*
- *Enterococcus* spp.
- *Candida albicans*
- Some Enterobacteriaceae members

Media

- Blood agar
- MacConkey agar
- Colistin nalidixic acid agar (CNA)
- CLED medium
- Vaginalis agar
- Cetrimide agar
- Mycoplasma isolation medium
- Sabouraud dextrose agar
- Chocolate agar
- Modified Thayer-Martin agar
- Listeria selective medium
- Mitis salivaris agar
- Todd-Hewitt broth
- Cell-free medium
- XLD medium

Specimen Collection

Female

- Catheter aspirate
- Bartholin gland aspirate
- Amniocentesis fluid
- Transcervical aspirate

- Fallopian aspirate
- Urethral discharge
- Swab of genital ulcer
- Swab of posterior vagina

Male

- Urethral swab
- Swab of genital ulcer
- Penial discharge
- Prosthetic secretions

For cases like syphilis, blood samples is collected and serological techniques are performed.

Microscopy

- Specimens was smeared on clean microscopic slide and allow to air-dry.
- Fix the specimen by using methanol.
- Stain by gram-staining procedure.
- Observe under 40× objective lens.

For the detection of syphilis, fluorescently labelled antibody technique, rapid plasma regain test, VDRL, RPR test, Khan test and Wasserman reaction are performed.

Culturing

- As per the observation under microscopy and instruction of the physician, select the media.
- Inoculate the specimen on the selective and enrichment media.
- Incubate all the media under aerobic, micro-aerobic and anaerobic environment based on the type of causative agent.

ISOLATION OF MICROORGANISMS FROM CSF

Bacterial meningitis is the result of infection of the meninges. Identification of the infecting agent is one of the most important functions of the diagnostic laboratory because acute meningitis is a very serious infection. Aerobic bacteria commonly cause meningitis. Inoculation of anaerobic bacteria is not recommended. CSF may contain very few microorganisms per ml of fluid, therefore, concentration of the specimen is recommended. Any positive finding on gram-staining or culture must be reported to the physician immediately.

Possible Pathogens

- *Escherichia coli*
- *Streptococcus agalactiae*

- *Listeria monocytogenes*
- *Haemophilus influenzae*
- *Neisseria meningitidis*
- *Streptococcus pneumoniae*

Specimen

- Lumbar puncture
- Brain abscess

The chance of recovery increases with the volume of the specimen.

Suggested volumes are

1 ml for bacterial culture

2 ml for fungus

Do not refrigerate the specimen.

Media

- Blood agar
- Chocolate agar
- Modified Thayer-Martin agar
- New York city agar medium
- GC medium

Processing

Record the gross appearance of CSF.

Centrifuge the specimen.

Aspirate the supernatant.

Vortex the sediment vigorously for at least 30 seconds.

Microscopy

- Prepare the smear by placing one or two drops of sediment on an alcohol-rinsed slide allowing the drop to form a large heap.
- Fix the smear by using methanol and perform gram-staining.
- Examine the slide and report to the physician immediately if there are any positive findings.

Culturing

- Using a sterile pipette, inoculate the media by placing one or two drops of the vortexed sediment on two plates.
- Incubate all plates at 35°C in 5 to 10% carbon dioxide containing environment for 48 to 72 hours.
- Examine all plates for growth.
- If growth is observed, perform the identification procedure.

ISOLATION OF ANAEROBES

Microbes exhibit great diversity in their ability to use free oxygen for growth. These variations in oxygen requirements reflect the differences in the bio-oxidative enzyme system present in the various species. The microbes can be classified into various types based on their oxygen requirements.

Aerobes require oxygen as a terminal electron acceptor and will not grow in the absence of oxygen.

Anaerobes do not use oxygen for growth and metabolism but rather obtain their energy from fermentative process.

Microaerophilic require oxygen as a terminal electron acceptor but fail to grow on the surface and no growth under anaerobic condition.

Facultative anaerobes can grow either oxidatively or anaerobically.

Capnophilic require carbon dioxide for growth.

Anaerobic metabolism is essentially fermentative and may include an electron transport system with an organic final electron acceptor. Products of anaerobic metabolism include short chain fatty acids, alcohols and amines. Oxygen toxicity is due to alteration of redox potential in the presence of oxygen rendering enzymes inactive and to the ability of anaerobes to deal with the toxic products of oxygen metabolism (hydrogen peroxide, superoxide, hydroxy free radicals, singlet oxygen).

Anaerobes can be grown on agar plates in an atmosphere without oxygen usually nitrogen 80%, carbon dioxide 10% and hydrogen 10%. Carbon dioxide stimulates growth and the hydrogen combines with oxygen and may be used to maintain anaerobic condition. This is achieved in anaerobic jars with sealed and clamped lids from which air is evacuated and replaced by the anaerobic gas mixtures. In all cases oxygen and hydrogen reaction is promoted by the presence of catalyst. Anaerobes may also be grown in liquid cultures in an anaerobic jar/chambers or even in deep broth with reducing agents.

All facultative anaerobes and aerobes have the following pathways for handling oxygen but most anaerobes do not have it.

Cytochrome metabolic pathways for oxygen

Superoxide dismutase which catalyses the reaction

$$H_2 + 2O_2 \qquad H_2O_2 + O_2$$

Catalase which catalyses the reaction

$$2\,H_2O_2 \qquad 2H_2O + O_2$$

Anaerobes cannot be cultivated in the presence of oxygen. An anaerobic environment is essential for the growth of strict anaerobes.

Anaerobic bacteria cause a variety of infections in humans. Anaerobic infections are generally endogenous. According to various reports 50–60% of important infections are caused by anaerobic bacteria. Anaerobes are very important because they can resist many antimicrobial agents.

Anaerobic Bacteria

Gram-positive bacilli

- *Clostridium*
- *Desulfotomaculum*
- *Acetobacterium*
- *Lactobacillus*
- *Propionibacterium acnes*
- *Eubacterium lentum*
- *Bifidobacterium*
- *Actinomycetes*

Gram-negative bacilli

- *Anaerobacter*
- *Bilophila*

Gram-positive cocci

- *Peptococcus*
- *Peptostreptococcus*
- *Ruminococcus*
- *Sarcina*

Gram-negative cocci

- *Veillonella*
- *Acidaminococcus*
- *Desulpholobus*
- *Wollinella*
- *Bacteroides*

Prevotella
Porphyromonas
Fusobacterium

Anaerobic Infections

- Appendicitis
- Otitis media
- Endocarditis
- Myocarditis
- Peritonitis
- Salpingitis
- Sinusitis
- Bacteraemia
- Cholecystitis
- Dental and oral infections
- Endometritis
- Osteomyelitis
- Emphyema
- Septic arthritis
- Trauma

Specimen

Anaerobes are often missed unless the specimen is properly collected, transported to the laboratory and then isolated properly.

Head and neck	Abscess and biopsy
Lungs	Transtracheal aspirate, lung puncture and biopsy
CNS	Abscess, biopsy, lumbar puncture
Abdomen	Peritoneal fluid and biopsy
Urinary tract	Suprapubic aspirate
Female genital tract culdoscopy specimen	Endometrial aspirate
Bone and joint	Aspirate
Soft tissue	Aspirate

Aspirates are transported with vials with anaerobic atmosphere.

All specimens should be transported within 30 minutes.

Media

Successful isolation of anaerobes depends on choosing the correct primary growth media and environmental condition. Most anaerobes require hemin and vitamin K for growth.

Brucella blood agar	Brucella
Phenylethyl alcohol agar	Inhibit facultative gram-negative rods
Bacteroides bile esculin agar	Bacteroides

Cycloserine cefoxitin fructose agar	*C difficile* (yellow ground glass colony)
Egg yolk agar	*Clostridium perfringens*
Robertson cooked meat	*Clostridium*
Chopped meat broth	*Clostridium*
Columbia blood agar	*Prevotella*
CDC anaerobic agar	*Prevotella, Fusobacterium*
Anaerobic kanamycin vancomycin blood agar	*Fusobacterium*
LKV medium	*Fusobacterium*, *Bacteroides* *Prevotella*

Processing

Both microscopic and culturing techniques are performed.

Microscopy

Modified gram-staining is a common microscopic technique (Kopeloff's modification)

- Prepared the smear.
- Fix by using methanol.
- Flood alkaline crystal violet and add 5 drops of sodium bicarbonate. Wait for 2–3 minutes.
- Wash with water.
- Add Kopeloff's iodine for 2 minutes.
- Decolorize with 3:7 acetone alcohol.
- Flood safranin for 10–30 seconds.

Culturing

- The specimen is streaked on selective and differential medium.
- Use capillary pipettes for the inoculation of primary inoculation media with liquid specimen.
- Inoculate liquid media near the bottom with 1 or 2 drops of inoculum.
- For streaking, place one drop of sample on each medium and then streak the drop with platinum or nichrome loop, using a quadrant streak technique.
- Mince the solid tissue specimens with sterile scissors. Add one part of enriched thioglycolate medium per volume of tissue and grind the mixture with sterile tissue grinder.
- All the plates should be incubated under anaerobic condition.

Available methods of anaerobiosis are

- Pyrogallol method or rolling tube or Wright's tube method
- Agar shake method
- Jar technique
- Evacuation and replacement method
- Gas pak method
- Anaerobic glove box technique
- Anaerobic holding jar method
- Pre-reduced anaerobically sterilized media (PRAS)

Rolling tube method

- It was developed by Hungate in 1950 to isolate strict anaerobes.
- Prepare medium and sterilize at 121°C for 15 minutes and make a slant.
- Inoculate the slant with the test specimen.
- Burn the cotton and insert a plug of burned cotton into the middle region of the tube.
- Add few pellets of pyrogallol and pour 0.5 ml of 1 M sodium hydroxide.
- Seal the tube with rubber bunk.
- Incubate the tubes at 37°C for 72 hours.

Agar shake method It is a simple method requiring simple equipment and yet it is extremely effective for the isolation of anaerobes. Van Niel pioneered this method in 1931.

Procedure

- Prepare 9 ml of 1% agar-containing media in a series of test tubes.
- Keep the media in test tube molten at 42–44°C.
- Inoculate the first tube in the dilution series with preferable amount of soil or water or sediment and shake well.
- Then perform dilution from first tube to end.
- After dilution cool the tubes and seal with a film of liquid paraffin.
- Then insert the absorbent cotton into the tubes and add a few pellets of pyrogallol and 0.5 ml of 1 M calcium carbonate prior to sealing.
- Finally seal the tubes with a rubber bung, an additional safeguard to prevent the entering of oxygen.
- Incubate all tubes at room temperature for 72 hours.
- Isolate the colonies from the tube by using sterile Pasteur pipette and purify the isolates.

Anaerobic jar technique An anaerobic jar is a cylindrical container made up of plastic, glass or metal. A metal or plastic lid is clamped to a flange at the top of the jar to create an airtight seal. Some jar lids have vents or valves through which air can be evacuated and an anaerobic gas mixture can be added.

Evacuation-replacement jar procedure (ER) A suitable gas mixture for the ER procedure is 10% hydrogen, 5% carbon dioxide and 85% nitrogen. There is no need to use commercial high pressure vaccum pump. To perform the procedure, replace the used catalyst in the lid of the jar with fresh one. Put the material to be incubated inside the jar. Place methylene blue indicator in the jar. After closing the jar with lid, connect the vent on the lid and evacuate the jar to 20–24 in. of mercury and fill the jar with commercial-grade nitrogen. Repeat the cycle. After the third evacuation, fill the jar with anaerobic gas mixture. Clamp the rubber tubing attached to the vented jar, disconnect the jar from the vacuum and gas line. Then place the jar in an incubator.

Anaerobic gas-pak jar method It uses the cold catalyst consisting of palladium-coated alumina pellets, which is active at room temperature. The disposable gas pak H_2–CO_2 generator consists of a sealed foiled envelope containing two tablets. One contains citric acid and sodium bicarbonate and the other contains sodium borohydride. When water is introduced into the envelope, the former tablet releases CO_2, while the other releases H_2. Hydrogen combines with oxygen and forms water through condensation process.

Procedure

- Replace the used catalyst with fresh one.
- Put the material to be incubated inside the jar.
- Place methylene blue indicator in the jar.
- Cut the corner of the gas-pak envelope.
- Place the envelope in upright position.
- Add 10 ml of water.
- Clamp the lid on the jar.
- Incubate the jar under appropriate environment.

Anaerobic glove box technique Anaerobic glove box is a self-contained anaerobic system. It consists of a gas-tight chamber with glove portals and an entry lock for the transfer of materials in or out of the chamber. The operator of the chamber uses gloves to handle the culture. A H_2-containing atmosphere is recirculated through palladium catalyst to remove O_2 from inside chamber. R.G.Freter and colleagues at the University of Michigan developed a flexible vinyl plastic glove box. Media are incubated in an incubator placed inside the chamber. Accessories of glove box are a rigid metal entry lock, vacuum pump, a gas mixture tank (85% N_2, 10% H_2 and 5% CO_2) and a tank of commercial-grade N_2. Relative humidity within glove box should be maintained at 70–85% moisture.

PRAS and the roll-streak tube method W.E.C. Moore and associates developed it. Their system is based on R.E. Hungate method. During preparationof the PRAS media, media components are combined, boiled to remove dissolved O_2 and then tubed, autoclaved and stored in butyl rubber stopper tubes under O_2-free environment. Commonly three types of commercially available CO_2 can be used for gassing roll tubes and liquid media.

1. Anaerobic-grade CO_2 which does not pass through catalyst.
2. Commercial-grade CO_2 which passes through copper catalyst.
3. A mixture of 97% CO_2 and 3% H_2.

Anaerboic disposable plastic bag The anaerobic bio-bag consists of a clear plastic bag, an H_2–CO_2 generator, palladium catalyst pellets, and a resazurin indicator, the generator is activated, and then the bag is sealed.

Anaerobic holding jar Three holding jars are used, the first to hold an inoculated media, the second for plates that are growing colonies to be subculture, and the third to receive freshly inoculated plates. Commercial-grade N_2 can be used in the holding jar system. Flow rate of N_2 to each jar is regulated by the needle valves on the many fold, and adjust the gas tank regulator to 416 psi/in. for 1 to 2 minutes to purge the jar of air. Then decrease the flow rate to 1 or 2 bubbles per seconds.

Anaerobic jar with hydrogen By removing most of the air from the anaerobic jar and replacing it with hydrogen or preferably hydrogen mixed with carbon dioxide and nitrogen anaerobic condition is obtained. In the presence of catalyst, hydrogen reacts with oxygen to form water. A common catalyst used is palladium.

Table 5.1 Summary of infective syndromes and of various organs in the human body and their diagnostic procedure

Site of infection	Infective syndrome	Possible aetiological agents	Specimen collection procedure	Culture media used for cultivation
Skin and subcutaneous connective tissue	Folliculitis Furuncles Carbuncles Scalded skin syndrome Impetigo Ecthyma Erysipelas Cellulitis Fasciitis Myoitis	Gram-positive cocci *Staphylococcus aureus* *Micrococcus* *Enterococcus sp.* *Streptococcus pyogenes* Gram-negative cocci *Neisseria* sp. Gram-negative bacilli *Pseudomonas* sp. *Aeromonas* sp.	Tissue obtained during surgical procedure Aspirated material from an abscess or deep wound Pus or exudates obtained from an infected site during surgery can be aspirated in to the syringe. Skin swab	Colistin–Nalidixic acid agar (CNA) Phenyl ethyl alcohol agar Chocolate agar Blood agar Bacteroides bile esculin agar MacConkey agar Thayer-Martin medium

(Contd.)

Table 5.1 (Continued)

Site of infection	Infective syndrome	Possible aetiological agents	Specimen collection procedure	Culture media used for cultivation
	Gas gangrene Lyme disease Piedra Tinea corporis Ringworm Rubeola Rubella Varicella (chickenpox) Warts Erythema infectiosum Measles German measles Herpes infection	*Plesiomonas* sp. *Pasteurella* sp. *Actinobacillus* sp. *Streptobacillus* sp. *Haemophilus* sp. *Francisella* sp. Fungi *Microsporum canis* *Trichophyton mentagrophytes* *Trichophyton verrucosum* *Epidermophyton* sp. *Candida albicans* *Geotrichum* sp. *Cephalosporium* sp. *Acremonium* sp. Virus Rubella virus Varicella-zoster virus Measles virus Herpes simplex virus	Skin scrapings Nail clippings Hair	Baird-Parker agar KF streptococcus agar Cetrimide agar Thioglycolate medium Sabouraud dextrose agar DTM medium Amies transport medium Viral transport medium Cell culture medium
Upper respiratory tract	Sinusitis Otitis media Pharyngitis Laryngitis Pertussis Epiglottitis Tracheitis Thyroiditis	*Streptococcus pyogenes* Other beta-haemolytic streptococci *Bordetella pertussis* *Streptococcus pneumoniae*	Nasopharyngeal swab Sodium alginate throat swab for pertussis	Blood agar plate Cystine tellurite blood agar Loeffler agar slant Chocolate agar Charcoal cephalexin blood agar (CCBA)

(Contd.)

Table 5.1 (Continued)

Site of infection	Infective syndrome	Possible etiological agents	Specimen collection procedure	Culture media used for cultivation
	Common cold Strep throat Diphtheria Suppurative-parotitis Tonsillitis Scarlet fever Parotitis Rheumatoid fever	*Haemophilus influenzae* *Corynebacterium diphtheriae* *Pseudomonas aeruginosa* *Borrelia* sp. *Bacteroides melaninogenicus* *Neisseria meningitidis* *Klebsiella* sp. *Staphylococcus aureus* Viruses Reovirus Coxsackie virus A Coxsackie virus B Rhinovirus Coronavirus Mastadenovirus HSV type 1 Fungus *Candida albicans*	Sinus washings Surgical biopcy Swab of posterior pharynx Swab of tonsil Pernasal swab Nasopharyngeal aspirates	Modified tinsdale medium Sabouraud dextrose agar Thayer-Martin agar Cetrimide agar New York City agar medium Baird-Parker agar MacConkey agar
Lower respiratory tract	Pneumonia Pertussis Tuberculosis Bronchitis Legionnaire's disease Influenza Histoplasmosis Lung abscess Plaque Aspergillosis Anthrax Cryptococcosis	*Streptococcus pyogenes* *Klebsiella pneumoniae* Group B Streptococci *Chlamydiae pneumoniae* *Pseudomonas aeruginosa* *Staphylococcus aureus* *Streptococcus pneumoniae* *Mycoplasma pneumoniae*	Needle aspiration method Transtracheal aspiration Lung aspiration Bronchial washing Blood collection	Blood agar Chocolate agar MacConkey agar Baird-Parker agar Cetrimide agar LJ Medium

(Contd.)

Table 5.1 (Continued)

Site of infection	Infective syndrome	Possible aetiological agents	Specimen collection procedure	Culture media used for cultivation
		Haemophilus influenzae	Early morning sputum	Buffered charcoal yeast extract agar
		Neisseria meningitidis		
		Legionella pneumophila		
		Moraxella catarrhalis		
		Yersinia pestis		Cell-free medium
		Bacillus anthracis		
		Histoplasma capsulatum		Bordet-Gengou medium
		Cryptococcus neoformans		Thayer-Martin medium
		Candida albicans		
		Nocardia asteroides		
		Blastomyces dermatidis		Charcoal yeast extract medium
		Aspergillus fumigatus		
		Torulopsis spp.		
		Fusarium spp.		
		Curvularia lunata		
		Francisella tularensis		
		Geotrichum spp.		
		Toxoplasma gondii		
		Cryptosporidium		
		Adenovirus		
		Influenza virus		
		Herpes simplex virus 1		
		Cytomegalovirus		
Gastro-intestinal tract	Dental caries	*Escherichia coli*	Stool	MacConkey agar
	Gum disorders	*Salmonella enteritidis*		
	Trench mouth	*Shigella* spp.	Rectal swab	GN broth HE agar
	Vincents disease	*Campylobacter*		
	Diarrhoea	*Vibrio* spp.	Gastric aspirate	TCBS agar
	Dysentery	*Plesiomonas* spp.		Alkaline peptone water
	Shigellosis	*Aeromonas* spp.	Gastric biopsy	Campylobacter isolation medium
	Salmonellosis	*Yersinia enterocolitica*		
	Gastritis	*Clostridium perfringens*	Intubation	
	Gastroenteritis	*Clostridium difficile*		Yersinia selective medium
	Enteric fever	*Staphylococcus aureus*		
	Liver abscess	*Streptococcus mutans*		Deoxycholate citrate agar
	Liver granuloma	*Bacillus cereus*		
		Helicobacter pylori		XLD agar
	Cholera			

(Contd.)

Table 5.1 (Continued)

Site of infection	Infective syndrome	Possible aetiological agents	Specimen collection procedure	Culture media used for cultivation
	Amoebisis	Rotavirus		Wilson and Blair medium
	Giardiasis	Coxsackievirus		Selenite broth
	Schistosomiasis			Salmonella-Shigella agar
	Ascariasis			
	Enterobiasis			
Gastro-intestinal tract	Dental caries	*Escherichia coli*	Stool	MacConkey agar
	Gum disorders	*Salmonella enteritidis*		GN broth
	Trench mouth	*Shigella* spp.	Rectal swab	HE agar
	Vincents disease	*Campylobacter*		TCBS agar
	Diarrhoea	*Vibrio* spp.		Alkaline peptone water
	Dysentry	*Plesiomonas* spp.	Gastric aspirate	Campylobacter isolation medium
	Shigellosis	*Aeromonas* spp.		Yersinia selective medium
	Salmonellosis	*Yersinia enterocolitica*		Deoxycholate citrate agar
	Gastritis	*Clostridium perfringens*	Gastric biopsy	XLD agar
	Gastroenteritis	*Clostridium difficile*		Wilson and Blair medium
	Enteric fever	*Staphylococcus aureus*		Selenite broth
	Liver abscess	*Streptococcus mutans*	Intubation	Salmonella-Shigella agar
	Liver granuloma	*Bacillus cereus*		MacConkey sorbitol agar
	Cholera	*Helicobacter pylori*		Cefsulodin irgasan novobiosin agar
	Amoebiasis	Rotavirus		Cycloserine cefoxitin fructose egg yolk agar
	Giardiasis	Coxsackievirus		Campy blood agar
	Schistosomiasis	Adenovirus		Blood agar
	Ascariasis	Echovirus		Blood agar with $10\ \mu g$ ampicillin
	Enterobiasis	Poliovirus		Inositol brilliant green bile salt agar
	Food poisoning	*Entamoeba histolytica*		Bismuth sulphite agar
	Peptic ulcer	*Giardia lamblia*		Vibrio agar
		Giardia intestinalis		Entamoeba isolation medium
		Trichomonas hominis		
		Balantidium coli		
		Isospora belli		
		Cryptosporidium spp.		
		Ascaris lumbricoides		
		Schistosoma spp.		
		Taenia spp.		
		Enterobius vermicularis		

(Contd.)

Table 5.1 (Continued)

Site of infection	Infective syndrome	Possible aetiological agents	Specimen collection procedure	Culture media used for cultivation
Urine	Bacteriuria Pyuria (pus cells in urine) Cystitis Pyelonephritis	*Staphylococcus saprophyticus* *Pseudomonas aeruginosa* *Escherichia coli* *Proteus* spp. *Klebsiella* spp. Haemolytic streptococci *Enterococcus* spp.	Mid-stream urine Cathetral suprapubic aspiration	Blood agar MacConkey agar Cetrimide agar CLED agar SS agar KF Streptococcus agar Nutrient agar
Blood	Septic shock Endocarditis Vascular encephalitis Pericarditis Myocarditis Septicaemia Tularemia Brucellosis Plaque Infectious mono-nucleosis Dengue Malaria Anthrax Typhoid fever Puerperal sepsis Rheumatic fever Relapsing fever Yellow fever	*Staphylococcus aureus* Viridans streptococci *Streptococcus pneumoniae* *Streptococcus pyogenes* *Salmonella typhi* *Escherichia coli* *Klebsiella pneumoniae* *Corynebacterium diphtheriae* *Yersinia pestis* *Leptospira* spp. *Brucella* spp. Group B streptococci *Proteus* spp. *Haemophilus influenzae* *Neisseria* spp.	Needle aspiration	Thioglycolate broth Tryptone soya diphasic broth Blood agar Chocolate agar MacConkey agar SS agar

(Contd.)

Table 5.1 (Continued)

Site of infection	Infective syndrome	Possible aetiological agents	Specimen collection procedure	Culture media used for cultivation
Genital tract	Female Gonorrhoea Syphilis Amnionitis Bartholinitis Cervicitis Endometritis Urithritis Vulvovaginitis Salpingitis Male Epididymitis Orchitis Prostatis Urithritis Gonorrhoea Syphilis	*Neisseria gonorrhoeae* *Treponema pallidum* *Haemophilus influenzae* *Listeria monocytogens* *Chlamydia trachomatis* *Ureaplasma urealyticum* *Bacteroides fragilis* *Peptostreptococcus* *Streptococcus pyogenes* *Streptococcus agalactiae* *Proteus mirabilis* Enterococci *Gardnerella vaginalis* *Candida albicans* *Pseudomonas aeruginosa*	Female Catheter aspirate Amniocentesis fluid Bartholin gland aspirate Transcervical aspirate Fallopian aspirate Swab of genital ulcer Urethral discharge Swab of posterior vagina Male Urethral swab Swab of genital ulcer Penial discharge Prosthetic secretions	Blood agar Chocolate agar MacConkey agar Modified Thayer-Martin agar Colistin Nalidixic acid agar Listeria selective medium CLED medium Mitis salivaris agar Vaginalis agar Todd-Hewitt broth Cetrimide agar Cell-free medium Mycoplasma isolation medium XLD medium Sabouraud dextrose agar
CSF	Meningitis	*Escherichia coli* *Streptococcus agalactiae* *Listeria monocytogens* *Haemophilus influenzae* *Neisseria meningitidis* *Streptococcus pneumoniae*	Lumbar puncture Brain abscess	Blood agar Chocolate agar Modified Thayer-Martin agar New York city agar medium GC medium

(Contd.)

Table 5.1 (Continued)

Site of infection	Infective syndrome	Possible aetiological agents	Specimen collection procedure	Culture media used for cultivation
Anaerobic infections	Appendicitis Cholecystitis Otitis media Dental and oral infections Endocarditis Endometritis Myocarditis Osteomyeletis Peritonitis Empyema Salpingitis Septic arthritis Sinusitis Trauma Bacteraemia	Gram-positive bacilli *Clostridium* *Desulfotomaculum* *Acetobacterium* *Eubacterium lentum* *Actinomycetes* Gram-positive cocci *Peptostreptococcus* *Ruminococcus* Gram-negative bacilli *Anaerobacter* *Bilophila* *Wollinella* *Bacteroides* *Prevotella* *Porphyromonas* *Fusobacterium* Gram-negative cocci *Veillonella* *Acidaminococcus*	Head and neck Abscess and biopsy Lung Transtracheal aspirate, lung puncture and biopsy CNS Abscess, biopsy, lumbar puncture Abdomen Peritoneal fluid and biopsy Urinary tract Subrapubic aspirate Female genital tract Culdoscopy, endometrial aspirate Bone and Joint Aspirate	Brucella blood agar Phenyl ethyl alcohol agar Bacteroides bile esculin agar Cycloserine cefoxitin fructose agar Egg yolk agar Robertson cooked meat medium Chopped meat broth Columbia blood agar CDC anaerobic agar (TSA + YE + haemin + vit. K, L-cystine) Anaerobic kanamycin-vancomycin blood agar LKV medium

1. Explain the basic steps involved in any diagnosis.
2. State the characteristic features of a sample.
3. How are samples transported to the laboratory?
4. What is a transport medium?
5. Explain the importance of microscopic examination in the diagnosis of infection.
6. Explain the basic techniques involved in bacterial identification.
7. What are the immunological techniques involved in microbial identification?
8. How is phage typing important in bacterial identification?
9. How is *S. aureus* isolated from skin?
10. Explain how bacterial pathogens are isolated from skin.
11. Explain KOH mount.
12. Define nasopharyngeal swab.
13. Describe Albert staining.
14. Explain the process involved in the identification of *S. pyogenes*.
15. How do you cultivate *C. diphtheria* from sputum?
16. Explain about Wayson staining technique.
17. How do you identify protozoans from stools?
18. Explain the processing of urine.
19. Describe the features of cultivation of bacteria from blood.
20. Describe the process of cultivation of bacteria from CSF.
21. Describe the various methods of isolation of anaerobic bacterial isolation.
22. Define
    i. Acid-fast staining
    ii. Bartlet grading
23. Describe the cultivation of
    i. *M. tuberculosis.*
    ii. *E. coli*
    iii. *Salmonella*
    iv. *Shigella*
24. Describe
    i. Suprapubic aspiration
    ii. Cathetral aspiration
    iii. Mid-stream urine collection

25. Write notes on
    i. Lumbar punture
    ii. Levine tube
    iii. Hektoen enteric medium
    iv. GC medium
    v. Rolling tube method
    vi. Gas-pak method
    vii. Griffith typing
    viii. Lancefeld grouping

Antimicrobial Chemotherapy

INTRODUCTION

Any chemical or biological agent that destroys or inhibits the growth of microorganisms is called an antimicrobial agent. Antibiotics are substances of microbial origin that have antimicrobial activity. The treatment of a disease with a chemical substance is known as chemotherapy.

The earliest evidence of successful chemotherapy is from ancient Peru, where the Indians used the bark from the cinchona tree to treat malaria. Paul Ehrlich is called as "Father of Chemotherapy". He discovered *p*-rosaniline, which has antitrypanosomal effects, and arsphenamine, which is effective against syphilis. Ehrlich postulated the term "magic bullet", a selective toxin for parasites but not toxic to humans. An important milestone in antimicrobial chemotherapy is the discovery of penicillin G in 1929 by Alexander Flemming. In 1939, Florey and colleagues at Oxford University tested antimicrobial activities of lysozyme and again isolated penicillin.

ANTIMICROBIAL AGENTS

Characteristic Features of an Ideal Antimicrobial Agent

An ideal antimicrobial agent should have certain characteristic features. It should

- have selective toxicity.
- be bactericidal in nature.
- be effective against a broad range of microorganisms.
- not cause allergy response.
- remain active in body fluids.
- be stable and water-soluble.
- have long shelf life.

- be cheap.
- have low level of resistance.
- have larger therapeutic index.

CLASSIFICATION OF ANTIMICROBIAL AGENTS

Bacterial cells grow and divide, replicating repeatedly to reach large numbers. To grow and divide, organisms must synthesize or take up many types of biomolecules. Antimicrobial agents interfere with specific processes that are essential for growth and/or division. Based on the mode of action, antimicrobials are classified into:

1. Inhibitors of bacterial and fungal cell wall synthesis
2. Inhibitors of cytoplasmic membrane functions
3. Inhibitors of nucleic acid synthesis
4. Inhibitors of ribosome function

Table 6.1 Classification of antimicrobial agents based on source

Natural antibiotics		Synthetic drugs	Semi-synthetic drugs
Name of antibiotics	Source		
Amphotericin B	*Streptomyces* spp.	Sulphonamide	Ampicillin
Nystatin, kanamycin, streptomycin neomycin, tobramycin, rifampin		Trimethoprim Chloramphenicol Ciprofloxacin Isoniazid	Carbenicillin
			Methicillin
Gentamicin	*Micromonospora*		Amoxycillin
Erythromycin	*Streptomyces erythraeus*		
Initial chloramphenicol	*Streptomyces venezuelae*		
Polymyxin and bacitracin	*Bacillus* spp.		
Penicillin and griseofulvin	*Penicillium* spp.		
Cephalosporin	*Cephalosporium*		

Antimicrobial agents may be either bactericidal, i.e., killing the target bacterium, or bacteriostatic, i.e., inhibiting its growth. Drugs act as antimetabolites, which block the metabolic pathway.

Drugs show different spectra of activity. Based on the spectrum of activity they are classified as:

1. Narrow-spectrum drugs which act only on limited groups of microorganisms.
2. Broad-spectrum drugs which act on different kinds of microorganisms.

Chemotherapeutic agents are again classified based on their source and preparations. They are:

1. Natural antimicrobial agents which are synthesized by microorganisms.
2. Semi-synthetic drugs which are modified form of natural drugs.
3. Synthetic drugs which are synthesized chemically (Table 6.1).

Chemotherapeutic agents are also classified based on microbial interaction. They are

1. Antibacterial drugs
2. Antifungal drugs
3. Antiviral drugs (Table 6.2)

Table 6.2 Classification of antimicrobial agents based on microbial interaction

Antibacterial drugs	Antifungal drugs	Antiviral drugs
Penicillin G	Amphotericin B	Ganciclovir
Erythromycin	Nystatin	Foscarnet
Cephalosporin	Griseofulvin	Amantadine
Methicillin	Polymyxin B	Rimantidine
Vancomycin	Nilkomycin	Idoxurdine
Ampicillin	Miconazole	Acyclovir
Doxycycline	Ketoconazole	Azidothymidine
Tetracycline	Clotrimazole	Famciclovir
Rifampicin	Fluconazole	Ribavirine
Gentamicin	Tolnaftate	Vidarabine
Chloramphenicol		

Inhibitors of Bacterial Cell Wall Synthesis

Gram-positive bacterial cell wall contains peptidoglycan and teichoic acid, and the bacterium may or may not be surrounded by a protein or polysaccharide envelope. Gram-negative bacterial cell wall contains peptidoglycan, lipopolysaccharide, lipoprotein, phospholipid and protein. The critical attack site of anti-cell wall agents is the peptidoglycan layer. This layer is essential for the survival of bacteria in hypotonic environments; loss or damage to this layer destroys the rigidity of the bacterial cell wall, resulting in death.

Penicillins and cephalosporins/cephamycins are widely used to inhibit both gram-positive and gram-negative bacilli (Figure 6.1).

β-Lactam antibiotics Penicillin was the first antibiotic used in medicine. Most penicillin group of antibiotics are derivatives of 6-aminopenicillanic acid. They differ from one another only with respect to the side chain attached to its amino group. The most common structure of penicillin group is the β-lactam ring. Penicillins inhibit the enzyme catalysing the transpeptidation reaction. They block the synthesis of a complete peptidoglycan and lead to osmotic lysis. Penicillins are a group of β-lactam antibiotics, e.g. penicillin G, penicillin V, ampicillin, carbenicillin, methicillin, ticarcillin, etc.

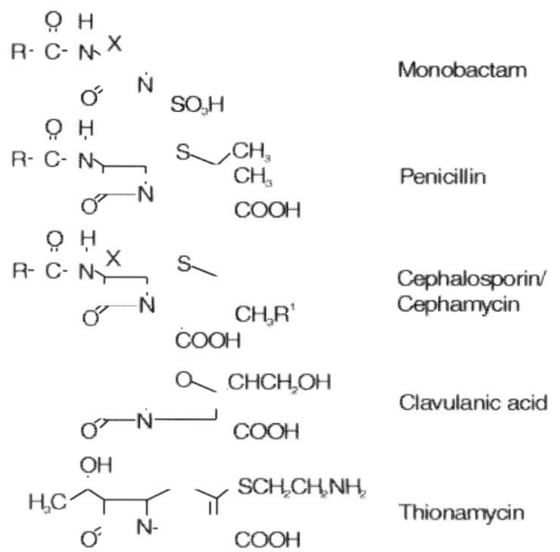

Figure 6.1 Basic structures of β-lactam antibiotics

Vancomycin interrupts cell wall synthesis by forming a complex with the C-terminal D-alanine residues of peptidoglycan precursors (Figure 6.2).

Inhibitors of Cytoplasmic Membrane Functions

Bacterial membranes Bacterial membranes are composed basically of lipid, protein and lipoprotein which act as a diffusion barrier for water, ions, nutrients and transport systems. Antimicrobial agents can cause disorganization of the membrane.

The best known compounds are polymyxin B and polymyxin E. Polymyxins disorganize membrane permeability so that nucleic acids and cations leak out and the cell dies. Gramicidins are also membrane-active antibiotics that appear to act by producing aqueous pores in the membranes. They are also used only topically.

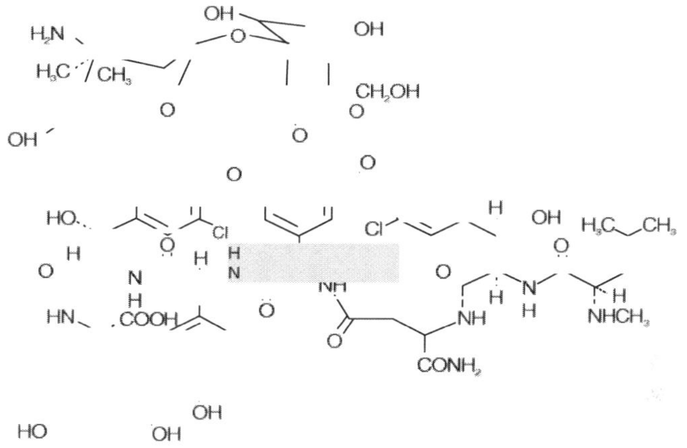

Figure 6.2 Vancomycin

Fungal membranes The polyene antibiotics act on sterol membrane by binding to rigid hydrophobic centre and a flexible hydrophilic section. Polyene interacts with fungal cells to produce a membrane–polyene complex that alters the membrane permeability, resulting in internal acidification of the fungus with exchange of K^+ and sugars, loss of phosphate esters, organic acids, nucleotides and eventual leakage of cell protein. Amphotericin B is used systemically (Figure 6.3). Nystatin is used as a topical agent and primaricin as an ophthalmic preparation.

Other agents that interfere with the synthesis of fungal lipid membranes are referred to as imidazoles: miconazole, ketoconazole, clotrimazole, and fluconazole.

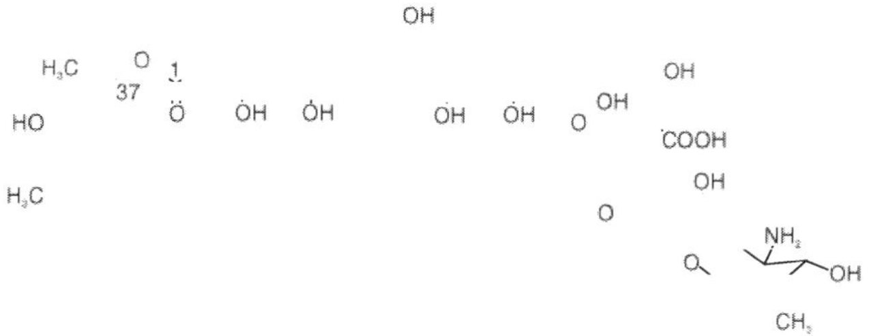

Figure 6.3 Amphotericin B

Inhibitors of Nucleic Acid Synthesis

Antimicrobial agents can interfere with nucleic acid synthesis at different levels. They can inhibit nucleotide synthesis and they can interfere with the polymerases involved in the replication and transcription of DNA.

Interference with nucleotide synthesis A large number of agents interfere with purine and pyrimidine synthesis. Flucytosine (5-fluorocytosine) is an antifungal agent that inhibits yeast species. Adenosine arabinoside inhibits viruses. Acyclovir is a nucleoside analogue that, after being converted to a triphosphate, inhibits the thymidine kinase and DNA polymerase of herpesviruses. Zidovudine (AZT) inhibits human immunodeficiency virus.

Agents that impair the template function of DNA A number of substances bind to DNA by intercalation. Chloroquine and miracil D (lucanthone) inhibit the further synthesis of DNA of plasmodia and schistosomes respectively.

Inhibition of DNA-*directed* DNA *polymerase* Rifamycins are a class of antibiotics that inhibit DNA-directed RNA polymerase (Figure 6.4).

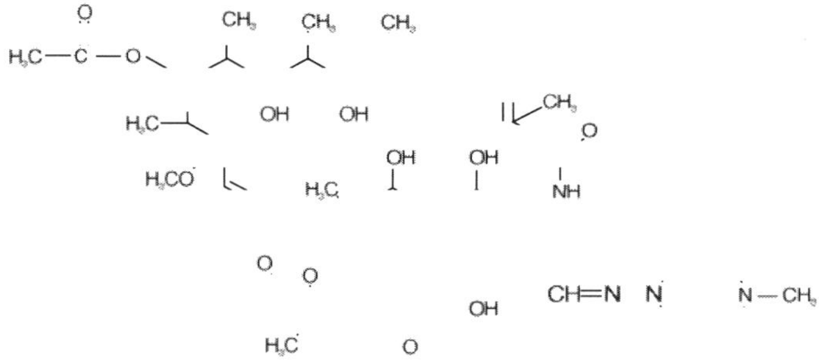

Figure 6.4 Rifampin

Inhibition of DNA *replication* The newer fluoroquinolones such as ciprofloxacin (Figure 6.5), norfloxacin, and ofloxacin interact with DNA gyrase and possess a broad spectrum of antimicrobial activity.

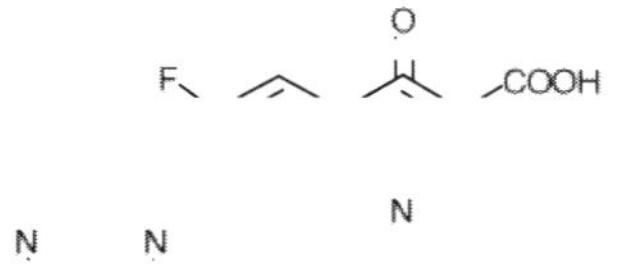

Figure 6.5 Ciprofloxacin

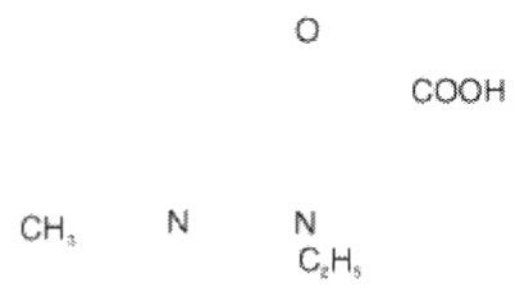

Figure 6.6 Nalidixic acid

Nalidixic acid (Figure 6.6) inhibits only aerobic gram-negative species. Ciprofloxacin acts on gram-positive species and cyclopropyl is active against *Pseudomonas* species. Nitroimidazoles such as etronidazole inhibit anaerobic bacteria and protozoa.

Metronidazole binds to DNA and causes DNA breakage (Figure 6.7).

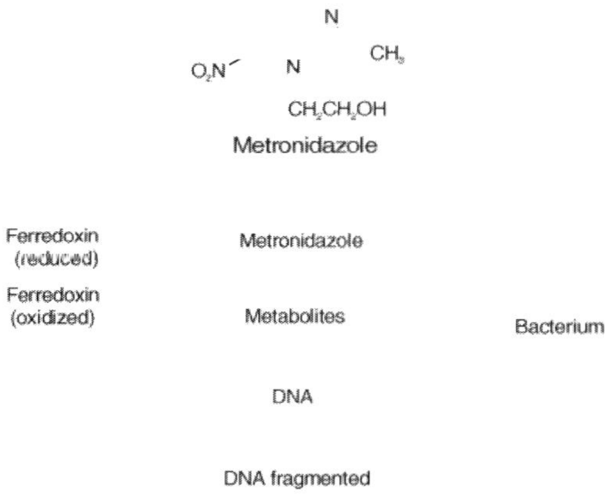

Figure 6.7 Mode of action of metronidazole

Antimicrobial Inhibitors of Ribosome Function

A number of antibacterial agents act by inhibiting ribosome function. Aminoglycosides act by binding to specific ribosomal subunits. Aminoglycosides are complex sugars connected by glycosidic linkage. Rifampin, tetracycline, streptomycin, gentamicin, chloramphenicol, erythromycin and clindamycin inhibit protein synthesis by binding ribosomes (Figure 6.8).

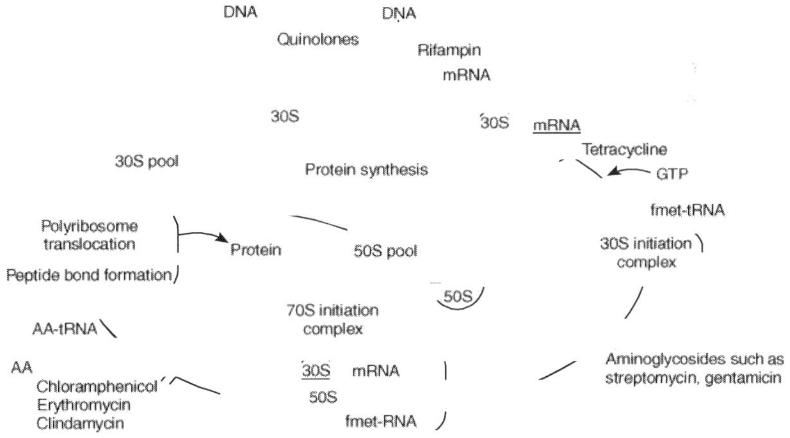

Figure 6.8 Diagrammatic representation of inhibition sites of protein biosynthesis by various antibiotics that bind to the 30S and 50S ribosomes

Tetracyclines are bacteriostatic broad-spectrum antibiotics. Based on the chemical structure at the major and secondary sites of modification (Figure 6.9), they are named as chlortetracycline, doxytetracycline, tetrecycline, doxycycline and minocycline.

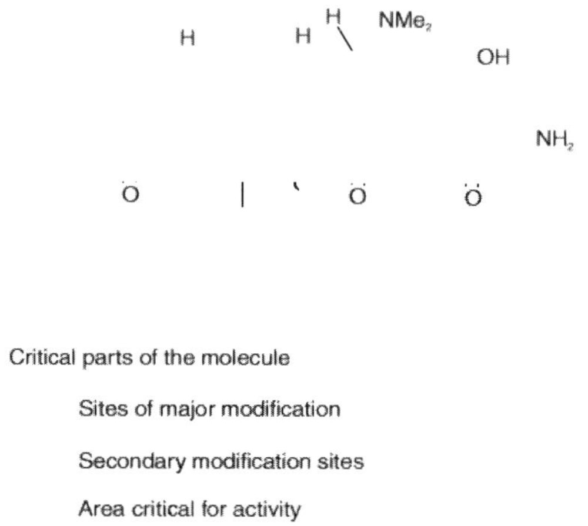

Figure 6.9 Structure of tetracycline showing the area critical for activity and major and minor points of modification

Chloramphenicol (Figure 6.10) is a bacteriostatic agent that inhibits both gram-positive and gram-negative bacteria. It inhibits peptide bond formation by binding to a peptidyltransferase enzyme on the 50S ribosome. Macrolides impair a peptidyltransferase reaction. Erythromycin (Figure 6.11) inhibits *Haemophilus*, *Mycoplasma*, *Chlamydia* and *Legionella*.

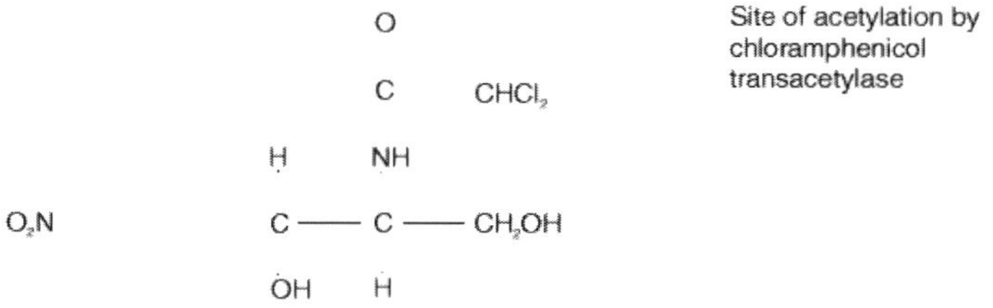

Figure 6.10 Structure of chloramphenicol (prototype or macrolide)

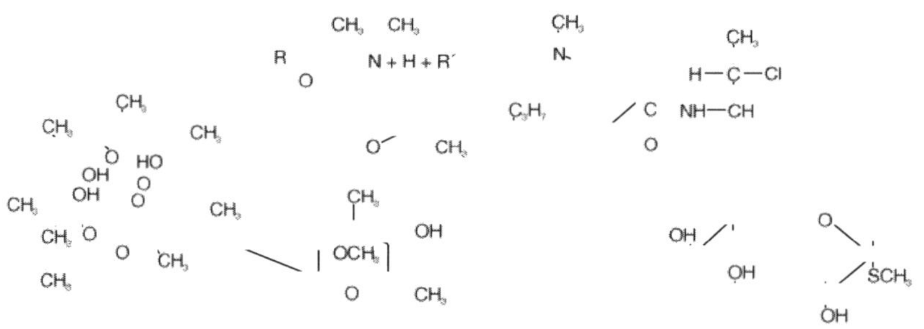

Figure 6.11 Drugs that inhibit other biochemical targets

Both trimethoprim and sulphonamides interfere with the biosynthesis of tetrahydrofolate, which is necessary for the ultimate synthesis of DNA, RNA and bacterial cell wall proteins (Figure 6.12).

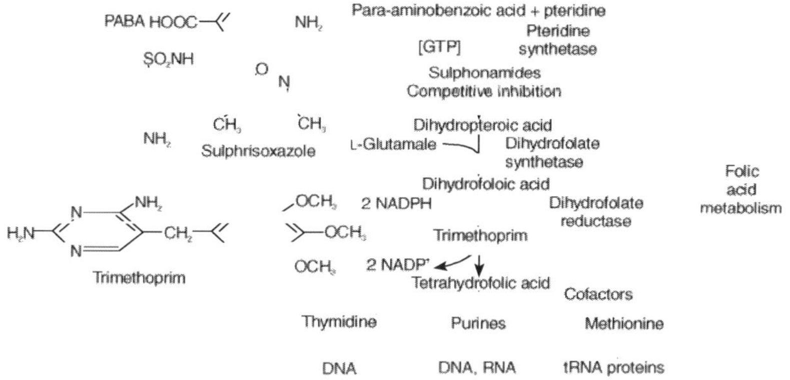

Figure 6.12 Structure of sulphonamide and trimethoprim with sites of inhibition of folic acid metabolism

Antibacterial Agents That Affect Mycobacteria

Isoniazid is a nicotinamide derivative that inhibits mycobacteria. It affects the synthesis of lipids. Ethambutol is mycostatic, whereas isoniazid is mycocidal. The other antituberculosis drugs, rifampin and streptomycin, affect mycobacteria in the same manner that they inhibit bacteria. Pyrazinamide is a synthetic analogue of nicotinamide. It is bactericidal.

Side Effects of Antimicrobial Agents

The following are some side effects caused by antimicrobial agents.

- Allergy
- Renal injury
- Depressed bone marrow function
- Gastric upset
- Diarrhoea
- Anaemia
- Hepatic injury
- Nausea
- Loss of hearing
- Hypotension
- Neutropenia
- Kidney damage

BACTERIAL RESISTANCE

There are a number of ways in which bacteria can become resistant. Most of the early studies of bacterial resistance is focused on single-step mutational events of chromosomal origin. Similarly, a single-step mutation that alters a ribosomal protein conferred resistance to streptomycin. In the late 1950s, Japanese workers found that enteric bacteria such as *Shigella dysenteriae* had become resistant not only to sulphonamides but also to tetracyclines and chloramphenicol. This resistance was not due to a chromosomal change, but rather due to the presence of extrachromosomal DNA that was transmissible. This type of resistance is called plasmid-mediated resistance.

Resistance-conferring plasmids are present in virtually all bacteria. For example, resistance to ampicillin appeared in *Haemophilus influenzae* in 1974 and in *Neisseria gonorrhoeae* in 1976.

Bacteria also contain transposons which can insert into plasmids and also into the chromosome. Transposon-mediated resistance to most of the major antibiotics has been found in the past few years.

Mechanism of Resistance

The basic mechanisms by which a microorganism can resist an antimicrobial agent are:

1. Alter the receptor for the drug (the molecule on which it exerts its effect)
2. Decrease the amount of drug that reaches the receptor by altering entry or increasing removal of the drug

Bacteria can possess one or all of these mechanisms simultaneously.

Resistance due to altered receptors

β-Lactam resistance Many bacterial pathogens resist attack by inactivating drugs through chemical modifications, e.g. hydrolysis of β-lactam ring by β-lactamase enzymes.

Vancomycin resistance Certain transposable genetic elements encode special cell wall-synthesizing enzymes which change the structure of the normal D-Ala-D-Ala side chain in the peptidoglycan assembly pathway. The altered side chain (D-Ala-D-Lac) does not bind vancomycin and allows normal peptidoglycan polymerization to occur in the presence of the drug.

Rifampin resistance The resistance of bacteria to rifampicin is caused by an alternation of one amino acid in DNA-directed RNA polymerase, which results in reduced binding of rifampin. The degree of resistance is related to the degree to which the enzyme is changed, but does not correlate strictly with enzyme inhibition.

Sulphonamide–trimethoprim resistance Sulphonamides can be rendered ineffective by altered or new dihydropteroic synthetase that has poor affinity for sulphonamides and preferentially binds *p*-aminobenzoic acid. Sulphonamide resistance of this type can result from a point mutation or from acquisition of a plasmid that causes synthesis of the new enzyme.

Quinolone resistance Resistance to quinolones can be caused by mutations in DNA gyrase subunits A or B, reduced outer membrane permeability in gram-negative cells or by active efflux transporters found in many bacteria.

Resistance due to decreased entry of a drug

Tetracycline resistance It is common in both gram-positive and gram-negative bacteria. In most cases, it is plasmid-encoded and inducible; however, chromosomal, constitutive resistance is found in some organisms such as *Proteus* species. Many plasmid-encoded specified tetracycline resistance determinants have been found in enteric bacteria. The most common of these determinants, TetB, is also present in *H. influenzae*.

Aminoglycoside resistance In the most important form of aminoglycoside resistance, the compound is modified outside the cell and resistance is partly due to poor uptake of the altered compound. Also, all aminoglycosides have free amino and hydroxy groups that are essential for binding to ribosomal proteins. A number of enzymes can acetylate the amino groups and phosphorylate or adenylate the hydroxyl groups.

Chloramphenicol resistance Many gram-positive and gram-negative bacteria, including some recently discovered *H. influenzae* strains, are resistant to chloramphenicol because they possess the enzyme chloramphenicol transacetylase, which acetylates hydroxyl groups on the chloramphenicol structure.

Antibiotics are frequently used in combination for the following reasons:

- To treat a life-threatening infection
- To prevent emergence of bacterial resistance
- To treat mixed infections of aerobic and anaerobic bacteria
- To enhance antibacterial activity (synergy)
- To use lower doses of a toxic drug

Mechanism to Reduce Bacterial Resistance

- Select new antibiotics which will be a major force in slowing the development of antimicrobial resistance
- Control, reduce or cycle antibiotic usage
- Improve hygiene in hospitals, hospital personnel
- Discover and develop new antibiotics
- Modify existing antibiotics
- Develop inhibitors of antibiotic-modifying enzymes
- Define agents that would "cure" resistance plasmids

DETERMINATION OF THE LEVEL OF ANTIMICROBIAL ACTIVITY

Dilution Susceptibility Test

This test can be used to determine minimum inhibitory concentration (MIC). In the broth dilution test, a series of broth tubes containing antibiotics in the concentration range of 0.1–128 micrograms per ml is prepared and inoculated with standard test organisms. The lowest concentration of antibiotics resulting in no growth after 16 to 20 hours of incubation is referred to as MIC.

Disc Diffusion Test

It is also called Kirby and Bauer test. Antibiotics are impregnated into a paper disc, placed in a bacterial lawn and observed for zone of inhibition.

E-test

Nowadays, to overcome the disadvantages of MIC (drug dilution test) on broth, strip-based method is used. It is called as E-test. Various concentrations of antibiotics

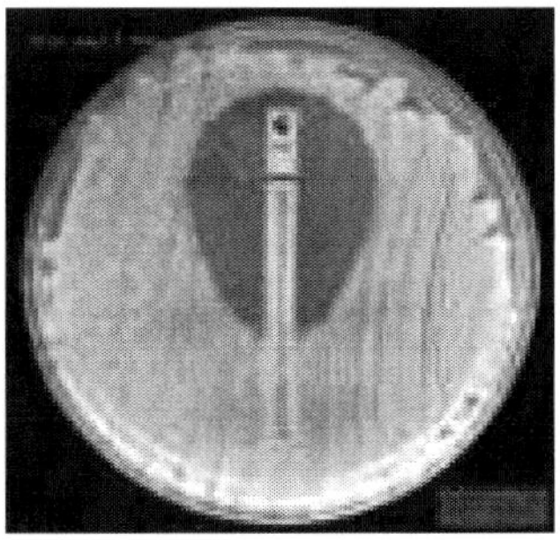

Figure 6.13 E-test

are impregnated into the strip and are placed on the medium seeded with bacterium. It is incubated for 24 hours at room temperature and observed for zone of inhibition (Figure 6.13).

1. Classify chemotherapeutic agents.
2. What are antimetabolites?
3. What are magic bullets?
4. Describe about antiviral drugs.
5. Discuss the way by which microorganisms develop resistance.
6. State the reason for combined therapy.

7. What are the attributes of ideal antimicrobial agents?
8. Write short notes on
   i. Paul Ehrlich
   ii. Alexander Flemming
9. Give a brief account of
   i. Polymyxin B
   ii. Nystatin
   iii. Amphotericin B

1. How do semi-synthetic drugs differ from synthetic drugs?
2. List drugs that create side effects in animals.
3. State reasons why people are turning to traditional system of medicine.
4. State the reason behind the development of multidrug resistance.
5. How many soil microorganisms are involved in the antibiotic synthesis process?

Ananthanarayan and Jayaram Paniker. *Textbook of Microbiology*, 4th edn. Orient Longman. 2000.
Mandel, G.L., Bennet, J.E. and Dolin, R. *Principles and Practice of Infectious Disease*, 4th edn. Churchill Livingstone, New York. 1995.
Murray, P.R. *Manual of Clinical Microbiology*, 6th edn. ASM Press, Washington DC. 1995.
Tripathi, K.D. *Essentials of Medical Pharmacology*, 5th edn. Jaypee Publishers, 2003.
aac.asm.org/
gsbs.utmb.edu/microbook
jac.oupjournals.org/contents-by-date.0.shtml
ww.cancerchemotherapy.org
www.8bestsites.com
www.bmb.leeds.ac.uk/mbiology
www.bsac.org.uk
www.bsac.org.uk
www.cancerchemotherapy.org
www.icaac.org/
www.ischemo.org
www.journals.asm.org
www.microbiology.ru/cmac/indexe.shtml
www-micro.msb.le.ac.uk

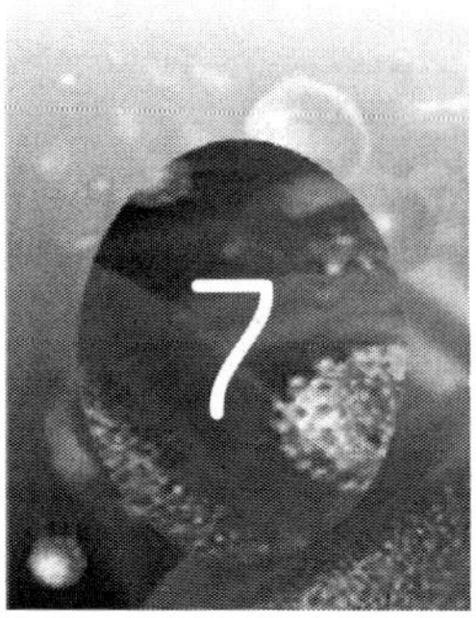

Epidemiology and Control of Community Infections

INTRODUCTION

Each disease affecting the body alters body structures and functions in particular ways, and these alterations are usually indicated by several kinds of evidence. For example, the patient may experience certain symptoms—change in body functions such as pain and malaise (a vague feeling of body's discomfort). These subjective changes are not apparent to the observers. The patient can also exhibit signs, which are objective changes that the physician can observe and measure. Frequently evaluated signs include lesions, swelling, fever and paralysis. Sometimes a specific group of symptoms or signs always accompanies a particular disease, such a group is called a syndrome. The diagnosis of a disease is achieved by evolution of the signs and symptoms together with the results of the laboratory tests.

Diseases are often classified in terms of how they bring about an effect within a host and within a given population. Any disease that spreads from one host to another either directly or indirectly is said to be a communicable disease. Those diseases that spread from one person to another very easily are called contagious diseases. A non-communicable disease is not spread from one host to another.

The science of epidemiology originated and evolved in response to the great diseases such as typhoid, cholera, smallpox and yellow fever. The practical goal of epidemiology is to establish effective control, prevention and eradication measures within a given population.

Epidemiology is defined as the science that evaluates the occurrence, determinants, distribution and control of health and disease in a defined population.

Epidemiology is very important in today's crowded, overpopulated world, where frequent travel and mass production and distribution of foods and goods are a way of life, and where disease can spread rapidly.

HISTORY

Hippocrates (460–361BC) is called the Father of Medical Science. Medical science and environmental science together form epidemiology.

Modern epidemiology began in 1854 when John Snow, a British physician reasoned out that cholera was transmitted by contaminated water. He is called as "Father of Epidemiology".

The most outstanding achievement of international surveillance was the development of a program for smallpox eradication. The multidisciplinary approach adopted by the WHO, in which programs were community-based with measurable goals and constant monitoring, resulted in the last endemic case being recorded in October 1977. Smallpox was officially declared as eradicated in December 1979.

RESPONSIBILITIES OF AN EPIDEMIOLOGIST

Any individual who practises epidemiology is called an epidemiologist.

The responsibilities of an epidemiologist include

- Analysing the aetiology of a disease
- Identifying the factors of a disease
- Assembling and analysing data such as
  - Age
  - Sex
  - Occupation
  - Personnel habits
  - Socio-economic status
  - History of immunization
  - Common history
  - Information about other diseases
- Analysing the time of disease occurrence
- Evaluating control measures
- Planning overall health care for the community

TYPES OF EPIDEMIOLOGY

Epidemiologists use three basic types of investigation when analysing the occurrence of a disease.

1. Descriptive epidemiology
2. Analytical epidemiology
3. Experimental epidemiology

Descriptive Epidemiology

It entails collecting all data that describe the occurrence of the disease under study. It provides relevant information about the affected persons and the place and time period in which the disease occurred. Such a study is generally retrospective. Information often reveals the factors responsible for the aetiology of the disease. The epidemiologist is aided by the knowledge that disease transmission frequency occurs via air, the ingestion of contaminated food or direct contact with infected individuals. These studies backtrack to the cause and source of the disease.

Analytical Epidemiology

It analyses a particular disease to determine its possible cause with case control method, the epidemiologist looks for factors that might have produced the disease. Cohert study is useful for this analysis. A group of persons who have the disease is compared with other groups which are free of the disease. These statistics are compared to determine all the possible factors—genetic, environmental and others.

Experimental Epidemiology

It begins with the hypothesis about a particular disease; experiments to test the hypothesis are then conducted with a group of people. Human volunteers are used, especially to establish routes of transfer and to establish that particular pathogens are actually responsible for a specific disease, e.g. analysis of common cold transmission route.

EPIDEMIOLOGICAL TERMINOLOGY

When a disease occurs occasionally, and at irregular intervals in a human population, it is a sporadic disease (e.g. typhoid). When it maintains a steady, low-level frequency at moderately regular intervals, it is an endemic disease (e.g. common cold).

Hyperendemic diseases gradually increase in occurrence frequency beyond the endemic level but not up to the epidemic level (e.g. common cold during winter season).

An epidemic is a sudden increase in occurrence of the disease above the expected level. The first case of an epidemic is called index case. An outbreak is a sudden, unexpected occurrence of a disease in a limited segment of a population (AIDS).

A pandemic is an increase in disease occurrence within a large population over a very wide region (AIDS).

The discipline that deals with the factors that influence the frequency of a disease in an animal population is known as epizootiology. Moderate prevalence of a disease in animals is termed enzootic, a sudden outbreak is termed epizootic and wide dissemination is called panzootic.

TOOLS OF EPIDEMIOLOGY

Statistics is the branch of mathematics dealing with the collection, organization and interpretation of numerical data. As a science particularly concerned with rates and comparison of rates, epidemiology was the first medical field in which statistics were extensively used.

A morbidity rate measures the number of individuals that become ill because of a specific disease within a susceptible population during a specific period. It is an incidence rate and reflects the number of new cases in a period.

Prevalence rate refers to the total number of individuals infected in a population at any time no matter when the disease begins.

Seroprevalence rate refers to the number of deaths in a population at any time.

Mortality rate is the relationship of the number of deaths caused by the given disease to the total number of cases of the disease.

The determination of morbidity, prevalence and mortality rates aids public health personnel in directing health care efforts to control the spread of infectious disease.

RECOGNITION OF AN INFECTIOUS DISEASE IN A POPULATION

Epidemiologists can recognize an infectious disease in a population by using various surveillance methods. Surveillance is a dynamic activity that includes gathering information on the development and occurrence of the disease, analysing the data, summarizing the findings and using the information to select control methods.

A combination of the following surveillance methods is used most often:

1. Generation of morbidity data from case reports.
2. Collection of mortality data from death certificates.
3. Investigation of actual cases.
4. Collection of data from reported epidemics.
5. Field investigation of epidemics.
6. Review of laboratory experiments: survey of population for antibodies against the agent and specific microbial serotypes, skin tests, cultures, stool analysis, etc.
7. Population surveys using valid statistical sampling to determine who has the disease.
8. Use of animal and vector disease data.
9. Collection of information on the usage of specific biologics—antibiotics, antitoxins, vaccines and other prophylactic measures.
10. Use of demographic data on population characteristics such as human movements during specific time of the year.

Correlation with a Single Causative Agent

After an infectious disease has been recognized in a population, epidemiologists correlate the disease outbreak with a specific organism—its exact cause must be discovered. At this point, diagnostic microbiology enters the investigation.

TYPES OF EPIDEMICS

Two types of epidemics were recognized on the basis of the outbreak. They are:

Common-source epidemic
Propagated epidemic

Common-source Epidemic

It is characterized by a sharp rise to a peak and then a rapid, but not as pronounced, decline of the number of individuals infected. This type of epidemic usually results from a single common contaminated source such as food or water, e.g. food poisoning.

Propagated Epidemic

It is characterized by a relatively slow and prolonged rise and then a gradual decline in the number of individuals infected. This type of epidemic usually results from the introduction of a single infected individual into a susceptible population, e.g. mumps and chickenpox.

Herd immunity is the resistance of a population to infection and pathogen spread because of the immunity of a large percentage of the population. Public health officials immunize large portions of the susceptible populations in an attempt to maintain high level of herd immunity. Any increase in the number of susceptible individuals may result in endemic diseases becoming epidemic. The proportion of immune individuals to susceptible ones must be constantly monitored because new susceptible individuals continually enter a population through migration and birth. In addition, pathogens can change so much through processes such as antigenic shift and drift.

CONTROL OF EPIDEMICS

There are three kinds of control measures.

First kind is directed towards reducing or elimination of the sources or reservoirs of infection. This is possible by

- Isolation of cases
- Destruction of animal reservoir
- Treatment of sewage
- Therapy that reduces or eliminates infectivity of the individual

The second kind is designated to break the connection between source and susceptible individual. This can be achieved by

- Chlorination of water supplies
- Pasteurization of milk
- Destruction of vectors by using insecticides

The third type reduces the number of susceptible individuals and raises overall immunity. This is done by

- Passive immunization
- Active immunization

1. How can epidemics be controlled?
2. What is an index case?
3. How do epidemiologists recognize an infectious disease within a given population?
4. Write short notes on:
   i. Common-source epidemic
   ii. Epidemic
   iii. Epidemiologists
   iv. Epidemiology
   v. Herd immunity
   vi. Index case
   vii. Cohert study
   viii. Morbidity rate
   ix. Mortality rate
   x. Endemic disease

1. Why is international cooperation needed to fulfil the requirements of epidemiologic study?
2. Is there any risk in eliminating a pathogen from a world?

Beaglehole, R., Bonita, R. and Kjellstrom, T. *Basic Epidemiology*. World Health Organization, Geneva, Switzerland. 1993.

Benenson, A. *Control of Communicable Disease Manual,* 16th edn. American Public Health Association, Washington, DC. 1995.

Bennett, J.V., Brachman, P.S. *Hospital Infections*, 3rd edn. Little Brown, Boston. 1992.

Brachman, P. *Bacterial Infections of Human*: *Epidemiology and Control*. Plenum, New York. 1991.

Lilienfeld, D.E. and Stolley, P. *Foundations of Epidemiology,* 3rd edn. New York. Cambridge University Press. 1994.

Evans, A.S. and Brachman, P.S. *Bacterial Infections of Humans*: *Epidemiology and Control*, 2nd edn. Plenum, New York. 1991.

Fox, J.P, Hall, C.E. and Elveback, L.R. *Epidemiology, Man and Disease*. Macmillan, New York. 1970.

Hennekens, C.H. and Buring, J.E. *Epidemiology in Medicine*. Little Brown, Boston. 1987.

Langmuir, A.D. "The Surveillance of Communicable Diseases of National Importance". *N. Engl. J. Med.* 268: 182. 1963.

MacMahon, B. and Pugh, T.F. *Epidemiology Principles and Methods*. Little Brown, Boston. 1970.

Mandell, G.L., Douglas, R.G. Jr., Bennett, and J.E. *Principles and Practice of Infectious Diseases*, 3rd edn. Churchill Livingstone, New York. 1990.

Morse, S. *Emerging Viruses*. Oxford University Press. 1993.

Mausner, J. S. and Kramer, S. *Epidemiology,* 2nd edn. WB Saunders, Philadelphia. 1985.

cat.inist.fr/?aModele=afficheN&cpsidt

research.medicine.wustl.edu

www.amazon.com/Hospital-Epidemiology-Infection-Control-Mayhall/

www.business-magazines.com/prd102150.php?siteid=global

www.cambridge.org/journals/journal_catalogue.asp?

www.hopkinsmedicine.org/heic

www.ingentaconnect.com/content

www.martindalecenter.com/PHealth_4_EPI.html

www.ncbi.nlm.nih.gov/entrez/query.fcgi?cmd=Retrieve

www.ovid.com/site/catalog/Journal/1339.

www.studiegidsen.vu.nl

www.vcu.edu/hospitalepi

www.zoonotics.net/part_iii_

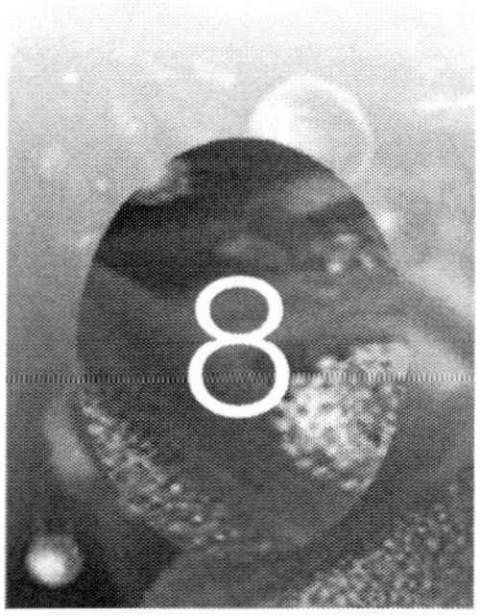

8 Collection of Various Specimens for Diagnosis

INTRODUCTION

In clinical microbiology, a specimen represents a portion or quantity of human material that is tested, examined or studied to determine the presence or absence of particular microorganisms. Several guidelines on the safety precautions to be used in selection, collection and handling of the specimens are listed.

Highly essential features of the sample collection are selecting the proper specimen and collecting adequate samples for examination. It is used for the confirmation that a microorganism is responsible for the infectious disease process.

A poorly collected specimen results in false positive results leading to incorrect or harmful therapy.

- The specimen must be a material from the actual infection site and must be collected with a minimum of contamination from adjacent tissues, organs or secretions.
- Optimal time and total number of specimens for collection must be established for the best chance of recovery of causative microorganisms.
- A sufficient quantity of specimen must be obtained to perform the culture techniques.
- Appropriate collection devices, specimen containers and culture media must be used to ensure proper isolation.
- Whenever possible, specimens should be obtained before the administration of antibiotics.
- The specimen container must be properly labelled.
- The following information should be transcribed along the specimen.

Name of the patient	:	
Hospital name	:	
Age of the patient	:	
Specimen type	:	
Clinical diagnosis	:	
Possible isolation	:	
Date and time of collection	:	

TRANSPORTATION OF SPECIMEN

Speed in transporting specimen is of prime importance because some laboratories refuse to accept specimens if they have been in transit too long. Certain specimens should be transported in a medium that preserves the microorganisms and helps to maintain the ratio of one organism to another. Special treatment is required for specimens when the organism is anaerobic. Anaerobic specimens are transported to the laboratory within 10 minutes. If any delay is expected, the specimen must be injected immediately into an anaerobic vial. Vials should contain a transport medium with an indicator like methylene blue. For anaerobic culturing, aspirates are more useful than swabs.

Commonly available transport medium are

- Amies transport medium
- Bile peptone transport medium
- Buffered glycerol saline base
- CVTR medium
- Specimen preservative medium
- Transport charcoal medium
- Transport medium
- Wangs semi-solid medium
- Cary-Blair medium
- Stuart's medium
- Alkaline peptone water

SPECIMEN COLLECTION

Skin

The most common specimen collected from skin is pus. The following methods are adopted to collect pus and its associated specimens.

- Tissue obtained during surgical procedure
- Aspirated material from an abscess or deep wound
- Pus or exudates obtained from an infected site during surgery can be aspirated into the syringe
- Skin swab
- Skin scrapings
- Nail clippings
- Hair

Before sample collection, skin surface should be disinfected with 70% alcohol.

Upper Respiratory Tract (URT)

URT infections are assessed with the help of any one of the following methods.

- Nasopharyngeal aspirates
- Nasopharyngeal swab
- Swab of posterior pharynx
- Sodium alginate throat swab for pertussis
- Sinus washings
- Surgical biopsy
- Swab of tonsil
- Pernasal swab

Nasopharyngeal aspirates Gently pass a sterile catheter through one tonsil as far as the nasopharynx. Attach a sterile syringe to the catheter, and aspirate the specimen of mucopus. Dispense the specimen into a sterile container.

Nasopharyngeal swab or throat swab Bright light from over the shoulder of the patient should be focused into the oral cavity. The patient is instructed to open the mouth at "aah" position and breathe deeply. The tongue is gently depressed with tongue blade to visualize the tonsillar fossae and posterior pharynx. The swab is extended between the tonsillar pillars and behind the vulva. The tonsillar areas and posterior pharynx should be firmly rubbed with the swab. After collection, the cotton swabs are placed into sterile Cary-Blair transport media to prevent desiccation during transit to the laboratory.

Lower Respiratory Tract (LRT)

For the isolation of pathogen from LRT, samples are obtained through needle aspiration method. In this method, a needle is inserted into the throat region and sputum specimen is collected. Needle aspiration includes transtracheal aspiration and lung aspiration. Other methods are bronchial washing and blood collection.

Early morning sputum should be collected because it contains pooled overnight secretions in which pathogenic bacteria are likely to be more concentrated. For collection, a sterile wide-mouthed jar with a tightly firmed screw cap lid can be used.

Intestinal Tract

Stool or rectal swab or gastric aspirate or gastric biopsy are the sample collection methods from gastrointestinal system.

The collection of diarrhoeal stool is not difficult. In cases of diarrhoea, stool specimen should be collected in clean, wide-mouthed containers that can be covered with tight-fitting lid. In some instances, collection of rectal swab rather than stool specimen may be necessary, particularly in neonates or in severely debilitated adults.

Gastric aspirate is collected with the help of intubation. Sample collection from the hollow tube is called intubation. Long sterile tube is attached with syringe and the tube is either swallowed or passed through a nostril into the patient's stomach. Specimens are withdrawn periodically. The most common intubation tube is Levin tube.

Urinary Tract

Pathogens of urinary tract are isolated and identified very specifically by making use of any one of the following samples.

Mid-stream urine It is collected in sterile, dry, wide-necked, leak-proof container. About 20 ml of sample should be collected. Clean catch method is used to collect mid-stream urine. First voided urine is not collected because it is contaminated with microbes from lower portion of the urethra. If immediate delivery to the laboratory is not possible, the urine should be refrigerated at 4°C. If a delay of more than 1 hour is anticipated, boric acid should be added to the urine. Specimens containing boric acid need not be refrigerated.

Catheter A catheter is a tubular instrument used for withdrawing fluids from the body cavity. Three types of catheters may be used—hard catheter, French catheter and Foely catheter. Multiple samples may be collected at a time using Foely catheter.

Suprapubic aspiration Suprapubic aspiration is performed only in neonates, small children and occasionally for adults with clinically suspected UTI, who fail to establish diagnosis. This technique is best performed when the bladder is full. The suprapubic skin overlaying the urinary bladder is disinfected and a tap is made. In the immediate site where the tap is to be made, about 1 ml of anaesthetic solution is injected subcutaneously. With the point of a sharply tapered surgical blade, a small lance wound incision is made through the epidermis. Through this wound, an 18-gauge needle is gently extended into the urinary bladder and 10 ml of urine is aspirated into the syringe.

BLOOD

Blood collection is performed by needle aspiration procedure. To reduce contamination during vein puncture, the following method should be followed.

1. Wash with green soap.
2. Rinse with clean water.
3. Apply 1–2% tincture of iodine and allow to dry for 1–2 minutes.
4. Remove iodine with 70% alcohol.
   i. Blood should be collected before antimicrobial treatment has been started and at the time when the patient's temperature begins to rise.
   ii. Blood for culture should be taken by vein puncture.
   iii. 10–20 ml of blood is to be collected from adult.
   iv. 1–2.4 ml should be collected from young infants.
   v. 2.4–5 ml should be collected from old infants.
   vi. Two blood cultures should be performed from each patient to confirm the causative agent.

Genital Tract

Genital samples are collected primarily during venereal infection. Methods include the following.

Female

- Catheter aspirate
- Amniocentesis fluid
- Bartholin gland aspirate
- Transcervical aspirate
- Fallopian aspirate
- Swab of genital ulcer
- Urethral discharge
- Swab of posterior vagina

Male

- Urethral swab
- Swab of genital ulcer
- Penial discharge
- Prosthetic secretions

For cases like syphilis, blood samples are collected and serological techniques are performed.

CSF

CSF is sterile in normal cases. CSF is collected during CNS infection.

- Lumbar puncture
- Brain abscess

The chance of recovery increases with the volume of the specimen.

Suggested volumes are

- 1 ml for bacterial culture
- 2 ml for fungus

Do not refrigerate the specimen.

Anaerobe Isolation

Anaerobes are often missed unless the specimen is properly collected, transported to the laboratory and then isolated properly. The following table gives the various methods of collection of specimen from different regions of the body.

Head and neck	Abscess and biopsy
Lungs	Transtracheal aspirate, lung puncture and biopsy
CNS	Abscess, biopsy, lumbar puncture
Abdomen	Peritoneal fluid and biopsy
Urinary tract	Suprapubic aspiration
Female genital tract	Endometrial aspirate
Bone and joint	Aspirate
Soft tissue	Aspirate

Aspirates are transported with vials with anaerobic atmosphere.

All specimens should be transported within 30 minutes.

1. What are the general guidelines that should be followed in collecting and handling specimens?
2. Write a short notes on
   i. Intubation
   ii. Catheter
   iii. Needle aspiration
   iv. Transport medium
   v. Suprapubic aspiration

Baron, E.J., Peterson, L.R. and Finegold, S.M. (eds.): *Bailey and Scott's Diagnostic Microbiology*, 9th edn. C.V. Mosby, St. Louis. 1994.

Flemming, D.O. *Lab Safety*, 2nd edn. ASM Press. 1995.

Forbes, B.A. *Bailey and Scotts Diagnostic Microbiology*, 10th edn. C.V. Mosby, St. Louis. 1998.

Isenberg, H.D. *Clinical Microbiology Procedures Handbook.* ASM Press. 1992.

Koneman, E.W., Allen, S.D., Schreckenberg P.C. and Winn, W.C. (eds.). *Atlas and Textbook of Diagnostic Microbiology*, 4th edn. J.B. Lippincott, Philadelphia. 1992.

Kunin, C.M. *Detection, Prevention and Management of Urinary Tract Infections*, 4th edn. Lea & Febiger, Philadelphia. 1987.

Miller, M.J. *A Guide to Specimen Management in Clinical Microbiology.* ASM Press. 1996.

Murray, P.R., Baron, E.J., PFaller, M.A., Tenover, P.C. and Yolken, R.H. (eds.) *Manual of Clinical Microbiology*, 6th edn. American Society for Microbiology, Washington, DC. 1995.

Pennington, J.E. (ed.) *Respiratory Infections: Diagnosis and Management*, 3rd edn. Raven Press, New York. 1994.

Woods, G.L. and Washington, J.A. "The Clinician and the Microbiology Laboratory." Mandell, G.L., Bennett, J.E. and Dolin, R. (eds.). *Principles and Practice of Infectious Diseases*, 4th edn. Churchill Livingstone, New York, 1995.

curator.jsc.nasa.gov/lunar/lunar

massfatality.dna.gov/Chapter11/SampleCollection

www.aphis.usda.gov/ppq/ep/soybean_rust/

www.compostingcouncil.org/pdf/Sample_Collection

www.dna.gov/lab_services/samplekit

www.toolkit.cch.com/tools/letter_m.

www.uga.edu/~sisbl/sampling.

www.utmb.edu/lsg/labsurvivalguide/sample_procurement.htm

www.vgl.ucdavis.edu/forensics/samplecollection.html

www.wada-ama.org/rtecontent/document/urine_testing_

Selective Cum Differential Media used for the Isolation of Bacteria

A selective medium is prepared by the addition of specific substances to a culture medium that will permit growth of one group of bacteria while inhibiting the growth of some other groups. Differential media is prepared by adding certain chemicals to differentiate bacteria. These media are used for the cultivation and differentiation of bacteria. The following table provides a list of selective cum differential media for the cultivation of different groups of bacteria.

Pathogen	Description	Selective, differential and enrichment media used for cultivation
Escherichia coli	Gram-negative rod, motile	Eosin methylene blue agar
		Xylose lysine deoxycholate agar
		Salmonella-Shigella agar
		Rajhans medium
		Hektoen enteric agar
		Violet red bile agar
		Tergitol 7 agar with TTC
		CLED with bromothymol blue
		Deoxycholate agar
		Endo agar
		MacConkey agar
Shigella spp.	Gram-negative rod, non-motile	Fluid selenite cystine broth*
		Fluid tetra thionate medium*
		Xylose lysine deoxycholate agar
		Salmonella-Shigella agar
		Hektoen enteric agar
		Deoxycholate citrate agar
		GN broth*

* enrichment medium

(Contd.)

Table (Continued)

Pathogen	Description	Selective, differential and enrichment media used for cultivation
Salmonella spp.	Gram-negative rod, motile	Hektoen enteric agar Deoxycholate citrate agar Fluid selenite cystine broth* Xylose lysine deoxycholate agar Salmonella-Shigella agar Violet red bile agar GN broth* Bismuth sulphite agar CLED with bromothymol blue Brilliant green agar
Klebsiella	Gram-negative rod, non-motile	McConkey agar Tergitol 7 agar with TTC CLED with bromothymol blue Eosin methylene blue agar
Yersinia enterocolitica	Gram-negative rod	PSB enrichment broth* Yersinia selective medium CIA agar PSTA enrichment broth*
Proteus spp.	Gram-negative rod, motile	McConkey agar Tergitol 7 agar with TTC CLED with bromothymol blue Deoxycholate agar
Vibrio	Gram-negative, comma-shaped bacilli, motile	Alkaline peptone water* Monsor medium TCBS agar Vibrio agar
Aeromonas	Gram-negative rod, motile	Aeromonas isolation medium Furunculosis agar Inositol brilliant green bile agar Rippey Cabelli agar Blood agar with 10 μl/ml ampicillin

* enrichment medium

(Contd.)

Table (Continued)

Pathogen	Description	Selective, differential and enrichment media used for cultivation
Pseudomonas aeruginosa	Gram-negative rod, motile	McConkey agar Tergitol 7 agar with TTC Cetrimide agar Pseudomonas isolation agar Malachite green agar
Campylobacter spp.	Gram-negative rod, motile	Doyle's enrichment broth* Preston agar Campy agar
Neisseria spp.	Gram-negative cocci, non-motile	Blood agar Chocolate agar GC medium Modified Thayer-Martin Agar New York City agar medium Columbia blood agar
Haemophilus spp.	Gram-negative rod, motile	Eugonic agar Levinthols medium Chocolate agar GC medium
Brucella spp.	Gram-negative rod	Schaedler broth Brucella selective medium
Mycobacterium	Acid-fast bacilli	Petragnani medium Lowenstein-Jensen medium Middle brook medium Pfizer TB medium Dubos oleic agar Kirschner medium
Staphylococcus	Gram-positive cocci, non-motile	Blood agar Mannitol salt agar Baird-Parker agar Vogel-Johnson agar DNase test medium Phenyl ethyl alcohol agar

* enrichment medium

(Contd.)

Table (Continued)

Pathogen	Description	Selective, differential and enrichment media used for cultivation
Streptococcus	Gram-positive cocci, non-motile	Blood agar Streptococcus enrichment broth* Edwards medium Neomycin blood agar Kanamycin esculin azide agar Synders test medium Todd–Hewitt medium
Enterococcus	Gram-positive cocci, non-motile	Azide dextrose broth KF Streptococcus agar with TTC SF broth Slanez and Bartly medium
Corynebacteria	Gram-positive rod, non-motile	Loeffler agar Cysteine tellurite blood agar Hoyle medium Tinsdale medium
Bordetella	Gram-negative rod	Bordet-Gengou medium Rogen-Lowey medium
Listeria	Gram-negative rod	Listeria enrichment broth Modified McBride Listeria agar Listeria selective medium
Legionella	Gram-negative rod	Buffered charcoal yeast extract agar Feeley-Gorman broth Legionella agar
Bacillus anthracis	Gram-positive rod	Columbia blood agar Eugonic broth Eugonic agar
Clostridium	Gram-positive rod	Cooked meat medium Forget-Fredett agar Wikins-Chalgren anaerobic agar Reinforced clostridial agar Clostridium difficile agar McClung agar

* enrichment medium

(Contd.)

Table (Continued)

Pathogen	Description	Selective, differential and enrichment media used for cultivation
Bacteroides	Gram-positive rod	Wikins-Chalgren anaerobic agar
		Bacteroides bile esculin agar
		LD esculin agar
		Veal infusion agar
Mycoplasma	Pleomorphic	Mycoplasma broth and agar
		Cell-free culture medium
		BYE agar

PART II

BACTERIOLOGY

10 General Characteristics of Bacteria

Bacteria are microscopic, unicellular prokaryotic organisms. The study of bacteria is known as bacteriology.

Bacteria are the most abundant of all organisms. They are ubiquitous in soil, water, air and as symbionts of other organisms. Many pathogens are bacteria. Their size ranges from 0.5 μm to 3 μm.

Bacteria were first observed by Antoni van Leeuwenhoek in 1676. The name "bacterium" was introduced much later by Ehrenberg in 1828.

MAJOR FEATURES

- Bacteria have distinctive cell wall, or unique cell envelopes, which contain a peptidoglycan layer. They lack a true nucleus, instead have a region called the nucleoid (i.e., DNA).
- They show absorptive mode of nutrition.
- They multiply by both sexual and asexual reproduction. Asexual reproduction includes binary fission, budding, fragmentation and spore formation. Sexual method includes transformation, transduction and conjugation.
- They produce spores under adverse conditions like heat, desiccation and chemicals.

Classification

Bacteria can be classified according to

- Morphology (Shape, cell wall structure and how they take up Gram stain)
- Physiological features (Cellular respiration and growth factors)

CLASSIFICATION BASED ON MORPHOLOGY

Shape

Bacterial cells vary in shape:

- Cocci are spherical, e.g. *Streptococcus, Staphylococcus*
- Bacilli are rods or cylindrical, e.g. *Bacillus, Clostridium*
- Spirilli are spiral or helical, e.g. *Treponema*
- Filamentous are complex forms, e.g. *Leptothrix*, *Crenothrix*

Cell Wall Structure

Bacteria can also be classified based on differences in the composition of cell walls. The difference becomes clear by means of a technique called gram staining, which identifies bacteria as either gram-positive or gram-negative. After staining, gram-positive bacteria hold the dye and appear purple, while gram-negative bacteria release the first dye used and appear red from the second (counter) dye. For example, *Staphylococcus* is gram-positive and *Escherichia coli* is gram-negative.

CLASSIFICATION BASED ON PHYSIOLOGY

Cellular Respiration

Bacteria can also be classified according to whether or not they need oxygen to survive.

- Aerobic bacteria require oxygen, e.g. *Bacillus cereus.*
- Anaerobic bacteria cannot tolerate oxygen, e.g. *Clostridium* spp.
- Facultative anaerobes are generally aerobes, but have the capacity to grow in the absence of oxygen, e.g. *Staphylococcus* spp.

Growth Factors

Bacteria can further be classified according to how they grow and metabolize.

Energy Source

- Chemotrophs utilize chemical compounds as energy source.
- Phototrophs utilize light as energy source.

Nutrient Source

- Heterotrophs derive carbon from preformed organic nutrients such as sugar.
- Autotrophs derive carbon from inorganic sources such as carbon dioxide.

BIOCHEMICAL CHARACTERS

Biochemical characters are very useful for identification because they are directly related to the nature and activity of microbial enzymes and protein transport. Indole, methyl red, and Voges-Proskauer tests are some of the tests used for the assessment of biochemical characters.

IMPORTANCE

Bacteria are important in every phase of human life. They are ideal for all kinds of genetic research because they have the ability to grow and multiply within a short span of time. They are both harmful and useful to the environment and animals, including humans.

Some of the beneficial aspects of bacteria include:

- Bacteria help in nitrogen fixation in soil.
- They degrade a variety of organic compounds and help in the disposal of organic matter.
- Bacteria, often in combination with yeasts and moulds, are used in the preparation of fermented foods, organic acids, antibiotics, etc.
- Using biotechnological techniques, bacteria can be bioengineered for the production of therapeutic drugs and for the bioremediation of toxic wastes.

But many bacteria also act as pathogens and cause several diseases.

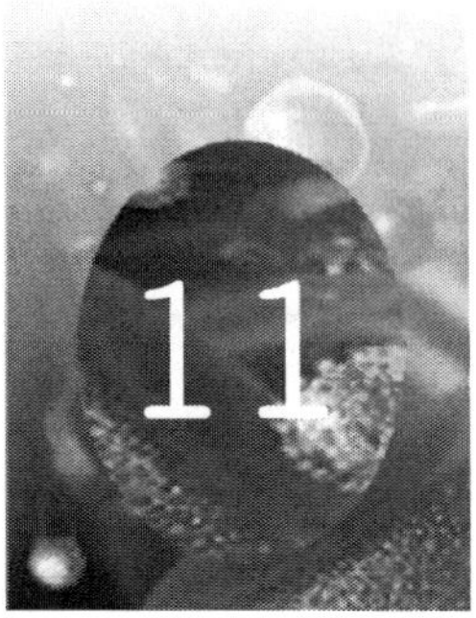

Classification of Pathogenic Bacteria

Classification refers to the arrangement of organisms into groups based on mutual similarity. Bergey's Manual provides the best classification of bacteria. The first edition of *Bergey's Manual of Systematic Bacteriology* describes 33 sections of bacteria within 4 volumes. The first edition uses morphological, physiological and metabolic characteristics as a criteria for classification. The second edition of systematic bacteriology uses 16S rRNA study and G + C content availability as major criteria. They classified bacteria in 30 sections within 5 volumes. Volume 1 describes archaea, cyanobacteria; volume 2 the proteobacteria; volume 3 gram-positive bacteria with low G + C; volume 4 gram-positive bacteria with high G + C; volume 5 spirochaetes, anaerobes, Planctomycetes. The classification described below is based on the *Bergey's Manual of Determinative Bacteriology*, 9th edition. This is a single book easily followed by all microbiologists. Morphology, physiology and biochemical characters are considered for classification. Out of 33 groups, only 13 groups from the total *Bergey's Manual* classification are pathogenic, and these are described one by one. The group number given is that from the original text.

GROUP 1: THE SPIROCHAETES

- Gram-negative helical cells that are highly flexible.
- Motile by periplasmic flagella.
- Chemoorganotrophic heterotrophs.
- Anaerobic, microaerophilic, facultative anaerobic or aerobic.
- Pathogenic to humans and animals.

Important genera are *Borrelia*, *Spirochaeta*, *Leptospira* and *Treponema*.

GROUP 2: AEROBIC/MICROAEROPHILIC, MOTILE, HELICAL/VIBRIOID GRAM-NEGATIVE BACTERIA

- Gram-negative helical or vibroid cells.

- Motile by polar flagella.
- Aerobic or microaerophilic, having a respiratory type of metabolism.
- Chemoorganotrophic heterotrophs.
- Pathogenic to animals and humans.
- Predatory on other microorganisms.

Important genera are *Campylobacter* and *Helicobacter.*

GROUP 4: GRAM-NEGATIVE AEROBIC/ MICROAEROPHILIC RODS AND COCCI

- Chemoorganotrophic heterotrophs.
- Occur in soil, fresh water or marine environments and in oral cavity of humans and animals.
- Pathogenic to animals and humans.

Important genera are *Alcaligenes, Alteromonas, Bordetella, Brucella, Flavobacterium, Legionella, Moraxella, Neisseria* and *Pseudomonas*.

GROUP 5: FACULTATIVELY ANAEROBIC GRAM-NEGATIVE RODS

- Chemoorganotrophic heterotrophs.
- Occur either free-living or in association with animal, human or plant hosts
- Some are pathogenic.

Important genera are *Citrobacter, Escherichia, Hafnia, Klebsiella, Proteus, Salmonella, Serratia, Shigella, Yersinia, Aeromonas, Plesiomonas, Photobacterium, Vibrio, Actinobacillus, Haemophilus, Pasteurella* and *Streptobacillus.*

GROUP 6: GRAM-NEGATIVE, ANAEROBIC, STRAIGHT, CURVED AND HELICAL RODS

- Chemoorganotrophic heterotrophs.
- Obtain energy by anaerobic respiration or by fermentation.

Important genera are *Bacteroides, Fusobacterium* and *Porphyromonas.*

GROUP 8: ANAEROBIC GRAM-NEGATIVE COCCI

- Chemoorganotrophic heterotrophs.
- Have a strictly fermentative type of metabolism.

The important genus is *Veillonella.*

GROUP 9: THE RICKETTSIAE AND CHLAMYDIAE

- Obligate intracellular parasites.
- May be rod-shaped, coccoid or pleomorphic.
- Many species are pathogenic.

Important genera are *Rickettsia*, *Chlamydia.*

GROUP 12: GRAM-POSITIVE COCCI

- Chemoorganotrophic.
- Mesophilic.
- Non-spore-forming cocci.
- Aerobic cocci that occur in pairs, clusters or tetrads.
- Facultatively anaerobic or microaerophilic cocci and strictly anaerobic cocci also occur in pairs, clusters or tetrads.

Important genera are *Enterococcus, Leuconostoc, Micrococcus, Peptococcus, Peptostreptococcus, Staphylococcus* and *Streptococcus.*

GROUP 13: ENDOSPORE-FORMING GRAM-POSITIVE RODS AND COCCI

- Bacteria that produce heat-resistant endospores.
- Mostly motile rods or filaments.
- Strict aerobes, facultative anaerobes, microaerophiles or strict anaerobes.

Important genera are *Bacillus* and *Clostridium.*

GROUP 14: REGULAR, NON-SPORING GRAM-POSITIVE RODS

- Rod-shaped cells.
- Non-sporing, non-pigmented, mesophiles.
- Chemoorganotrophic heterotrophs.
- Fermentative, saccharolytic microaerophiles.
- Aerobes or facultative anaerobes.
- Strict aerobes.

Important genera are *Lactobacillus, Erysipelothrix* and *Listeria.*

GROUP 15: IRREGULAR, NON-SPORING GRAM-POSITIVE RODS

- Irregular rods.
- Some may exhibit club-shaped forms, branched filamentous elements, or mixtures of rods or filamentous and coccoid forms.
- Aerobes or facultative anaerobes to microaerophilic or strict anaerobes.
- Pathogens of animals and plants.

Important genera are *Acetobacterium, Actinomyces*, *Bifidobacterium*, *Brevibacterium* and *Corynebacterium*.

GROUP 16: THE MYCOBACTERIA

- Aerobic.
- Non-motile, non-sporing, slow-growing rod-shaped bacteria.
- Acid-fast bacterium.
- Branched filaments formed occasionally.
- No aerial mycelium formed.

Important genus is *Mycobacterium.*

GROUP 26: NOCARDIFORM ACTINOMYCETES

Subgroup 1: Mycolic acid-containing bacteria (e.g. *Rhodococcus, Nocardia*).

Subgroup 2: Pseudonocardia and related genera (e.g. *Pseudonocardia*).

Subgroup 3: Nocardioides (e.g. *Terrabacter*).

Subgroup 4: *Jonesia*

GROUP 30: MYCOPLASMAS

- Pleomorphic cells devoid of cell walls.
- Growth on agar shows characteristic "fried egg" appearance.
- Requires sterols for growth.
- May show gliding motility.
- Facultative anaerobes or obligatory anaerobes.

Important genera are *Mycoplasma, Spiroplasma* and *Ureaplasma.*

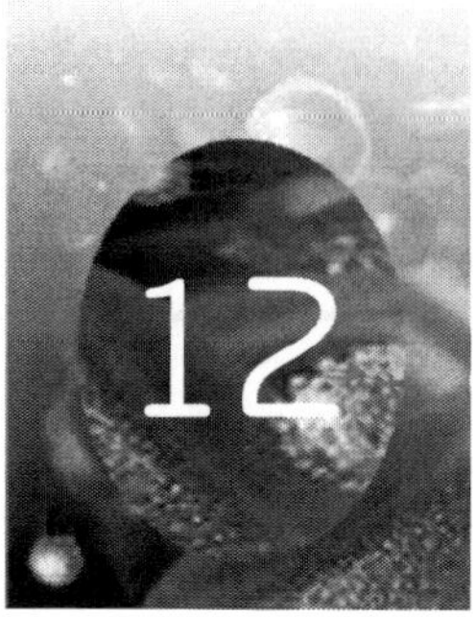

Staphylococcal Infections

INTRODUCTION

Staphylo means "grape-like clusters" which is due to three planar cell divisions (Figure 12.1). Staphylococci are gram-positive spherical bacteria that occur in 1 μm microscopic clusters. It was first observed in human pyogenic lesions by Von Reckling Heusen in 1871. Sir Alexander Ogston demonstrated the causative role of *Staphylococcus* in 1880. The genus *Staphylococcus* is placed under the family Micrococcaceae, but the staphylococci are phylogenetically unrelated to any other genera in the family. The most significant pathogen of this genus is *S. aureus.* At least 30 species of staphylococci have been recognized by biochemical analysis and in particular by DNA–DNA hybridization. Eleven of these can be isolated from humans as commensals. *S. aureus* (nares) and *S. epidermidis* (nares, skin) are common commensals and also have the greatest pathogenic potential. Staphylococci are the most resistant bacteria among non-sporing bacteria, and have high resistant capability to withstand high temperature, drying, etc.

Staphylococcus aureus

Clusters (Staphylococci)

Figure 12.1 Cell division in staphylococci

Some of the important species of staphylococci include *S. aureus, S. auricularis, S. capitis, S. caprae, S. epidermidis, S. haemolyticus, S. felis, S. intermedius, S. vitulus, S. hominis, S. lugdunensis, S. saprophyticus, S. schleiferi, S. warneri* and *S. xylosus*.

CLASSIFICATION

Staphylococcus is the member of the family Micrococcaceae. It is located in Section 12 of the first edition of *Bergey's Manual of Systematic Bacteriology* (Volume 2) and 9th edition of *Bergey's Manual of Determinative Bacteriology*. The second edition of *Bergey's Manual of Systematic Bacteriology* describes staphylococcus in a different way and given in the section 22 of Volume 3 based on G+C content. This section comprises aerobic and facultative aerobic rods and cocci (Box 12.1).

Kingdom	Bacteria
Phylum	Firmicutes
Class	Bacilli
Order	Bacillales
Family	Staphylococcaceae
Genus	*Staphylococcus* (Rosenbach, 1884)

CHARACTERISTIC FEATURES

Staphylococci are gram-positive cluster-forming cocci (Figures 12.2 and 12.3). They are non-motile, non-spore-forming, facultative anaerobes. They produce golden yellow colonies on nutrient agar, beta haemolytic colonies on blood agar, black colonies on Baird-Parker agar, yellow colonies on mannitol salt agar, pink-coloured zoned colonies on DNase test agar, black colonies on Vogel-Johnson medium.

Figure 12.2 Grape-like clusters of *Staphylococcus*

1. Few strains produce microcapsule or slime
2. Optimum pH for growth is 7–7.5
3. Optimum temperature is 30–37°C
4. *S. epidermidis* produces white colonies on nutrient agar and non-β-haemolytic colony on blood agar. *S. aureus* utilizes potassium tellurite and converts it into telluramine. It is a black-coloured compound. Biochemical features of *S. aureus* in given in Box 12.2.

Gram-positive cocci, non-motile, non-spore-former, indole-negative, methyl-red positive, VP-negative, urease-positive, nitrate-positive, gelatin-hydrolysis-positive, phosphatase-positive, coagulase-positive, fermentation of glucose-positive, mannitol-positive.

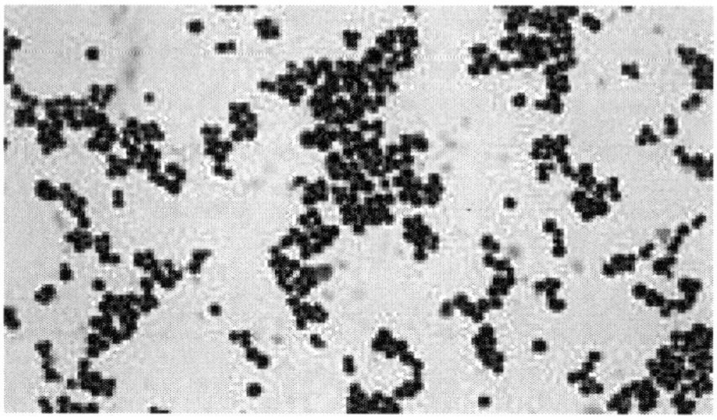

Figure 12.3 Gram-positive cluster of *S. aureus*

VIRULENCE FACTORS

The microbial produce that contributes to virulence or pathogenicity is called the virulent factor. The following are some of the virulent factors of *S. aureus*.

1. Surface protein that promote colonization, e.g. protein A, microcapsule, etc.
2. Invasins that promote bacterial spread, e.g. leucocidin, kinases, hyaluronidase
3. Surface factors that inhibit phagocytic activity, e.g. capsule, protein A
4. Biochemical properties that enhance their survival in phagocytes, e.g. carotenoids, catalase
5. Immunological disguises, e.g. coagulase, protein A
6. Membrane-damaging toxins, e.g. haemolysins, leucotoxin, leucocidin
7. Exotoxins that damage host tissue, e.g. staphylokinase
8. Inherent and acquired resistance to antimicrobial agent

Virulence factors	Effect
Leucocidin	Damages WBC
Catalase	Reduces oxidative burst
Coagulase	Causes plasma to clot
Haemolysins	Lyse RBC
Hyaluronidase	Reduces nerve impulse conduction
β-lactamase	Inactivates penicillin
Exfoliative toxin	Causes skin damage

Enterotoxin	Food poisoning

Figure 12.4 shows the various virulent factors.

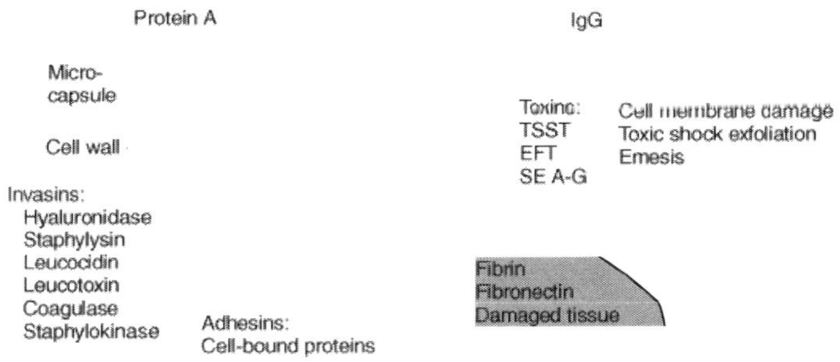

Figure 12.4 Virulent factors of *S. aureus*

EPIDEMIOLOGY

- It spreads from person to person by aerosols.
- Babies colonize staphylococci from their surrounding.
- Carriers are also responsible for transmission.
- Most of the staphylococci are carried by surgeons.

PATHOGENESIS

Staphylococcus attaches to the host epithelial and endothelial cells (laminin and fibronectin) with the help of protein. Most strains express a fibrin/fibrinogen-binding protein (clumping factor) which promotes attachment to blood clots and traumatized tissues. Adhesins promote attachment to collagen.

Invasion of host tissues by staphylococci apparently involves the production of a huge array of extracellular proteins, some of which may occur as cell-associated proteins.

Membrane-damaging toxins like alpha toxin, beta toxin, delta toxin, gamma toxin, coagulase, leucocidin, staphylokinase, DNase, lipase, protease and fatty acid modifying enzyme (FAME) are present at the site of invasion.

Susceptible cells have a specific receptor for alpha toxin, which allows the toxin to bind and cause small pores through which monovalent cations can pass. In humans, platelets and monocytes are sensitive to alpha toxin. Secondary reaction of the activity of this toxin causes release of cytokines that trigger production of inflammatory responses.

Beta toxin is a sphingomyelinase, which damages membranes rich in lipid. A lysogenic phage is known to encode the toxin. Delta toxin is a very small peptide toxin produced by *S. aureus*.

Beta toxin (leucotoxin) and leucocidin are the two components of protein toxin that damage membranes of susceptible cells.

Leucotoxin has a haemolytic activity.

Coagulase is an extracellular protein which binds to prothrombin in the host to form a complex called staphylothrombin, resulting in the conversion of fibrinogen to fibrin.

Most strains express plasminogen activator called staphylokinase, which is involved in fibrinolysis and aid in bacterial spreading.

FAME is responsible for abscess formation; lipases, proteases, DNases are used for providing nutrients to bacteria after attachment. Bacteria must be able to resist host defence mechanism. Capsular polysaccharide and catalase enzymes prevent lysis of cells through phagocytosis. Protein A, which binds IgG by their Fc region, disturbs opsonization and phagocytosis. Leucocidin acts on PMN.

Once the bacterium has entered inside the tissue, survival of *Staphylococcus* depends on several factors:

- Number of entering organisms
- Site of infection
- Inflammatory response
- Immunological history

Based on the site of infection and specific and non-specific defence mechanisms the following diseases are observed on the human body.

Folliculitis—a small red bump develops at hair follicle, hair can be pulled from its follicle.

Furuncle—infection in follicle spreads to adjacent tissues causing localized redness, swelling.

Carbuncle—a large area of redness, swelling and pain.

Cellulitis—infection in subcutaneous tissue with inflammation.

Stye—infection in eyelid.

Other infections are:

- Sinusitis
- Pneumonia
- Impetigo
- Osteomyelitis
- Scalded skin syndrome (SSS)
- Toxic shock syndrome (TSS)

- Bacteraemia
- Endocarditis
- Cystitis
- Diarrhoea

LABORATORY DIAGNOSIS

Specimen

Skin abscesses, blood, skin swab, pus, mucus, sputum, throat swab, tissue obtained during surgical procedure, aspirated material from deep wound, pus or exudates obtained from an infected site during surgery are the specimens used to isolate *S. aureus*.

Transport Medium

Amies transport medium

Culture Media used for Cultivation

- Phenyl ethyl alcohol agar
- Blood agar
- Mannitol salt agar
- Baird-Parker agar

Processing of the Specimen

- All the specimens should be emulsified in broth to create a homogeneous suspension for the preparation of gram-stained smear.
- A drop of aspirated purulent material is placed on a slide.
- Clear aspirated material should be concentrated by centrifugation.
- Both impression smear and homogenized smear are performed while using tissues.
- Swabs should be vortexed in 0.5–1 ml of broth to extract as much suspension as possible. Each swab should be squeezed against the side of the broth tube to expel remaining fluid and discarded.

Microscopy

Gram Staining

- A drop of homogenized suspension is used for smearing.
- All smears are air-dried.
- The smear is flooded with 90–100% alcohol and allowed to air-dry for fixing.
- The smear is stained by gram-staining technique.
- The slide is then observed under oil immersion lens and the appropriate result reported.

Culture

- Both liquid and solid media are used for cultivation.
- Homogenized tissue materials are inoculated into blood agar.
- The sample is inoculated into Baird-Parker agar.

Examination of the Media

- All broths and plates are examined. The broth is gram-stained and subcultured if there is no growth on plates but it is evident in broth.
- *S. aureus* produces beta haemolytic colonies on blood agar.
- *S. aureus* produces black-coloured colonies on Baird-Parker agar.

PREVENTION AND TREATMENT

For cutaneous infection, cloxacillin and dicloxacillin are highly effective. Erythromycin may also be used.

For serious systemic *Staphylococcus* infections, parenteral administration of nafcillin or oxacillin is recommended. Vancomycin or cephalosporins are suitable for allergic patients.

There was no vaccine available for prevention of human staphylococcal infection up to February 2002. In February 2002, an experimental bivalent vaccine against *Staphylococcus aureus* was reported to be safe and immunogenic for approximately 40 weeks in patients with end stage renal disease undergoing haemodialysis. The vaccine called StaphVAX is composed of *Staphylococcus* type 5 and 8 capsular polysaccharide conjugated with non-toxic recombinant *Pseudomonas aeruginosa* exotoxin A.

1. How does pathogenic *Staphylococcus* differ from non-pathogenic *Staphylococcus*?
2. Why is there no vaccine for *Staphylococcus*?
3. Why should we not use human blood for blood agar preparation?
4. What are the media used for the cultivation of *Staphylococcus*?
5. Why are pathogenic microbes resistant to different antibiotics?
6. Why is pathogenic *Staphylococcus* resistant to β-lactam antibiotics?

Holt, J.G. (ed.). *Bergey's Manual of Determinative Bacteriology*, 9th edn. Williams & Wilkins. 1994.

Madigan, Michael, Martinko and John (eds.). *Brock Biology of Microorganisms*, 11th edn. Prentice Hall. 2005.

Ryan, K.J. and Ray, C.G. (eds.). *Sherris Medical Microbiology*, 4th edn. McGraw-Hill. 2004.

www.cdc.gov/ncidod/hip/ARESIST/visa.htm

en.wikipedia.org/wiki/Staphylococcus

gsbs.utmb.edu/microbook/ch012.htm

kidshealth.org/parent/infections/ bacterial_viral/staphylococcus.html

medic.med.uth.tmc.edu/path/00001456.htm

ohioline.osu.edu/hyg-fact/5000/5564.html

textbookofbacteriology.net/staph.html

www.bact.wisc.edu/bact330/lecturestaph

www.ccohs.ca/oshanswers/biol_hazards/methicillin.html

www.cfsan.fda.gov/~mow/chap3.html

www.emedicine.com/med/topic2166.

www.hpa.org.uk/infections/topics_az/staphylo/menu.htm

www.molbio.princeton.edu/ courses/mb427/2001/projects/02/staph.htm

www.sanger.ac.uk/Projects/S_aureus/

www-micro.msb.le.ac.uk/video/Staphylococcus.html

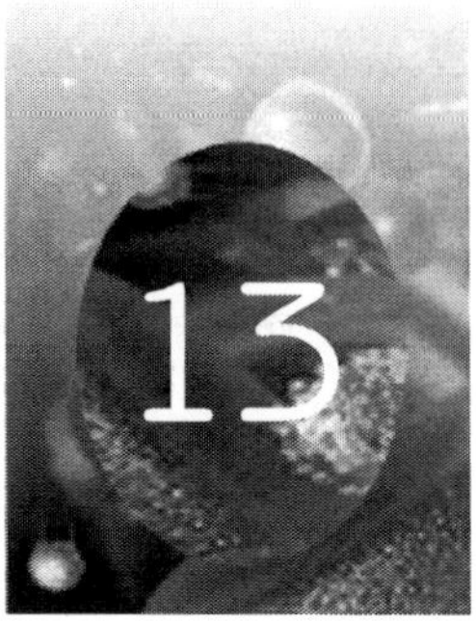

Streptococcal Infections

INTRODUCTION

The genus *Streptococcus* comprises a large and biologically diverse group of gram-positive cocci that grow in pairs and chains (Figures 13.1. and 13.2). Streptococci were first described by Billroth in 1874 in exudates from erysipelas and wound infections.

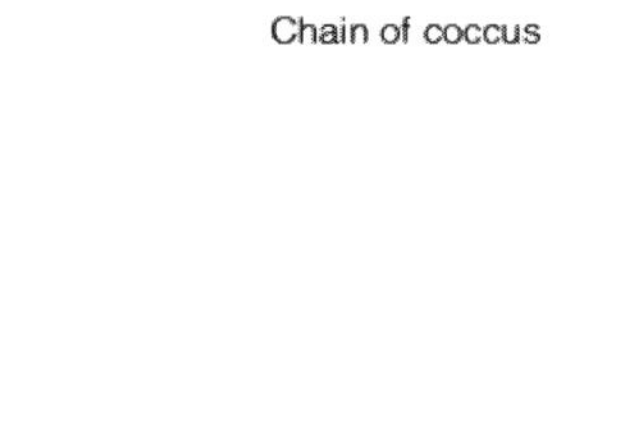

Figure 13.1 Streptococcus

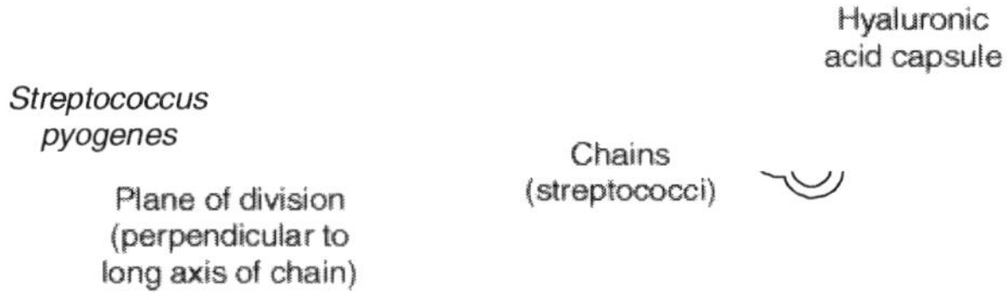

Figure 13.2 Cell division in Streptococci

Rosenbach (1884) isolated the cocci from human suppurative lesions and named them *Streptococcus pyogenes*. The genus *Streptococcus* currently includes 27 recognized species. Many of the species exist as commensals. Some are highly pathogenic while others are saprophytic.

CHARACTERISTIC FEATURES

- *Streptococcus* is a gram-positive bacterium, spherical to ovoid in shape and less than 2 μm in diameter.

- It is a facultative anaerobe with fermentative metabolism. This bacterium is catalase-positive and oxidase-negative.

CLASSIFICATION

Streptococci belong to the family Streptococcaceae and the order Lactobacillales. Various strategies are adopted to classify streptococcus. The genus *Streptococcus* is large and complex. The first edition of *Bergey's Manual of Systematic Bacteriology* describes 38 species clustered into four groups namely pyogenic streptococci (*S. pyogenes*), oral streptococci (*S. oralis*), anaerobic streptococci (lactococci) and other streptococci (*S. bovis*). Box 13.1 shows the recent classification of streptococcus.

Kingdom	Eubacteria
Phylum	Firmicutes
Class	Bacilli
Order	Lactobacillales
Family	Streptococcaceae
Genus	*Streptococcus* (Rosenbach, 1884)

Classification based on Haemolysis

Schottmuller first proposed haemolysis as a criterion for classification but it was established on a firm basis by Brown (1919). This classification is based on haemolysis pattern on blood agar.

1. Alpha haemolytic streptococci exhibit incomplete haemolysis, which imparts green appearance to the haemoglobin. These species are called viridans streptococci.
2. Beta haemolytic streptococci produce a sharply defined, clear, colourless zone of haemolysis. They account for the majority of diseases.
3. Gamma haemolytic streptococci are non-haemolytic. The streptococci included in this group are non-pathogens.

Classification based on Antigens

The haemolytic streptococci were classified by Rebecca Lancefield (1933) serologically into groups based on the nature of carbohydrate antigen on the cell wall. These are known as Lancefield groups. About 19 serogroups are available and are named from A to U (without I and J). Serogroups are further subdivided into subtypes based on the protein (M, T and R) antigens and is called Griffith typing. About 80 types of *S. pyogenes* have been recognized.

Streptococci

Alpha haemolysis (Viridans group)

Gamma haemolysis (Non-pathogenic group)

Beta haemolysis (Haemolytic group)

b haemolytic group classified into 19-Lancefield groups
A,B,C,D,E,F,G,H,K,L,M,N,O,P,Q,R,S,T &U

Griffith typing

MEDICALLY IMPORTANT STREPTOCOCCI

Various medically important streptococci are given in Table 13.1.

Table 13.1 Classification and grouping of streptococci

Group	Haemolysis	Example
A	Beta	*S. pyogenes*
B	Beta	*S. agalactiae*
C	Beta	*S. equii*
D	Beta	*S. bovis*
F	Beta	*S. milleri*
G	Beta	*S. canis*
–	Alpha	*S. pneumoniae*
Oral streptococci	Alpha	*S. sanguis* *S. oralis*
Viridans streptococci		*S. gordonii* *S. mitis* *S. mutans* *S. salivaris*

VIRULENCE FACTORS

The following virulent factors of streptococcus are involved in pathogenesis.

Virulence factors	Effect
Hyaluronic acid (capsule)	Prevents opsonization
Peptidoglycan	Induces fever, cardiac necrosis
M Protein	Prevents phagocytosis
C5a peptidase	Destroys chemotactic signals
Lipoteichoic acid	Improves colonization
Haemolytic haemolysin	Lyses RBC
Cytolytic haemolysin	Lyses RBC and epithelial cells
Pyogenic exotoxin	Induces pus formation
Esterases	Induces pus formation
Streptokinase	Induces pus formation
Phosphatase	Induces pus formation
Proteinase	Induces pus formation
DNase	Cleaves DNA
ATPase	Cleaves ATP
Nuclease	Lyses nucleus
Cardiohepatic toxin	Kills or affects heart and liver cells
Streptolysin O	Oxygen-sensitive haemolysin
Streptolysin S	Oxygen-resistant haemolysin

PATHOGENESIS

The pathogenicity of *S. pyogenes* (Group A) is multifactorial. A large number of virulent factors are produced by the pathogen during the infection.

The organism enters into the human body by various direct and indirect mechanisms. They include:

- Scratching of skin
- Wound
- Droplets
- Aerosols
- Blood transfusion

From the portal of entry, organisms adhere to epithelial cells via lipoteichoic acid and enhance colonization. Once adherence has taken place, those strains that are able to resist phagocytosis and killing by leucocytes can proliferate and begin to invade local tissues. Local pharyngeal or cutaneous infection may ensue or organism may invade adjacent tissues or distant tissues via bloodstream. Once antibody response is induced, organisms may be readily engulfed and killed by phagocytes. The cell wall reacts with IgG in a non-immune manner. Cell wall antigens also have the capability to activate alternative complement pathway.

During infection, *Streptococcus* produces various types of virulent factors which may induce both suppurative (tonsillar abscess, otitis, septicaemia, osteomyelitis) and non-suppurative (acute rheumatic fever, glomerulonephritis) infections.

DISEASES CAUSED BY STREPTOCOCCI

- Pharyngitis
- Scarlet fever
- Septicaemia
- Erysipelas
- Impetigo
- Rheumatic fever
- Acute glomerulonephritis
- Puerperal sepsis
- Meningitis
- Endocarditis
- Urinary tract infections

Pharyngitis

Streptococcus attaches to the buccal epithelial cells through lipoteichoic acid. Streptococci are rapidly killed after ingestion. Some virulent strains resist the immune response and induce inflammatory response. Peptidoglycan also activates complement and induces inflammatory response.

Pharyngeal infection may be symptomatic or may be associated with strep throat, fever, chills, headache, malaise, nausea and vomiting. Pharynx may be mildly erythematous or beefy red with greyish yellow exudates and there may be bleeding from the throat. Streptococci may spread to the surrounding tissues and lead to suppurative complications. It may also lead to meningitis.

Scarlet means "strep throat with a red skin rash". It is closely related to erythrogenic toxin. It can accompany pharyngitis. A fine papular, red rash disseminates throughout the body.

Skin Infection

Group A streptococci produce impetigo, cellulitis, erysipelas, wound infection and gangrene.

Impetigo—superficial infection that usually begins as small vesicles, progressing to sweeping lesions with amper crust and slightly cloudy purulent exudates.

Erysipelas—spreading infection of the skin or mucous membrane, are usually seen in face. Lesions are characterized by erythema, oedema and induration.

Cellulitis—an inflammation of the skin and underlying connective tissue in diabetic patients. It may also lead to gangrene.

Puerperal sepsis or child bed fever—a common cause of maternal death. It occurs following child birth.

Acute Rheumatic Fever

It is a non-suppurative inflammatory response. It is manifested as arthritis, carditis, chorea, erythema marginatum or subcutaneous nodules. Symptoms occur within 2–3 weeks. Pathogenesis is poorly understood. Antigenic cross-reactivity between streptococcal antigens and heart tissue, direct toxicity due to exotoxins, actual invasion of heart by *Streptococcus* are the reasons for rheumatic fever. Prolonged bed rest is needed for recovery. Salicylates and corticosteroids are used for treatment.

Glomerulonephritis

It usually follows cutaneous infections. Kidney damage was originally believed to be secondary to the deposition of immune complexes. Symptoms are oedema, oliguria, hypertension, congestive heart failure and proteinuria.

Group B *Streptococcus (Streptococcus agalactiae)* causes bacteraemia, neonatal sepsis, meningitis and urinary tract infection.

Group C *Streptococcus* causes pharyngitis.

Group D *Streptococcus* causes genito-urinary infection and endocarditis.

Group G *Streptococcus* causes pharyngitis.

LABORATORY DIAGNOSIS

Specimen

- Skin swab
- Throat swab
- Nasopharyngeal aspiration
- Blood
- Sputum
- Pus

Microscopy

A loopful of specimen is smeared on a glass slide and fixed with methanol. The smear is gram-stained and observed under high-power magnification.

Culture

Most streptococci grow aerobically and anaerobically. Temperature range for growth is 22–42°C. Specimen is streaked on blood agar and incubated at 37°C. Colonies that appear are less than 1 mm in diameter, they appear greyish white or colourless, dry or shining and usually irregular. *S. pyogenes* produces beta haemolytic colonies. Kanamycin blood agar is recommended for group B *Streptococcus*. When grown on serum starch agar *S. agalactiae* produces an orange pigment. Isolates are identified by making use of biochemical parameters (Table 13.2 and Box 13.2).

Serology

ASO *test* Antistreptolysin O test is used to detect streptolysin O in blood.

Dick *test* Erythrogenic toxin produces erythematous reaction in susceptible individuals. For the detection of *S. pyogenes*, 0.1 ml of toxin is injected into humans by intradermal injection, which produces an erythematous reaction. This is called Dick test described by Dick in 1924.

Table 13.2 Biochemical characterization of different streptococci

Group	Haemolysis	Vancomycin sensitivity	Bacitracin sensitivity	SXT	CAMP	Hippurite	PYR	Bile Esculin	Growth at 6.5% NaCl	Optochin	Bile solubility
A	β	S	S	R	–	–	+	–	V	R	–
B	β	S	R	R	+	+	–	–	–	R	–
C	β	S	V	S	–	–	–	–	–	R	–
F	β	S	V	S	–	–	–	–	–	R	–
G	β	S	V	S	–	–	–	–	–	R	–
Enterococcus	α	S	R	R	–	V	+	+	+	R	–
Non-enterococcus	α	S	R	S	–	–	–	+	–	R	–
Viridans streptococci	α	S	V	S	–	V	V	–	–	R	–
S. pneumoniae	α	S	V	S	–	–	–	–	–	S	+

*CAMP—Christe Atkins Munch Pieterson Test, Va —Vancomycin, V—Variable, S—Sensitive, R—Resistant, +—positive, – —Negative

Gram-positive cocci, β-haemolytic colonies, non-motile, optochin-resistant, catalase- and oxidase-negative, vancomycin-sensitive, bacitracin-sensitive, SXT-resistant, CAMP-negative, hippurite-hydrolysis -negative, PYR-positive, esculin-hydrolysis-positive, bile-solubilization-negative.

TREATMENT

Antibiotics like erythromycin, clindamycin, cephalexin, penicillin, vancomycin, streptomycin, etc. are used for treatment.

1. Describe the virulence factors of streptococci.
2. What are the major groups of streptococci?
3. How can you classify streptococci based on haemolysis?
4. Write short notes on
   i. Lancefield serogrouping
   ii. Griffith typing
   iii. ASO test
   iv. Dick test
   v. CAMP

biology.kenyon.edu/Microbial_Biorealm/bacteria/gram-positive/streptococcus/
en.wikipedia.org/wiki/Streptococcus
gsbs.utmb.edu/microbook/ch013.htm
homepages.pavilion.co.uk/tetrix/streptococcus
medic.med.uth.tmc.edu/path/00001457.htm
textbookofbacteriology.net/streptococcus.html
www.bact.wisc.edu/Bact330/lecturespneumo
www.bact.wisc.edu/bact330/lecturespyo
www.cdc.gov/ncidod/dbmd/diseaseinfo/streppneum_t.htm
www.cdc.gov/node.do/id/0900f3ec800075ca
www.cfsan.fda.gov/~mow/chap21.html
www.childbirth.org/articles/GBS.html

www.dhpe.org/infect/strepa.html
www.dhpe.org/infect/strepb.html
www.drgreene.com/21_1196.html
www.emedicine.com/med/topic2184.html
www.emedicine.com/med/topic2185.html
www.genome.ou.edu/smutans.html
www.genome.ou.edu/strep.html
www.med.sc.edu:85/fox/strep-staph.html
www.niaid.nih.gov/factsheets/strep.
www.pbrc.hawaii.edu/microangela/strept.
www.rockefeller.edu/vaf/catalog.html
www.rockefeller.edu/vaf/strep.html
www.sanger.ac.uk/Projects/S_pyogenes/
www.wpro.who.int/media_centre/ fact_sheets/fs

Dental Caries

INTRODUCTION

Dental diseases are among the most important widespread diseases around the globe. Although not an important cause of mortality, these may sometimes have serious complications on the general health of people. Due to the increased concern for oral health, prevalence of dental caries and periodontal disease, in 1994 WHO has dedicated World Health day (27th February) to oral health.

India is a vast country having diversity in its geography and culture, which contributes to oral disorders. The study undertaken by Urban Health Training Centre of Government Medical College, Nagpur stated the following reasons for the ill oral conditions:

1. Differences in eating habits
2. Oral cleaning habits
3. Fluoride content of water

ANATOMY OF TEETH

The tooth is composed of a crown that protrudes above the gum and the root that is inserted into a long socket. A dense coating called enamel protects the outer surface of the crown. The enamel is an extremely hard, non-cellular material composed of tightly packed rods of calcium phosphate that cannot be replaced once the root has matured.

A layer of cementum, which anchors the tooth to the periodontal membrane that lines the socket, surrounds the root. The major portion of the tooth inside the crown and the root is composed of a highly regular calcified material called dentin. The core contains the pulp cavity, which has blood vessels and nerves. The places where the bone protrudes from the crown are covered by connective tissue and a mucous membrane called gingiva or gum.

Dental pathology generally affects both hard and soft tissues. Microbes initiate both categories of disease after adhering to the tooth surface.

CAUSATIVE AGENTS

Different organisms like *S. mutans, S. gordonii*, *Actinomyces, Lactobacillus acidophilus, Bacteroides, Neisseria* and *Treponema* are involved in creating dental caries. But the most prevalent and important organisms are *S. mutans* and *Lactobacillus acidophilus.*

Streptococcus mutans

- The cells are spherical or ovoid, 0.5–2 μm occurring in pairs or in chains.
- They are sometimes lancet-shaped.
- They are non-motile, non-sporing facultative anaerobes and are chemoorganotrophs.
- They require rich media for growth.
- They show fermentative type of metabolism and produce lactate but no gas.
- Esculin hydrolysis is positive.
- Optochin sensitivity is negative.
- No haemolysis.
- Acid is produced from inulin, lactose, mannitol, raffinose, sorbitol.
- They are Voges-Proskauer-positive and catalase-negative. Classification of streptococcus is already given in previous chapter (Box 14.1).

Lactobacillus acidophilus

- They are gram-positive rods, non-sporing and motile.
- They are facultative anaerobes.
- Growth is enhanced by 5% CO_2.
- They are chemoorganotrophs and produce non-pigmented colonies.
- The organisms ferment sugar and produce lactate.
- Nitrate is not reduced.
- They are gelatin-negative, catalase-negative and cytochrome-negative.

  Section 22 of the 2nd edition of *Bergey's Manual* describes this bacterium (Box 14.1).

Kingdom	Bacteria
Division	Firmicutes
Class	Bacilli
Order	Lactobacillales
Family	Lactobacillaceae
Genus	*Lactobacillus* (Beijerinck, 1901)
species	*L. acidophilus, L. brevis*

DENTAL CARIES

It is the most common human disease. It is a complex mixed infection of the dentition that gradually destroys the enamel and often lays the groundwork for the destruction of deeper tissues.

Caries development (Figure 14.1) occurs in many phases and requires multiple interactions involving the anatomy, physiology, diet and bacterial flora of the host.

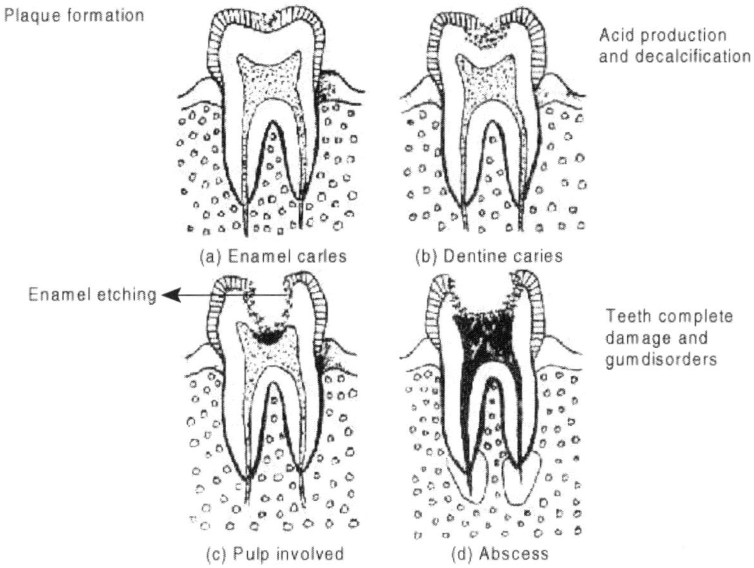

Figure 14.1 Formation of dental caries

Steps Involved in Caries Development

1. Pellicle formation
2. Plaque formation
3. Acid production
4. Localization
5. Enamel etching

Pellicle formation Within few moments after cleaning of teeth, a thin film of mucous coat develops, which is made up of adhesive salivary proteins. This acts as a substratum upon which bacteria first gain a foothold.

Plaque formation The most predominant microbes involved are the gram-positive organisms belonging to the genera *Streptococcus* and *Actinomyces*. These bacteria have structures like fimbriae and slime layer that allows them to cling to the tooth surface and to each other. This results in the formation of a dense mass of whitish substance called plaque. Dextran helps in the attachment of microbes to the enamel.

After plaque formation, secondary bacteria start to invade the plaque. The microbes invading are *Lactobacillus, Bacteroides, Fusobacterium, Neisseria* and *Treponema*.

Acid production If the plaque is not removed from sites that readily trap food, it usually evolves into caries lesion.

Initiation of caries formation by *Streptococcus mutans*

Attachment to the enamel and plaque formation

Bacteria hydrolyse sucrose into its monosaccharide

Enzyme glucose transferase assembles glucose into dextrans

Fermentation and acid production

Acid softens enamel

Secondary infection

Bacteria reach the pulp cavity and cause the ache

Further damage

Localization and enamel etching *Streptococcus* and *Lactobacillus* are directly involved in acid production by fermenting dietary carbohydrate. If this acid is removed by flushing action of mouth or diluted, there is less chance of getting caries.

If the acids are not removed, it will bind to the enamel. The environment will be acidic. This causes dissolution or decalcification of the calcium phosphate of the enamel. This initial lesion remains localized in the enamel which is known as first-degree caries. Filling can repair this. Once the deterioration has reached the level of dentin, this is formed as the second-degree caries. Exposure of pulp is considered as third-degree caries. Severe tenderness and toothache follow this. In this case the tooth cannot be repaired. This may follow periodontal disease.

Periodontal Disease

It is a soft tissue infection, commonly found in 97–100% of people. Gingivitis, bleeding of gums, is seen.

Periodontitis is responsible for nearly 10% of the tooth loss in older adults. Gums are inflamed and bleed easily. Sometimes pus forms in the pocket surrounding the teeth.

Acute necrotizing ulcerative gingivitis is termed as Vincent's disease or Trench mouth. This disease causes severe pain during chewing, foul smell from mouth. The bacteria usually associated are *Prevotella intermedia*. If this condition is not treated, the tooth may fall out of its socket. This may lead to irregular arrangement of teeth. Sometimes broken teeth irritate walls of oral cavity. Continuous irritation initiates abnormal growth. It is considered as oral cancer.

EPIDEMIOLOGY

- Worldwide distribution
- Occurs in all age groups
- Commonly seen in 15–30 age groups
- In India, males have higher incidence than females

LABORATORY DIAGNOSIS

- Direct examination
- Microscopic examination

Sample is transported with the help of Amies transport medium, inoculated into *Streptococcus* selection medium and also streaked on blood agar. Biochemical tests and carbohydrate fermentation tests are then performed to differentiate *Streptococcus* spp.

PREVENTION

- Minimal ingestion of sucrose
- Brushing
- Professional cleaning to remove plaque
- Use of fluorinated water
- Use of mouthwash
- Use of toothpaste
- Use of starch in diet

Natural ayurvedic treatment will be more effective than antibiotics.

SOME IMPORTANT DEFINITIONS

Caries Bone or tooth decay.

Dental plaque A thin film on the surface of teeth consisting of bacteria embedded in the matrix of bacterial polysaccharide, salivary glycoprotein and other substances.

Periodontal disease It is a disease located around the teeth or in the periodontium, the tissue supporting the teeth. It includes the cementum, periodontal ligament, alveolar bone and gingiva.

Periodontosis A degenerative, non-inflammatory condition of the periodontium, which is characterized by destruction of tissue.

1. Name some common dental diseases.
2. What are the three elements that cause tooth decay?
3. What diseases affect the gums? What is trench mouth?
4. What will happen if you do not look after your gums?
5. What are the causes for irregular arrangement of teeth?
6. What is oral cancer? Name its causes.

1. Why are there no dental caries in cattle?
2. How can caries and periodontal disease be prevented?
3. What is the role of gum?

en.wikipedia.org/wiki/Dental_caries
health.allrefer.com/health/dental-caries-info.html
oralhealth.dent.umich.edu/VODI/html/00/toc-dc.html
www.caobisco.com/english/pdf/Dental.pdf
www.cdc.gov/mmwr/preview/mmwrhtml/ss5403a1.htm
www.db.od.mah.se/car/data/cariesser.html
www.dent.ucla.edu/ce/caries/
www.dental.mu.edu/oralpath/lesions/caries/caries.htm
www.lalecheleague.org/NB/NBSepOct02p164.
www.nlm.nih.gov/medlineplus/ency/article/001055.htm
www.nutritionaustralia.org
www.pediatriconcall.com/forpatients/CommonChild/DentalCaries.asp
www.uic.edu/classes/peri/peri343/
www.umm.edu/ency/article/001055.htm
www.usc.edu/hsc/dental/ opath/Chapters/Chapter08

Pneumonia

Pneumonia is an infection of one or both lungs, usually caused by bacteria, viruses and fungi. It is an endogenous infection, because causative agents are contracted from normal flora of the respiratory tract. An infection of the lung that involves the small air sacs or alveoli and the tissue around them is called pneumonia.

CAUSATIVE AGENTS

Bacteria that cause pneumonia include *Klebsiella pneumoniae, Mycoplasma pneumoniae, Streptococcus pneumoniae, Legionella pneumophila.*

Viruses that cause pneumonia include respiratory syncytial virus, adenovirus and coxsackievirus.

The most common causative agent of pneumonia is *Streptococcus pneumoniae*. It causes epidemic or even pandemic forms of pneumonia. It remains a leading cause of morbidity and mortality in humans of all ages. Currently, over 3 million people develop pneumonia each year in the United States.

Streptococcus pneumoniae was first isolated by Pasteur and Sternbery independently during the year 1881 from human saliva.

CHARACTERISTIC FEATURES

- It is a facultative gram-positive coccus.
- Also called diplococcus but often they are called pneumococcus (Figure 15.1).

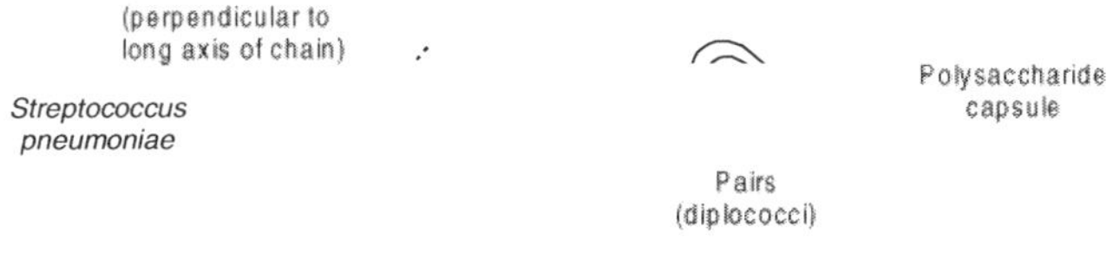

Figure 15.1 Cell division and arrangement of *Streptococcus pneumoniae*

- They are capsulated and non-motile, sometimes lancet shaped.
- The colonies are alpha haemolytic.
- *S. pneumoniae* ferments inulin and solubilizes bile.

ANTIGENIC STRUCTURE

Capsular Antigen

It is made up of a complex polysaccharide that forms the hydrophobic gel on the surface of the organism. On the basis of capsular antigen, 84 serotypes were recognized (Figure 15.2).

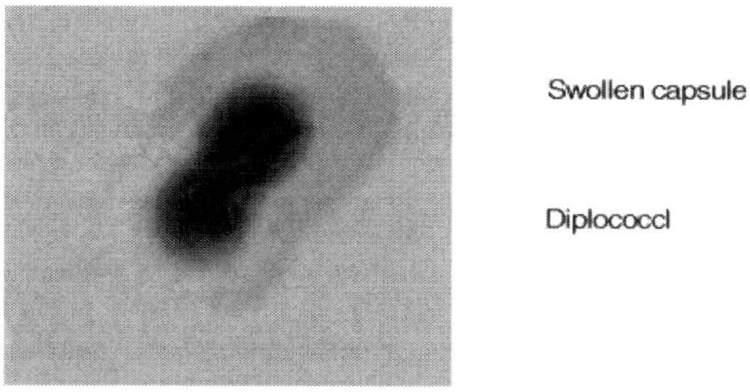

Figure 15.2 Appearance of *Streptococcus pneumoniae*

Cell Wall Antigen

- Species-specific.
- Structural component of cell wall.
- Phosphocholine is a major antigenic determinant.
- Galactosamine, glucose phosphate, ribitol and trideoxydiaminohexoses are the other components.

F-antigen or Forssman Antigen

This antigen cross-reacts with Forssman series of mammalian cell-surface antigen. It is a lipoteichoic acid, consisting of the C polysaccharide covalently linked to a lipid moiety. Distributed on the outer surface of the cell membrane, protein F is involved in attachment to fibronectin.

M Protein

It is type-specific.

It is similar to M protein of *S. pyogenes*. M protein is a fibrillar surface protein. The M proteins are associated with both colonization and resistance to phagocytosis. M proteins have been identified on the basis of antigenic specificity.

VIRULENCE FACTORS

The virulence is directly related to capsular antigen. Adhesins promote colonization. It binds specifically on *N*-acetyl hexosamine galactose polysaccharide component of cell wall.

Capsule

It makes the pathogen resistant to phagocytosis and facilitates colonization.

Neuraminidase

It attacks the glycoprotein and glycolipid component of the cell membrane. It cleaves the *N*-acetyl muramic acid residue of the cell wall.

Protease

It specifically cleaves the immunoglobulin.

Toxin of Pneumococci

- Pneumolysin is a toxin released during cell lysis.
- It is a 52.8-kDa, thio-activated cytotoxin.
- It binds with the cholesterol component of the cell membrane and disturbs them by forming pores, thus killing the ciliated epithelial cells.
- It makes a contribution to the inflammatory response.
- It activates the complement pathway directly.
- It limits the phagocytic action of PMNs and monocytes by decreasing oxidative burst.

Hydrogen Peroxide

It is produced under micro-aerophilic conditions. It causes lung tissue damage.

Lysozyme

It mediates autolysis of bacteria and releases inflammatory cell materials.

PATHOGENESIS

Everyone carries bacteria in their nose and throat as normal flora. 10–40% of healthy population have *S. pneumoniae*. It is prevalent during winter seasons. The bacterium is passed from person to person through coughing and sneezing.

In human beings about 80% of lobar pneumonia and 60% bronchopneumonia are caused by *S. pneumoniae*. Death occurs about 1–3 days after onset of symptoms.

Breathing of droplet nuclei allows the entry of bacteria to upper respiratory tract. It is inhibited by host defence mechanism. Pneumococci resist the defence mechanism and enter into the lungs. They usually settle in the air sac of the lung where they rapidly grow. During growth, capsule mediates colonization and also prevents phagocytosis.

Peptidoglycan fragments and teichoic acid of the cell wall activate the alternate pathway of complement and elicit the production of IL-1 and TNF-α. As a result, opsonins like C3b and IgG, IgM antibodies are formed against cell wall antigens. These opsonins diffuse through the porous matrix of the capsule to the bacterial surface, where they bind and activate the classical complement pathway. This contributes to the further production of C5a.

Complement activation attracts more PMNs but does not aid phagocytosis because C3b and antibodies are bound to cell wall. As a result of expanding inflammatory response, tissue damage occurs, but it does not clear the bacteria.

Inflammatory response leads to blood vessel damage, which results in increase of vascular permeability, leakage of blood, vasodilation, extravasation, excess mucous secretion, accumulation of mucus, lung tissue damage, and disturbance in gas exchange that leads to suffocation.

Damaging lung tissue not only provides the bacteria with source of nutrients but also creates a protected area for the bacterium. Then the bacteria spread throughout the body and create bacteraemic and toxaemic conditions.

CLINICAL MANIFESTATION

Symptoms of pneumonia include the following:

- Shortness of breath
- Fever
- Chills
- Cough
- Chest pain
- Pain during inhalation
- Fluid around lungs—pleural effusion
- Abscess
- Emphyema
- Rust-coloured sputum

Complications

Complications of pneumonia are

- Local destruction of lung tissue
- Frank cavitation
- Respiratory failure
- Respiratory distress syndrome
- Bronchiectasis
- Ventilator dependence

- Pulmonary abscess
- Emphyema
- Superinfection and
- Death

LABORATORY DIAGNOSIS

Specimen

Sputum or throat swab is taken to isolate *S. pneumoniae*.

Macroscopy

The sputum appears rust coloured.

Microscopy

The specimen is gram-stained and observed under the high-power objective lens. If lancet-shaped diplococci are observed, it indicates *Streptococcus pneumoniae* infection.

Culture

Specimen is streaked on blood agar plate and incubated at 37°C for 24 hours under aeration. Alpha haemolytic colonies are observed. Isolated bacteria are identified by making use of morphological and biochemical tests (Box 15.1).

Spherical, diplococci, capsulated, gram-positive, α-haemolytic on blood agar, bile-solublizer, non-motile, inulin-fermentor, vancomycin-sensitive, SXT-sensitive, optochin-sensitive, bacitracin-variable

TREATMENT

Cell wall inhibitors like penicillin and ampicillin are the drugs of choice. Erythromycin and chloramphenicol are also used in some instances.

1. Define.
   i. Virulent factors of *S. pneumoniae*
   ii. Role of lysozyme in pathogenesis of *S. pneumoniae*
2. Explain pathogenesis of pneumococci.
3. Describe the morphological and biochemical features of *S. pneumoniae*.

bmj.bmjjournals.com/cgi/content/full/328/7433/203
en.wikipedia.org/wiki/Pneumococcus
focus.hms.harvard.edu/2005/Apr22_2005/immunology.shtml
www.amazon.com/exec/obidos/ tg/detail/-/155581297X?v=glance
www.answers.com/topic/pneumococcus
www.answers.com/topic/streptococcus-pneumoniae
www.childrensvaccine.org/html/v_pneumo_qf.htm
www.chop.edu/consumer/jsp/division/generic.jsp?id=75726
www.drreddy.com/shots/pneumococcus.html
www.gov.mb.ca/health/publichealth/
www.healthsystem.virginia.edu/ uvahealth/peds_growth/pnmco
www.ibs.fr/content/ibs_eng/ presentation/lab/lim/cellwall.htm
www.immunize.org/pneumopoly/
www.keepkidshealthy.com/prevnar.html
www.keepkidshealthy.com/WELCOME/ immunizations/pneumococcus.html
www.medterms.com/script/main/art.asp?articlekey
www.pneumococcus.com/
www.vaccinealliance.org/General_Information/Immunization_informa/Diseases_Vaccines/pneumo.php
www.vaccineinformation.org/pneumchild/photos.asp
www.who.int/vaccines/en/pneumococcus.shtml
www.wrongdiagnosis.com/p/pneumococcus
ymghealthinfo.org/content.asp?pageid=P02269

Diphtheria

INTRODUCTION

Center for Disease Control (CDC) describes diphtheria as "an upper respiratory tract illness characterized by sore throat, low-grade fever and an adherent membrane of the tonsil(s), pharynx and/or nose." Diphtheria is a rapidly developing, acute, febrile infection, which involves both local and systemic pathology. A local lesion develops in the upper respiratory tract and involves necrotic injury to epithelial cells. As a result of this injury, blood plasma leaks into the area and a fibrin network forms, which is interlaced with rapidly growing *Corynebacterium diphtheriae* cells. This membranous network covers the site of local lesion and is referred to as the pseudomembrane.

HISTORY AND BACKGROUND

The bacterium that causes diphtheria was first described by Klebs in 1883, and was cultivated by Loeffler in 1884, who applied Koch's postulates and properly identified *Corynebacterium diphtheriae* as the causative agent of the disease.

In 1884, Loeffler concluded that *C. diphtheriae* produced a soluble toxin, and thereby provided the first description of a bacterial exotoxin.

In 1888, Roux and Yersin demonstrated the presence of the toxin in the cell-free culture fluid of *C. diphtheriae* which, when injected into suitable lab animals, caused the systemic manifestation of diphtheria.

In 1913, Schick designed a skin test as a means of determining susceptibility or immunity to diphtheria in humans. Diphtheria toxin will cause an inflammatory reaction when very small amounts are injected intracutaneously. The Schick test involves injecting a very small dose of toxin under the skin of the forearm and evaluating the injection site after 48 hours. A positive test (inflammatory reaction) indicates susceptibility (non-immunity). A negative test (no reaction) indicates immunity (antibody neutralizes toxin).

CAUSATIVE AGENT

Corynebacteria are gram-positive, aerobic, non-motile, rod-shaped bacteria related to the actinomycetes. They do not form spores or branch as do the actinomycetes, but they have the characteristic of forming irregular-shaped, club-shaped or V-shaped arrangements in normal growth. They undergo snapping movements just after cell division, which brings them into characteristic arrangements resembling Chinese letters.

Three strains of *C. diphtheriae* were recognized by McLeod on the basis of growth and other characteristics. They are gravis, intermedius and mitis.

The gravis strain has a generation time (*in vitro*) of 60 minutes; the intermedius strain has a generation time of about 100 minutes; and the mitis strain has a generation time of about 180 minutes. The faster growing strains typically produce a larger colony on most of the growth media.

CLASSIFICATION

Corynebacterium is the only genus of the family Corynebacteriaceae. [Section 15, volume 2, first edition of *Bergey's Manual* and Section 23, volume 4 (high G + C gram-positives), second edition of *Bergey's Manual* (Box 16.1)].

Kingdom	Bacteria
Phylum	Actinobacteria
Order	Actinomycetales
Suborder	Corynebacterineae
Family	Corynebacteriaceae
Genus	*Corynebacterium*
species	*Corynebacterium diphtheriae*

PATHOGENESIS

The incubation period is usually two to five days. Transmission occurs by droplet spread through contact with a patient or carrier or articles soiled with discharges from infected lesions.

Humans are the primary reservoir for *C. diphtheriae*. Sources of infection include carriers who have recovered from infection. The organisms are inhaled and they establish infection in the upper respiratory tract, usually in the nose or throat.

Diphtheria bacilli do not tend to invade tissues below or away from the surface of the epithelial cells at the site of the local lesion. At this site they produce the toxin that is absorbed and disseminated through lymph channels and blood to the susceptible tissues of the body. Degenerative changes in these tissues, which include heart, muscle, peripheral nerves, adrenals, kidneys, liver and spleen, result in the systemic pathology of the disease.

The pathogenicity of *C. diphtheriae* includes two distinct phenomena:

1. *Invasion* Invasion of the local tissues of the throat, which requires colonization and subsequent bacterial proliferation.
2. *Toxigenesis* The diphtheria toxin causes the death of eukaryotic cells and tissues by inhibition of protein synthesis. Although the toxin is responsible for the lethal symptoms of the disease, the virulence of *C. diphtheriae* cannot be attributed to toxigenicity alone, since a distinct invasive phase apparently precedes toxigenesis.

Toxigenicity

Two factors have great influence on the ability of *C. diphtheriae* to produce the diphtheria toxin:

1. Low extracellular concentrations of iron
2. The presence of a lysogenic prophage in the bacterial chromosome

The role of iron in artificial culture The most important factor controlling yield of the toxin is the concentration of inorganic iron (Fe^{++} or Fe^{+++}) present in the culture medium.

The role of β-phage Only those strains of *C. diphtheriae* that are lysogenized by a specific β-phage produce diphtheria toxin.

Mode of Action of the Diphtheria Toxin

The diphtheria toxin is a two-component bacterial exotoxin synthesized as a single polypeptide chain containing an A (active) domain and a B (binding) domain. Proteolytic nicking of the secreted form of the toxin separates the A chain from the B chain.

$$\text{EF-2} + \text{NAD}^+ \longrightarrow \text{ADP-Ribose-Diphthamide-EF-2} + \text{Nicotinamide} + \text{H}^+$$

The toxin binds to a specific receptor (now known as the HB-EGF receptor) on susceptible cells and enters by receptor-mediated endocytosis. Acidification of the endosome vesicle results in unfolding of the protein and insertion of a segment into the endosomal membrane. Apparently as a result of activity on the endosome membrane, the A subunit is cleaved and released from the B subunit as it inserts and passes through the membrane. Once in the cytoplasm, the A fragment regains its conformation and its enzymatic activity. Fragment A catalyses the transfer of ADP-ribose from NAD to the eukaryotic elongation factor 2 (EF-2), which inhibits the function of the latter in protein synthesis. Ultimately, inactivation of all the host cell EF-2 molecules causes death of the cell (Figure 16.1).

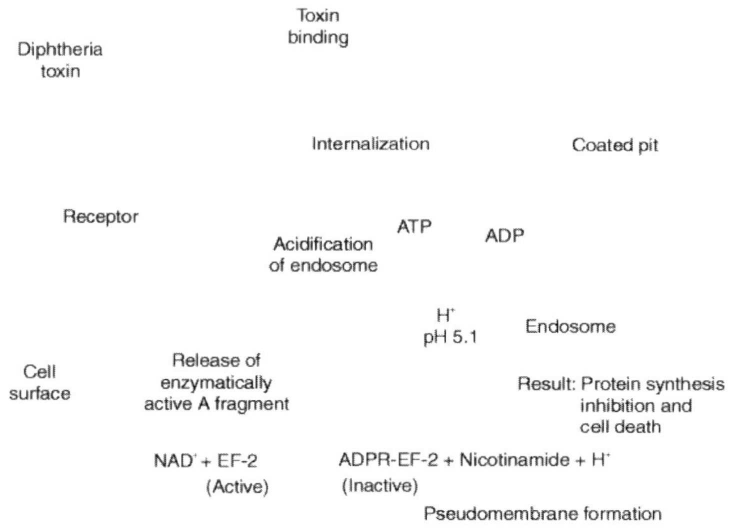

Figure 16.1 Pathophysiology of diphtheria toxin

Susceptibility and Resistance

- Infants born from immune mothers are relatively immune.
- Passive immunity is usually lost in six months.
- Clinical attack does not always lead to lasting immunity.
- Immunity may be acquired through inapparent infection.
- Prolonged active immunity is induced by immunization with diphtheria toxoid.
- Antitoxic immunity protects against systemic disease but not against local infection in the nasopharynx.

CLINICAL MANIFESTATION

The infection manifests itself as acute bacterial disease of tonsils, pharynx, larynx, nose and occasionally other mucous membranes, skin, conjunctiva and genitalia. The throat is sore in facial or pharyngo-tonsillar diphtheria with cervical lymph nodes enlarged and tender. In severe cases, there is marked swelling and oedema of the neck.

Cutaneous diphtheria also occurs, generally without systemic symptoms. Laryngeal disease is serious in infants and young children. Case fatality rate is 5–10 per cent for non-cutaneous diphtheria.

LABORATORY DIAGNOSIS

Collection of Specimens

- Nasopharyngeal cultures should be obtained with a flexible alginate swab that reaches deep into the posterior nares.
- Throat cultures should be taken with a cotton swab that is firmly applied to any area with a membrane or inflammation.

Presumptive Diagnosis

It is based on observing pseudo-whitish membrane on tonsils. Dextrose proteose peptone agar, Hoyle's medium, Tinsdale agar are used for cultivation. *C. diphtheriae* produces grey-coloured colonies on blood agar and tellurite blood agar. Other features of identification of *C. diphtheriae* are given in Box 16.2.

Gram-positive, Chinese letter arrangement, Klebs bacilli, metachromatic granules present, catalase-positive, oxidase-negative, growth observed in Loeffler serum slope, black coloured colony in tellurite blood agar

SPECIFIC TREATMENT

Administration of antitoxin is the most important factor in treatment and must not be delayed till bacteriological confirmation. Antibiotics are commonly used. These are used to eradicate *C. diphtheriae* from disease sites, and include oral erythromycin (preferred to penicillins), IM penicillin and benzathine penicillin (single dose).

PREVENTION

Diphtheria toxoid is recommended for all persons at 2, 4, 6, 18 months and 5 years (given as DPT (Triple antigen)) and at 15 years and every 10 years thereafter (given as ADT).

Booster doses of DPT vaccine should be given to high risk population and health workers for at least 10 years.

1. What are the symptoms of diphtheria?
2. List the media used for the cultivation of *C. diphtheriae*.
3. Describe the mode of action and pathogenic process of diphtheria toxin.
4. Describe the different strains of *C. diphtheriae*.
5. Write notes on Schick test.

1. Can *Corynebacterium* survive within a host cell?
2. How can individuals develop lifelong immunity against diphtheria?
3. How are metachromatic granules detected?
4. Describe the role of β-phage in *C. diphtheriae* in pathogenesis.

Brooks, R. and Joynson, D.H.M. "Bacteriological diagnosis of diphtheria." *J. Clin. Pathol.* 43, 576–80. 1990.

Elmer, W. Koneman, Stephen, D. Allen, William, M. Janda, Paul, C. Schreckenberger and Washington, C. Winn, Jr. *Introduction to Diagnostic Microbiology*. J.B. Lippincott Company, New York. 1994.

Elmer, W. Koneman, Stephen, D. Allen, William, M. Janda, Paul, C. Schreckenberger and Washington, C. Winn, Jr. *Color Atlas and Textbook of Diagnostic Microbiology*, 5th edn. J.B. Lippincott, New York. 1997.

Facklam, R.R., Lawrence, D.N. and Sottnek, F.O. "Modified culture techniques for *Corynebacterium diphtheriae* isolation for desiccated scrabs." *J. Clin. Microbiol.* 7, 1378. 1978.

Flint Enguist, S.J., Krug R.M., Racaniello L.W. and Skalka, A.M. *Principles of Virology*. A.S.M. Press, Washington. 2000.

Henry, D. Isenberg. *Essential Procedure for Clinical Microbiology*. A.S.M. Press, Washington. 1998.

John, L. Ingraham and Catherine, A. Ingraham. *Introduction to Microbiology*, 2nd edn. Brooks/Cole. 2000.

Kisley, S.R., Price, S. and Ward, T. "*Corynebacterium ulcerans*—a potential cause of diphtheria." *Commun. Disease Rev.* 4, No.5, R63. 1994.

PathogenDescriptions/Corynebacterium.htm www.sanger.ac.uk/Projects/C_diphtheriae/

wonder.cdc.gov/wonder/prevguid/m0051752/m0051752.asp

www.bact.wisc.edu/Bact330/lecturediphth

www.bio.nite.go.jp/dogan/MicroTop?GENOME_

www.biology.kenyon.edu/Microbial_Biorealm/bacteria/corynebacterium/

www.cehs.siu.edu/fix/medmicro/coryn.htm

www.ebi.ac.uk/2can/genomes/bacteria/Corynebacterium_glutamicum.
www.emedicine.com/emerg/topic138.htm

www.emedicine.com/MED/topic459.htm
www.gsbs.utmb.edu/microbook/ch032.
www.homepages.pavilion.co.uk/tetrix/corynebacterium.html
www.life.umd.edu/classroom/bsci424/
www.medic.med.uth.tmc.edu/path/00001439.htm
www.monografias.com/trabajos5/coryne/coryne.shtml
www.phac-aspc.gc.ca/msds-ftss/msds42e
www.textbookofbacteriology.net/diphtheria.html
www.thefreedictionary.com/corynebacterium
www.thelabrat.com/protocols/corynebacteriumagar.shtm
www.wrongdiagnosis.com/medical/corynebacterium_minutissimum.htm
www.suite101.com/page.cfm/380/dip.htm

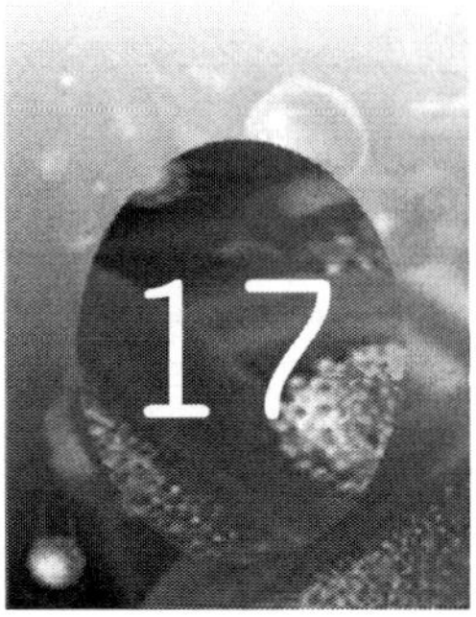

Meningitis

INTRODUCTION

Meningitis is the infection of meninges of brain or spinal cord, during which accumulation of fluid is observed. It is caused by a variety of bacteria and viruses which includes *Haemophilus influenzae, Neisseria meningitidis, Escherichia coli, Staphylococcus aureus, Streptococcus pneumoniae* and *Streptococcus pyogenes*. Although a variety of cocci cause meningitis, the term "meningococcus" is reserved for *N. meningitidis. H. influenzae* is another important organism that causes meningitis in children of age from 5 months to 5 years. Bacterial meningitis is called septic meningitis, others are called aseptic meningitis.

Haemophilus influenzae

H. influenzae is a small, non-motile, gram-negative bacterium of the family Pasteurellaceae. Encapsulated strains of *H. influenzae* isolated from cerebrospinal fluid are coccobacilli, with their size ranging from 0.2 to 0.8 μm.

H. influenzae is highly adapted to its human host. It is present in the nasopharynx of approximately 75 per cent of healthy children and adults. *Haemophilus* meaning "loves haem", specifically requires a precursor of haem in order to grow. Nutritionally, *H. influenzae* prefers a complex medium and requires preformed growth factors that are present in blood, specifically X factor (i.e., haemin) and V factor (NAD or NADP). In the laboratory, it is usually grown on chocolate blood agar. The bacterium grows best at 35–37°C and has an optimal pH of 7.6. *H. influenzae* is generally grown in the laboratory under aerobic conditions or under slight CO_2 tension (5% CO_2), although it is capable of glycolytic and respiratory growth using nitrate as the final electron acceptor. Satellism is a major feature of *H. influenzae.* It means that when *H. inflenzae*-containing blood agar is cross streaked with *S. aureus*, it produces large adjacent colonies and small distant colonies. This is due to haemolytic activity of *S. aureus*. Haemolytic activity releases X and V factors from RBCs (Figure 17.1).

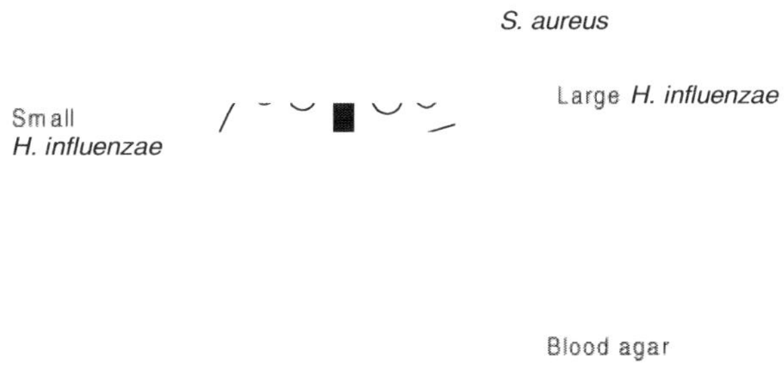

Figure 17.1 Satellism

Neisseria meningitidis

It is also called meningococcus. It is a gram-negative coccus, usually seen in pairs. *N. meningitidis* strains are grouped on the basis of their capsular polysaccharide into 12 serogroups. They are A, B, C, H, I, K, L, X, Y, Z, 29E and W135. A, B, C, Y and W135 are involved in human infections.

N. meningitidis is usually cultivated in a peptone blood-based medium in a moist chamber containing 5–10% CO_2. New York city agar medium and modified Thayer-Martin agar act as selective media (Box 17.1).

Figure 17.2. *N. meningitidis*

Kingdom	Bacteria
Phylum	Proteobacteria
Class	Beta Proteobacteria
Order	Neisseriales
Family	Neisseriaceae
Genus	*Neisseria* (Trevisan, 1985)

PATHOGENESIS OF *H. influenzae*

H. influenzae infection usually begins in the upper respiratory tract as nasopharyngitis and may be followed by sinusitis and otitis, possibly leading to pneumonia. In severe cases, bacteraemia may occur which frequently results in joint infections or meningitis.

The pathogenesis of *H. influenzae* infection is not completely understood, although the presence of the type b polysaccharide capsule is known to be the major factor in virulence. Encapsulated organisms can penetrate the epithelium of the nasopharynx and invade the blood capillaries directly. From blood, organisms enter into the central nervous system in an unknown way. Non-typable (non-encapsulated) strains are less invasive, but they are apparently able to induce an inflammatory response that causes disease.

In infants and young children (under 5 years of age), *H. influenzae* type b causes bacteraemia and acute bacterial meningitis. Occasionally, it causes epiglottitis (obstructive laryngitis), cellulitis, osteomyelitis and joint infections. Non-typable *H. influenzae* causes ear infections (otitis media) and sinusitis in children, and is associated with respiratory tract infections (pneumonia) in infants, children and adults.

PATHOGENESIS OF *N. meningitidis*

The human nasopharynx is the only known reservoir of *N. meningitidis*. Meningococci are spread via respiratory droplets, and transmission requires aspiration of infective particles. Meningococci attach to the non-ciliated columnar epithelial cells of the nasopharynx. Fimbriae and other outer membrane proteins mediate attachment. Invasion of mucosal cells occurs by a mechanism called parasite-directed endocytosis. During endocytosis, *Neisseria* along with vacuole is transported to the base of the cell, where the bacteria are released by exocytosis into the sub epithelial cells. The bacteria resist phagocytosis. The bacterium then enters the central nervous system via blood and lymphatics.

CLINICAL MANIFESTATION

In infants	Irritability, refusal to take food and vomiting
Neonates	Apnea, coma
Adults	Fever, headache, vomiting, photophobia, chills, myalgia and weakness

VIRULENCE FACTORS

Neisseria meningitidis

- Lipooligosaccharide
- Antiphagocytic capsule
- Fimbriae

Gram-negative, fastidious bacteria, coccobacilli, non-motile, catalase-positive, oxidase-positive, requires X and V factor, grey-coloured colony on chocolate agar, satellism

H. *influenzae*

- Endotoxin
- Lipooligosaccharide
- Neuraminidase

- IgA protease
- Fimbriae
- Polyribosyl ribitol phosphate (PRP) capsule
- Outer membrane proteins designated as P1 and P2

H. influenzae is susceptible to lysis by antibody and complement. During meningitis, phagocytosis is probably the main host defence mechanism since complement rarely occurs in the cerebrospinal fluid.

Gram-negative, diplococci (also called as meningococci), coffee bean-shaped, catalase-positive, oxidase-negative, non-motile

LABORATORY DIAGNOSIS

One ml of cerebrospinal fluid (CSF) is collected by using lumbar puncture method.

Processing

Sample is subjected to microscopy after concentration (centrifugation). Gram-stained smear is observed to find out the type of infection. Sample is inoculated on chocolate agar, New York city agar medium, modified Thayer-Martin agar. All the plates are incubated microaerobically in the presence of 5–10% CO_2. Identifying features of *H. inflenzae* and *Neisseria meningitidis* are given in Boxes 17.1 and 17.2.

TREATMENT AND PREVENTION

The recommended treatment for *H. influenzae* meningitis is ampicillin and a third-generation cephalosporin or chloramphenicol. Third-generation cephalosporins, such as ceftriaxone or cefotaxime, are effective against *H. influenzae*. Tetracyclines and sulpha drugs remain effective in treating sinusitis or respiratory infection. Amoxicillin plus clavulanic acid (augmentin) is effective against β-lactamase-producing strains.

Current recommendations for vaccination of infants require parenteral administration of three different vaccines, Diphtheria-Pertussis-Tetanus (DPT), Hib conjugate and Hepatitis B. Tetramune is the first licensed combination vaccine that provides protection against diphtheria, tetanus, pertussis and Hib disease.

1. Name two different types of meningitis.
2. List four bacteria that cause meningitis in humans.

3. List the differences between *N. meningitidis* and *Haemophilus influenzae*.
4. Describe about *H. influenzae* meningitis.
5. Classify *N. meningitidis*.

Brooks, G.F. and Donegan, E.A. *Meningococcal infection*. Edward Arnold, London. 1985.

Conde-Glez, C.J., Morse, S., Rise, P. *et al*. (eds.). *Pathobiology and Immunobiology of Neisseriaceae*. Instituto Nacional De Salud Publica, Cuernavaca. 1994.

Holmes, K. K. Mardh, P. A. and Sparling, P. F. (eds.). *Medical Microbiology*, 2nd edn. McGraw-Hill, New York. 1990.

Salyers, A. A. and Whitt, D. D. *Bacterial Pathogenesis: A Molecular Approach*. American Society for Microbiology. Washington DC. 1994.

www.cdc.gov/ncidod/dbmd/diseaseinfo/meningococcal_g

www.en.wikipedia.org/wiki/meningitis

www.kidshealth.org/parent/infections/lungs/meningitis

www.mayoclinic.com/health/meningitis

www.meningitidis.org.uk

www.meningitis.org

www.meningitis-trust.org

www.nlm.nih.gov/medlineplus/meningitidis

www.who.int/topics/meningitis/en

Whooping Cough

INTRODUCTION

Pertussis or whooping cough is a highly contagious disease involving the respiratory tract. It is caused by a bacterium that is found in the mouth, nose and throat of an infected individual. Pertussis means "intensive cough". Pertussis is primarily spread by direct contact with discharges from the nose and throat of an infected individual.

CAUSATIVE AGENT

- This disease is caused by *Bordetella pertussis.*
- *Bordetella pertussis* was first isolated in 1906 by Bordet and Gengou from the sputum of children.
- Formerly it was called *Haemophilus pertussis.*
- It is a small, ovoid, coccobacillus, non-motile and non-sporing.
- It is capsulated, but tends to lose the capsule on repeated cultivation.
- Its bipolar metachromatic granules are demonstrated by toluidine blue staining.
- The bacteria are nutritionally fastidious and are usually cultivated on rich media supplemented with blood. On blood agar the organism grows slowly and requires 3–6 days to form pinpoint colonies.
- The organisms are strict aerobes and grow best at 35–36°C.
- It does not ferment sugars, form indole, reduce nitrates, utilize citrate, hydrolyse urea, and is catalase- and oxidase-positive.
- It is a delicate organism, killed readily by heating, drying and disinfectants.
- Freshly isolated strains have fimbriae.
- The organism is sensitive to unsaturated fatty acids, sulphides and peroxides.

CLASSIFICATION

B. pertussis belongs to the family Alcaligenaceae available in group 4 gram-negative aerobic rods and cocci (volume 1, first edition, *Bergey's Manual*) or in volume 2 proteobacteria, section XVI (β-proteobacteria) of second edition of *Bergey's Manual*) (Box 18.1).

Kingdom	Bacteria
Phylum	Proteobacteria
Class	Beta Proteobacteria
Order	Burkholderiales
Family	Alcaligenaceae
Genus	Bordetella (Moreno-Lopez, 1952)
species	*B. parapertussis, B. pertussis*

Differential characteristics of *B. pertussis* and *B. parapertussis* are given in the following table.

Feature	*B. pertussis*	*B. parapertussis*
Motility	–	+
Growth on blood-free peptone agar	–	+
Pigment production	–	+
Nitrate reduction	+	–
Urea hydrolysis	+	+
Oxidase reaction	+	–

– Negative, + Positive

VIRULENCE FACTORS

Adherence factors and toxins are involved in pathogenicity. These virulent factors are controlled by at least two sets of genes. The genes may be inactivated at 25°C and activated at 37°C.

Adherence mechanisms of *B. pertussis* involve a "filamentous haemagglutinin" (FHA), which is a fimbria-like structure on the bacterial surface, and a cell-bound pertussis toxin (PTx). Short-range effects of soluble toxins play a role as well in invasion during the colonization stage (Figure 18.1).

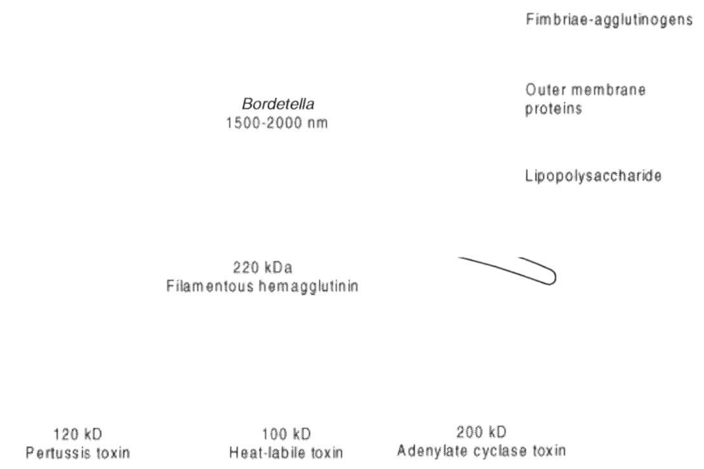

Figure 18.1 Virulent factors of *Bordetella pertussis*

Toxins Produced by B. *pertussis*

B. pertussis produces a variety of substances with toxic activity.

It secretes its own invasive adenylate cyclase, which enters mammalian cells. This toxin acts locally to reduce phagocytic activity and probably helps the organism to initiate infection. This toxin is a 45-kDa protein that may be cell-associated or released into the environment.

It produces a highly lethal toxin (formerly called dermonecrotic toxin), which causes inflammation and local necrosis of adjacent sites where *B. pertussis* is located. The lethal toxin is a 102-kDa protein composed of four subunits, two of which have a molecular weight (MW) of 24 kDa and the other two have a MW of 30 kDa (Figure 18.2). It causes necrotic skin lesions when low doses are injected subcutaneously in mice. It is lethal at higher doses. The role of the toxin in whooping cough is not known.

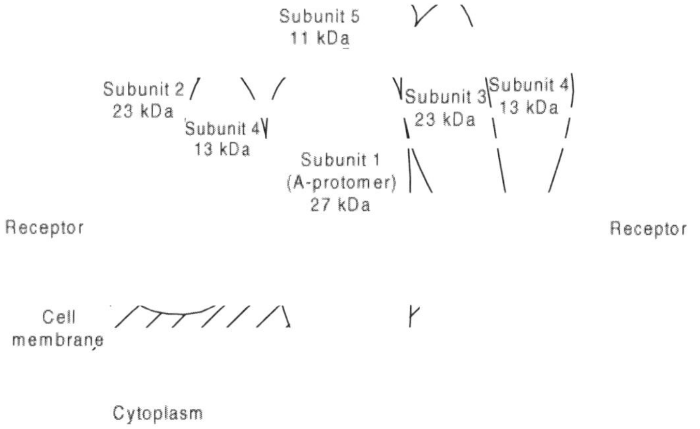

Figure 18.2 Attachment of pertussis toxin on cell membrane

It produces a substance called the tracheal cytotoxin, which is toxic for ciliated respiratory epithelium and which will stop the ciliary action of cells. It also stimulates the release of cytokine IL-1, and so causes fever.

It produces the pertussis toxin, PTx, a protein that mediates both colonization and toxaemic stages of the disease. PTx is a two-component, A + B bacterial exotoxin. The A subunit (S1) is an ADP ribosyl transferase. The B component, composed of five polypeptide subunits (S2 through S5), binds to specific carbohydrates on cell surfaces. The A subunit gains enzymatic activity and transfers the ADP ribosyl moiety of NAD to the membrane-bound regulatory protein Gi that normally inhibits the eukaryotic adenylate cyclase. The Gi protein is inactivated and cannot perform its normal function to inhibit adenylate cyclase. The conversion of ATP to cyclic AMP cannot be stopped and intracellular levels of cAMP increase. This will disrupt the cellular function, and in the case of phagocytes, decrease their phagocytic activities such as chemotaxis, engulfment, oxidative burst and bactericidal killing. Systemic effects of the toxin include lymphocytosis and alteration of hormonal activities that are regulated by cAMP, such as increased insulin production (resulting in hypoglycaemia) and increased sensitivity to histamine (resulting in increased capillary permeability, hypotension and shock). This alters both AMI and CMI responses and may explain the high frequency of secondary infections that accompany pertussis (Figure 18.3).

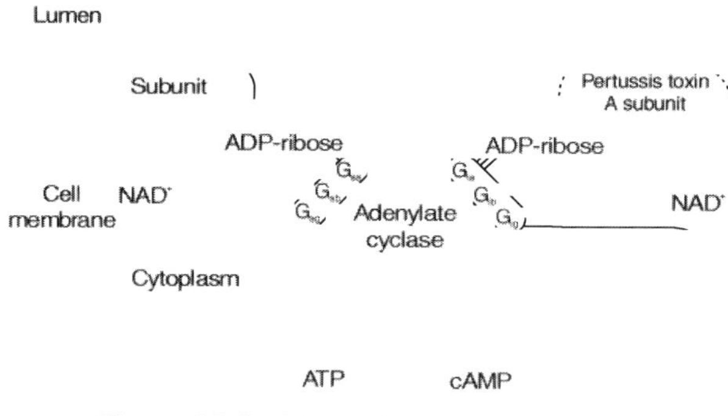

Figure 18.3 Mode of action of pertussis toxin

Lipopolysaccharide

As *B. pertussis* is gram-negative, it possesses lipopolysaccharide (endotoxin) in its outer membrane, but its LPS is unusual. It is heterogeneous, with two major forms differing in the phosphate content of the lipid moiety. The alternative form of lipid A is designated as lipid X. The unfractionated material elicits the usual effects of LPS (i.e., induction of IL-1, activation of complement, fever, hypotension, etc.).

Agglutinogens

The agglutinogens are surface antigens responsible for agglutination of the bacterial cells in the presence of their corresponding antibodies. To date, 14 different agglutinogens (AGG 1 through AGG 14) have been distinguished. AGG1 is specific for *B. pertussis*.

Outer Membrane Proteins

At least four different outer membrane protein structures are distinguished on *B. pertussis*; they are designated OMP 15, OMP 18, OMP 69 and OMP 91. They are believed to be protective antigens.

PATHOGENESIS

The agent of whooping cough is transmitted primarily via droplets. Infection results in colonization and rapid multiplication of the bacteria on the mucous membranes of the respiratory tract.

Colonization

B. pertussis colonizes the cilia of the mammalian respiratory epithelium (Figure 18.4).

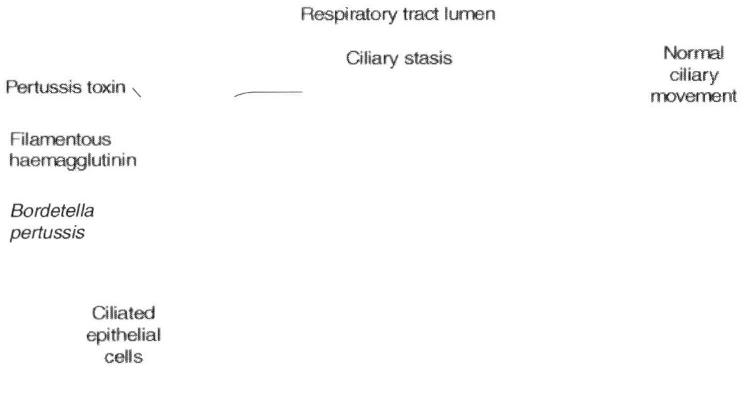

Figure 18.4 Inhibition of ciliary activity by *Bordetella*

Filamentous haemagglutinin is a large (220 kDa) protein that forms filamentous structures on the cell surface. FHA binds to galactose residues on the surface of ciliated cells. One of the toxins of *B. pertussis*, the pertussis toxin (PTx), is also involved in adherence to the tracheal epithelium.

Antibodies against PTx components prevent colonization of ciliated cells by the bacteria and provide effective protection against infection. Thus, pertussis toxin is clearly an important virulence factor in the initial colonization stage of infection.

The second or toxaemic stage of pertussis follows relatively non-specific symptoms of the colonization stage. It begins gradually with prolonged and paroxysmal coughing that often ends in a characteristic inspiratory gasp (whoop).

During the second stage, *B. pertussis* can rarely be recovered, and antimicrobial agents have no effect on the progress of the disease. This stage is mediated by a variety of soluble toxins.

On the basis of symptoms produced by the organisms, the disease stage is divided into three

1. Catarrhal stage
2. Paroxysmal stage
3. Convalescent stage

Catarrhal Stage

It is characterized by low-grade fever, rhinorrhoea, progressively worsening cough and also mucous membrane inflammation.

Paroxysmal Stage

The patient tries to cough up the mucus by making 5–15 rapidly continuous cough followed by the characteristic whoop—a hurried deep inspiration. The total blood leucocyte levels may resemble those of leukemia (> 100,000/mm^3), with 60 to 80 per cent being lymphocytes.

Convalescent Stage

It is the recovery stage.

Stages	Incubation period	Catarrhal	Paroxysmal	Convalescent
Duration	7–10 days	1–2 weeks	2–4 weeks	3–4 weeks
Symptoms	None	Low-grade fever, rhinorrhoea, progressively worsening cough and also mucous membrane inflammation	Rapidly continuous cough followed by the characteristic whoop	Diminished paroxysmal cough, pneumonia
Culturing	None	Possible	Possible	Rare

CLINICAL MANIFESTATION

- Pertussis begins as a mild upper respiratory infection.
- Symptoms resemble those of common cold, such as running nose, slight fever and a mild cough.

- Within two weeks, the cough becomes more severe and is characterized by rapid coughing followed by a high pitched whoop.
- Thick, clear mucus may be discharged.

The incubation period is usually 5 to 10 days, but may be as long as 21 days.

A person can transmit pertussis from seven days following exposure, to three weeks after the onset of coughing episodes.

COMPLICATIONS

Complications of pertussis may include pneumonia, middle ear infection, loss of appetite, dehydration, seizures (disorder of the brain) and death.

LABORATORY DIAGNOSIS

In the early stages of infection, the organism is present in enormous numbers in the patient's respiratory secretions and is rarely detected in smears.

Sample Collection

Samples are collected by the following methods:

Cough plate method Here, the culture plate is held about 10–15 cm in front of the patient's mouth during induced coughing, so that droplets of respiratory exudate impinge directly on the medium.

Post-nasal swab Secretions of the posterior pharyngeal wall are collected with an alginate swab on a bent wire passed through mouth.

Pernasal swab Here a swab on flexible wire is passed along the floor of the nasal cavity and material is collected from the pharyngeal wall. This method yields highest percentage of isolation.

All specimens should be immediately plated on to Bordet-Gengou medium and incubated at 35–36°C for 48–72 hours. Colonies are small, dome-shaped, smooth, opaque, grayish-white, refractile and glistening, resembling mercury drops. A hazy zone of haemolysis surrounds colonies. For routine use, charcoal–blood agar (Regan–Lowe medium) is most widely used. A (2,6-*o*-dimethyl)-β-cyclodextrin-supplemented Stainer–Scholte broth can be used as an enrichment medium. *Bordetella* species does not need factors X and V (NAD^+ and haemin). Identifying features of *B. pertussis* are given in Box 18.2.

Gram-negative, non-motile, catalase-positive, oxidase-positive, glistening mercury-like colonies on Bordet-Gengou medium, indole-positive, VP-negative, H_2S-negative, MR-positive, C-positive, urease-positive, nitrate-positive, capsule present, ash-coloured colony on Regan–Lowe medium

Modern serological techniques, such as enzyme-linked immunosorbent assay (ELISA) have been used to detect IgG, IgM, IgA and IgE antibodies.

TREATMENT

Treatment with erythromycin will eliminate viable *B. pertussis* organisms from the respiratory tract within a few days.

Whooping Cough Vaccine

Several new acellular vaccines have been developed from purified components of *B. pertussis*.

For decades, the pertussis vaccine has been given in combination with vaccines against diphtheria and tetanus. The combination is known as the DPT vaccine. Recently, infants have been receiving a vaccine that combines the DPT vaccine with the vaccine against *Haemophilus influenzae* type b meningitis (Hib). This vaccine is called DTPH. The Diphtheria-Pertussis-Tetanus vaccine using acellular pertussis is known as DTaP. The Diphtheria-Pertussis-Tetanus vaccination is given in five doses: at 2, 4, 6, 12–18 months and 4–6 years of age.

1. Describe the different stages of pertussis.
2. How do you differentiate between the two species of *Bordetella*?
3. How can you identify β-pertussis bacteriologically and immunologically?
4. Write short notes on
   i. DPT vaccine
   ii. Pertussis toxin
   iii. FHA
   iv. Media used for the cultivation of *Bordetella*

1. If natural infection of pertussis develops lifelong immunity, how does pertussis toxin induce fever?

Alough, J. E. and Freer, J. H. *Sourcebook of Bacterial Protein Toxins*. Academic Press, London. 1991.

Munoz, J.J. and Bergman, R.K. *Bordetella Pertussis*. Vol 4. Marcel Dekker, New York. 1997.

Robinson, A., Duggleby, C.J., Gorringe, A. R. *et al*. Antigenic variation in *Bordetella pertussis.* p 147. In Birbeck T. H. and Penn C. W. (eds.). *Antigenic Variations and Infectious Diseases*. Society for General Microbiology. IRL Press. Oxford. 1986.

Wardlaw, A.C. and Parton, R. *Pathogenesis and Immunity in Pertussis*. John Wiley & Sons, New York. 1988.

www.en.wikipedia.org/wiki/pertussis

www.home-remedies-for-you.com/remedy/whooping-cough.html

www.kidshealth.org/parent/infections/lung/whooping_cough.html

www.mayoclinic.com/health/whooping-cough/DS00445

www.medinfo.co.uk/conditions/whoopingcough.html

www.metrokc.gov/health/prevcont/pertussis.html

www.nlm.nih.gov/medlineplus/whoopingcough

www.whoopingcough.net

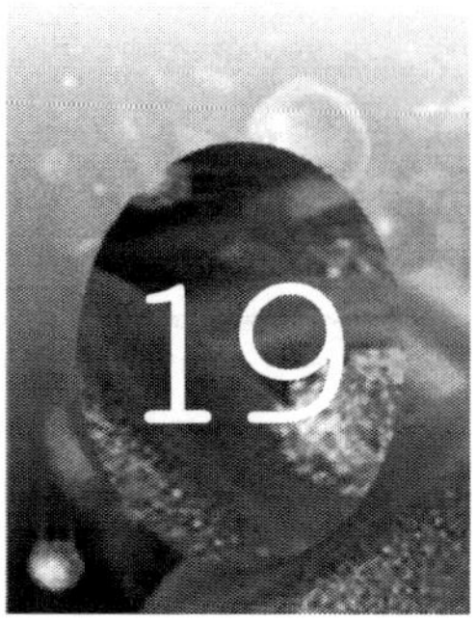

Tuberculosis

INTRODUCTION

Tuberculosis is one of the lower respiratory tract infections. It is a progressive granulomatous disease of the lungs. It is an ancient disease with evidence seen in skeletons from the Stone Age and in bones from some of the early Egyptian mummies. Although the infectious nature of tuberculosis was established by Villemin (1865), the causative agent was first described by Robert Koch in the year 1882 and who named it as "mammalian tubercle bacilli". It is caused by very closely related species of *Mycobacterium* namely *Mycobacterium tuberculosis* (in humans) and *Mycobacterium bovis* (in animals).

TUBERCULOSIS TODAY

- Someone is infected with tuberculosis every second.
- One-third of the world's population is infected with tuberculosis complex bacteria.
- If left untreated, one person with tuberculosis will infect 10–15 people per year.

CAUSATIVE AGENT

- The causative agent is *Mycobacterium tuberculosis*. (Figure 19.1)
- It is an acid-fast bacillus. It is due to mycolic acid content of the cell wall.
- It is a fairly large, non-motile, rod-shaped bacterium.

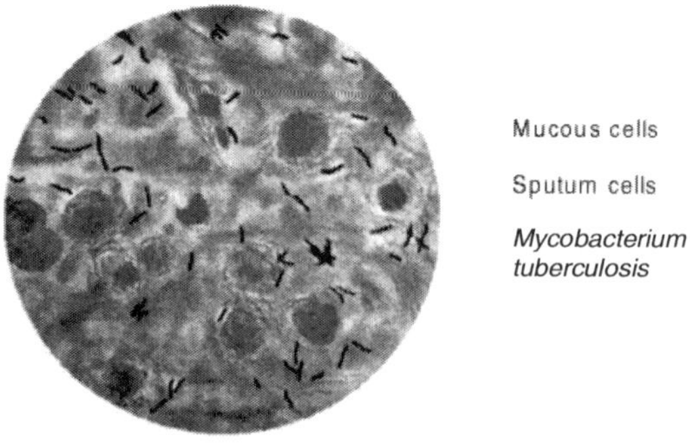

Figure 19.1 *Mycobacterium tuberculosis*

- It is non-sporing and non-capsulated.
- In tissues, it occurs as thin, straight rods, occurring singly, in pairs or in clumps.
- It is a weakly gram-positive, obligate aerobe and is a slow grower with a generation time of 6–12 hours.
- Optimum temperature for growth is 37°C, pH is 6.4–7.
- It can grow only in media containing egg, asparagine, potato and serum.

CLASSIFICATION

Mycobacterium is a member of the family *Mycobacteriaceae*. Section 1, first edition of *Bergey's manual* and section 18, second edition of *Bergey's Manual* describe the characters of *Mycobacterium* (Box 19.1).

Kingdom	Bacteria
Phylum	Actinobacteria
Order	Actinomycetales
Suborder	Corynebacterineae
Family	Mycobacteriaceae
species	*Mycobacterium tuberculosis, M. leprae* (Lehmann and Neumann, 1896)

VIRULENCE FACTORS

Mycobacterium is composed of large amount of lipids. It includes mycolic acids (long-chain fatty acid C78–C90) waxes and phosphatides. Muramyl dipeptide complexed with mycolic acid can cause granuloma formation. Phosphatide induces tubercle formation. Lipid can cause accumulation of macrophages and neutrophils.

Virulent strains of tubercle bacilli form microscopic serpentine cords in which bacilli are arranged in parallel chains. Cord formation is correlated with virulence. A cord factor (Trehalose 6,6 dimycolate) inhibits migration of leucocytes, causing chronic granuloma. Polysaccharides induce hypersensitivity reaction. Protein found along with wax (mycolic acid) induces antibody formation.

The advantages of high concentration of lipids in *M. tuberculosis* are:

- Impermeability to stains and dyes.
- Resistance to many antibiotics.
- Resistance to killing by acidic and alkaline compounds.
- Resistance to osmotic lysis via complement deposition.
- Resistance to lethal oxidation and survival inside the macrophages.

PATHOGENESIS

M. tuberculosis infections occur by airborne transmission of droplet nuclei containing a few viable cells of virulent organisms. Coughing generates about 3000 droplet nuclei, talking for 5 minutes generates 3000 droplet nuclei and singing generates 3000 droplet nuclei per minute.These droplet nuclei are the major agents responsible for tuberculosis.

After the inhalation of droplet nuclei, the bacilli are deposited in the alveolar spaces of the lungs, where they are non-specifically engulfed by alveolar macrophages. However the macrophages are not activated and so are unable to destroy the intracellular organisms.

A portion of the infectious inoculum resists intracellular destruction and persists, eventually multiplying and killing the macrophages. Other macrophages also begin to extravasate from the peripheral blood. These macrophages also phagocytoze *M. tuberculosis* but they are not activated and hence cannot destroy the pathogen.

At this stage, lymphocytes begin to infiltrate. The lymphocytes, specifically T cells, recognize the antigen processed and presented through MHC molecule. This results in T-cell activation and release of cytokines and other factors (Figure 19.2).

$CD4^+$ cells produce gamma interferons that activate macrophages and enhance mycobactericidal capabilities of CD4 cells. These cells can limit the replication of intracellular *M. tuberculosis* and may kill the bacilli.

$CD8^+$ cells attack infected macrophages expressing mycobacterial antigens and lyse the cells by releasing them from the protective niche and exposing them to activated macrophages.

Activation of macrophages can result in bacterial killing while cytotoxicity may release bacteria from phagocytes and allow their engulfment and destruction by activated macrophages.

T_H1 cells produce IL-2 and gamma interferon, which promote the inflammatory reactions and CMI. T_H2 cells produce IL-4, IL-5 and IL-10 and promote antibody production. Cord factor may be directly cytotoxic to macrophages. Most of the tissue destruction associated with tuberculosis results from cell-mediated hypersensitivity.

Accumulation of mycobacteria stimulates an inflammatory focus which matures into a granulomatous lesion characterized by a mononuclear cell infiltrate surrounding a core of degenerating epithelioid and multinucleated giant (Langerhans') cells. This lesion (called a tubercle) may become enveloped by fibroblasts, and its centre often progresses to caseous necrosis. Liquefaction of the caseous material and erosion of the tubercle into an adjacent airway may result in cavitation and the release of massive numbers of bacilli into the sputum. In the resistant host, the tubercle eventually becomes calcified.

Early in infection, mycobacteria may spread distally either indirectly through the lymphatics to the hilar or mediastinal lymph nodes via the thoracic duct into the bloodstream, or directly into the circulation by erosion of the developing tubercle into a pulmonary vessel. This is called disseminated tuberculosis. Extrapulmonary haematogenous dissemination results in the seeding of other organs (e.g. spleen, liver and kidneys) and eventually, reinoculation of the lungs. This is often considered as reactivated tuberculosis.

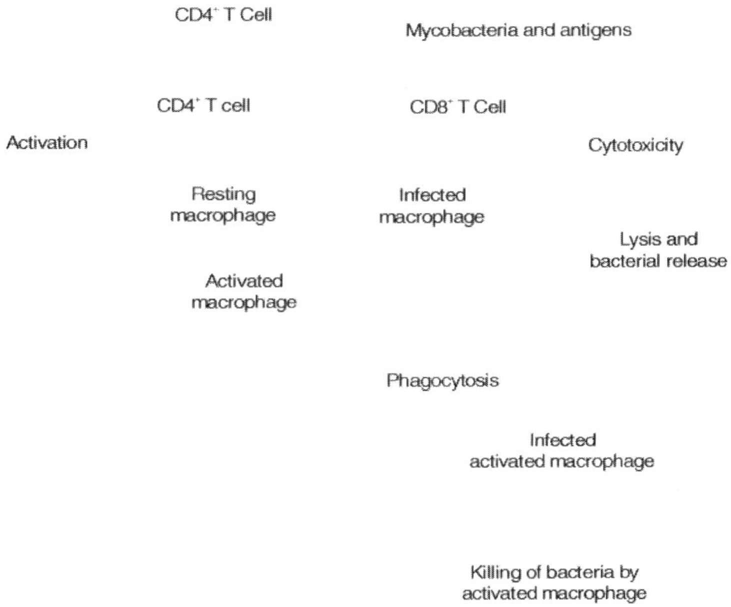

Figure 19.2 Pathophysiology of tubercle bacilli

Miliary lesions, which are small granulomas, resembling millet seeds that spread throughout the lung fields is known as miliary tuberculosis . Secondary lesions caused by miliary tuberculosis can occur almost in any anatomical location but usually involve the genitourinary system, joints, lymph nodes. The lesions are of two types.

1. *Exudative lesion* It results from the accumulation of PMNs around *M. tuberculosis*. It results in soft tubercle.
2. *Productive lesion (granulomatous)* It occurs when hosts become hypersensitive to tuberculoprotein. This situation gives rise to the formation of hard tubercle.

Usually the host will begin to control the infection at some point. When primary lesion heals, it becomes fibrous and calcifies. When this happens, the lesions are referred to as Ghon complex. If the complex contains viable cells, these foci are referred to as Simon foci.

CLINICAL MANIFESTATION

Clinical signs and symptoms develop in only a small proportion (5–10 per cent) of infected healthy people. These patients usually contract pulmonary disease; prominent symptoms are:

- Chronic productive cough
- Low-grade fever
- Night sweats
- Fatigue
- Weight loss

Tuberculosis may manifest itself with extrapulmonary complications including lymphadenitis, kidney, bone or joint involvement, meningitis, or disseminated (miliary) disease. Lymphadenitis and meningitis are more common among normal infants with tuberculosis, and all extrapulmonary manifestations increase in frequency among immunocompromised individuals such as patients on chronic renal dialysis and elderly, malnourished or HIV-infected individuals.

EPIDEMIOLOGY

Tuberculosis is particularly common in groups such as the elderly, the chronically malnourished, alcoholics and the poor. The prevalence of clinical tuberculosis among the homeless in the United States may be up to 300 times higher than the national average rate. In recent years, the incidence of disease in racial minorities in the United States has been more than five times than that observed in the whites.

LABORATORY DIAGNOSIS

Infection in an asymptomatic individual can be diagnosed with the help of the intradermal purified protein derivative (PPD) skin test. Intradermal introduction of PPD into a previously infected, hypersensitive person results in the delayed (48–72 hours) appearance of an indurated (raised, hard) reaction with or without erythema. The Mantoux test requires the intradermal injection of a measured volume (0.1 ml) containing a specified quantity (5 tuberculin units) of PPD. The transverse diameter of induration is measured 48 to 72 hours later. Interpretation varies.

Sample Collection

In a person with symptoms suggestive of tuberculosis, clinical specimens such as sputum, bronchial washings, pleural fluid, urine or cerebrospinal fluid should be collected. Sputum should be collected in the early morning. Sputum is processed for microscopy and culturing as per methodology given in chapter 5.

Diagnostic Methods

Commercial chemiluminescent DNA probes, gas-liquid chromatography, high-performance liquid chromatogrphy and thin-layer chromatography allow identification of a few species of mycobacteria within hours after sufficient growth is present on solid or in a liquid medium.

In the future, nucleic acid amplification methods may prove useful for the detection of mycobacteria directly in clinical material within 24 hours or less of specimen receipt. Currently, standardized guidelines for susceptibility testing of *Mycobacterium tuberculosis* has been developed. BACTEC TB vial (indirect test) is a test used to detect the antibiotic sensitivity pattern of *Mycobacterium tuberculosis.*

TREATMENT

First-line anti-mycobacterial drugs are isoniazid, rifampin, pyrazinamide and ethambutol. Previously streptomycin was one in the combination.

Combined drugs are used for the treatment of tuberculosis. It is also used to reduce the resistant nature of the pathogen. This combined therapy is administered for 12–24 months.

CONTROL

A vaccine against *M. tuberculosis* is available. It is called BCG (Bacillus of Calmette and Guerin, after the two Frenchmen who developed it). BCG consists of a live attenuated strain derived from *Mycobacterium bovis*. This strain of *Mycobacterium* has remained avirulent for over 60 years. The vaccine is not 100% effective. Studies suggest a 60–80% effective rate in children.

1. Describe the pathogenesis of tuberculosis.
2. Discuss the reason behind symptoms and complications of tuberculosis.
3. Explain the diagnostic procedures of tuberculosis.
4. Write short notes on
    i. Cord factor
    ii. Tubercle
    iii. Caseous necrosis
    iv. Miliary tuberculosis
    v. Reactivated tuberculosis
    vi. PPD
    vii. Mantoux
    viii. LJ medium

1. What is the current status of tuberculosis?
2. How is immunity developed against tuberculosis?
3. What are the uses of mycolic acid content of tubercle bacilli?

Bloom, B.R. (ed.). *Tuberculosis—Pathogenesis, Protection and Control*. ASM Press, Washington, DC. 1994.

Johnson, J.L. Elner, J.J. and Shiratsuchi, H. "Monocyte–*Mycobacterium avium* complex interactions: Studies of potential virulence factors for humans." *Immunol. Ser.* 60:263. 1994.

Krahenbuhl, J.L. and Adams, L.B. "The role of the macrophage in resistance to the leprosy bacillus." *Immunol. Ser.* 60:281. 1994.

Modlin, R.L. "Th1-Th2 paradigm: insights from leprosy." *J. Invest. Dermatol.* 102:828. 1994.

Reichman, L.B. and Hershfield, E.S. (eds.). *Tuberculosis—A Comprehensive International Approach*. Marcel Dekker, New York. 1993.

Rom, W.N. and Garay, S. (eds.). *Tuberculosis*. Little Brown and Co., New York. 1995.

Shinnick, T. (ed.). "Tuberculosis." *Curr. Top. Microbiol. Immunol.* Springer-Verlag, Heidelberg. 1995.

lrsitbrd.nic.in

textbookofbacteriology.net/tuberculosis.html

www. ntiindia.kar.nic.in

www.hpa.org.uk/infections/topics_az/tb/menu.html

www.lungusa.org/site/pp.asp

www.osha.gov/SLTC/tuberculosis/index.html

www.stoptb.org

www.tbcindia.org/

www.topics.nytimes.com/diseasesconditionsandhealthtopics/tuberculosis/

www.umdnj.edu/ntbcweb

Leprosy

INTRODUCTION

Leprosy is a chronic granulomatous disease of man involving primarily the skin, peripheral nerves and nasal mucosa but capable of affecting any tissue or organs. As per WHO it is defined as a hypopigmented or reddish skin lesion with definite loss of sensation. It is one of the nervous system infections. It is a disease of great antiquity, having been recognized from Vedic times in India and from biblical times in the Middle East. It probably originated in the tropics and spread to rest of the world. Leprosy is also called Hansen's disease because Hansen first observed lepra bacilli in 1868. The first successful propagation of the leprosy bacillus in the laboratory did not occur until 1960. In 1970 it was found that *Mycobacterium leprae* causes a systemic infection in the nine-banded armadillo.

CAUSATIVE AGENT

M. leprae causes leprosy. It was the first bacillus isolated from humans. It is one of the least understood bacterium. Hansen, a Norwegian physician first observed it from lepra cells of human beings. It does not grow on artificial media. It reaches the environment from the nose and upper respiratory tract of persons with leprosy.

It is a straight or slightly curved rod showing considerable morphological variations. Organisms are found singly or in large masses termed globi. Large numbers of bacilli may be packed in the cells in an arrangement that suggests packets of cigars. It is one of the acid-fast bacilli. Acid-fastness of the pathogen is removed by using pyridine extraction procedure.

The presence of a phenolase in *M. leprae* obtained from lepromatous skin nodules provides the simplest test for separating *M. leprae* from other *Mycobacterium*. (Phenolase converts 3, 4 dihydroxyphenylalanine (DOPA) to a coloured product).

It grows well on foot pads of nine-banded armadillo at 30°C. This temperature is obtained by controlling air temperatures at 20–25°C. Different levels of antibodies to *M. leprae* are associated with different forms of leprosy, but at present there is little definitive information on the molecular structure of the specific antigens. Twenty different antigens can be detected by immunodiffusion. Some of these antigens are common to all mycobacteria. Only 2 determinants specific to *M. leprae* have been identified and purified. One of these is a phenolic glycolipid and the second is a lipoarabinomannon. It is a dominant antigen.

CLASSIFICATION

Classification is similar to *M.tuberculosis* (Box 19.1).

PATHOGENESIS

The organism enters the human body through respiratory route or through skin. Some authors say transmission of lepra bacilli requires close contact with infected patients. It mainly attacks nerve cells and grows very slowly in mononuclear macrophages, especially the histiocytes of skin and Schwann cells of the nerves. Rate of infection depends on the status of the human immune system. Attack of immune cells against affected nerve cells produces nerve damage leading to deformity. On the basis of immunological findings, histopathology and clinical findings, Ridley and Jopling have established a classification scheme consisting of five types of leprosy (Table 20.1).

Table 20.1 Classification of leprosy

Type of disease	Immune response	Bacilli in skin	Bacilli in nasal mucosa
Tuberculoid (TT)	Good	None	None
Borderline Tuberculoid (BT)	Less good	None or few	None
Borderline (BB)	Partial	Few	None
Borderline Lepromatous (BL)	Poor	Moderate	Few
Lepromatous (LL)	Nil	Many with globi	Many with globi

Among these, two forms are stable. They are tuberculoid and lepromatous forms.

Tuberculoid

In this, skin biopsy specimens show mature granuloma formation in the dermis that consists of epitheloid cells, giant cells and marked infiltration of lymphocytes. Acid-fast bacilli usually cannot be demonstrated. The organisms invade the nerves and selectively colonize the Schwann cells. The larger nerves are swollen and destroyed by granulomas or inflammatory cells. The nerve damage is non-specific and arises as a consequence of cell-mediated immune response.

Lepromatous

In this there is no cellular immune response. The lesions are small and many. They are shiny with no loss of feeling. Skin and nasal smears contain many bacteria. 10^9 bacilli are observed per gram of tissue. Organisms tend to invade vascular channels, which results in continuous bacteraemia and consistent involvement of the reticuloendothelial system. The nerves are also infected but are less compared to tuberculoid type.

Borderline

The lesions clinically resemble tuberculoid leprosy but bacteriologically and immunologically resemble the lepromatous type. Immunological response ranges from less good to poor.

OTHER TERMS USED IN THE CLASSIFICATION

In the early stages of leprosy, a small macule may form, which does not sufficiently develop for it to be classified into one of the recognized clinical forms already described. The histological appearance of the lesion is not specific and consists of diffuse round cell infiltration in the dermis. In many persons, the lesion heals spontaneously and completely and in others it may progress to tuberculoid or lepromatous types.

According to World Health Organization report (1982) leprosy is divided into two groups: Paucibacillary (lesion contains few bacteria) which includes all cases of tuberculoid types and some cases of borderline type (BT, TT), and multibacillary which (they contain large numbers of bacilli) includes all cases of lepromatous types and some cases of borderline (BB, BL and LL).

Common symptoms include progressive nerve damage, chronic skin lesions and ulcerative lesions of mucous membrane, deformed faces, loss of fingers and toes.

EPIDEMIOLOGY

There are approximately 12 million persons with leprosy in the world currently. Among them 4 million infections are seen in India. This disease is prevalent in tropical areas. In endemic areas of Africa, 20–50 persons in every 1000 may be infected. Within the US, leprosy is endemic. Humans appear to be the only natural hosts for *M. leprae*. The route of infection is probably by inhalation and sometimes through skin. The incidence of the disease in contacts is low and only 5% of the contacts suffer from the disease. In some cases like lepromatous type, bacilli may be transmitted through bloodsucking insects.

LABORATORY DIAGNOSIS

Specimen

Skin biopsy

Scrapings from lesions and nasal mucosa

Direct Slit Skin Smear

Slit skin smears made from skin scrapings from patches and ear lobes and punctured material from nodules and nasal mucosal scrapings are stained by modified Ziehl–Neelson method using 4–5% sulphuric acid as a decolorizing agent. Smear shows acid-fast bacilli arranged in parallel bundles within macrophages (lepra cells) and this confirms lepromatous leprosy. The living cells stain uniformly but dead cells look fragmented and irregular.

On the basis of bacteriological index of skin smear, the severity of infections is assessed. Bacteriological index is defined as the number of viable bacilli in a lesion, which is assessed from stained smear by oil immersion lens.

1–10 bacilli in 100 fields	=	1+
1–10 bacilli in 10 fields	=	2+
1–10 bacilli per field	=	3+
10–100 bacilli per field	=	4+
100–1000 bacilli per field	=	5+
More than 1000 bacilli, clumps and groups per field	=	6+

On the basis of morphological index, the effect of antibiotic treatment is assessed. It is the percentage of uniformly stained bacilli out of the total bacteria present in tissue.

Skin and Nerve Biopsy

Skin biopsy is collected from active edge of the patches and nerve biopsy from thickened nerve for histological confirmation of tuberculoid leprosy when AFB cannot be demonstrated by direct smear.

Animal Inoculation

The culture of *M. leprae* on artificial media has not yet been achieved. It is possible, however to produce the growth of the organism in certain animals such as the footpads of mouse, rat and nine-banded armadillo (Figure 20.1).

Large number of bacilli are obtained from armadillo. This type of cultivation was successfully completed during the year 1971. Body temperature of the animal that favours the growth of the organism is 30°C.

Figure 20.1 Nine-banded armadillo

Indirect Tests

Lepromin test Lepromin is a boiled extract of lepromatous tissue in isotonic saline. When 0.1 ml of lepromin (contain 40 million dead cells) is injected intradermally to an individual, it produces an area of nodular infiltration at the site of injection in skin, which reaches its maximum size in 3–5 weeks. This is a non-specific skin test, which can be helpful in classifying leprosy and assessing the future course of the disease. One of its main values is in confirming lepromatous leprosy.

TREATMENT

Dapsone (4,4 diamino diphenyl sulphone) was an effective monotherapy for all types of leprosy till 1982. As per *Indian Journal of Leprosy* the following drugs are recommended for leprosy patients.

Rifampicin, dapsone, clofazimine, ethionamide, quinolones (pefloxacin, ofloxacin, sparfloxacin), brodimoprim, aminoglycosides (streptomycin, kanamycin), prothionamide, mimocycline, clathromycin are the drugs used to treat leprosy.

Future drugs are sparfloxacin, clarithromycin, mimocycline, ofloxacin.

1. Give the classification of leprosy.
2. Is there any way to cultivate lepra bacilli?
3. Explain the different diagnostic parameters of leprosy.
4. Write short notes on
    i. Lepra bacilli
    ii. Hansen's disease
    iii. Paucibacillary
    iv. Multibacillary
    iv. Lepromin test
    vi. Bacteriological index
    vii. Morphological index

1. Why is there a difficulty in cultivating lepra bacilli?
2. Why is there no specific pathology described for leprosy?

Modlin, R.L. "Th1-Th2 paradigm: Insights from leprosy." *J. Invest. Dermatol.* 102:828. 1994.

Reichman, L.B. and Hershfield, E.S. (eds.). *Leprosy—A Comprehensive International Approach*. Marcel Dekker, New York. 1993.

Rom, W.N. Garay, S. (eds.). *Leprosy*. Little Brown and Co., New York. 1995.

Shinnick, T. (ed.). "Tuberculosis." *Curr. Top. Microbiol. Immunol*. Springer-Verlag, Heidelberg. 1995.

dir.yahoo.com/Health/Diseases_and_Conditions/Leprosy/

www.emedicine.com/derm/topic223.htm -

www.hpa.org.uk/infections/topics_az/leprosy/menu.htm

www.lepra.org.uk/ - 12k

www.leprosy.ca/

www.leprosy.org/LEPinfo.html

www.leprosymission.org.uk/

www.leprosymission.org/ -

www.merck.com/mmhe/sec17/ch194/ch194a.html -

www.neuro.wustl.edu/neuromuscular/nother/infect.htm

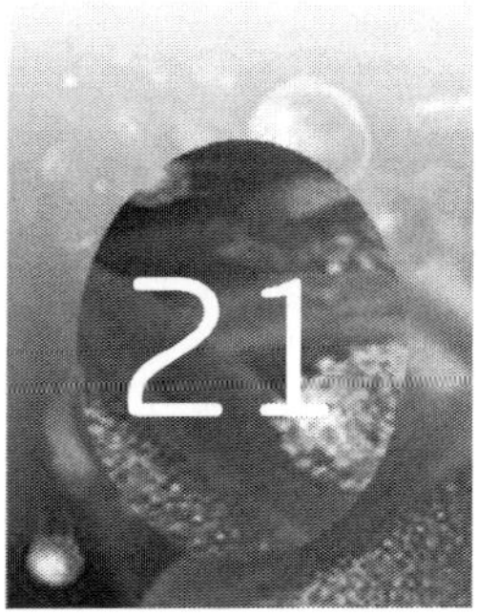

Diarrhoea

Diarrhoea is the most common illness caused by a variety of microorganisms. It affects a large number of people. It is a symptom but not a disease. Diarrhoea is the leading cause of illness and death among children in developing countries including India. This is because children tend to eat less during an episode of diarrhoea. Also, diarrhoea can adversely affect the digestion of food. As a result, the body may not be able to utilize the food effectively. The body requires additional nutrients during any infection in order to fight the germs that cause disease. Inadequate diet and poor digestion together adversely affect the nutritional status of a child.

Diarrhoea is defined as a condition in which there is change in the consistency and character of stool, and three or more water stools are passed.

There are three main types of diarrhoea; they are acute watery diarrhoea, dysentery and persistent diarrhoea.

1. *Acute watery diarrhoea* has three main characteristics.
   - i. Symptoms start suddenly.
   - ii. Stools are loose and watery.
   - iii. Recovery normally takes place within 3 to 7 days.
2. *Dysentery* has two main characteristics. These include:
   - i. Passage of blood in stools.
   - ii. It may be associated with cramps in the abdomen, loss of appetite and rapid loss of weight.
3. *Persistent diarrhoea* has three main characteristics. These include:
   - i. Passage of watery stool along with blood.
   - ii. Symptoms persist for more than 14 days.
   - iii. Weight loss.

If there is any continuous diarrhoea it is called repeated diarrhoea.

BACILLARY DYSENTERY

Dysentery is characterized by the daily loss of 200 to 300 ml of serum protein in faeces. This loss of serum protein results in depletion of nitrogen stores that exacerbates malnutrition and growth stunting. Bacillary dysentery constitutes a significant proportion of acute intestinal disease in children and adults of developing countries, and this infection is a major contributor to stunted growth in children.

CAUSATIVE AGENT

Gram-negative facultative anaerobes of the genus *Shigella* are the principal agents of bacillary dysentery.

The genus *Shigella* is differentiated into four species:

- *S. dysenteriae* (serogroup A, consisting of 12 serotypes);
- *S. flexneri* (serogroup B, consisting of 6 serotypes);
- *S. boydii* (serogroup C, consisting of 18 serotypes); and
- *S. sonnei* (serogroup D, consisting of a single serotype).

Serogroups A, B and C are very similar physiologically while *S. sonnei* can be differentiated from the other serogroups by positive β-D-galactosidase and ornithine decarboxylase biochemical reactions. The identification of *Shigella* by species in the clinical laboratory is usually accomplished by slide agglutination using commercially available, absorbed rabbit antisera.

In DNA hybridization studies, *Escherichia coli* and *Shigella* species cannot be differentiated at the polynucleotide level. Enteroinvasive *E. coli* (EIEC) are very similar to *Shigella* biochemically and they also evoke diarrhoea and/or dysentery. Some EIEC are also serologically related to *Shigella*. For example, EIEC serotype O124 agglutinates with *S. dysenteriae* serotype 3 antiserum.

CLASSIFICATION

Shigella is a sensitive bacterium that belongs to the family Enterobacteriaceae (section 5, volume 1, first edition, *Bergey's Manual*/volume 3, second edition, *Bergey's Manual*). Taxonomic position is given in Box 21.1.

Kingdom	Bacteria
Phylum	Proteobacteria
Class	Gamma Proteobacteria
Order	Enterobacteriales
Family	Enterobacteriaceae
Genus	*Shigella* (Castellani and Chalmers, 1919)

BIOCHEMICAL CHARACTERISTICS

Test	*Shigella sonnei*	*S. dysentriae*, *S.flexneri* and *S. boydii*
Indole	+	+/–
Methyl red	+	+
Voges-Proskauer	–	–
Citrate	–	–
TSI	K/A	K/A
Gas	–	–
H_2S	–	–
Motility	–	–
Urease	–	–
Phenylalanine deaminase	–	–
Lysine decarboxylase	–	–
Arginine decarboxylase	–	–
Ornithine decarboxylase	+	–
ONPG	+	–
Nitrate	–	+
Carbohydrate fermentation		
Glucose	+	+
Mannitol	+	–
Lactose	–	–
Maltose	+	+
Sucrose	–	–
Xylose	–	–
Trehalose	+	+

PATHOGENESIS

Humans are the primary reservoirs of *Shigella* spp. The excreta of infected individuals via direct faecal–oral contamination transmit the infection most often. After entry of virulent bacilli into the intestine, it will cause severe illness. About 100–200 active cells of *Shigella* are enough to cause shigellosis in healthy individuals. After entry, bacteria bind to host cell surface; filamentous actin condenses into microfilaments beneath the site of adherence. Myosin (component of cytoskeleton involved in movement of organelles) also accumulates in this area.

Microfilament rearrangements lead to formation of pseudopods. Normally non-phagocytic mucosal cells ingest attached bacteria. Pseudopods engulf bacteria, bring them inside host cell within phagocytic vesicle.

After uptake, extensive reorganization of actin filaments continues around phagocytic vesicle. Ingested bacteria escape endocytic vesicle and multiply in cytoplasm.

A complex of two plasmid-encoded determinants, designated invasion plasmid antigens (IPA) B and C, trigger the bacterial cell attachment. This complex also induces the endocytic uptake by M cells, epithelial cells and macrophages. IPA-B also mediates lysis of endocytic vacuoles in epithelial cells and macrophages. In the latter case, IPA proteins also cause release of IL-1 cytokine and macrophage apoptosis. The organism then secretes intracellular spread protein (ICSA), which elicits polymerization of filamentous actin.

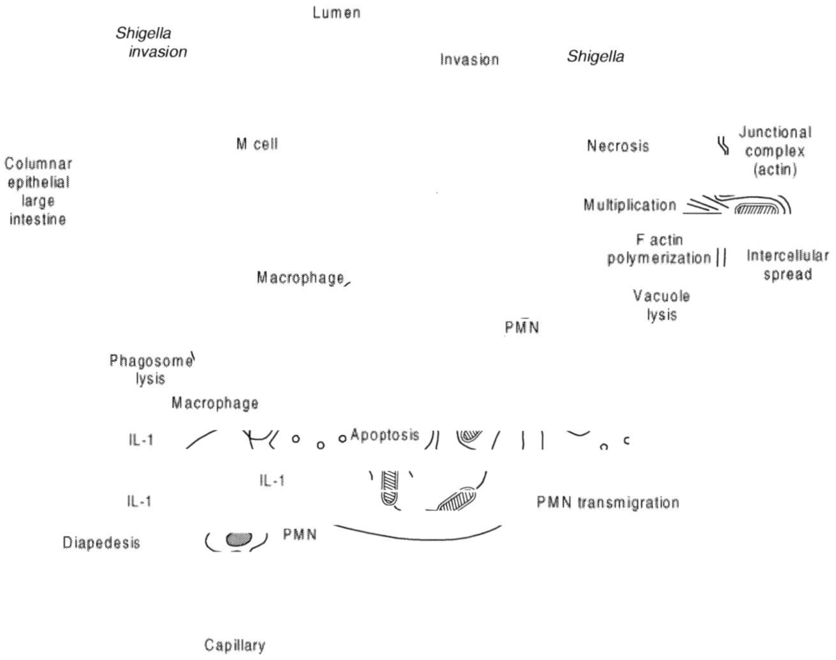

Figure 21.1 *Shigella* pathogenesis

There are two types of spread that are observed:

1. Organelle-like movement (OLM)—movement along with actin filament.
2. Intracellular spread—actin filaments polymerize and elicit the formation of actin tail, which provides a motive force for *Shigella* impinging on the plasma membrane of the infected cell. The resulting protrusions deform the

plasma membrane of host cells. ICSB plasmid-encoded protein then lyses the plasma membranes, resulting in intercellular spread.

IL-1 release evokes an influx of PMN. Eventually infected cells undergo apoptosis and this leads to inflammatory response and lysis of infected cells (Figure 21.1).

CLINICAL MANIFESTATION

Shigellosis has two basic clinical presentations:

1. Watery diarrhoea is associated with vomiting and mild to moderate dehydration.
2. Dysentery is characterized by a small volume of bloody, mucoid stools and abdominal pain (cramps and tenesmus).

Symptoms include vomiting, fever, mucous and blood stool, mild dehydration, weight loss, haemolytic ureamic syndrome (HUS) and Reiter syndrome.

Reiter's Syndrome

It is an arthritis that appears after the intestinal infection. In this condition autoimmune response triggered by *Shigella* antigens affect joint tissues, leading to inflammation of the joints.

Haemolytic Uraemic Syndrome (HUS)

- HUS strains produce unusually high levels of Shiga toxin.
- Primary effect may be damage to blood vessels.
- Vascular damage that causes renal failure explains paralysis, occasional neurological complication and kidney failure.
- Enhances LPS-mediated release of cytokines such as IL-1 and TNF.
- Acts synergistically with TNF or LPS to kill cultured human endothelial cells.

Complications

Possible complications of shigellosis include bacteraemia, convulsions and other neurological complications, reactive arthritis, and haemolytic uraemic syndrome (HUS).

Toxin

During multiplication, *Shigella dysenteriae* serotype 1 produces an enterotoxin called shiga toxin which is an extremely potent, ricin-like, cytotoxin that inhibits protein synthesis in susceptible mammalian cells. Shiga toxin is associated with the haemolytic uraemic syndrome, a complication of infections with *S. dysenteriae* 1.

Shiga Toxin

- A-B (1 : 5) toxin is released during cell lysis.
- B subunits bind to host cell glycolipids.
- A subunit internalized by endocytosis, nicking and translocation of A occurs inside the host cell.
- It stops protein synthesis by inactivation of 60S subunit of ribosome.

Toxic Activity

- It acts as an enterotoxin (provokes fluid loss).
- It acts as a neurotoxin (causes paralysis).
- It sometimes acts as a cytotoxin (induces apoptosis).

IMMUNITY TO SHIGELLOSIS

Shigella is an intracellular parasite and it has the tendency to survive within macrophages.

Humoral Response

- In serum, immunoglobulins found are of IgM type.
- In the intestinal region, IgA is found as secretory product.

EPIDEMIOLOGY

Humans are the primary reservoirs of *Shigella* species, with captive subhuman primates as accidental hosts. In developing countries with prevailing conditions of inadequate sanitation and overcrowded housing, the infection is transmitted most often by the excreta of infected individuals via direct faecal–oral contamination.

LABORATORY DIAGNOSIS

Clinical

Patients presenting with watery diarrhoea and fever should be suspected of having shigellosis. The diarrhoeal stage of the infection cannot be distinguished clinically from other bacterial, viral and protozoan infections. Nausea and vomiting can accompany shigella diarrhoea, but these symptoms are also observed during infections with non-typhoidal salmonellae and enterotoxigenic *E. coli*. Bloody, mucoid stools are highly indicative of shigellosis.

Laboratory

Diagnosis is dependent upon the isolation and identification of *Shigella* from the faeces. Positive cultures are most often obtained from blood-tinged plugs of mucus in freshly passed stool specimens obtained during the acute phase of the disease. Rectal swabs may also be used to culture *Shigella*. Isolation of *Shigella* in the clinical laboratory typically involves an initial streaking for isolation on differential/selective media with aerobic incubation to inhibit the growth of the anaerobic normal flora.

Commonly used primary isolation media include McConkey agar, Hektoen-enteric agar and Salmonella-Shigella (SS) agar. These media contain bile salts to inhibit the growth of other gram-negative bacteria and pH indicators to differentiate lactose fermenters (coliforms) from non-lactose fermenters such as *Shigella*. A liquid enrichment medium (Hajna gram-negative broth) may also be inoculated with the stool specimen and subcultured onto the selective/differential agarose media after a short growth period (Box 21.2). On triple sugar iron agar (TSI), *Shigella* species produce an alkaline slant and an acid butt with no bubbles of gas in the agar (Table 21.1). Slide agglutination tests with antisera for serogroup and serotype confirm the identification.

Hektoen enteric agar	Colourless colony
SS agar	Colourless colony
McConkey agar	Non-lactose-fermenting colony
XLD agar	Colourless colony
Rajhans medium	Colourless colony

TREATMENT

Absorbable drugs such as ampicillin (2 g/day for 5 days), trimethoprim (8 mg/kg/day), sulfamethoxazole(40 mg/kg/day), ciprofloxacin (1 g/day for 3 days) are effective against multiple drug resistant strains.

PREVENTION

- Exclusive breast-feeding till 4 to 6 months and continuation of breast-feeding for at least one year.
- Avoidance of bottle-feeding.
- Use of freshly cooked food.
- Cleaning of feeding bottles every day.
- Washing hands with soap in running water.
- Disposal of stools passed by a child in sanitary toilet immediately.
- Immunization for measles also reduces intestinal burden.
- Ensuring clean child play area.

1. Describe the nature of *Shigella*.
2. How do you cultivate *Shigella*?
3. Describe the pathogenesis of *Shigella*
4. Discuss the epidemiology, laboratory diagnosis and prevention of dysentery.
5. Write short notes on
   i. Shiga toxin
   ii. HUS
   iii. Reiter syndrome

1. How is *Shigella* different from other enteric pathogens?
2. How does *Shigella* invade tissues?
3. How can Shiga toxin act as a neurotoxin, cytotoxin and enterotoxin.

Hale, T.L. "Genetic basis of virulence in *Shigella* species." *Micro.* Rev. 55:206, 1991.

Keusch, G.T. and Bennish, M.L. Shigellosis. p. 593. In Evans, A. S., Brachman, P.S. (eds.). *Bacterial Infections of Humans, Epidemiology and Control*. Plenum, New York. 1991.

Perdomo, O.J.J., Cavaillon, J.M. and Huerre, M. *et al*. "Acute inflammation, causes epithelial invasion and mucosal destruction in experimental shigellosis." *J. Exp. Med.* 180:1307. 1994.

rehydrate.org/dd/su55.htm

www.bustedtees.com/shirt/dysentery

www.cdc.gov/ncidod/dbmd/diseaseinfo/waterbornediseases_t.htm

www.dysentery.net/ -

www.health.state.ny.us/diseases/communicable/shigellosis/fact_sheet.htm - - www.mwra.state.ma.us/germs/dysentery.htm

www.m-w.com/cgi-bin/netdict?dysentery

www.nhsdirect.nhs.uk/articles/article.aspx?articleId=

www.wrongdiagnosis.com/d/dysentery/intro. www.hpa.org.uk/infections/topics_az/shigella/menu.htm

Cholera

INTRODUCTION

Cholera (frequently called asiatic cholera or epidemic cholera) is a severe diarrhoeal disease caused by the bacterium of the family Vibrionaceae. It is an acute illness. The infection is often mild or without symptoms, but sometimes it can be a severe disease characterized by profuse watery diarrhoea, vomiting and leg cramps, resulting in acidosis and hypervolumic shock.

CAUSATIVE AGENT

- *Vibrio cholerae* (Figure 22.1) is the causative agent of cholera.
- It is a gram-negative comma-shaped bacillus with single polar flagellum.
- It was first isolated in pure cultures by Robert Koch in 1883.
- Most vibrios have relatively simple growth factor requirements and will grow in synthetic media with glucose as the sole source of carbon and energy. However, since vibrios are typically marine organisms, most species require large amount of NaCl or seawater base for optimal growth.
- In liquid media vibrios are motile by polar flagella that are enclosed in a sheath continuous with the outer membrane of the cell wall.
- Generation time for *V. cholerae* is less than 30 minutes.

Slightly curved
single polar
flagellum

Figure 22.1 Vibrio cholerae

Kingdom	Bacteria
Phylum	Proteobacteria
Class	Gamma Proteobacteria
Order	Vibrionales
Family	Vibrionaceae
Genus	*Vibrio* (Pacini,1854)
species	*V. cholerae, V. fischeri, V. harveyi, V. parahaemolyticus, V. vulnificus*

BIOCHEMICAL CHARACTERISTICS

Indole-positive, methyl red-positive, Voges-Proskauer-positive, citrate-positive, TSI-A/A gas, H_2S, motile, urease-negative, catalase-positive, oxidase-positive, phenylalanine-negative, lysine decarboxylase-positive, arginine-positive, ornithine-positive, ONPG-positive, nitrate-positive

ANTIGENIC TYPES

Vibrios that cause epidemic infection have been subdivided into two biotypes: Classic and EI Tor. Both biotypes (EI Tor and Classic) contain two major serotypes, Inaba and Ogawa. These serotypes are differentiated by agglutination and vibriocidal antibody tests on the basis of their dominant heat-stable lipopolysaccharide somatic antigens.

The antigenic determinants of *Vibrio cholerae* are as follows:

Serotype	O Antigens
Ogawa	A, B
Inaba	A, C
Hikojima	A, B, C

PATHOGENESIS

Cholera is a serious epidemic disease that has killed millions of people and continues to be a major health problem worldwide.

Cholera is acquired by drinking water that has been contaminated with human faeces or by consuming food that has been washed in contaminated water. *V. cholerae* persists in environment because it can grow both in salt water and fresh water.

Adhesins, neuraminidase, motility, chemotaxis and toxin production are the important virulence factors of *V. cholerae*.

It also has the capability to survive the gastric secretions and low pH of the stomach. They are well adapted to survive in the small intestine. *V. cholerae* is resistant to bile salts and can penetrate the mucous layer of the small intestine, possibly aided by the secretion of neuraminidase and proteases.

They withstand propulsive gut motility by their own swimming ability and chemotaxis directed against the gut mucosa.

Toxin co-regulated pili (TCP pili) mediates attachment to the intestinal mucosa.

Two other possible adhesins in *V. cholerae* are a surface protein that agglutinates red blood cells (haemagglutinin) and a group of outer membrane proteins, which are products of the ACF (accessory colonization factor) genes.

Non-fimbrial adhesins mediate a tighter binding to host cells.

After attachment, the cholera bacilli colonize on mucosal surface and produce an enterotoxin, cholera toxin, that is responsible for cholera symptoms.

Cholera toxin activates the adenylate cyclase enzyme in cells of the intestinal mucosa leading to increased levels of intracellular cAMP, and the secretion of H_2O, Na^+, K^+, Cl^-, and HCO_3^- into the lumen of the small intestine.

The toxin contains 5 binding (B) subunits of 11,500 daltons, an active (A_1) subunit of 23,500 daltons and a bridging piece (A_2) of 5,500 daltons that links A_1 to the five B subunits.

V. cholerae enterotoxin is a product of *ctx* genes. *ctx*A encodes the A subunit of the toxin, and *ctx*B encodes the B subunit. The genes are a part of the same operon. The transcript (mRNA) of the *ctx operon* has two ribosome-binding sites.

The components of cholera toxin are synthesized in the cytoplasm and are assembled in the periplasm after translation. Extra B subunits can be excreted by the cell, but A must be attached to 5B in order to exit the cell. A_1 and A_2 subunits are linked by a disulphide bond.

After the release of cholera toxin into the lumen, it is bound to the oligosaccharide moiety of G_{M1} ganglioside receptor on host cells.

A_1 subunit is released from the toxin, presumably by reduction of the disulphide bond that links it to A_2, and enters the cell by an unknown translocation mechanism. One hypothesis is that the 5B subunits form a pore in the host cell membrane through which the A_1 unit passes (Figure 22.2).

Once fragment A_1 enters inside the cell, it enzymatically catalyses the transfer of the ADP-ribosyl moiety of NAD to a component of the adenylate cyclase system. The process is complex. Adenylate cyclase (AC) is activated normally by a regulatory protein (GS) and GTP. Hydroxylation of GTP by regulatory protein (G_1) leads to inactivation of adenylate cyclase enzyme.

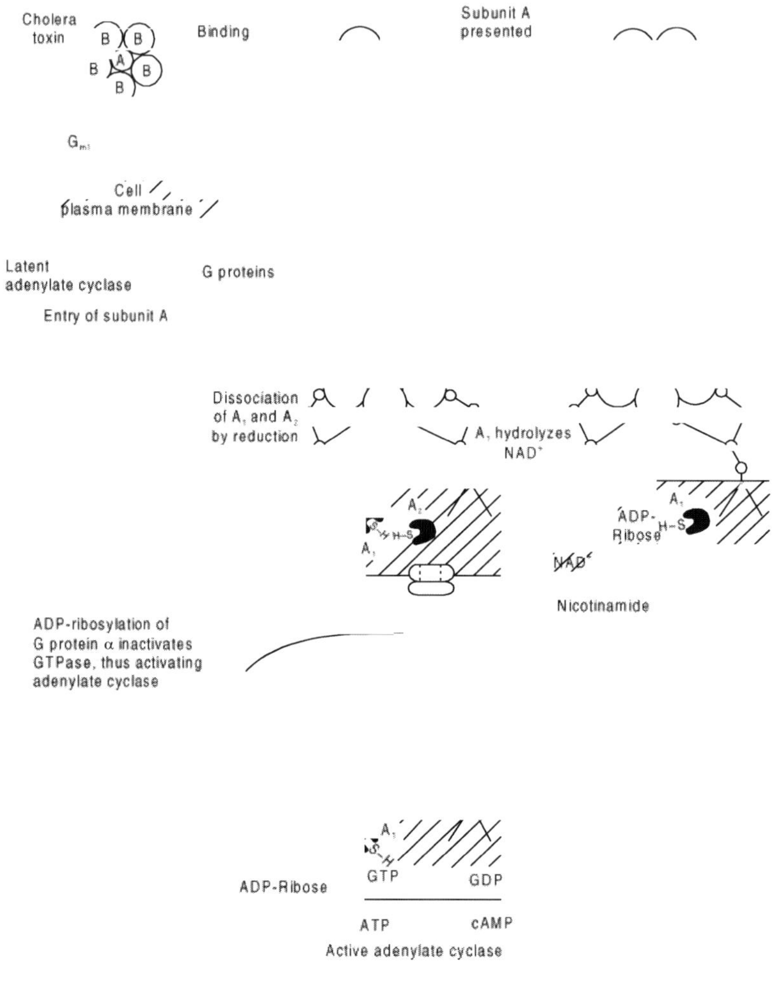

Figure 22.2 Mode of action of cholera toxin

In case of cholera, A_1 fragment catalyses the attachment of ADP-ribose (ADPR) to the regulatory protein forming GS-ADPR from which GTP cannot be hydrolysed. This leads to the activation of the adenylate cyclase (Figure 22.3).

The net effect of the toxin is to cause cAMP to be produced at an abnormally high rate, which stimulates mucosal cells to pump large amounts of Cl^- into the intestinal contents. H_2O, Na^+ and other electrolytes flow due to the osmotic and electrical gradients caused by the loss of Cl^-. The lost H_2O and electrolytes in mucosal cells are replaced from the blood. Thus, the toxin-damaged cells become pumps for water and electrolytes causing diarrhoea, loss of electrolytes and dehydration that are characteristic of cholera.

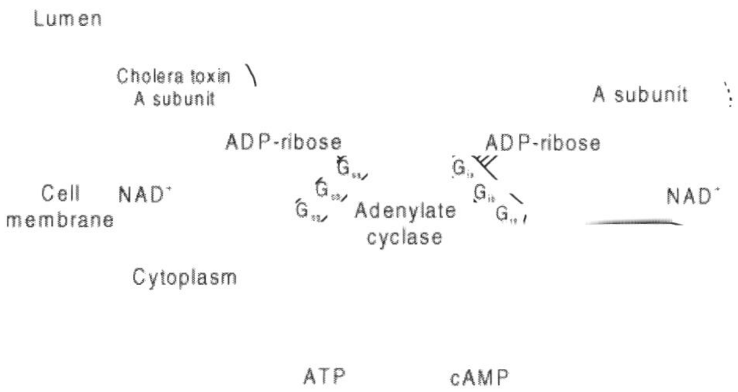

Figure 22.3 *Activation of adenylate cyclase*

CLINICAL MANIFESTATION

The clinical description of cholera begins after an incubation period of 6–48 hours with sudden onset of massive diarrhoea. The patient may lose gallons of protein-free fluid and associated electrolytes, bicarbonates and ions within a day or two. This loss of fluid leads to acidosis and shock. The watery diarrhoea is also called "rice-water stool" and contains enormous numbers of vibrios. The loss of potassium ions may result in cardiac complications and circulatory failure. Untreated cholera frequently results in high (50–60%) mortality rates.

LABORATORY DIAGNOSIS

1. Culture

Specimen	Stool
Transport medium	Alkaline peptone water and Venkatraman–Ramakrishna medium
Selective medium	Trisodium citrate bile salt sucrose agar (TCBS)
Colony morphology on TCBS	Circular yellow coloured colonies

2. Serological identification is also performed.

TREATMENT

Treatment of cholera involves the rapid intravenous replacement of the lost fluid and ions. Following this replacement, administration of isotonic maintenance solution should continue until the diarrhoea ceases. If glucose is added to the maintenance solution it may be administered orally, thereby eliminating the need for sterility and intravenous administration. Most antibiotics and chemotherapeutic agents have no value in cholera therapy, although a few (e.g. tetracyclines) may shorten the duration of diarrhoea and reduce fluid loss.

VACCINES

The first attempts at a vaccine in 1960s were directed at whole cell preparations injected parenterally. At best, 90% protection was achieved and this immunity waned rapidly to the baseline within one year. Purified LPS fractions from different biotypes have also been given as vaccines with variable success. The cholera toxin can be converted to toxoid in the presence of formalin and glutaraldehyde.

1. Briefly discuss various features of cholera.
2. Explain the pathogenesis and symptoms related to diarrhoea.
3. Write short notes on
   i. Rice-water stool
   ii. Cholera toxin
   iii. *Vibrio cholerae*
   iv. Ogawa
   v. TCP
   vi. cAMP
   vii. TCBS

1. Why does not *V. cholerae* satisfy Koch's postulates?
2. Is *V. cholerae* an invasive bacterium?
3. Why is antibiotic therapy not useful for the treatment of cholera?

Barua, D. and Greenough, III W.B. *Cholera*. Plenum Book Company, New York and London. 1992.

Finkelstein, R.A. "Cholcra." In: Germanier R. (ed.). *Bacterial Vaccines*. Academic Press, San Diego. 1984.

Van Heyningen, W.E. and Seal, J.R. *Cholera: The American Scientific Experience, 1947-1980*. Westview Press, Boulder, CO. 1983.

Wachsmuth, I.K. Blake, P.A. and Olsvik, O. *Vibrio cholerae and Cholera: Molecular to Global Perspectives.* ASM Press, Washington, D.C. 1994.

Wadstrom, T. (ed.). *Bacterial Toxins and Cell Membranes*. Academic Press, San Diego. 1978.

World Health Organization: Diarrhoeal diseases control programme. Report of the tenth meeting of the technical advisory group (Geneva, March 1317, 1989). WHO/ D/89 32:1, 1989.

en.wikipedia.org/wiki/Cholera.

www.who.int/topics/cholera/

textbookofbacteriology.net/cholera.html

toyotabxof.satublog.com/

www.cdc.gov/node.do/id/

www.health.state.ny.us/diseases/communicable/cholera/fact_sheet.htm

www.nlm.nih.gov/medlineplus/ency/article/

Gastroenteritis

INTRODUCTION

Gastroenteritis is a condition in which there is an inflammation of the intestine, along with diarrhoea. Different types of microorganisms cause gastroenteritis in humans. They are *E. coli*, *Campylobacter*, *Yersinia*, *Citrobacter* and *Aeromonas*. Gastroenteritis due to *Escherichia coli* plays a major role because of new strain development.

E. coli is an important cause of acute watery diarrhoea in adults and children. *E. coli* is a common member of the normal flora of the large intestine. As long as these bacteria do not acquire genetic elements encoding for virulence factors, they remain benign commensals.

E. coli is the head of the large bacterial family, Enterobacteriaceae, the enteric bacteria, which are facultative anaerobic gram negative rods; it is versatile and well adapted to its characteristic habitats. Its nutritional equivalent is very simple, it grows very fast and its genome is well characterized. *E. coli* can respond to environmental signals such as chemicals, pH, temperature, osmolarity, etc.

CLASSIFICATION

Phylum	Proteobacteria
Class	Gamma Proteobacteria
Order	Enterobacteriales
Family	Enterobacteriaceae
Genus	*Escherichia*
species	*E. coli* (T. Escherich, 1885)

Indole-positive, methyl red-positive, Voges-Proskauer-negative, citrate-negative, TSI—Acid slant : acid butt : gas-positive : H_2S-negative, urease-negative, catalase-positive, oxidase-negative, phenylalanine-negative, lysine-decarboxylase-positive, nitrate-positive

VIRULENCE FACTORS

Adhesins, CFA-I/CFA-II, type1 fimbriae, P fimbriae, S fimbriae, Intimin (non-fimbrial adhesin), invasins, haemolysis, motility/chemotaxis and flagella are the major virulence factors.

Toxins of *E. coli* are LT toxin, ST toxin, shiga-like toxin, cytotoxins and endotoxin (LPS).

Factors providing antiphagocytic surface properties are capsules, K antigens and LPS. LPS and K antigens provide defence against serum bactericidal reaction.

Defence against immune response is provided by capsules, K antigens, LPS and antigenic variation.

Genetic attributes of *E. coli* are due to genetic exchange by transduction and conjugation, transmissible plasmids, R factors and drug-resistant plasmids, toxin and other virulent plasmids.

SEROTYPING OF E. *coli*

E. coli is serogrouped according to the presence or absence of specific heat-stable somatic antigens (O antigens) composed of polysaccharide chains linked to the core polysaccharide.

O specificity is determined by sugar or amino sugar composition. More than 170 different O-specific antigens have been defined since Kauffmann began this method of typing *E. coli* in 1943.

56 types of H antigens are available. H antigen typing is important for *E. coli* associated with diarrhoeal disease.

TYPES OF *E. coli*

As a pathogen, *E. coli*, of course, is best known for its ability to cause intestinal diseases. Five classes (virotypes) of *E. coli* that cause diarrhoeal diseases are now recognized: enterotoxigenic *E. coli* (ETEC), enterohaemorrhagic *E. coli* (EHEC), enteropathogenic *E. coli* (EPEC), enteroinvasive *E. coli* (EIEC) and enteroaggregative *E. coli* (EAggEC).

PATHOGENESIS

ETEC

E. coli diarrhoeal diseases are contracted orally by ingestion of food or water contaminated with a pathogenic strain. ETEC occurs in all age groups, but mortality is most common in infants, particularly in the most undernourished or malnourished infants in developing nations. The disease varies from minor discomfort to a severe cholera-like syndrome.

The disease requires colonization and elaboration of one or more enterotoxins. Both traits are plasmid-encoded ETEC adhesins and fimbriae, which are species-specific, e.g. K-88 antigen is found in pigs, K-99 is found in calves and lambs. CFA-I and CFA-II are found in humans.

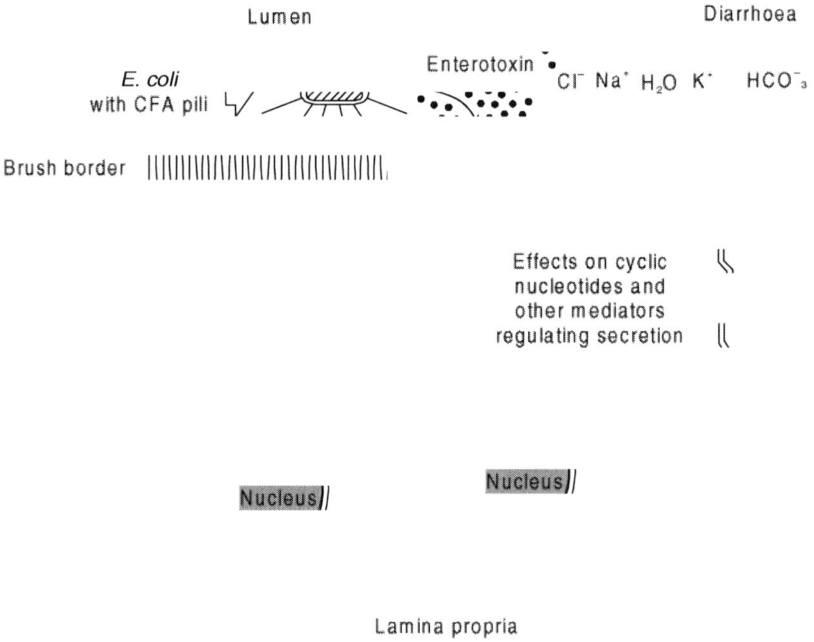

Figure 23.1 Pathogenesis of ETEC on humans

ETEC possesses specialized antigens unrelated to common pili, which act as ligands to bind the bacterial cells to specific complex CHO receptors on the epithelial cell surface of the small intestine, since this is on the gut surface, these pili are termed colonization factor antigens (CFAg). After colonization, *E. coli* produces enterotoxins, it induces the LT (heat-labile) and ST toxins (heat-stable). ST is actually a family of toxic peptides. STa can stimulate intestinal guanylate cyclase, the enzyme that converts GTP to cGMP. Increased intracellular cGMP inhibits intestinal fluid uptake, resulting in net fluid secretion. STb do not seem to cause diarrhoea by the same mechanism (Figure 23.1). EIEC pathogenesis is similar to bacillary dysentery (Figure 23.2) (Also refer Chapter 21).

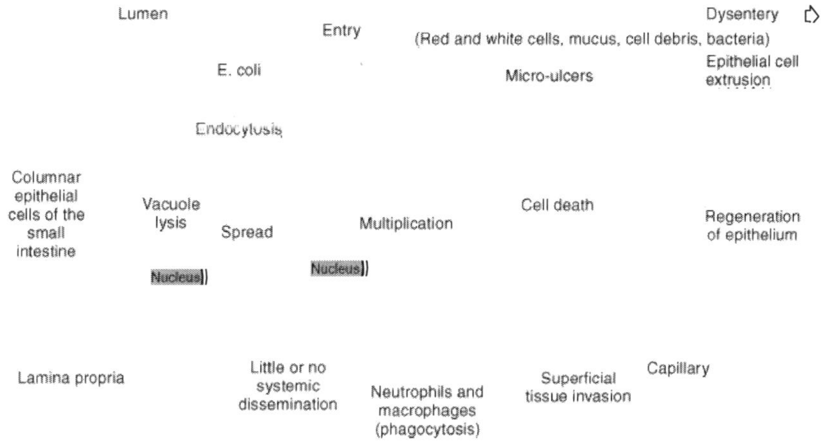

Figure 23.2 Diagrammatic representation of EIEC diarrhoea

EPIDEMIOLOGY

The disease is transmitted from person to person.

The incidence of *E. coli* diarrhoea is clearly related to hygiene, food processing and general sanitization.

The geographic frequency of *E. coli* is inversely proportional to the sanitation standards.

EMB agar	Metallic sheen colonies
XLD agar	Yellow coloured colonies
SS agar	Pink coloured colonies
Rajhan's medium	Greenish blue colonies
CLED agar	Bluish colonies
MacConkey agar	LF colonies (pink)
Hektoen enteric agar	Salmon-coloured colonies

LABORATORY DIAGNOSIS

- Stool specimen examination
- Culturing of stool
- Serological identification of specific CFA
- Demonstration of LT and ST toxin
- Identification of genes for virulent factors

Table 23.1 Differentiating features of different *E. coli pathogens*

	ETEC	EIEC	EPEC	EaggEC	EHEC
Attachment	Fimbrial adhesions, e.g. CFA-I, CFA-II, K-88, K-99	Non-fimbrial adhesins, possibly outer membrane protein	Non-fimbrial adhesin (intimin)	Adhesins not characterized	Adhesins not characterized, probably fimbriae
Invasive/non-invasive	Non-invasive	Invasive (penetrate and multiply within epithelial cells)	Moderately invasive (not as invasive as shigella or EIEC)	Non-invasive	Moderately invasive
Toxin	Produce LT and/or ST toxin	Does not produce Shiga toxin	No specific toxin	Produce ST-like toxin (EAST) and a haemolysin	Does not produce LT or ST but does produce shiga toxin
Symptoms	Watery diarrhoea in infants and travellers; no inflammation, no fever. Activity is like cholera toxin.	Dysentery-like diarrhoea (mucus, blood), severe inflammation, fever	Usually infantile diarrhoea; watery diarrhoea similar to ETEC; some inflammation, no fever Symptoms probably result from invasion rather than toxigenesis	Persistent diarrhoea in young children without inflammation, no fever	Paediatric diarrhoea, copious bloody discharge (haemorrhagic colitis), intense inflammatory response, may be complicated by haemolytic uraemic syndrome.

TREATMENT

Treatment is by fluid replacement procedure as in cholera and antibiotic treatment as in shigella infections.

1. How do you differentiate ETEC and EIEC?
2. What is the method adopted to serotype *E. coli*?
3. Discuss the characters of different *E. coli* serogroups.
4. Explain the pathogenesis of *E. coli.*
5. Describe the laboratory diagnosis and treatment of *E. coli* gastroenteritis.
6. Write short notes on
   i. Gastroenteritis
   ii. Toxins of *E. coli*

1. How and why does *E. coli* change its characteristics?

Dytoc, M. Sone and R. Cockerill, F, III *et al*. "Multiple determinants of verotoxin-producing *Escherichia coli* O157:H7 attachment-effacement." *Infect. Immun*. 61:3382. 1993.

Evans, D.J. Jr. and Evans, D.G. "Colonization factor antigens of human pathogens." *Current Topics Microbiol. Immunol.* 151:129. 1990.

Giron, J.A., Ho ASY, and Schoolnik, G.K. "An inducible bundle-forming pilus of enteropathogenic *E.coli*." Science 254:710. 1991.

Giron, J.A., Levine, M.M. and Kaper, J.B. "Longus: a long pilus ultrastructure produced by human enterotoxigenic *Escherichia coli*." *Molec. Microbiol*. 12:71. 1994.

Spangler, B.D. "Structure and function of cholera toxin and the related *Escherichia coli* heat-labile enterotoxin." *Microbiol. Rev*. 56:622. 1992.

Tesh, V.L. O'Brien, A.D. Adherence and colonization mechanisms of enteropathogenic and enterohemorrhagic *Escherichia coli*. "*Microb. Pathogenesis*" 12:245. 1992.

Wenneras, C., Svennerholm, A.M., Ahren, C. and Czerkinsky, C. "Antibody-secreting cells in human peripheral blood after oral immunization with an inactivated enterotoxigenic *Escherichia coli* vaccine." *Infect. Immun.* 60:2605. 1992.

www.cdc.gov/ncidod/dvrd/gastro.html

www.emedicinehealth.com/gastroenteritis/article_em.html

www.nlm.nih.gov/medlineplus/ency/article/000252.html

www.mayoclinic.com/health/first-aid-gastroenteritis/FA00030

www.mckinley.uiuc.edu/handouts/gastroenteritis.html

www.merck.com/mmhe/sec09/ch122/ch122a.html hcd2.bupa.co.uk/fact_sheets/html/Gastroenteritis.html

www.patient.co.uk/showdoc/23068743/

www.uct.ac.za/depts/mmi/jmoodie/gastro2.html

www.webmd.com/content/article/5/1680_51287.html

Typhoid Fever

INTRODUCTION

Enteric fever is a collective term used for invasive infections caused by a small group of bacteria called *Salmonella*. The term "invasive" is used to describe the tendency to spread, penetrate and intrude. Enteric fever is an important public health problem worldwide. An estimated 14–27 lakh people are affected by enteric fever in India every year. It is prevalent all over the world. Louis gave the name typhoid. Budd pointed out that the disease is transmitted through the excreta of patients. Typhoid fever occurs only in humans.

CAUSATIVE AGENT

Typhoid fever is caused by

- *Salmonella typhi*
- *Salmonella paratyphi A*
- *Salmonella paratyphi B*
- *Salmonella paratyphi C*

GENERAL CHARACTERISTICS OF SALMONELLA

Salmonella are gram-negative, straight rods, motile by peritrichous flagella. They are facultative anaerobes with respiratory and fermentative type metabolism. The organisms are oxidase-negative and catalase-positive. Some strains produce hydrogen sulphide. The differentiating features of the four species of *Salmonella* that cause typhoid are listed in Table 24.1.

Lactose fermentation-negative, indole-negative, sucrose-negative, mannitol-positive, malonate-positive, glucose-positive, citrate utilization-positive

Table 24.1 Differentiating characteristics of *Salmonella* spp.

Character	*S. typhi*	*S. paratyphi A*	*S. paratyphi B*	*S. paratyphi C*
Glucose	Acid	Acid and gas	Acid and gas	Acid and gas
Hydrogen sulphide	+	–	–	–
Lysine	+	–	+	–
Ornithine	–	+	+	–
Rhamnose	–	+	+	–
Xylose	D	–	Acid and gas	Acid and gas
Citrate	–	–	+	–
D-Tartrate	A	–	–	Acid and gas
Mucate	D	–	Acid and gas	–

CULTURAL CHARACTERISTICS

- *Salmonella* grows on ordinary media.
- On MacConkey agar it produces small, circular, translucent, NLF colonies.
- On Wilson and Blair, bismuth sulphite medium, the colonies are jet black with metallic sheen colonies.
- On Salmonella-Shigella agar black-centred colonies are observed.
- On XLD agar slight pink colonies with black centre are observed.
- On deoxycholate citrate agar, black colonies are formed.

ANTIGENIC CHARACTERISTICS

The Kauffmann-White system used to classify *Salmonella* is based on identifying the O (somatic) and H (flagellar) antigens possessed by the different serovars. The detection of Vi antigen is also used in the identification of *S. typhi* and others.

O-antigens

O-antigens are cell wall heat-stable antigens responsible for antigenicity. *Salmonella* are grouped by their O-antigens. The groups are designated A–Z, 51–61 and 64–66. Many of the medically important *Salmonella* belong to groups A–G. Each O antigen has group factor.

H-antigens

These are flagellar heat-labile antigens. *Salmonella* are serotyped by H antigens. Many salmonellae are diphasic, that is, they can occur in two antigenic forms referred to as phase I and phase II. Phase I antigens are given alphabetical letters and phase II antigens are either numbered or given any other letters.

Vi-antigens

The surface (K) antigens are found on *S. typhi* and *S. paratyphi C*. It is associated with virulence Vi antigens which can interfere with O antigen test. If therefore an isolate agglutinates Vi antigen serum, but not O antigen, interference from Vi antigen should be suspected. A saline suspension of the organism should be heated in a container or boiling water for 20 minutes and after being allowed to cool, the bacterial cell should be re-tested with O antiserum.

Table 24.2 Differentiating *Salmonella* species based on antigens

Character	*S. paratyphi A*	*S. paratyphi B*	*S. paratyphi C*
Serogroup	A	B	C
Type	A	B	C
O antigen	1, 2, 12	1, 4, 5, 12	6, 7(Vi)
H-antigen Phase I	A	B	C
H-antigen Phase II	-	1, 2	1, 5

VIRULENCE FACTORS

- Endotoxin
- Invasins
- Catalase
- Superoxide dismutase
- Cationic proteins
- Factors involved in resistance to acidic pH
- Vi factor
- Aromatic amino acid synthesizing factors

PATHOGENESIS

Enteric fever is transmitted by faecal–oral route through contaminated food and water. The ingested organisms are mostly destroyed in the stomach. Sufficient number of bacilli passes through the gastric acid barrier and reaches the duodenum, where they multiply in alkaline medium.

In the small intestine, the bacilli attach themselves to the surface of epithelial cells of the villi, pass through them to the submucosal coat, where they are phagocytozed by neutrophils and macrophages. The virulent bacilli resist intracellular killing and multiply within these cells. These cells enter the mesenteric lymph nodes, where after a period of multiplication, the bacilli invade the bloodstream and primary bacteraemia follows. Cells rapidly clear bloodstream by mononuclear phagocytic system in the liver, bone marrow, spleen, lung and lymph nodes. Thus the internal organs are infected during primary bacteraemia in the first 7–10 days (Figure 24.1).

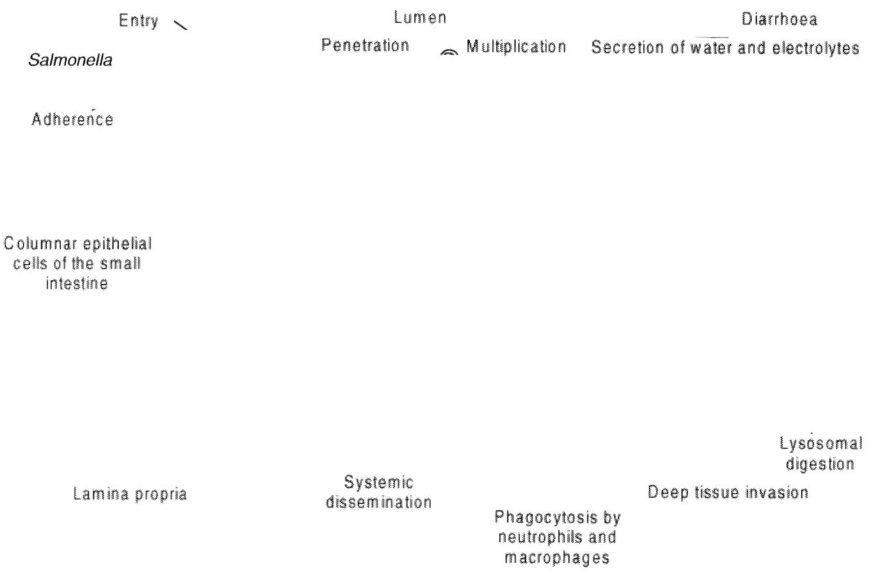

Figure 24.1 Pathogenesis of *Salmonella*

After primary bacteraemia, the organisms enter into the intestinal lymph nodes. During the 10th day of infection, parasitized cells undergo necrosis and the bacilli pass into the blood leading to secondary and heavier bacteraemia, which corresponds with the onset of clinical illness at about the 14th day after ingestion. During this period, some organisms undergo lysis, thereby liberating endotoxin in the circulation. The bacteraemia and toxaemia cause pyrexia and other symptoms.

From the bloodstream, some organisms localize in organs such as gall bladder, liver, spleen, etc. Some organisms are discharged from the gall bladder into the intestine which cause the inflammation of Peyer's patches of intestine produces necrosis and sloughing of the infected follicles with resultant typhoid ulcers which may lead to haemorrhage and perforation.

The febrile illness often causes severe mental cloudiness and headache. Skin rashes known as rose spots may appear on the chest and abdomen in the second and third week. The organism appears in stool during the second and third week and during the third to fourth week of infection.

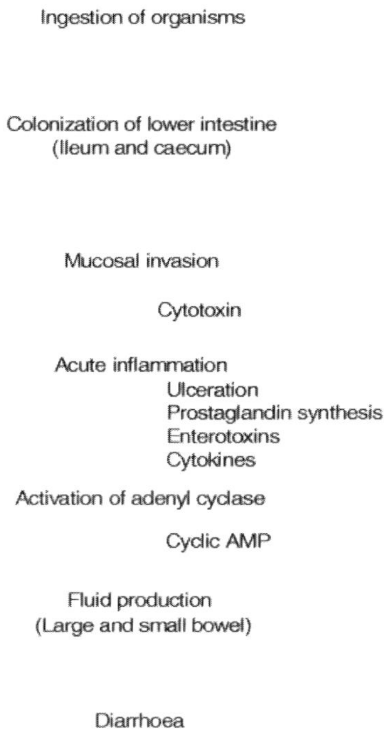

Figure 24.2 Steps involved in pathogenesis

Normally duration of infection varies from 7–14 days for typhoid fever and 4–5 days for paratyphoid fever. Typhoid fever is more severe than paratyphoid fever.

CLINICAL MANIFESTATION

- Fever
- Rash
- Relatively lower pulse rate
- Coated tongue
- Diffuse tenderness in the abdomen
- Enlarged liver and spleen
- Headache
- Pain in the abdomen
- Vomiting
- Diarrhoea
- Bleeding from the nose
- Dry cough

Complications

Bleeding from intestine, perforation of the intestine, enteric encephalopathy, kidney failure, meningitis, osteomyelitis, loss of weight and mental confusion are some of the complications.

Typhoid fever requires immediate diagnosis and treatment.

EPIDEMIOLOGY

S. typhi is maintained in nature by human carriers. There are three main types of typhoid carriers—convalescent, temporary and chronic. One notorious carrier of *S. typhi* was Mary Malan, a young Irish cook who lived in New York city in the early 1900s. She was responsible for 53 cases of typhoid during a 15-year period. It is mostly transmitted through faecal–oral route. Human is the only natural host. It can occur at any time during the year but is common during July to September.

LABORATORY DIAGNOSIS

Collection of specimen

Blood (or bone marrow or clot), stool and urine are clinical samples.

Isolation of the Organism

Blood culture The organisms can be best isolated during the first 7–10 days of illness, but in at least half the cases, the organisms can still be isolated in the second and third weeks.

A minimum of 10 ml of venous blood from an adult patient should be added to each of 50 ml of sodium taurocholate broth and glucose broth. From children at least 5 ml blood should be inoculated in each of the bottles.

The inoculated media should be incubated at 37°C overnight and subcultured on to MacConkey agar and blood agar.

The incubation of blood culture bottle should be extended up to 7 days if subcultures fail to yield any bacterial growth. It should be subcultured again on day 7 and if no growth is obtained blood culture may be declared as negative for enteric fever bacilli.

Periodic subcultures are made after day 2, 5 and 7 on MacConkey agar. No growth after seven days may be regarded as negative. Very rarely, organisms may appear as colonies after 30 days of incubation.

Clot culture The clot culture has the following advantages over blood culture.

- Serum is available for antibody titration.
- Special culture bottles are not required; so it is useful under field conditions.

Culture of faeces In at least 50% of the cases, culture of faeces is positive in the first week, so faecal culture should be attempted in the first week of illness itself. The isolation of *S. typhi* greatly increases in the second or third week. Rectal swabs give inferior results compared to faecal culture.

The faeces should be cultured on deoxycholate citrate agar (DCA) and Wilson and Blair bismuth sulphite brilliant green agar using Selenite F as enrichment medium. XLD agar gives better results in isolating *Salmonella* from faeces. It is recommended to use two enrichment media and two selective media to improve the isolation rate.

Urine culture Urine culture is positive in a quarter to one-third of cases, but there seems to be no regular excretion pattern in the urine. Bacteriuria is usually limited to the first few weeks of the illness and in great majority of the patients, the urine is free from organisms, weeks before the faeces becomes negative.

Stages of diagnosis are as follows:

Stage	Examination	Result
1st week	Blood culture	Leucopenia
	Blood culture	95%
2nd week	Blood culture	40–50%
	Widal	Low antibody titre
3rd week	Widal	100%
	Blood	15–20%
	Stool	80%
	Urine	80%
4th week	Widal	100%
	Stool	90%
	Urine	90%
	Blood	5–10%

Serodiagnosis

Detection of antibodies A large number of serological tests have been devised to detect and titrate the antibodies against common agents of enteric fever. The tube agglutination test (Widal test) has great historical but little diagnostic importance unless four-fold rise is exhibited in two samples drawn at intervals of 10–14 days from the start of the illness.

WIDAL TEST

In this test the patient's serum is tested by tube agglutination test for its titres of antibodies against H, O and Vi suspensions of enteric fever bacteria.

Procedure

A series of serum dilutions are made for each antigen to be tested. Tubes with 0.5 ml saline should be included as controls for each antigen to be used.

Perfectly clean and dry test tubes should be used and dilutions beginning with 1 : 10 and doubling through 1 : 320 or so should be prepared.

0.1 ml of serum is added to 0.9 ml of physiological saline and then diluted serially by mixing 0.5 ml diluted serum with 0.5 ml saline and 0.5 ml from the last tube should be discarded.

From specimen submitted to detect possible rise in titre, a series of 10 dilutions, ending with 1 : 5120 are prepared.

PREPARATION OF ANTIGEN

The killed standardized suspension of *S. typhi* O, *S. typhi* H, *S. paratyphi* A, H and *S. paratyphi* A, O are used. These are commercially available and their preparation should be undertaken as per the instructions of the manufacturers.

Procedure

- The antigen suspension should be shaken to ensure even distribution.
- 0.5 ml is added to each serum dilution and to saline for controls.
- The test tubes are incubated overnight at 37°C and are allowed to stand at room temperature for 15–20 minutes before reading.

PREVENTION

Sanitary measures and vaccines are the two main ways to prevent enteric fever. Key sanitary measures include:

- Provision of safe drinking water.
- Safe disposal of human sewage.
- Maintenance of high standards of food hygiene.
- Care in detecting, monitoring and treating chronic carriers.

Individual protection measures include:

- Drinking boiled, iodinated or chlorinated water.
- Care in consumption of food prepared outside.
- Immunization with typhoid vaccine.
- Avoidance of raw or partially cooked eggs.
- Washing hands with soap and water after using toilet and before preparing or handling food.
- People who have recovered from fever to use separate towels till they get completely cured.

VACCINATION

Three main types of vaccines are available:

1. Older heat-inactivated, phenol-preserved whole cell vaccines
   - Crude *S. typhi*, provide 50–60% protection

- 0.5 ml, 2 injections one in muscle and the other in skin
- Booster once in 3 years

2. Newer Vi polysaccharide vaccine
   - Muscular injection, 70–80% protection
   - 5 ml quantity, single dose
   - Booster dose during 3rd year
3. Oral typhoid vaccine
   - Oral vaccine has to be given once in three days

TREATMENT

The following drugs are used:

- Chloramphenicol
- Ampicillin
- Amoxicillin
- Clotrimoxazole
- Ciprofloxacin
- Ofloxacin
- Norfloxacin
- Pefloxacin
- Ceftriaxone
- Cefoperazone
- Cofotaxine
- Trimethoprim

1. Describe the features used to differentiate different *Salmonella* species.
2. How do you cultivate *Salmonella* sp.?
3. Explain the pathogenic attributes of *Salmonella typhi.*
4. Can you control the spread of typhoid? If so, how?
5. Discuss the diagnosis of typhoid serologically.
6. Describe the various stages of enteric fever diagnosis.
7. What are the complications of typhoid?

1. How do you detect H_2S production in TSI medium? Explain critically the process involved in it.
2. How do you differentiate enteric fever from other types of fever?
3. Why is typhoid fever also called as enteric fever?
4. Why is typhoid fever invasive infection?

Black, P.H., Kunz, L.J. and Swartz, M.N. "Salmonellosis. A review of some unusual aspects." *N. Engl. J. Med.* 262:811, 864, 921. 1960.

Chopra, A.K., Peterson, J.W. and Chary, P. *et al*. "Molecular characterization of an enterotoxin from *Salmonella typhimurium*." *Microb. Pathogen.* 16:85. 1994.

Finlay, R.B., Heffron, F. and Falkow, S. "Epithelial cell surfaces induce *Salmonella* proteins required for bacterial adherence and invasion." *Science.* 243:940. 1989.

Finlay, B.B., Leung, K.Y. and Rosenshine, I. *et al*: "*Salmonella* interactions with the epithelial cell: A model to study the biology of intracellular parasitism." *ASM News*. 58:486. 1992.

Galan, J.E. and Curtiss, R. "Cloning and molecular characterization of genes whose products allow *Salmonella typhimurium* to penetrate tissue culture cells." *Proc. Natl. Acad. Sci.* USA. 86:6383. 1989.

Giannella, R.A., Gots, R.E. and Charney, A.N. *et al*: "Pathogenesis of *Salmonella*-mediated intestinal fluid secretion: activation of adenylate cyclase and inhibition by indomethacin." *Gastroenterology.* 69:1238. 1975.

Rubin, R.H. and Weinstein, L. *Salmonellosis: Microbiologic, Pathologic and Clinical Features*. Stratton Intercontinental Medical Book Corp., New York. 1977.

Stephen, J.,Wallis, T.S. and Starkey, W.G. *et al*: "Salmonellosis: in retrospect and prospect." p. 175. In Evered, D. and Whelan, J. (eds.). *Microbial Toxins and Diarrhea Disease*. Ciba Foundation Symposium 112. Pitman Press, London. 1985.

www.mayoclinic.com/health/typhoid-fever

www.medicinenet.com/typhoid_fever

healthlink.mcw.edu

www.merck.com/mmhe/sec17/ch190

www.emedicine.com/oph/topic686.html

www.who.int/vaccines-documents

www.healthatoz.com

www.metrokc.gov/health/prevcont/typhoid.html

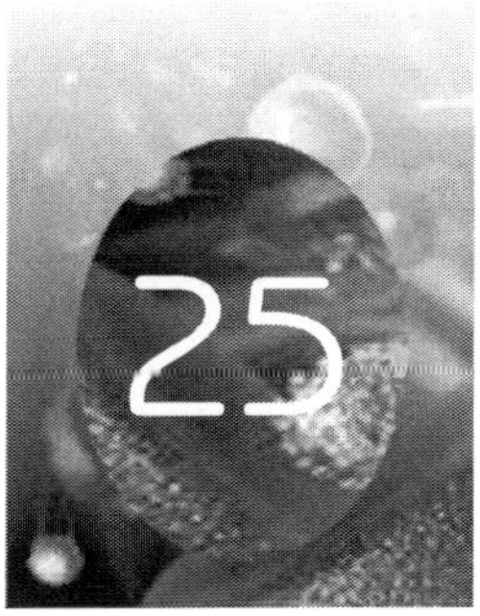

Gonorrhoea

INTRODUCTION

Gonorrhoea is a venereal disease, which has been known from ancient times. Galen first coined the name Gonorrhoea in 130 AD. The word Gonorrhoea means "flow of seed". The disease is acquired by sexual contact.

CAUSATIVE AGENT

- Gonorrhoea is caused by *Neisseria gonorrhoeae*.
- It is also called Gonococcus.
- Neisser first described it in gonorrhoeal pus in 1879.
- Bumm in 1885 cultured the coccus and proved its pathogenicity.

Soyabean-shaped cells

Figure 25.1 *Vibrio cholerae*

- It is a gram-negative coccus.
- It is a fastidious organism, 20% of strains require glutamine for primary isolation.
- The organism is aerobic and glucose is the principal carbon and energy source.
- Growth occurs best at pH 7.2–7.6 and at a temperature of 35–36°C.
- Some strains require 5–10% carbon dioxide.
- They grow well on chocolate agar and Mueller-Hinton agar with 5% blood.
- A popular selective medium is the Thayer-Martin medium. The medium contains vancomycin, colistin, nystatin and trimethoprim lactate.
  - Vancomycin inhibits gram-positive bacteria.
  - Colistin inhibits gram-negative bacteria except *Proteus.*
  - Nystatin inhibits yeast cells.
  - Trimethoprim lactate inhibits *Proteus*.

- Colonies are small, round, translucent, convex or slightly umbonate with finely granular surface and lobate margin.
- Four types of colonies have been recognized and named as T1, T2, T3 and T4.
  - T1 and T2 form small brown colonies.
  - T3 and T4 form larger granular non-pigmented colonies.
- Gonococci ferment only glucose and not maltose.
- Gonococci are antigenically heterogeneous; they are capable of changing their surface structures.
- Gonococcus is a very delicate organism readily killed by heat, drying and antiseptics.
- It is highly susceptible to sulphonamides, penicillin, etc.

ANTIGENIC CHARACTERISTICS

Pili, Protein I, Protein II and Protein III, LOS, Tbp1, Tbp2, Lbp are the major antigens responsible for pathogenicity.

PILI

It is a non-flagellar surface appendage, constructed from identical protein subunits linked in tandem to form long thin polymers, serologically heterogeneous. Molecular weight ranges from 17,000– 22,000 daltons. Piliated organisms predominate in T1 and T2 colony types. Pili undergo both phase variation and antigenic variation. A single gonococcal cell has genetic capacity to produce at least 8 antigenically variant pili. Pilus has 3 parts—a constant region in which amino acids are highly conserved, semivariable region and highly variable region. Pili are made up of pilin protein. This protein is composed of two major gene products of pil E (major pilin protein) and pil C (involved in pilin maturation). It is used for initial binding of *Gonococcus* to non-ciliated cells.

Protein I

- It is an outer membrane protein.
- It accounts for about 60% of the total weight of the outer membrane.
- It has a molecular weight of 32,000–36,000 daltons.
- It combines with protein III and form porins.
- Nature of this protein variation is observed between mild infective strains and highly infective strains.
- P1A is the antigenic form found in most invasive strains.
- P1B is the antigenic form found in strains that cause localized infections.
- It may prevent phagolysosome fusion in polymorphonuclear leucocytes and also reduces oxidative burst.

Protein II

- It is an outer membrane protein.
- PII proteins are a series of heat-modifiable proteins that are subject to phase variation.
- They have subunit molecular weight from 24,000–30,000 daltons.
- It is also called opa (opacity) protein, because it is responsible for colony opacity.
- At least 6–8 antigenic distinct PII proteins have been described.
- It is used for intimate attachment, contributes to invasion of host cell and is involved in microcolony formation.

Protein III

- It is an outer membrane protein.
- It combines with PI and forms porins.
- It interferes with complement-mediated killing.
- Protein H8 is a recently detected antigen found in all strains. Because of its surface location and widespread distribution among all pathogenic *Neisseria*, the H8 antigen has been suggested as a possible candidate for a vaccine against both gonococcal and meningococcal infection.

LOS (Lipooligosaccharide)

- LOS is an outer-membrane lipopolysaccharide.
- It mediates bacterium–bacterium attachment.
- It elicits inflammatory response and triggers TNF-α release that causes scarring of fallopian tubes.
- It contributes to serum resistance (organism is resistant to complement and serum antibody activity).
- Changes in the length of carbohydrate portion of LPS can render the bacterium serum resistant.
- Some strains can take an activated form of *N*-acetyl muramic acid (sialic acid) from human blood and covalently attach with galactose residues on LOS (sialic acid is a host molecule. It does not activate complement).
- Serum-resistant strains produce systemic infections; serum-sensitive strains produce localized infection.

Tbp1, Tbp2 and Lbp

Organism requires iron for better growth, which is acquired from humans, with the help of Tbp1 and Tbp2 surface receptors. Lbp acquires ions from lactoferrin and is transferred to transferrin, which is utilized by *Neisseria* through Tbp1 and Tbp2. Gonococci produce IgA protease, which cleaves IgA.

PATHOGENESIS

N. gonorrhoeae is a fragile organism, which does not survive outside the human host. It is only transmitted from person to person by direct contact. It can colonize on mucosal surfaces and columnar epithelial cells of cervix and urethra. Within one hour after contact with the mucosal surface, the infection is established. During infection, gonococci are anchored to the surface of urethral cells with the help of pili. Lipopolysaccharide mediates colonization of bacteria, protein II initiates attachment, and protein I prevents phagolysosome formation. At the same time LOS triggers inflammatory response. This is responsible for the symptoms. Activities of inflammatory response lead to lysis of macrophages and form purulent discharge. Local production of TNF-α elicited by LOS is thought to be the main cause of tissue damage. Protein I makes pores on the surface of macrophage and epithelial cells that also mediate tissue damage.

Disease in Men

An unprotected male has an approximately 22% chance of acquiring gonorrhoea during intercourse with an infected women. Incubation period is about 2–8 days. The patient experiences with burning sensation on urination and yellow purulent discharge (urethral discharge) that indicates urethritis. This leads to urinary tract infection, sterility and arthritis.

Disease in Women

About 20–80% of women with gonorrhoea are asymptomatic. The organism thrives in the cervices, fallopian tubes as well as in vaginal glands and other areas of female genital tract. Symptomatic infections usually begin with painful urination and vaginal discharge. It leads to pelvic inflammatory disease (PID). Scar tissue formed as a result of infection in fallopian tubes may block normal passage of ova through the tubes causing sterility. Scarring of fallopian tube can also lead to a dangerous complication—ectopic pregnancy in which the ovum is fertilized and develops in the fallopian tube and leads to life-threatening internal haemorrhage.

Sometimes gonococcus causes disseminated gonococcal infection. The gonococcus arthritis–dermatitis syndrome is the most common manifestation. It is the result of gonococcal bacteraemia. Symptoms are fever, chills, malaise, and arthritis.

Development of typical skin lesions is a type of infection occurring in about 1% of the cases. Generally patients take medicines at this stage. If symptoms do not disappear after three days it may lead to septic joint disease. Other symptoms are endocarditis and meningitis.

Ophthalmia neonatorum is one of the eye infections that occurs during childhood. This disease is transmitted at birth from mother to infant during passage through the birth canal.

Gonococcal vulvovaginitis occurs in girls at 2–8 years of age. The alkaline pH of prepubescent vagina is cited as one factor favouring the gonococcal disease in this age group.

LABORATORY DIAGNOSIS

Sample

Urethral discharge, vaginal discharge and blood are the samples collected to diagnose gonorrhoea.

Gram-staining

The smear is gram-stained. The bean-shaped gonococci are observed inside of PMNs.

Culture

Samples can be transported with the help of Stuart's medium and inoculated on chocolate agar, Mueller-Hinton agar, Thayer-Martin agar and New York city agar medium. The plates are incubated in an environment of 5–10% CO_2 at 35°C. Small, grey, entire, sticky colonies are formed on chocolate agar.

Serological tests are also available for the diagnosis of gonorrhoea.

TREATMENT

- 1935—Sulphonamides were used for treatment.
- 1946—Most of the strains showed resistance to sulphonamides and so penicillin was introduced for treatment.
- 1957—Gonococci required high concentration of penicillin.
- 1976—Organism produced beta lactamase and became resistant to penicillin.
- 1989—CDC recommended a single dose of ceftriaxone (250 mg) intramuscularly plus doxycycline (100 mg orally for 7 days).

Erythromycin (500 mg) may be substituted for doxycycline in pregnant women.

1. Explain pathogenesis process of *N. gonorrhoeae*.
2. Explain the virulence factors of *N. gonorrhoeae.*
3. Discuss the laboratory diagnosis of *N. gonorrhoeae*.

4. Write short notes on
    i. Thayer-Martin medium
    ii. Gonococcal vulvovaginitis
    iii. Ophthalmia neonatorum
    iv. Gonococci
    v. Pelvic inflammatory disease
    vi. Ectopic pregnancy
    vii. Disseminated gonococcal infection
    viii. Gonococcal arthritis–dermatitis

Brooks, G.F. and Donegan, E.A. *Gonococcal Infection*. Edward Arnold, London. 1985.

Conde-Glez, C.J., Morse, S. and Rice, P. *et al*. (eds.). *Pathobiology and Immunobiology of Neisseriaceae*. Instituto Nacional de Salud Publica, Cuernavaca. 1994.

DeVoe I.W. "The meningococcus and mechanisms of pathogenicity." *Microbiol. Rev.* 46:162. 1982.

Holmes, K.K., Mardh, P.A. and Sparling, P.F. (eds.). *Sexually Transmitted Diseases*, 2nd edn. McGraw-Hill, New York. 1990.

Morse, S.A., Broome, C.V. and Cannon, J. *et al.* (eds.). Perspectives on pathogenic Neisseriaceae. *Clin. Microbiol. Rev*. 2:S1. 1989.

Salyers, A.A. and Whitt, D.D. *Bacterial pathogenesis: A molecular approach*. American Society for Microbiology, Washington, DC. 1994.

Centers for Disease Control and Prevention: Sexually transmitted diseases treatment guidelines. MMWR, Vol. 42 (No. RR-14), 1993.

www.health.state.ny.us/nysdoh/consumer/gonor.html

www.encarta.msn.com/index/consiseindex

http://www.umr.edu/~umrshs/gonorrhea.html

http://www.homestead.com/chrisloga/stdhealth

Syphilis

INTRODUCTION

Syphilis was an epidemic in the late fifteenth century in Europe. Previously syphilis was also called great pox because lesions of syphilis are larger than the lesions of smallpox. The disease received its present name from the poem by Fracastoro in 1530 about the afflicted shepherd, "Syphilus".

From the sixteenth century until the nineteenth century, most doctors assumed that gonorrhoea and syphilis were types of the same disease.

Only in 1837 did the eminent French venereologist Phillipe Ricord discover the specificity of the two diseases through a series of experimental inoculations from syphilitic chancres.

Ricord was also the first physician to differentiate primary, secondary and tertiary syphilis, the three stages of infection.

CAUSATIVE AGENT

Syphilis is caused by the spirochaete *Treponema pallidum*. *Treponema* is a gram-negative, thin, motile, spiral-shaped bacterium belonging to the order Spirochaetales. The name "Spirochaete" is derived from the Greek word "coiled hair". Their spiral cellular shape is approximately 16 to 18 bends consisting of an outer sheath, periplasmic space with periplasmic flagella and a peptidoglycan layer (Figure 26.1).

Figure 26.1
Treponema pallidum

The approximate size of *Treponema* is 10 μm × 0.2 μm. Therefore, dark-field microscopy, which takes advantage of the Tyndall effect, must be used to observe *Treponema* cells. The cells can also be observed microscopically after staining with specific anti-treponemal antibodies labelled with fluorescent dyes. The spirochaetes are able to swim in viscous environments (e.g. oral cavity, intestinal tract), but are only able to spin in water due to the lack of friction. Any contact with air, antiseptics or sunlight will kill the microbe.

T. pallidum requires the presence of tissue culture cells, a microaerobic environment and serum components for growth. Its fastidious nature may account for its obligate parasitism and rapid death outside the host. If kept outside the body in a moist dark place it will live not more than two hours.

This microorganism is responsible for natural disease in humans only, although rabbits can be infected experimentally. *T. pallidum* cannot be grown in cell-free cultures. Limited growth has been achieved in cultured rabbit epithelial cells, but replication is slow (doubling time, 30 hours) and can be maintained for only a few generations.

CLASSIFICATION

T. pallidum belongs to the family Spirochaetaceae (section 1, volume 1, first edition, *Bergey's Manual*). Box 26.1 describes the taxonomic position of *T. pallidum*.

Kingdom	Eubacteria
Phylum	Spirochaeta
Class	Spirochaetes
Order	Spirochaetales
Family	Treponemataceae
Genus	*Treponema*
species	*T. pallidum*
	(Schaudinn and Hoffmann, 1905)

VIRULENCE FACTORS

The outer membrane proteins are associated with adherence to the surface of host cells, and virulent spirochaetes produce hyaluronidase, which may facilitate perivascular infiltration. Virulent spirochaetes are also coated with host cell fibronectin, which can protect against phagocytosis.

DISEASE IN HUMANS

Syphilis results in the formation of lesions throughout the body. The bacteria usually enter the body during sexual intercourse, through the mucosal membranes of the vagina or urethra, but may rarely be transmitted through scars, wounds or scratches.

Bacteria may also pass from an infected pregnant women across the placenta to the developing foetus resulting in congenital syphilis. Hence, the disease is passed through infectious blood, mucosa or sexual contact.

T. pallidum tends to invade the interstitial spaces of tissue at the site of infection and move rapidly to other locations. Both of these activities are aided by the active, "cork-screw" motility of *T. pallidum*, which allows it to bore its way through tissues.

Following inoculation, *T. pallidum* penetrates intact mucous membranes or broken skin, begins to divide slowly and disseminates. Syphilis has three distinct stages and a latency stage between the second and third stages. In all these stages the principal pathogen is *T. pallidum*.

Primary Syphilis

About 10–60 days after exposure to syphilis, the area of infection is marked by the appearance of a hard, slightly elevated, round, ulcerous development called a "chancre." (French *canker*, a destructive sore). The chancre represents the primary site of initial replication. The chancre usually appears on the penis, labia, cervix, ano-rectal region or around the mouth. A chancre, however, can form anywhere on exposed skin. The chancre will heal within 4–6 weeks, even without treatment, but may leave a scar. The lymph nodes also become inflamed. Histological examination of the lesions reveal endarteritis (inflammation of the inner wall of an artery) and periarteritis (inflammation of membranous sac surrounding the heart), characteristic of syphilitic lesions at all stages, and infiltration of the ulcer with polymorphonuclear leucocytes and macrophages. Ingestion of the spirochaete by the phagocytic cells is often seen, but the organisms frequently survive.

Secondary Syphilis

One to six weeks after the healing of chancre, a pale red rash appears usually on the palms or soles of feet, but may occur over the entire body. This rash is accompanied by fever, sore throat, headache, joint pain, poor appetite, weight loss, and hair loss. Sores form around the genitals or anus that secrete extremely infectious fluids. These symptoms usually last 3–6 months, but may then re-appear at any time. Without treatment, syphilis will continue to progress throughout the body.

The Latency Stage

There are no obvious symptoms during this stage, yet *T. pallidum* will be present in the host's body, lodging itself in the host's tissue. This stage can last from a few months to a lifetime, during which the infected person tests positive for the disease. Whether this stage is contagious or not is debatable. It is generally believed that the individual is not infectious, yet still has a chance of spreading the disease to those in sexual contact with the infectious person. However, this stage almost always passes the disease to the foetus. Approximately 50–70% of carriers in this stage will not progress to tertiary syphilis.

Tertiary Syphilis

Death in adults is due to the variable occurrence of later complications of syphilis in the skin, bones, central nervous system, heart and blood vessels. Gummy lesions are also present in this stage of syphilis, and if left untreated, they frequently lead to destruction of soft tissue or bone. The pathogenesis of gummas is uncertain; what is known is that spirochaetes can be demonstrated in lesions. Lesions of the late benign syphilis sometimes occur for an extended period from as few as 2 to over 40 years from the onset of infection. Gummas of critical organs like the heart, brain and liver, can be fatal. This stage of syphilis can lead to cardiovascular syphilis, neurosyphilis or death.

Congenital Syphilis

In uterine infections it can lead to significant foetal disease, with death, multiorgan deformities, or latent infections. Most infected infants are born without clinical evidence of disease, but then develop rhinitis followed by a widespread papular rash. Late bony destruction and cardiovascular syphilis are common in untreated infants who survive the initial course of the disease. Infants can also be treated, but antibiotics may be harmful due to their effects on the normal flora that may otherwise antagonize the growth of *T. pallidum*.

EPIDEMIOLOGY

Syphilis is found worldwide, and is the third most common sexually transmitted disease in the United States (next to gonorrhoea and chlamydial infection). More than 35,000 cases of primary and secondary syphilis were reported in the United States in 1987, with an incidence of 14.6 cases per 100,000 persons between 15 to 64 years of age. In 1995, the incidence of primary and secondary syphilis in the United States declined to 6.3 cases per 100,000 persons.

LABORATORY DIAGNOSIS

Table 26.1 *Summary of syphilis diagnosis*

Diagnostic test	Method of examination
Microscopy	Dark-field illumination
	Direct immunofluorescence antibody staining (FA)
Culture	Not used
Serology	Non-treponemal VDRL, RPR
Treponemal tests	FTA-ABS, MHA-TP

Culture

Efforts to culture *T. pallidum* have been generally unsuccessful and should not be attempted.

Serology

The diagnosis of syphilis in most patients is made by serological testing. The two general types of tests are biologically non-specific (non-treponemal) tests and the specific treponemal tests.

Non-treponemal tests measure IgG and IgM antibodies (reagin antibodies) developed against lipids from damaged cells during the early stage of the disease. The antigen used for the non-treponemal tests is cardiolipin, derived from beef heart. The two tests used commonly are the venereal disease research laboratory (VDRL) test and the rapid plasma reagin (RPR) test. Both tests measure coagulation of cardiolipin antigen by the patient's serum. Both tests are rapid, although complement in serum must be inactivated for 30 minutes before the VDRL test can be performed. Only the VDRL test can be used to test cerebrospinal fluid from patients with suspected neurosyphilis.

Treponemal tests are specific antibody tests used to confirm positive reactions with the VDRL or RPR tests. The treponemal tests can be positive before the non-treponemal tests become positive in early syphilis. Treponemal tests may also remain positive when the non-specific tests revert to negative in some patients who have late syphilis. The tests most commonly used are fluorescent treponemal antibody absorption (FTA-ABS) test and the microhaemagglutination test for *T. pallidum* (MHA-TP). The MHA-TP is technically easier to perform and interpret than the FTA-ABS tests.

TREATMENT, PREVENTION AND CONTROL

Penicillin is the drug of choice; long-acting benzathine penicillin is used for the early stages of syphilis and penicillin G is recommended for congenital and late syphilis. Tetracycline, erythromycin and chloramphenicol can be used as alternative antibiotics for patients allergic to penicillin. Only penicillin or chloramphenicol can be used for patients with neurosyphilis.

1. Why was the name syphilis given?
2. Describe the characters of *T. pallidum.*
3. How do you demonstrate *Treponema* and its antigens from clinical specimens?
4. Describe the pathological features of *T. pallidum.*
5. Explain the various features of syphilis.
6. Write short notes on
   i. FTA–ABS
   ii. *Treponema pallidum*
   iii. RPR

iv. Congenital syphilis
v. Gumma lesions
vi. VDRL
vii. Chancre

1. Why is there no artificial media for the cultivation of *T. pallidum*?

Chiu, M.J., Cockerell, C.J. and Houpt, K.R., *et al.* "Spirochaetal infections of the skin." In: *Atlas of Infectious Diseases.* Mandell, G.E. and Stevens, D.L. (eds.) Churchill Livingstone, Philadelphia. 1994.

Crissey, J.T. and Denenholz, D.A. "Syphilis." *Clin. Dermatol.* 2:1. 1984.

Dunn, R.A. and Rolfs, R.T. "The resurgence of syphilis in the United States." *Curr. Opin. Infect. Dis.* 4:3. 1991.

Jackman, J.D. and Radolf, J.D. "Cardiovascular syphilis." *Am. J. Med.* 87:425. 1989.

Koff, A.B. and Rosen, T. Nonvenereal Treponematoses: yaws, endemic syphilis, and pinta. *J. Am. Acad. Dermatol.* 29:519. 1993.

Larsen, S.A., Steiner, B.M. and Rudolph, A.H. Laboratory diagnosis and interpretation of tests for syphilis. *Clin. Microbiol. Rev.* 8:1. 1995.

Lukehart, S.A. and Holmes, K.K. "Syphilis." In: *Harrison's Principles of Internal Medicine* (Braunwald, E. *et al.* (eds.). McGraw-Hill Book Company, New York. 1994.

Miller, J.N. "Value and limitations of non-treponemal and treponemal tests in the laboratory diagnosis of syphilis." *Clin. Obstet. Gynecol.* 18:191. 1975.

Musher, D.M. Biology of *Treponema pallidum*. In: *Sexually Transmitted Diseases*, 2nd edn. Holmes, K.K. *et al.* (eds.). McGraw-Hill Book Company, New York. 1990.

dermatlas.med.jhmi.edu/derm/result.cfm?Diagnosis=

dpalm.uth.tmc.edu/treponema/tpall.html

microbewiki.kenyon.edu/index.php/Treponema

www.4woman.gov/faq/stdsyph.htm -

www.emedicine.com/EMERG/topic563.htm

www.emedicine.com/med/topic2224.htm

www.gsbs.utmb.edu/microbook/ch036.

www.kidshealth.org/parent/infections/std/syphilis.

www.nlm.nih.gov/medlineplus/ency/article/001327

www.nlm.nih.gov/medlineplus/syphilis.html

www.phac-aspc.gc.ca/msds-ftss/msds154e.html

www.tigr.org/tdb/CMR/gtp/htmls/Background.html...

www.bacteriamuseum.org/species/Tpallidum.shtml

www.wadsworth.org/databank/treponem.

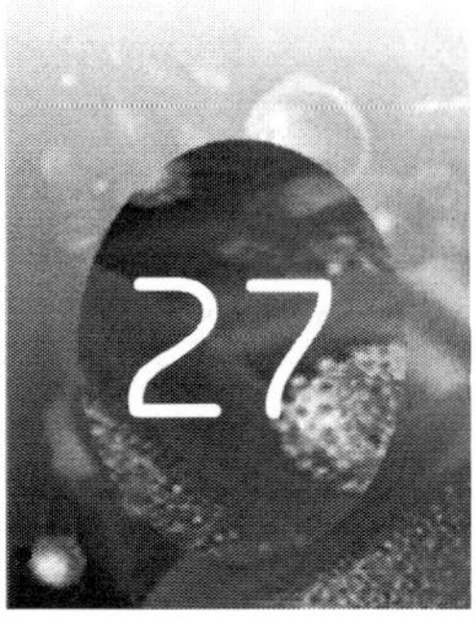

Gas Gangrene

INTRODUCTION

Clostridial wound infections may be divided into three categories: gas gangrene or clostridial myonecrosis, anaerobic cellulitis and superficial contamination. Oakky in 1954 has stated that gas gangrene can have a rapidly fatal outcome, dermataceous myonecrosis and requires prompt, often severe, treatment.

Bacteriology of gas gangrene is varied. Rarely it is due to infections with single species of clostridia. In gas gangrene and anaerobic cellulitis, the primary pathogen can be any one of various clostridial species including *C. perfringens*, *C. novyi*, *C. septicum* and occasionally, *C. bifermentans, C. histolyticum*, or *C. fallax*.

Characteristics of C. *perfringens*

- They are usually shorter gram-positive rods that seldom sporulate *in vitro*.
- Capsules may be observed by direct observation of smear from wounds but are not uniformly demonstrable in culture.
- Spores are central or subterminal but are rarely seen in artificial cultures.
- It is a non-motile, aerotolerant anaerobe.
- On blood agar plate, a double zone of haemolysis is produced.
- In litmus milk, fermentation of lactose leads to formation of acid. The acid coagulates the casein and the clotted milk is disrupted due to the vigorous gas production (stormy fermentation) (historical).
- It produce invasive enzymes like proteases, DNases, collagenases, etc.
- Alpha toxin is the most important toxin (lecithinase, phospholipase C). It is synthesized by phage-encoded genes.
- Lecithinase converts lecithin into diglyceride and phosphorylcholine.
- Lecithin is a major lipid in cell membranes.
- Lecithinase causes capillary endothelial cell leakage, oedema and myonecrosis.

PATHOGENESIS

All clostridial wound infections occur in an anaerobic tissue environment caused by an impaired blood supply secondary to trauma, surgery, foreign bodies or malignancy.

Contamination of the wound by clostridia from the external environment or from the host's normal flora causes the infection. Gas gangrene is an acute disease with a poor prognosis and often fatal outcome. Initial trauma to host tissue damages muscle and impairs blood supply. This lack of oxygenation causes the oxidation–reduction potential to decrease and allows the growth of anaerobic clostridia. Initial symptoms are generalized fever and pain in the infected tissue. As the clostridia multiply, various exotoxins (including haemolysins, collagenases, proteases and lipases) are liberated into the surrounding tissue, causing more local tissue necrosis and systemic toxaemia. Infected muscle is discoloured (purple mottling) and oedematous and produces a foul-smelling exudate; gas bubbles form from the products of anaerobic fermentation. As capillary permeability increases, the accumulation of fluid increases, and venous return eventually is curtailed. As more tissue becomes involved, the clostridia multiply within the increasing area of dead tissue, releasing more toxins into the local tissue and the systemic circulation. Because ischaemia plays a significant role in the pathogenesis of gas gangrene, the muscle groups most frequently involved are those in the extremities served by one or two major blood vessels.

Symptoms

Bloody exudate, oedema, muscle liquefaction. Gases (CO_2/H_2)—Gas gangrene. Lack of PMN due to ischaemia.

CNS manifestation

Delerium and disorientation are the common CNS manifestation.

HOST DEFENCES

Host defence against gas gangrene and other clostridial wound infections are mostly ineffective. Even repeated episodes of clostridial wound infection do not seem to produce effective immunity.

LABORATORY DIAGNOSIS

Biopsy from wounds form the sample. Primary diagnosis is made by observing gram-positive rods along with clinical presentation.

Diagnosis of clostridial wound infections is based on clinical symptoms coupled with gram-staining and bacterial culture of clinical specimens. Characteristic lesions and the presence of large numbers of gram-positive bacilli (with or without spores) in a wound exudate provide strong presumptive evidence. A commonly used laboratory test for presumptive identification of *C. perfringens* is the Nagler reaction which detects the presence of alpha toxin (phospholipase C), one of the most prominent toxins produced by *C. perfringens*.

On egg yolk media, *Clostridium* produces a halo of diglyceride from alpha toxin (Nagler–egg yolk with antitoxin), on blood agar plate (BAP), a double zone of haemolysis due to several haemolysins (one is alpha toxin).

Robertson's cooked meat medium is also used for cultivation. *C. perfringens* produce proteolytic reaction (black).

CONTROL

Correction of the anaerobic conditions combined with antibiotic treatment form the basis for therapy. Penicillin is the drug of choice for all clostridial wound infections; Chloramphenicol is a second-choice antibiotic. Prevention of wound contamination is the most important factor in controlling clostridial wound infections.

1. How do you control clostridial wound infection?
2. Explain the diagnosis of gas gangrene.
3. Describe the pathophysiology of gas gangrene.
4. Write short notes on
   i. Nagler reaction
   ii. *Clostridium perfringens*
   iii. Toxins of *C. perfringens*

Adelman *et al*. (2006). *Current Medical Diagnosis & Treatment*, 45th edn. McGraw-Hill.

Ryan, K.J. and Ray, C.G. (eds.) (2004). *Sherris Medical Microbiology*, 4th edn. McGraw-Hill.

Warrell *et al*. (2003). *Oxford Textbook of Medicine*, 4th edn. Oxford University Press.

Wells, C.L. and Wilkins, T.D. (1996). "Clostridia: Sporeforming Anaerobic Bacilli." In: *Barron's Medical Microbiology*, 4th edn. Univ. of Texas Medical Branch.

eMedicine - Clostridial Gas Gangrene : Article by Don R Revis, Jr, MD

http://en.wikipedia.org/wiki/Clostridium_perfringens"

microbewiki.kenyon.edu/index.php/Clostridium -

seafood.ucdavis.edu/haccp/compendium/chapt13.htm -

www.agen.ufl.edu/~foodsaf/sf205.html - 6k

www.azdhs.gov/phs/edc/edrp/es/clostridiumpf.html
www.emedicine.com/med/topic394.html
www.extension.iastate.edu/foodsafety/pathogens
www.foodlink.org.uk
www.inspection.gc.ca/english/fssa/concen/cause/perfrine.shtml
www.merck.com/mmhe/sec09/ch122
www.uky.edu/ag/foodscience

Tetanus

INTRODUCTION

The word "Tetanus" comes from the Greek word *tetanos*, which means "to stretch". This disease originates from wounds and selectively impairs the activity of inhibitory chemicals that control the nerve impulse conduction. This disease is worldwide in distribution. Tetanus has been known from very ancient times, having been described by Hippocrates.

Arthur Nicholaire discovered the tetanus bacillus in 1884. The organism was isolated in pure form by Kitasato in 1889.

Tetanus results from the infection of a wound or raw surface with tetanus bacilli. Most frequently this disease follows injury. Puncture wounds are particularly vulnerable as they favour the growth. Rarely it may follow surgical operation, usually due to lapses in asepsis. Sometimes the disease may be due to suppurations, such as otitis media. It is one of the complications in septic abortion. Tetanus may also be caused by unsterile injection.

It was a serious disease with a high rate of mortality, 80–90% death observed before the start of treatment. Even with proper treatment the case fatality rate varies from 15–50%.

Tetanus is more common in the developing countries with warm climate and rural areas where the soil is fertile and highly cultivated. In India, tetanus is estimated to be 4th commonest cause of death. It is worldwide in distribution. Non-immunized rural population develops tetanus at higher rates than urban population. The disease is very low in well-developed countries.

CAUSATIVE AGENT

- It is caused by *Clostridium tetani* (Figure 28.1).
- It is a quite long and thin organism.

- It is gram-positive and produces terminal spore, so it appears like a drumstick (Figures 28.2 and 28.3).
- It is motile by peritrichous flagella.
- It is an obligate anaerobe.
- Optimum temperature and pH for growth are 37°C and 7.4.
- The organism is moderately fastidious requiring vitamins and amino acids for growth.
- It is non-capsulated.
- It grows in cooked meat medium with turbidity and some gas formation within 48 hours; there is no digestion of meat and black colour is formed during prolonged incubation.
- It does not ferment any carbohydrate.
- In horse blood agar, alpha haemolysis is observed first, then beta haemolysis is observed due to prolonged incubation and production of haemolysins.
- The organism is indole-positive, methyl red-negative, Voges-Proskauer-negative, and liquefies gelatin very slowly.
- It is resistant to disinfections and withstands heat.
- Flagellar and somatic antigens have been demonstrated. Ten types of organisms were identified on the basis of flagellar antigens and have a single type of somatic antigen. Type IV is non-flagellated.
- The organism produces 2 distinct toxins:
  i. Tetanolysin

     It disturbs the activity of RBCs and WBCs.

     It is oxygen-labile and heat-labile.

  ii. Tetanospasmin
     - It is responsible for the symptoms of tetanus.

Terminal spore

Figure 28.1 *Clostridium tetani*

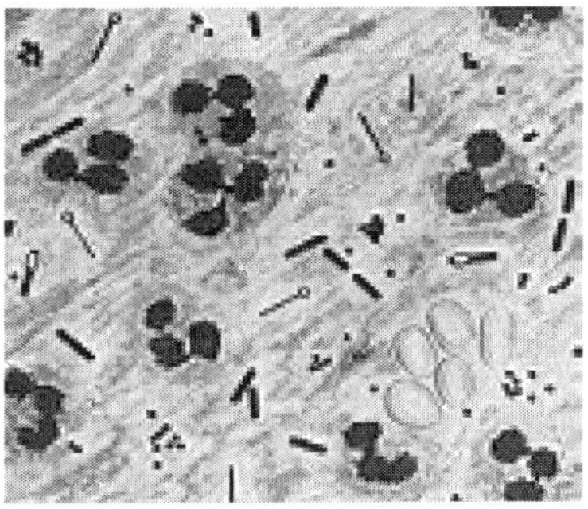

Figure 28.2 *Clostridium* in gram-stained pus

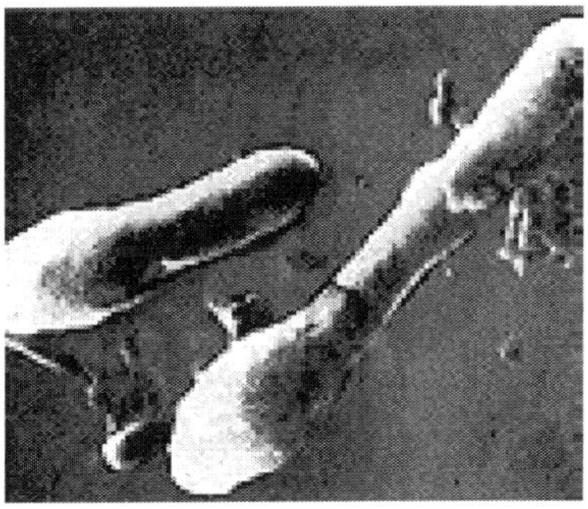

Figure 28.3 Electron microscopic view of terminal spore of *C. tetani*

- Toxins are released during autolysis of cell.
- Heat-labile protein is inactivated by heating for 20 minutes at 60°C.
- Its molecular weight is about 150,000 daltons.
- It has two non-identical polypeptide chains having molecular weight of 93,000 and 52,000 daltons both held together by disulphide bridges (Figure 28.4).
- It selectively binds to ganglioside.

Ganglioside-binding region is found on heavy chain region whereas light chain is responsible for toxic activity.

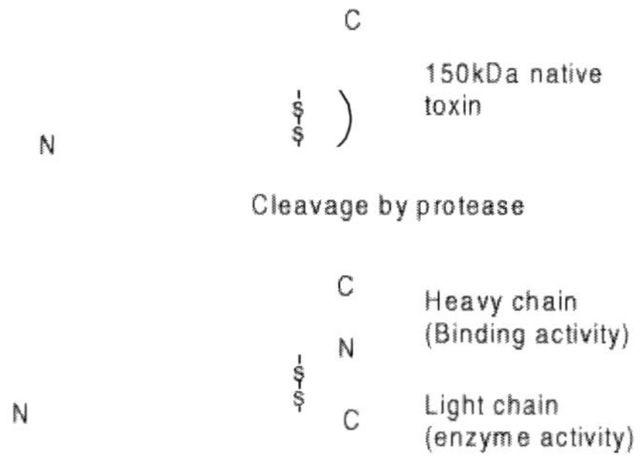

Figure 28.4 Tetanospasmin

CLASSIFICATION

Kingdom	Bacteria
Phylum	Firmicutes
Class	Clostridia
Order	Clostridiales
Family	Clostridiaceae
Genus	*Clostridium*
species	*C. tetani, C. perfringens*

PATHOGENESIS

- The spores of *C. tetani* are ubiquitous in nature. They have been found in 20–64% of soil samples. Spores are generally implanted at the site of contamination. Under favourable conditions the bacilli germinate.
- Organism alone does not cause disease. It has little invasive power. It requires the help of anaerobic bacterium, which is responsible for removal of oxygen and

creation of anaerobic environment. After implantation of spores into an appropriate environment it causes localized infection and later, generalized infection. After a specific incubation period (4–10 days) toxins are released from the cell by autolysis (Figure 28.5).

- Toxin spreads throughout the body and causes various complications. The toxins are taken up by internalization and transported intra-axonally. When it reaches the region of nucleus, it is transported to the interneurons where it inhibits the inhibitory transmitters.

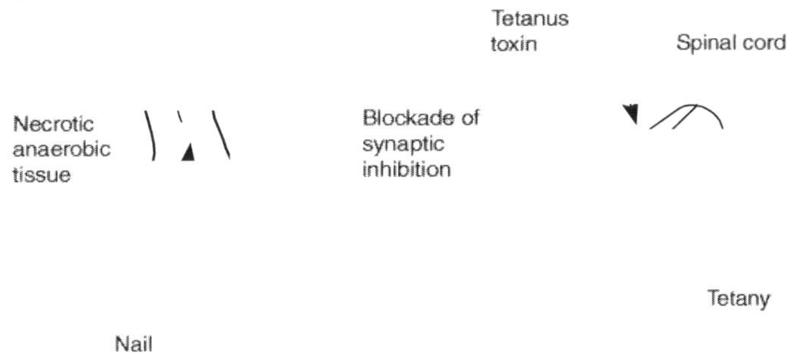

Figure 28.5 Role of tetanus toxin

CLINICAL MANIFESTATION

The earliest manifestation is muscle stiffness followed by the spasm of the master muscles, lockjaw. As the disease progresses, it causes clenching of the jaw, arching of the back, flexion of the arms, hypertension, hypotension, cardiac disturbance, respiratory problems and bone fractures.

EPIDEMIOLOGY

C. tetani can be isolated from the soil in almost every environment throughout the world. The organism can be found in the gastrointestinal flora of humans, horses and other animals. Isolation of *C. tetani* from the intestinal flora of horses, coupled with the high frequency of equine tetanus, led to the erroneous assumption that the horse was the animal reservoir of *C. tetani*.

LABORATORY DIAGNOSIS

Wound exudate and pus form the sample.

Microscopic examination is useful to demonstrate the presence of bacilli. Gram-staining is performed. Cultivation is with the help of Robertson's cooked meat medium. On this medium *C. tetani* produce saccharolytic reaction.

Based on the saccharolytic and proteolytic reactions, clostridia are differentiated.

Saccharolytic—formation of red colour

Proteolytic—formation of black colour

CONTROL

The nature of prophylaxis depends largely on the type of wound and immune status of the patient. The available methods are surgical attention, antibiotics and immunization.

Surgical prophylaxis aims at removal of foreign bodies, necrotic tissue and blood clots to avoid providing an anaerobic environment favourable for tetanus bacillus.

Injections of tetanus toxoid are prophylactic. Currently, booster doses are recommended only every 10 years by the CDC. An antibody titre above 0.01 international units (IU) per ml is usually considered protective. Human tetanus immunoglobulin (HTIG) in a dose of 250 IU intramuscularly is administered.

Routine immunization with tetanus toxoid should begin at 1–3 months of age with DPT vaccine. Three doses of DPT is given at intervals of 3–4 weeks, with booster doses 1 and 4 years later. Immunity to tetanus can be maintained by a single booster dose of toxoid every 10 years.

1. Describe the pathogenesis and laboratory diagnosis of tetanus.
2. How do you differentiate *C. tetanis* from *C. perfringens.*
3. Write short notes on
   i. Tetanopasmin
   ii. *C.tetani*
   iii. Prophylaxis of tetanus

Wells, C.L. and Wilkins, T.D. (1996). Clostridia: Sporeforming Anaerobic Bacilli." In: *Baron's Medical Microbiology,* 4th edn., Univ of Texas Medical Branch.

"Diphtheria, tetanus, and pertussis: recommendations for vaccine use and other preventive measures. Recommendations of the Immunization Practices Advisory Committee (ACIP).". *MMWR Recomm Rep* 40 (RR-10): 1–28 1991.
World Health Organization 1998 Maternal and Neonatal Tetanus
www.babybag.com/articles/cdc_tetn.html
www.bbc.co.uk/health/conditions/tetanus2.shtml
www.emedicinehealth.com/tetanus/article_em.html
www.fda.gov/Fdac/features/696_tet.html
www.health.state.ny.us/diseases/communicable/tetanus/fact_sheet.html
www.nlm.nih.gov/medlineplus/ency/article/000615.html
www. tetanus.org/
www.who.int/topics/tetanus/en
www2.ncid.cdc.gov/travel/yb/utils/
www.cyberhorse.net.au/csl/tetanus.html

Leptospira

INTRODUCTION

Leptospira are very thin, tightly coiled, obligate aerobic spirochaetes characterized by a unique flexuous type of motility. The genus is divided into two species: the pathogenic leptospire *L. interrogans* and the free-living leptospire *L. biflexa.* Serotypes of *L. interrogans* are the agents of leptospirosis, a zoonotic disease. The primary hosts for this disease are wild and domestic animals and the disease is a major cause of economic loss in the meat and dairy industries. Humans are accidental hosts. The first human case of leptospirosis was described in 1886 as a severe icteric illness and was referred to as Weil's disease .

CHARACTERISTICS

Leptospira has the general structural characteristics that distinguish spirochaetes from other bacteria (Figure 29.1). The cell is encased in a three- to five-layer outer membrane or envelope. Beneath this outer membrane are the flexible, helical

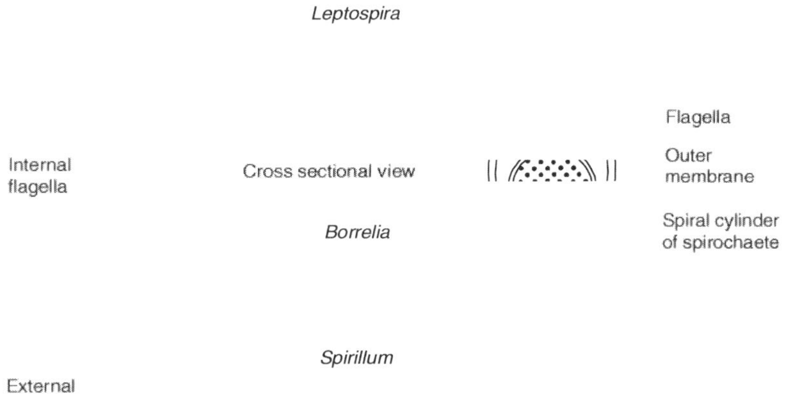

Figure 29.1 Morphological comparison of *Leptospira, Borrelia,* and *Spirillum*

peptidoglycan layer and the cytoplasmic membrane; these encompass the cytoplasmic contents of the cell. The structures surrounded by the outer membrane are collectively called the protoplasmic cylinder. An unusual feature of the spirochaetes is the location of the flagella, which lie between the outer membrane and the peptidoglycan layer. They are referred to as periplasmic flagella. Leptospires are too slender to be visualized in the bright-field microscope but are clearly seen by dark-field or phase-contrast microscopy. *Leptospira* differs from other spirochaetes in lacking glycolipids and having diaminopimelic acid rather than ornithine in its peptidoglycan.

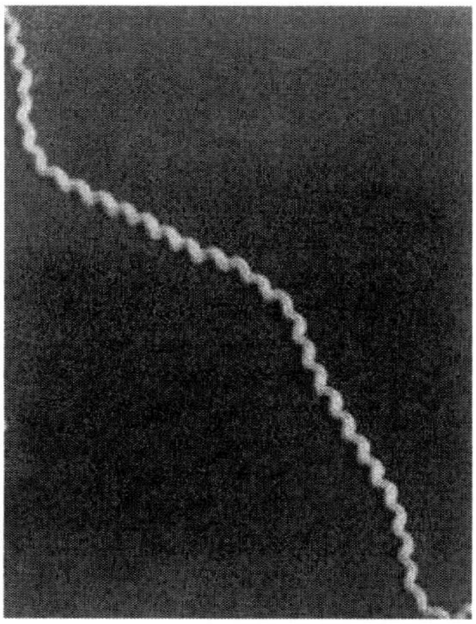

Figure 29.2 Electron micrograph of *Leptospira interrogans* serovar *icterohaemorrhagiae*

The leptospires are most readily cultivated in medium. Long-chain fatty acids and vitamins B_1 and B_{12} are the only organic compounds known to be necessary for growth of leptospires. When cultivated in media with pH 7.4 at 30°C, their average generation time is about 12 hours. Aeration is required for maximal growth. They can be cultivated in plates containing soft (1 per cent) agar medium, in which they form primarily subsurface colonies.

PATHOGENESIS

The mucosa and broken skin are the most likely sites of entry for the pathogenic leptospires. A generalized infection leads to bacteraemia. It occurs during the acute, leptospiraemic phase of the disease. The host responds by producing antibodies and it kills leptospira in combination with complement. Leptospires surviving in the brain, kidney and eyes multiply slowly and are shed in the urine (the leptospiruric phase). The leptospires may persist in the host for weeks to months; in rodents they may be shed in the urine for the lifetime of the animal. Leptospiruric urine is the vehicle of transmission of this disease.

The mechanism by which leptospires cause disease remains unresolved, as neither endotoxin nor exotoxins have been associated with them. Damage to the endothelial lining of the capillaries and subsequent interference with blood flow appear responsible for the lesions associated with leptospirosis. The most notable feature of severe leptospirosis is the progressive impairment of hepatic and renal function. Renal failure is the most common cause of death. The lack of substantial cell destruction in leptospirosis is reflected in the complete recovery of hepatic and renal function in survivors.

CLINICAL MANIFESTATION

The three organ systems most frequently involved are the central nervous system, kidneys, and liver. After an average incubation period of 7 to 14 days, the leptospiraemic acute phase is evidenced by abrupt onset of fever, severe headache, muscle pain, and nausea; these symptoms persist for approximately 7 days. Jaundice occurs during this phase in more severe infections. The more severe form of leptospirosis is frequently associated with infections having the serotype icterohaemorrhagiae and is often referred to as Weil's disease.

EPIDEMIOLOGY

Leptospirosis is a worldwide zoonosis with a broad spectrum of animal hosts. The primary reservoir hosts are wild animals such as rodents, which can shed leptospires throughout their lifetime. Domestic animals are also an important source of human infections. Direct or indirect contact with urine containing virulent leptospires is the major means by which leptospirosis is transmitted. Urine-contaminated soil can remain infective for as long as 14 days. In humans, leptospirosis has occurred in an infant being breast-fed by a mother with the disease.

LABORATORY DIAGNOSIS

The microscopic agglutination test is most frequently used for serodiagnosis. The organisms can be isolated from blood or urine on commercially available media, but the test must be requested specifically because special media are needed. Isolation of the organisms confirms the diagnosis.

CONTROL

Human leptospirosis can be controlled by reducing its prevalence in wild and domestic animals. Doxycycline has been used successfully as a chemoprophylactic agent for military personnel training in tropical areas.

Johnson, R.C. (ed.). *The Biology of Parasitic Spirochetes*. Academic Press, San Diego. 1976.

Johnson, R.C. (ed.). "Lyme disease." *Rheum. Dis. Clin. North Am*. 15:627. 1989.

Kaufmann, A.E. and Weyant, R.S. "Leptospiraceae." In Murray, P.R., Baron, E.J., Pfaller, M.A, Tenover, F.C. and Yolken, R.H. (eds.). *Manual of Clinical Microbiology*, 6th edn. American Society for Microbiology, Washington DC. 1995.

Rahn, D.W. and Malawista, S.E. Lyme disease: Recommendations for diagnosis and treatment. *Ann. Int. Med.* 114:472. 1991.

Schwan, T.G., Burgdorfer, W. and Rosa, P.A. *Borrelia.* In Murray P.R., Baron, E.J., Pfaller, M.A., Tenover, F.C. and Yoken, R.H. (eds.). *Manual of Clinical Microbiology*, 6th edn. American Society for Microbiology, Washington DC, 1995.

Steere, A.C., Grodzicki, R.L. and Kornblatt, A.N. *et al.* "The spirochaetal etiology of Lyme disease." *N. Engl. J. Med.* 308:733. 1983.

www.zoologix.com/dogcat/Datasheets/Leptospira.html

www.ijmm.org/article.asp

www.icmr.nic.in/ijmr/2006/November/1112

www.cehs.siu.edu/fix/medmicro/lepto.html

www.answers.com/topic/leptospira

www.leptospirosis.org/bacteria/serovars.php

www.med.monash.edu.au/microbiology/staff/adler/leptolab.html

www.bacterio.cict.fr/l/leptospira.html

medinfo.ufl.edu/year2/mmid/bms5300/bugs/leptint.html

1. Explain how leptospira is cultivated.
2. Leptospirosis—is it a zoonotic infection? Explain.
3. Explain how rat urine is responsible for the transmission of leptospirosis.
4. Explain various features of *Leptospira* and its infection.
5. Write short notes on
   i. Weils disease
   ii. Periplasmic flagella

Borrelia

INTRODUCTION

Borrelia has morphological characteristics similar to those of *Leptospira,* except that cells average 0.2 to 0.5 μm by 4 to 18 μm and have fewer coils (Figure 30.1). Seven to twenty periplasmic flagella originate at each end and overlap at the centre of the cell. *Borrelia* contains ornithine whereas diaminopimelic acid is found in *Leptospira*.

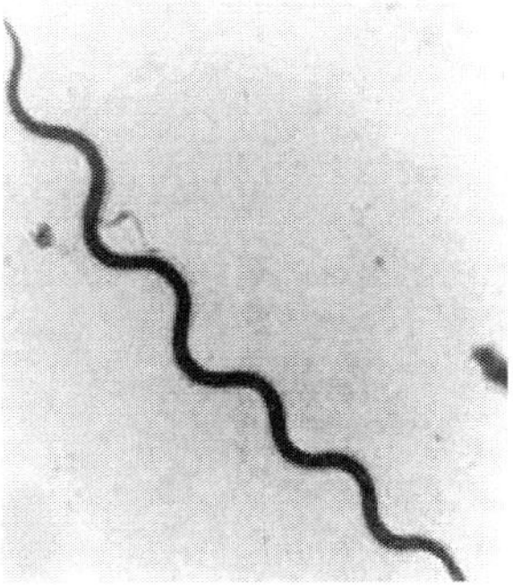

Figure 30.1 Electron micrograph of Borrelia

The nutritional requirements of borreliae are more complex than those of leptospires. Glucose, amino acids, long-chain fatty acids, *N*-acetyl glucosamine, and several vitamins are some of their required organic nutrients. The borreliae are microaerophilic organisms. *Borrelia hermsii* has a generation time of 12 hours when cultivated in artificial media at 35°C compared with only 6 to 10 hours in the mouse.

Borrelia species are responsible for relapsing fever and Lyme disease. The organisms are transmitted to humans primarily by lice or ticks. Relapsing fevers are acute recurrent illnesses characterized by febrile episodes that recede spontaneously but generally reappear with decreasing intensity and duration. *Borrelia recurrentis* is responsible for the louse-borne or epidemic type of relapsing fever with humans serving as the reservoir host. Lyme disease is another tick-borne illness and is caused by *B. burgdorferi.*

Borreliae must rely on an insect vector to transmit the organisms through the epidermis. The site of entry is usually not prominent as the organisms are not clinically recognized until they enter the blood. The mechanisms by which they reach the bloodstream are unknown. The relapses are due to the ability of borreliae to undergo multiple cyclic antigenic variations. As antibodies for the predominant antigenic type multiplying within the host appear, these organisms "disappear" from the peripheral blood and are replaced by a different antigenic variant within a few days. This process may occur several times in an untreated host, depending on the infecting *Borrelia* strain.

CLINICAL MANIFESTATION

The mechanism by which borreliae cause Lyme disease has not been elucidated. Early Lyme disease is characterized by an expanding annular red rash, erythema migrans, the characteristic "bull's-eye" rash with central clearing. It is frequently accompanied by fever, fatigue, headache, and muscle and joint pain. Arthritis, neuritis, and carditis may also be present during early Lyme disease. Persistent neurologic and arthritic infections (lasting for months to years) may occur in some patients.

EPIDEMIOLOGY

Borrelia is transmitted to humans by the body louse or ticks. *Borrelia recurrentis,* the cause of louse-borne (epidemic) relapsing fever, is carried by the human louse *Pediculus humanus*. *Borrelia burgdorferi,* the agent of Lyme disease, is transmitted by members of the *Ixodes ricinus* complex (hard ticks).

LABORATORY DIAGNOSIS

Diagnosis is based primarily on demonstration of the spirochaetes in blood during febrile episodes by dark-field examination, use of stained blood smears or mouse inoculation. Antibody detection by indirect immunofluorescence assay is available.

CONTROL

Relapsing fevers and Lyme disease are prevented by avoiding the vectors. Early Lyme disease can be effectively treated with oral tetracyclines and semisynthetic penicillins. Arthritic and neurologic disorders are treated with high-dose intravenous penicillin G or ceftriaxone. Patients who have failed to respond to penicillin or tetracycline therapy have been effectively treated with ceftriaxone.

1. Differentiate *Borrelia* from *Leptospira*.
2. Explain the characteristic symptoms produced by *Borrelia.*
3. Explain *Borrelia* is transmitted in between hosts.
4. Differentiate lyme disease from relapsing fever.

Burgdorfer, W. Barbour, A.G. and Hayes, S.F. *et al*. "Lyme disease: a tick-borne spirochaetosis." *Science*. 216:1317. 1982.

Felsenfeld, O. *Borrelia: Strains, Vectors, Human and Animal Borreliosis*. Warren Green, St. Louis. 1971.

G. Postic, D. and Saint Girons, I. *et al.* "Delineation of *Borrelia burgdorferi* sensu stricto, *Borrelia garinii* sp nov and group VS 461 associated with Lyme borreliosis." *Int. J. Syst. Bacteriol.* 42:378. 1992.

Johnson, R.C. (ed.). "Lyme disease". *Rheum. Dis. Clin. North Am.* 15:627. 1989.

Johnson, R.C. (ed.). *The Biology of Parasitic Spirochaetes*. Academic Press, San Diego. 1976.

Kaufmann, A.E. and Weyant, R.S. "Leptospiraceae." In Murray, P.R., Baron, E.J., Pfaller, M.A., Tenover, F.C. and Yolken, R.H. (eds.). *Manual of Clinical Microbiology*, 6th edn. American Society for Microbiology, Washington DC. 1995.

Rahn, D.W. and Malawista, S.E. "Lyme disease: recommendations for diagnosis and treatment." *Ann. Int. Med.* 114:472. 1991.

Schwan, T.G., Burgdorfer, W. and Rosa, P.A. *Borrelia.* In Murray, P. R., Baron, E. J., Pfaller, M.A., Tenover, F.C. and Yoken, R.H. (eds.). *Manual of Clinical Microbiology*, 6th edn. American Society for Microbiology, Washington DC. 1995.

Steere, A.C., Grodzicki, R.L. and Kornblatt, A.N. *et al.* "The spirochaetal etiology of Lyme disease." *N. Engl. J. Med* 308:733. 1983.

es.wikipedia.org/wiki/Borrelia

medind.nic.in/iau/t04/i2/iaut04i2p92.pdf

www.arupconsult.com/Topics/Infectious_Disease/Bacteria/Borrelia.html

www.biomedcentral.com/1471-2180/6/95.bioinfo.hku.hk/LeptoList/ -

www.infecto.edu.uy/revisiontemas/tema25/B.burdorferi.html

www.ncbi.nlm.nih.gov/entrez/query.fcgi

www.pbrc.hawaii.edu/microangela/lepto.html

www.ucmp.berkeley.edu/bacteria/bacteria.html

www.vetcontact.com

Helicobacter pylori

INTRODUCTION

Helicobacter is widely distributed in the animal kingdom. They have been known as animal pathogens for nearly 100 years. However, because they are fastidious and slow-growing in culture, they have been recognized as human gastrointestinal pathogens only during the 1990s. They can cause chronic superficial gastritis, peptic ulcer disease and can lead to gastric carcinoma.

CHARACTERISTICS

H. pylori are microaerophilic, non-sporulating, gram-negative curved rods, 3.5 μm long and 0.5 to 1 μm wide, with a spiral periodicity in fresh cultures and spherical (coccoid) forms present in older cultures. *H. pylori* have multiple, polar, sheathed flagellae (Figure 31.1), a unique composition of cell wall fatty acids, and a smooth surface. *H. pylori* produces urease and does not grow when incubated below 30°C. Growth is best on chocolate or blood agar plates after incubation for 2 to 5 days. Either a blood or a haemin is needed for better growth on liquid media.

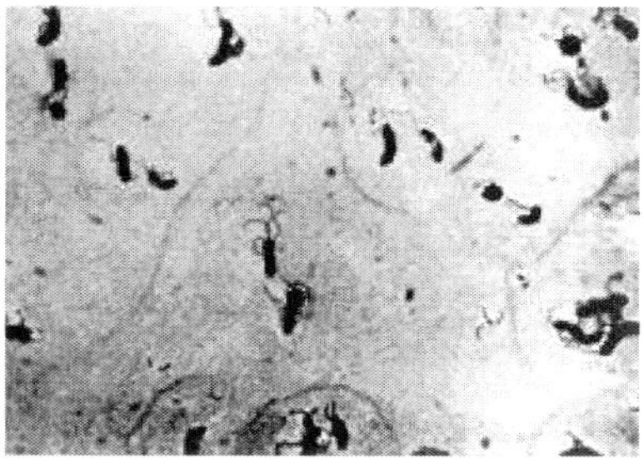

Figure 31.1 *H. pylori* showing typical thin, comma- or S-shaped forms

Pathogenesis

Helicobacter pylori are readily killed by brief exposure to hydrochloric acid solutions with pH below 4.0. This is paradoxical for an organism whose primary residence is the gastric lumen. However, several factors may explain this phenomenon. Firstly, *H. pylori* lives in the mucous layer overlaying the gastric mucosa, a niche protected against gastric acid. This mucus is relatively thick and viscous and maintains a pH gradient from approximately pH 2 adjacent to the gastric lumen to pH 7.4

immediately above the epithelial cells. Secondly, *H. pylori* is among the most efficient producers of urease. An important effect of this metabolic activity is the release of ammonia, which buffers acidity. Thirdly, *H. pylori* is highly motile even in very viscous mucus. This motility may allow the organism to migrate to the most favourable pH gradient. *Helicobacter* does not invade tissue but it produces extracellular products such as exotoxins. An 87-kDa cytotoxin that induces vacuolation of eukaryotic cells is expressed *in vitro* by about 50% of strains. Proteolytic activity affects the ability of mucus to retard diffusion of hydrogen ions. Mucus depletion over inflamed tissues is characteristic of *H. pylori*-associated gastritis. Ammonia, produced by urease, is known to be toxic to eukaryotic cells and may potentiate mucosal injury.

CLINICAL MANIFESTATION

Helicobacter pylori has repeatedly been shown to be associated with chronic superficial gastritis (CSG), which involves the antrum and the fundus of the stomach. Chronic gastritis is a well known risk factor for the development of gastric carcinoma.

EPIDEMIOLOGY

H. pylori is found worldwide and affects persons from diverse socioeconomic groups. *H. pylori* has been isolated from dental plaque, and DNA products may be detected in saliva by PCR. *H. pylori* has been isolated from faeces. These data indicate potential routes of transmission of *H. pylori*.

DIAGNOSIS

H. pylori can be presumptively identified in freshly prepared gastric biopsy smears by phase-contrast microscopy, based on the characteristic motility of the microorganisms, and by staining histologic sections from gastric biopsies with Gram (carbol fuchsin

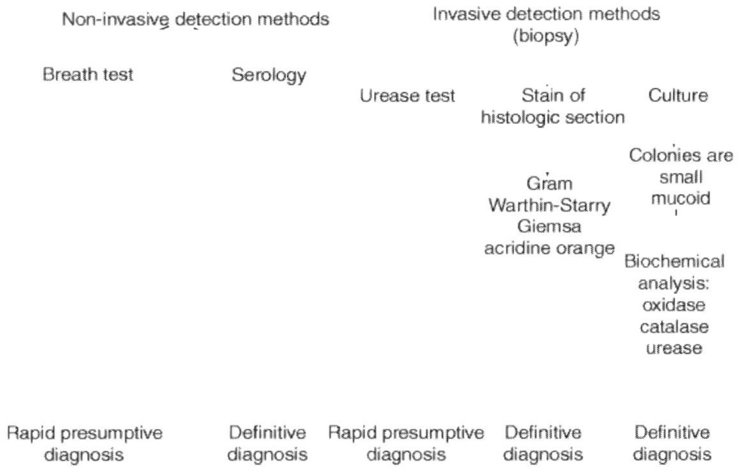

Figure 31.2 Detection methods for *H. pylori*

counterstain), Warthin–Starry silver, Giemsa, or acridine orange stains. *H. pylori* may be isolated from gastric tissue or from biopsies of oesophageal or duodenal tissue containing gastric metaplasia (Figure 31.2) using non-selective media, such as chocolate agar, or antibiotic-containing selective media, such as those of Skirrow or Goodwin. Spiral organisms that are oxidase-, catalase-, and urease-positive can be identified as *H. pylori*.

All of the above tests require endoscopy and biopsy. A non-invasive technique known as the urea breath test has been developed to diagnose *H. pylori* infection.

CONTROL

Antimicrobial therapy for treatment of this infection has emerged as the most important means to resolve *H. pylori* infection. Antimicrobial therapy is now one of the primary therapies for duodenal ulceration.

1. *Helicobacter pylori* is an acid-surviving bacteria? Is it true?
2. How does *H. pylori* overcome stomach acid for survival.
3. Describe the character of *H. pylori*.
4. Explain the pathogenic properties of *H. pylori*.
5. Does peptic ulcer lead to cancer? Explain.

Blaser, M.J. and Parsonnet, J. "Parasitism by the "slow" bacterium *Helicobacter pylori* leads to altered gastric homeostasis and neoplasia." *J. Clin. Invest*. 94: 4–8. 1994.
Dooley, C.P., Cohen, H. and Fitzgibbons, P.L. *et al*. "Prevalence of *Helicobacter pylori* infection and histologic gastritis in asymptomatic persons." *N. Engl. J. Med*. 321: 1562. 1989.
Goodwin, C.S., Armstrong, J.A. and Chilvers, T. *et al.* "Transfer of *Campylobacter pylori* and *Campylobacter mustelae* to *Helicobacter mustelae* comb. nov. respectively." *Int J. Syst. Bacteriol*. 39: 397. 1989.

bmj.bmjjournals.com/cgi/content/full/320/7226/31
de.wikipedia.org/wiki/Helicobacter_pylori
www.cdc.gov/ncidod/dbmd/diseaseinfo/hpylori_g.htm
www.emedicine.com/med/topic962.htm
www.gicare.com/pated/ecdgs30.htm
www.hpylori.com.au/
www.medicinenet.com/helicobacter_pylori/article.
www.tigr.org/tigr-scripts/CMR2/GenomePage3.spl?database=ghp
www.vianet.net.au/~bjmrshll/table1.htm
www.webmd.com/hw/digestive_problems/hw1531.asp

Campylobacter

INTRODUCTION

Campylobacter jejuni is a microaerophilic, non-fermentative, gram-negative organism. The name *Campylobacter*, meaning "curved rod," describes the appearance of the organisms (Figure 32.1). In young cultures, organisms are comma-shaped, spiral, S-shaped, or gull-wing-shaped; as cultures age, round or coccoid forms appear when subjected to atmospheric or temperature stresses.

CHARACTERISTICS

C jejuni, which is structurally similar to other gram-negative bacilli, is motile, with a single flagellum at one or both poles of the cell. The cell envelope has an inner bipolar lipid cell membrane, a thin peptidoglycan layer, an outer bipolar lipid layer with the lipid moiety of a lipopolysaccharide layer embedded in it, and the carbohydrate portion extending to the surface of the cell. Interspersed in the outer membrane layer are membrane proteins, some of which are exposed to the surface and are antigenic for infected hosts.

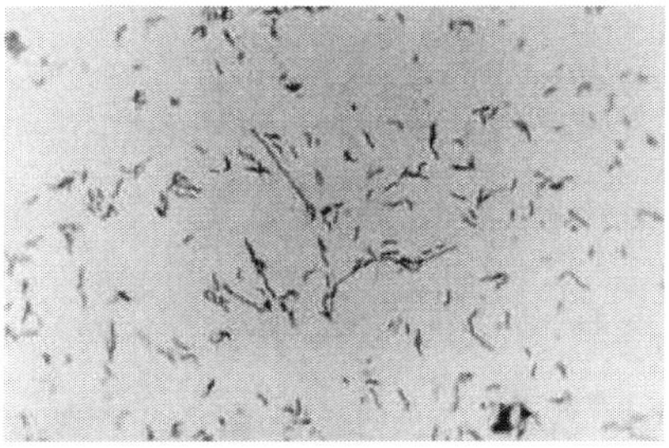

Figure 32.1 Forty-eight-hour culture of *C jejuni* pathogenesis

PATHOGENESIS

Campylobacter have been known as animal pathogens for nearly 100 years. However, because they are fastidious and slow-growing in culture, they have been recognized as human gastrointestinal pathogens only during 1980s. They can cause diarrhoeal illnesses. *Campylobacter jejuni*, and, less often, *C. coli* and *C. lari* are the most common bacteria causing acute diarrhoeal illnesses in developed countries.

The incubation period has ranged from 1–7 days, with most illness developing 2–4 days after infection. Infection leads to multiplication of organisms in the intestines. Patients shed 10^6 to 10^9 campylobacter per gram of faeces. The sites of tissue injury include the small and large intestines. Patients frequently have colic involvement consisting of inflammation of the lamina propria with neutrophils, eosinophils, and mononuclear cells. Destruction of epithelial glands with crypt abscess formation occurs in severe cases. The pathologic lesions seen in *Campylobacter* colitis are difficult to distinguish from those in ulcerative colitis. That most campylobacter enteritis in developed countries is associated with fever and the presence of faecal leucocytes and blood in the stool is consistent with the invasive characteristics of the organisms.

CLINICAL MANIFESTATION

Symptoms range from mild gastrointestinal distress lasting 24 hours to a fulminating or relapsing colitis that mimics ulcerative colitis or Crohn's disease. The predominant symptoms experienced by individuals in developed countries are diarrhoea, abdominal pain, fever, nausea, and vomiting. Guillain–Barré syndrome is an uncommon consequence of *C. jejuni* infection that is present 2–3 weeks after the diarrhoeal illness.

EPIDEMIOLOGY

In developed countries, *C. jejuni* is an important cause of diarrhoea, particularly in children and young adults. Between 3 and 14 per cent of patients with diarrhoea who seek medical attention are infected with *C. jejuni*. The ultimate reservoir for *C. jejuni* is the gastrointestinal tract of many wild animals, and a variety of domestic animals, including food animals (cattle, sheep, poultry, swine, and goats).

DIAGNOSIS

Campylobacter enteritis is hard to distinguish from enteritis caused by other pathogens. The presence of neutrophils or blood in the faeces of patients with acute diarrhoeal illnesses is an important indication of *Campylobacter* infection. Darting motility in a fresh faecal specimen observed by dark-field or phase-contrast microscopy or characteristic vibrio forms visible after gram-staining permit a presumptive diagnosis. The diagnosis is confirmed by isolating the organism from a faecal culture (Figure 32.2) or, rarely, from a blood culture. Because of its growth requirement for microaerobic atmosphere, special laboratory methods are needed to isolate *C jejuni*. Plating methods must be selective to inhibit the growth of competing microorganisms in the faecal flora. Campylobacter organisms have been isolated by filtration methods that do not use antibiotic-containing media. Polymerase chain reaction (PCR)-based techniques have been developed for rapid detection, culture confirmation and for typing of *C. jejuni* strains.

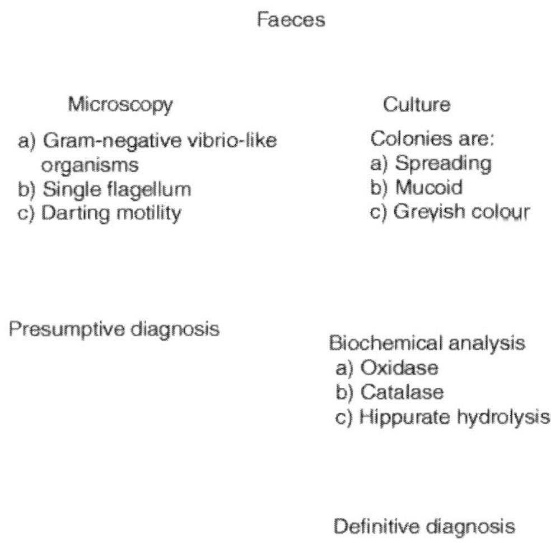

Figure 32.2 Detection of *C. jejuni* and related enteric bacteria

CONTROL

Control of campylobacter enteritis depends largely on interrupting the transmission of the organism to humans from farm and domestic animals, food of animal origin, or contaminated water. Individuals can reduce the risk of campylobacter infection by properly cooking and storing meat and dairy products, avoiding contaminated drinking water and unpasteurized milk, and washing their hands after contact with animals or animal products.

TREATMENT

Fluid and electrolyte replacement are the cornerstones for treatment. Erythromycin or ciprofloxacin appear to be the drugs of choice.

1. Characterize campylobacter morphologically and biochemically.
2. Is campylobacter a zoonotic infection?
3. How do you prevent campylobacter gastroenteritis?
4. Explain the process of diagnosis of campylobacter in humans.

Allos-Mishu, and B. Blaser, M.J. "*Campylobacter jejuni* and the expanding spectrum of related infections." *Clin. Infect. Dis.* 20: In press. 1995.

Graham, D.Y., Klein, P.D. and Evans, D.J. Jr, *et al.* "*Campylobacter pylori* detected non-invasively by the 13C-urea breath test." *Lancet I*. 1174. 1988.

Penner, J.L. "The genus *Campylobacter*: A decade of progress." *Clin. Microbiol. Rev*. 1:157. 1988.

Perez-Perez, G.I., Dworkin, B.M. and Chodos, J.E. *et al*. "*Campylobacter pylori* antibodies in humans." *Ann. Intern. Med*. 109:11. 1988.

Nachamkin, I., Blaser, M.J. and Tompkins, L.S. (eds.). *Campylobacter jejuni*: *Current strategy and future trends*. American Society for Microbiology, Washington DC. pp. 3–296. 1992.

www.cdc.gov/ncidod/eid/vol5no1/altekruse.html

www.who.int/mediacentre/factsheets/fs255/en/

en.wikipedia.org/wiki/Campylobacter

www.cfsan.fda.gov/~mow/chap4.html

www.kidshealth.org/parent/infections

Pseudomonas aeruginosa

Pseudomonas aeruginosa is an opportunistic pathogen of humans. It is a gram-negative, aerobic rod belonging to the family Pseudomonadaceae. These bacteria are common inhabitants of soil and water. Almost all strains are motile by means of a single polar flagellum. *P. aeruginosa* strains produce two types of soluble pigments, the fluorescent pigment pyoverdin and the blue pigment pyocyanin. Pyocyanin (from "pyocyaneus") refers to "blue pus" which is a characteristic of suppurative infections caused by *Pseudomonas aeruginosa*. *Pseudomonas aeruginosa* is notorious for its resistance to antibiotics.

The bacterium is naturally resistant to many antibiotics due to the permeability barrier afforded by its outer membrane LPS. *Pseudomonas* maintains antibiotic resistance plasmids, both R-factors and it is able to transfer these genes by means of the bacterial processes of transduction and conjugation.

PATHOGENESIS

The pathogenesis of pseudomonas infections is multifactorial. The ultimate pseudomonas infection may be seen as composed of three distinct stages:

1. Bacterial attachment and colonization
2. Local invasion
3. Disseminated systemic disease

COLONIZATION

The fimbriae of pseudomonas will adhere to the epithelial cells of the upper respiratory tract. Colonization may produce protease enzyme that degrades fibronectin in order to expose the underlying fimbrial receptors on the epithelial cell surface. Tissue injury may also play a role in colonization of the respiratory tract.

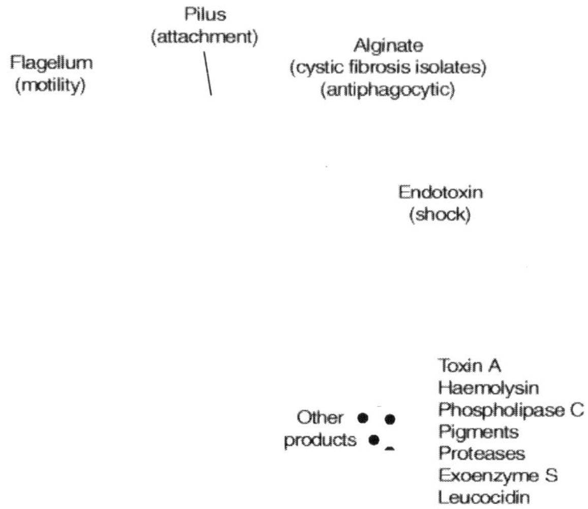

Figure 33.1 Virulence factors of *P. aeruginosa*

INVASION

The ability of *Pseudomonas aeruginosa* to invade tissues depends upon production of extracellular enzymes and toxins that break down physical barriers and damage host cells, as well as resistance to phagocytosis and the host immune defenses. Two extracellular proteases have been associated with virulence that exert their activity at the invasive stage: elastase and alkaline protease. The elastase cleaves collagen, IgG, IgA, and complement. Elastase disrupts the respiratory epithelium and interferes with ciliary function. Alkaline protease interferes with fibrin formation and will lyse fibrin. *P. aeruginosa* produces three other soluble proteins involved in invasion: a cytotoxin (MW 25 kDa) and two haemolysins. The cytotoxin is a pore-forming protein. It was originally named leucocidin because of its effect on neutrophils, but it appears to be cytotoxic for most eukaryotic cells. Of the two hemolysins, one is a phospholipase and the other is a lecithinase. They appear to act synergistically to break down lipids and lecithin. The cytotoxin and hemolysins contribute to invasion through their cytotoxic effects on eukaryotic cells.

The blue pigment, pyocyanin, impairs the normal function of human nasal cilia, disrupts the respiratory epithelium, and exerts a pro-inflammatory effect on phagocytes. A derivative of pyocyanin, pyochelin, is a siderophore that is produced under low-iron conditions to sequester iron from the environment for growth of the pathogen. No role in virulence is known for the fluorescent pigments.

TOXIGENESIS

P. aeruginosa produces two extracellular protein toxins, exoenzyme S and exotoxin A. These toxins induce the infections.

CLINICAL MANIFESTATION

Infection associated with *Pseudomonas aeruginosa* are endocarditis, pneumonia, bacteraemia, septicaemia, meningitis and brain abscesses, otitis, keratitis, neonatal ophthalmia, osteomyelitis and urinary tract infections, gastrointestinal infections like paediatric diarrhoea, typical gastroenteritis, and necrotizing enterocolitis, skin infections like burns, trauma or dermatitis; folliculitis, acne vulgaris.

DIAGNOSIS

Diagnosis of *P. aeruginosa* infection depends upon isolation and laboratory identification of the bacterium. Sample collection depends on type of infection associated with it. On MacConkey agar it produces non-lactose fermenting colony (NLF-Yellow) and on Cetrimide agar, it produces greenish blue colonies. It is identified on the basis of its Gram morphology, inability to ferment lactose, a positive oxidase reaction, its fruity odour, and its ability to grow at 42° C. Fluorescence under ultraviolet light is helpful in early identification of *P. aeruginosa* colonies.

EPIDEMIOLOGY AND CONTROL

Pseudomonas aeruginosa is a common inhabitant of soil, water, and vegetation. It is found on the skin of some healthy persons and has been isolated from the throat (5 per cent) and stool (3 per cent) of non-hospitalized patients.

Pseudomonas aeruginosa is primarily a nosocomial pathogen. The spread of *P. aeruginosa* can best be controlled by observing proper isolation procedures, aseptic technique, and careful cleaning and monitoring of respirators, catheters, and other instruments.

Topical therapy of burn wounds with antibacterial agents such as silver sulphadiazine, coupled with surgical debridement, dramatically reduces the incidence of *P. aeruginosa* sepsis in burn patients.

Pseudomonas aeruginosa is frequently resistant to many commonly used antibiotics. Although many strains are susceptible to gentamicin, tobramycin, colistin, and amikacin, resistant forms have developed. The combination of gentamicin and carbenicillin is frequently used to treat severe *Pseudomonas* infections.

1. Discuss the characteristics of *Pseudomonas aeruginosa*.
2. Explain the diagnosis of *P. aeruginosa*
3. Write notes on the control of *P. aeruginosa* infections.

Brown, M.R.W. (ed.). *Resistance of a Pseudomonas aeruginosa*. John Wiley & Sons, New York. 1975.

Clarke, P.H. and Richman, M.N. (eds). *Genetics and Biochemistry of Pseudomonas*. John Wiley & Sons, New York. 1975.

Coburn, J., Wyatt, R.T., Iglewski, B.H. and Gill, D.M. "Several GTP-binding proteins, including 24 C-H-ras, are preferred substrates of *Pseudomonas aeruginosa* exoenzyme." *S.J. Biol. Chem.* 264:9004. 1989.

Cross, A.S., Sadoff, J.C., Iglewski, B.H. and Sokol, P.A. "Evidence for the role of toxin A in the pathogenesis of infections with *Pseudomonas aeruginosa* in humans." *J. Infect. Dis.* 142:538. 1980.

Dunn, M. and Wunderink, R.G. "Ventilator-associated pnemonia caused by Pseudomonas infection." (Review) *Clinics of Chest Medicine*. (16):95. 1995.

Hancock, R.E.W., Mutharia, L.M. and Chan, L. *et al*. *Pseudomonas aeruginosa* isolates from patients with cystic fibrosis: a class of serum-sensitive, nontypable strains deficient in lipopolysaccharide O side chain. *Infect. Immun.* 42:170. 1983.

Liu, P.V. "Extracellular toxins of *Pseudomonas aeruginosa*." *J. Infect. Dis., Suppl.* 130:S95. 1974.

Mutharia, L.M., Nicas, T.I. and Hancock, R.E.W. "Outer membrane proteins of *Pseudomonas aeruginosa* serotyping strains." *J. Infect. Dis.* 46:770. 1982.

Poole, K. "Bacterial multidrug resistance emphasis on efflux mechanisms and *Pseudomonas aeruginosa*." (Review) *J. Antimicrobial Chemotherapy*. 34(4):453. 1994.

Pritchard, A.E. and Vasal, M.L. Possible insertion sequences in a mosaic genome organization upstream of the exotoxin A gene in *Pseudomonas aeruginosa*. *J. Bacteriol.* 172:2020. 1990.

Woods, D.E. and Iglewski, B.H. "Toxins of *Pseudomonas aeruginosa*: new perspectives." *Rev Infect Dis. Suppl.* 5:S715. 1983.

ccforum.com/content/10/4/R114

en.wikipedia.org/wiki/*Pseudomonas_aeruginosa*

textbookofbacteriology.net/pseudomonas.html

www.pseudomonas.com/

wordnet.com.au/Products/topics_in_infectious_diseases_Aug01.

www.cdc.gov/node.do/id

www.emedicine.com/med/topic1943.html

www.sciencedaily.com/releases

Chlamydia

INTRODUCTION

The chlamydiae are a small group of non-motile coccoid bacteria that are obligate intracellular parasites of eukaryotic cells. Chlamydial cells are unable to carry out energy metabolism and lack many biosynthetic pathways; therefore they are entirely dependent on the host cell to supply them with ATP and other intermediates. Because of their dependence on host biosynthetic machinery, the chlamydiae were originally thought to be viruses; however, they have a cell wall and contain DNA, RNA, and ribosomes and therefore are now classified as bacteria. The group consists of a single genus, *Chlamydia* (order Chlamydiales, class Chlamydiaceae). This genus contains the species *C. trachomatis* and *C. psittaci*, as well as a new organism, the TWAR organism, which has recently been proposed as a third species (*C. pneumoniae*). All three species cause disease in humans. *Chlamydia psittaci* infects a wide variety of birds and a number of mammals, whereas *C. trachomatis* is limited largely to humans. *Chlamydia pneumoniae* (TWAR organism) has been found only in humans.

STRUCTURE AND MULTIPLICATION

The chlamydiae exist in nature in two forms:

1. A non-replicating, infectious particle called the elementary body (EB), 0.25 to 0.3 μm in diameter, that is released from ruptured infected cells and can be transmitted from one individual to another (*C. trachomatis*, *C. pneumoniae*) or from infected birds to humans (*C. psittaci*), and
2. An intracytoplasmic form called the reticulate body (RB), 0.5 to 0.6 μm in diameter, that engages in replication and growth. The elementary body is covered by a rigid cell wall and contains a DNA genome. It contains an open reading frame for a gene involved in DNA replication. The elementary body also contains an RNA polymerase responsible for the transcription of the DNA genome after entry into the host cell cytoplasm and the initiation of the growth cycle.

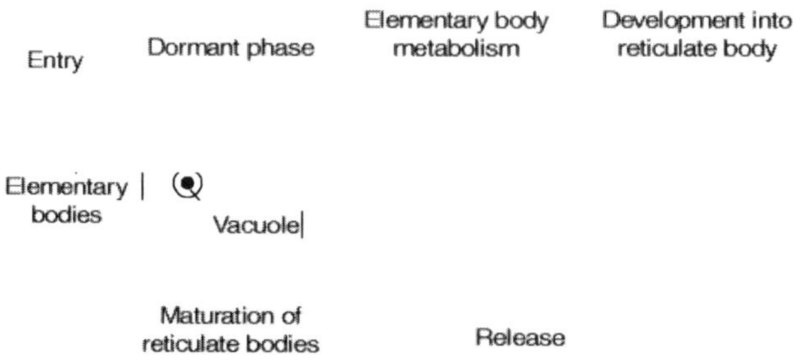

Figure 34.1 Development cycle of chlamydiae

A complex series of events occurs during the developmental cycle of chlamydiae. Studies on the growth cycle of *C. trachomatis* and *C. psittaci* in cell cultures *in vitro* revealed that the infectious elementary body develops into a non-infectious reticulate body (RB) within a cytoplasmic vacuole in the infected cell. There is an eclipse phase of about 20 hours after entry of the elementary body into the infected cell, during which the infectious particle develops into a reticulate body. In these structures the chlamydial genome is transcribed into RNA, proteins are synthesized, and the DNA is replicated. The reticulate body divides by binary fission to form particles which, after synthesis of the outer cell wall, develop into new infectious elementary body progeny. The yield of chlamydial elementary bodies is maximal 36 to 50 hours after infection (Figure 34.1). Table 34.1 describes the events of *Chlamydia* replication.

Table 34.1 Developmental cycle of the chlamydiae

Agent	Host cell
Stage 1 *Dormant phase* Elementary bodies (EBs) have no or little metabolic activity, contain a DNA genome (ca. 600 genes), plasmid DNA (ca 8 kbp), ribosomal subunits, cytoplasm, cell membrane, and cell wall containing a major outer membrane protein (MOMP), three cysteine-rich outer membrane proteins (60 to 62, 15 and 74 kDa), and two cell-binding proteins (31 and 18 kDa).	Host cell can be in any phase of its growth cycle at infection.
Stage 2 *Initiation of EB metabolism*	Duration: 0–12 hours after infection.
EBs adsorb to host cell membrane. The organization of the EB cell wall changes (deficiency of the cysteine-rich proteins; no change in MOMP).	Cell phagocytozes EBs. Cell forms vacuole around EBs.

(Contd.)

Table 34.1 (Continued)

Agent	Host cell
EBs absorb to host cells and utilize glucose 6-phosphate as substrate.	Mitochondria support EB development. Cellular enzyme in glucose metabolic pathway is utilized by EB.
EBs utilize mitochondrial functions.	Host cell nucleic acid metabolism continues.
DNA in EBs changes conformation from compact organization to loose arrangement, possibly for transcriptional processes.	
DNA-dependent RNA polymerase molecules attached to DNA genome, are activated, and transcribe the genome.	
At this stage, EB development becomes sensitive to rifampin. Protein synthesis with existing ribosomes begins in EBs. Low level of DNA synthesis in EB is detected since development is sensitive to 5×10^{-4} M hydroxyurea.	Nucleotides in host cell pool are available for the prokaryotic parasite.
EB mass increases because of macromolecule synthesis.	
Stage 3 *Development of EB into reticulate body*	Duration: 12–35 hours after infection.
Rate of DNA biosynthesis increases.	
DNA is replicated by a DNA-dependent DNA polymerase encoded by chlamydial and plasmid DNAs. Replication of DNA is semiconservative, with synthesis of short Okazaki-like fragments on the DNA template.	Cell responds to enlargement of EB by increasing the size of vacuole in which the agent develops.
DNA synthesis is accompanied by endogenous synthesis of thymidine, probably from deoxycytidine precursor.	
DNA synthesis is inhibited by folic acid analogs and hydroxyurea.	
DNA-dependent RNA polymerase transcribes ribosomal genes in chlamydial DNA. Precursors of rRNA are synthesized and processed to yield two RNA species, of 1.1 and 0.55 Mda. rRNA species assemble into 50S and 30S ribosomal subunits by combining with ribosomal proteins synthesized according to genetic information in the chlamydial DNA.	
Glucose 6-phosphate is catabolized by enzymes in the agent. In *C. trachomatis* only, glycogen is synthesized and deposited in vacuole, where RBs develop.	

(Contd.)

Table 34.1 (Continued)

Agent	Host cell
Synthesis of DNA, ribosomes, and proteins is coupled with formation of new RBs by a binary fission-like process. At this stage, the cytoplasmic vacuoles are filled with RBs.	Duration: 20–48 hours after infection. Inhibition of protein synthesis of cytoplasmic polyribosomes by chloramphenicol does not affect chlamydial life cycle. The cytoplasmic vacuole containing the RBs is markedly enlarged. In *C. psittaci* the vacuole fills most of the cell cytoplasm. In *C. trachomatlis* the vacuole is more rigid and defined.
Chlamydial enzymes are involved in the biosynthesis of the RB-limiting membranes.	
Chlamydial enzymes synthesize RB cell walls.	
Stage 4 Maturation of RBs, formation of EBs.	
DNA synthesis in RBs is coupled with division of particles into two smaller daughter cells (PRE-EBs),	
Internal organization in pre-EB particles the DNA genome is condensed in the centre of the particle, with cell wall loosely arranged around the particle.	
RBs start the synthesis of cysteine-rich proteins.	Vacuole is disrupted, and host cell death is inevitable.
Cell wall synthesis leads to rigid EB particles. All this stage, EBs are formed and the life cycle is completed. EBs are released from the ruptured cells.	

(Modified from Becker Y: "Chlamydia: Molecular Biology of prokaryotic obligate parasites of eukaryocytes." *Microbiol. Rev.* 42: 299 1978, with permission.)

PATHOGENESIS

Human diseases caused by chlamydiae can be divided into two types:

1. Chlamydial agents transmitted by direct contact (*C. trachomatis* genital and ocular infections, *C. pneumoniae* ocular infection) and

2. Chlamydial agents that are transmitted by the respiratory route (*C. psittaci* and *C. pneumoniae*.)

The spread of *C. trachomatis* from person to person may cause trachoma, inclusion conjunctivitis, or lymphogranuloma venereum. Transmission of *C. trachomatis* from the urogenital tract to the eyes and vice versa occurs via contaminated fingers, towels, or other fomites and, in neonates, by passage through an infected birth canal. These diseases appear in an epidemic form in populations with low standards of hygiene. *Chlamydia trachomatis* genital infections are sexually transmitted. *Chlamydia psittaci* is transmitted from infected birds or animals to humans through the respiratory tract. *Chlamydia pneumoniae* spreads from infected individuals by respiratory tract infections but is not sexually transmitted.

CLINICAL MANIFESTATION

Chlamydia trachomatis infections Trachoma, a *C. trachomatis* infection of the conjunctival epithelial cells, results in subepithelial infiltration of lymphocytes, leading to the development of follicles. The infected epithelial cells contain cytoplasmic inclusion bodies. As a result of damage to the epithelial cells, fibroblasts and blood vessels invade the infected area, a pannus forms, and the cornea becomes vascularized and clouded. The eyelids become scarred and malformed, causing trichiasis, an abnormal inward growth of the eyelashes.

Chlamydia trachomatis also causes sexually transmitted genital and rectal infections. The frequency of *C. trachomatis* infections in men may equal or exceed the frequency of gonorrhoea. Non-gonococcal urethritis, epididymitis, and proctitis in men can result from infection with *C trachomatis*. Superinfection of gonorrhoea patients with *C. trachomatis* also occurs. Acute salpingitis and cervicitis in young women can be caused by a *C. trachomatis* infection ascending from the cervix.

Neonates exposed to *C. trachomatis* in an infected birth canal may develop acute conjunctivitis within 5 to 14 days. The disease is characterized by marked conjunctival erythema, lymphoreticular proliferation, and purulent discharge.

Recently, *C. trachomatis* has been suspected of causing lower respiratory tract infections in adults. Evidence also indicates that *C. trachomatis* may cause pneumonia or bronchopulmonary infections in immunocompetent persons.

The initial lesions, or vesicles, appear in the urogenital tract in men and women. If the disease does not heal spontaneously, regional lymph nodes become involved.

Chlamydia psittaci Infections

Chlamydia psittaci infects birds through the respiratory tract. Humans exposed to dead or living infected birds may develop fever, a mild influenza-like disease, or toxic fulminating pneumonitis after an incubation period of 2 to 4 weeks. *Chlamydia psittaci* can cause pneumonia in cats and sheep as well as in humans.

TWAR Organism Infections

Recently, a new *Chlamydia* strain (designated *C. pneumoniae* serovar TWAR organism) that spreads from person to person in human populations was reported to cause outbreaks of respiratory tract infections in immunocompetent persons.

EPIDEMIOLOGY

The disease flourishes in hot, dry areas where there is a shortage of water and where standards of hygiene are low. The agent is spread to the eyes by flies, dirty towels, fingers, or cosmetic eye pencils. *Chlamydia trachomatis* also resides in the genital tract, cervix, and urethra of adults, and genital infection is spread sexually.

DIAGNOSIS

Most diseases caused by the chlamydiae are diagnosed on the basis of their clinical manifestations. *Chlamydia trachomatis* can be identified microscopically in scrapings from the eyes or the urogenital tract. Inclusion bodies in scraped tissue cells are identified by iodine staining of glycogen present in the cytoplasmic vacuoles in infected cells. To isolate the agent, cell homogenates that contain the chlamydial elementary bodies are centrifuged onto the cultured cells (e.g. irradiated McCoy cells). After incubation, typical cytoplasmic inclusions are seen in the cells stained with Giemsa stain or iodine. Staining with iodine can distinguish between inclusion bodies of *C. trachomatis* and *C. psittaci*, as only the former contain glycogen. Each chlamydial agent can also be identified by using specific immunofluorescent antibodies prepared against either *C. trachomatis* or *C. psittaci*. Sera and tears from infected humans are used to detect anti-*Chlamydia* antibodies by the complement fixation or microimmunofluorescence tests. It is possible to diagnose *C. trachomatis* in tissue biopsy specimens by *in situ* DNA hybridization with cloned *C. trachomatis* DNA probes.

CONTROL

Tetracycline and erythromycin are the antibiotics commonly used to treat chlamydial infections in humans. Penicillin is not effective. Patients with trachoma have been treated effectively with erythromycin, rifampin, sulphonamides, chloramphenicol, and tetracyclines. Repeated treatment cycles of long-acting sulphonamides also have been used in local or systemic treatment of trachoma infections. Tetracyclines or sulphonamides sometimes are effective in patients with lymphogranuloma venereum, but treatment does not always improve the condition.

1. How is chlamydia developed as a pathogen.
2. Describe various developmental phases of chlamydia.
3. How will you differentiate chlamydia from bacteria?

4. Does chlamydia synthesize ATP?
5. Differentiate RB from EB.
6. Define EB.
7. Define RB.

Moulder, J.W. "Order II Chlamydiales Storz and Page 1971,334AL." p. 729. In Krieg, N.R. and Holt, J.G. (eds.). *Bergey's Manual of Systematic Bacteriology*. Vol 1. Williams & Wilkins, Baltimore. 1984.

www.emedicine.com/emerg/topic925.htm, www.gsbs.utmb.edu/microbook/ch039.htm www.kidshealth.org/teen/sexual_health/stds/std_chlamydia.html www.ashastd.org/learn/learn_chlamydia_facts.cfm

www.plannedparenthood.org/sexual-health/std/chlamydia.htm en.wikipedia.org/wiki/Chlamydia_trachomatis

www.idph.state.il.us/public/hb/hbchlam.htm

www.herpes-coldsores.com/std/chlamydia_pictures.htm

www.stdservices.on.net/std/chlamydia/

www.nhsdirect.nhs.uk/he.asp?articleID=

Rickettsiae

INTRODUCTION

Rickettsiae are small, gram-negative bacilli. The common features of rickettsiae are their epidemiology, their obligate intracellular lifestyle and the laboratory procedure. Some organisms in the family Rickettsiaceae are closely related genetically (e.g. *Rickettsia rickettsii*, *R. akari*, *R. prowazekii*, and *R. typhi*); some are related partially (e.g. *Ehrlichia* and *Bartonella*); and others not related to *Rickettsia* species (e.g. *C. burnetii*). *Rickettsia* species include two antigenically defined groups that are closely related genetically but differ in their surface-exposed protein and lipopolysaccharide antigens. These are the spotted fever and typhus groups. The cell wall contains lipopolysaccharides, a major component that differs antigenically between the typhus group and the spotted fever group. These rickettsiae also contain major outer membrane proteins. Rickettsiae are able to synthesize ATP via metabolism of glutamate. Majority of strains show capsule.

Rickettsioses is the common term used to describe diseases caused by Rickettsia. It is a zoonotic infection except for Q fever. They are usually transmitted to humans by arthropods (tick, mite, flea, louse, or chigger). Rocky Mountain spotted fever, Q fever, murine typhus, sylvatic typhus, human monocytic ehrlichiosis, human granulocytic ehrlichiosis, and rickettsialpox are some of the diseases associated with Rickettsia. Rocky Mountain spotted fever is among the most severe of human infectious diseases, with a mortality of 20 to 25 per cent unless treated with an appropriate antibiotic.

The rickettsiae are maintained in nature principally by transovarial transmission from infected female ticks to infected ova that hatch into infected larval offspring (Figure 35.1).

The clinical feature of Rocky Mountain spotted fever is due to severe damage to blood vessels by *R. rickettsii*. These organisms have the ability to spread and invade vascular smooth muscle cells as well as endothelium. Damage to the blood vessels leads to visible haemorrhages and thrombocytopenia.

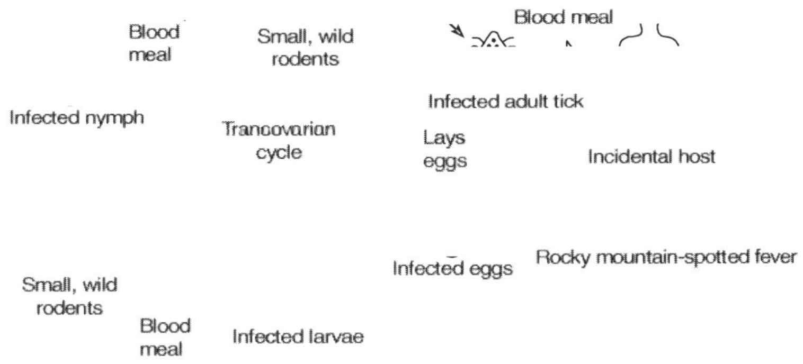

Figure 35.1 Transovarian passage of *R. rickettsii* in the tick vector

EPIDEMIC TYPHUS AND BRILL–ZINSSER DISEASE

Epidemics of louse-borne typhus fever have had important effects on the course of history; for example, typhus in one army but not in the opposing force has determined the outcome of wars. Populations have been decimated by epidemic typhus. During and immediately after World War I, 30 million cases occurred, with 3 million deaths. Unsanitary, crowded conditions in the wake of war, famine, flood, and other disasters and in poor countries today encourage human louse infestation and transmission of *R. prowazekii.* Epidemics usually occur during the cold months in poor highland areas, such as the Andes, Himalayas, Mexico, Central America, and Africa. Lice live in clothing, attach to the human host several times daily to take a blood meal and become infected with *R. prowazekii* if the host has rickettsiae circulating in the blood. Scratching inoculates rickettsiae into the skin. Between epidemics *R. prowazekii* persists as a latent human infection. Years later, when immunity is diminished, some persons suffer recrudescent typhus fever (Brill–Zinsser disease). These milder sporadic cases can ignite further epidemics in a susceptible louse-infested population.

MURINE TYPHUS

Murine typhus is prevalent throughout the world, particularly in ports, countries with warm climates and other locations where rat populations are high. *Rickettsia typhi* is associated with rats and fleas, particularly the oriental rat flea, although other ecologic cycles (e.g. opossums and cat fleas) have been implicated. Fleas are infected by transovarian transmission or by feeding on an animal with rickettsiae circulating in the blood. Rickettsiae are shed from fleas in the faeces, from which humans acquire the infection through the skin, respiratory tract, or conjunctiva. The incidence declined coincident with increased utilization of the insecticide DDT. Although the infection and clinical involvement affects the brain, lungs, and other visceral organs in addition to the skin, mortality in humans is less than 1 per cent.

PATHOGENESIS

Rickettsiae are transmitted to humans by the bite of infected ticks and mites and by the faeces of infected lice and fleas. They enter via the skin and spread through the bloodstream to infect vascular endothelium in the skin, brain, lungs, heart, kidneys, liver, gastrointestinal tract, and other organs. Rickettsial attachment to the endothelial cell membrane induces phagocytosis, soon followed by escape from the phagosome into the cytosol (Figure 35.2). Rickettsiae divide inside the cell. *Rickettsia prowazekii* remains inside the apparently healthy host cell until massive quantities of intracellular

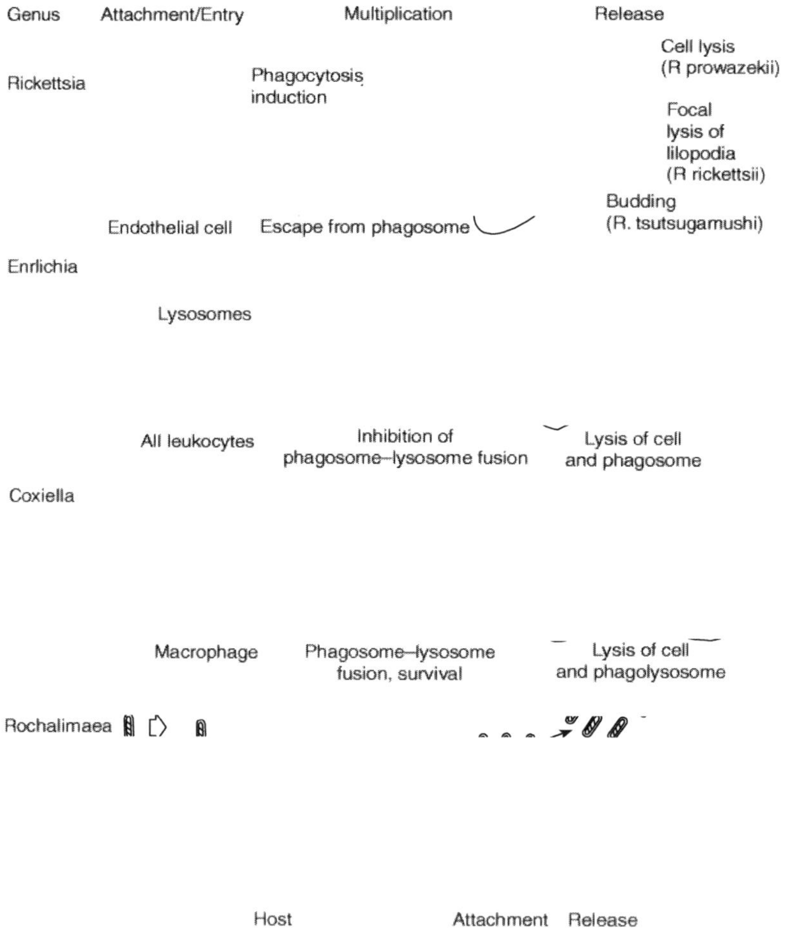

Figure 35.2 Pathogenesis of the rickettsial agents

rickettsiae accumulate and the host cell bursts, releasing the organisms. In contrast, *R. rickettsii* leaves the host cell via long, thin cell projections (filopodia) after a few cycles of binary fission. Hence, relatively few *R. rickettsii* organisms accumulate inside any particular cell, and rickettsial infection spreads rapidly to involve many other cells. Perhaps because of the numerous times the host cell membrane is traversed, there is an influx of water that is initially sequestered in cisternae of cytopatheically dilated rough endoplasmic reticulum in the cells more heavily infected with *R. rickettsii.*

The bursting of endothelial cells infected with *R. prowazekii* is a dramatic pathologic event. The mechanism is unclear, although phospholipase activity, possibly of rickettsial origin, has been suggested. Injury to endothelium and vascular smooth muscle cells infected by *R. rickettsii* seems to be caused directly by the rickettsiae, possibly through the activity of a rickettsial phospholipase or rickettsial protease or through free-radical peroxidation of host cell membranes. Host immune, inflammatory, and coagulation systems are activated and appear to benefit the patient. Cytokines and inflammatory mediators account for an undefined part of the clinical signs. Rickettsial lipopolysaccharide is biologically relatively non-toxic and does not appear to cause the pathogenic effects of these rickettsial diseases.

The pathologic effects of these rickettsial diseases originate from the multifocal areas of endothelial injury with loss of intravascular fluid into tissue spaces (oedema), resultant low blood volume, reduced perfusion of the organs, and disordered function of the tissues with damaged blood vessels (e.g. encephalitis, pneumonitis, and haemorrhagic rash).

DIAGNOSIS

Diagnosis of rickettsial infections is often difficult. The clinical signs and symptoms (e.g. fever, headache, nausea, vomiting, and muscle aches) resemble many other diseases during the early stages when antibiotic treatment is most effective. Rickettsiae are both fastidious and hazardous, in nature. They are the intracellular parasites. Only few laboratories undertake their isolation and diagnostic identification (Figure 35.3). Some laboratories are able to identify rickettsiae by immunohistology

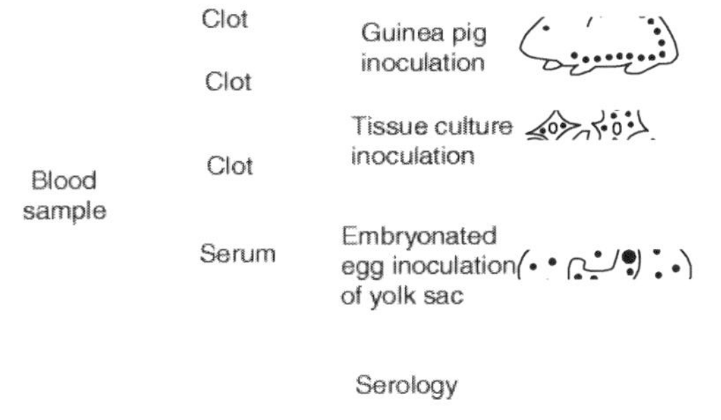

Figure 35.3 Laboratory methods used in confirming diagnosis of rickettsial infection

in skin biopsies as a timely, acute diagnostic procedure, but to establish the diagnosis physicians usually rely on serologic demonstration of the development of antibodies to rickettsial antigens in serum collected after the patient has recovered. The indirect fluorescence antibody test is also useful.

CONTROL

Doxycycline, tetracycline and chloramphenicol are effective in controlling the infection in the individual patient. Human infections are prevented by controlling the vector and reservoir hosts. Vaccines against spotted fever and typhus group rickettsiae have been developed to prevent infection.

COXIELLA BURNETII AND Q FEVER

Coxiella burnetii is sufficiently different genetically from the other rickettsial agents that it is placed in a separate group. It is very resistant to chemicals and dehydration. Its transmission to humans is by the aerosol route but to animals by a tick vector. *Coxiella burnetii* is an obligately intracellular bacterium with some peculiar characteristics. It is small, generally 0.25 μm by 0.5 to 1.25 μm. It is an intracellular bacterium showing pleomorphism.

CLINICAL MANIFESTATIONS

Q fever is a highly variable disease, ranging from asymptomatic infection to fatal chronic infective endocarditis. Some patients develop an acute febrile disease that is a non-specific influenza-like illness or an atypical pneumonia. Other patients are diagnosed after identification of granulomas in their liver or bone marrow.

PATHOGENESIS

Human Q fever follows inhalation of aerosol particles derived from heavily infected placentae of sheep, goats, cattle, and other mammals. *Coxiella burnetii* proliferates in the lungs, causing atypical pneumonia in some patients. Haematogenous spread occurs, particularly to the liver, bone marrow, and spleen. The disease varies widely in severity, including asymptomatic, acute, subacute, or chronic febrile disease, granulomatous liver disease and chronic infection of the heart valves. The target cells are macrophages in the lungs, liver, bone marrow, spleen, heart valves, and other organs. *Coxiella burnetii* is phagocytozed by Kupffer cells and other macrophages and divides by binary fission within phagolysosomes. Apparently it is minimally harmful to the infected macrophages. Different strains have genetic and phenotypic diversity. The lipopolysaccharides are relatively non-endotoxic. Host-mediated pathogenic mechanisms appear to be important, especially immune and inflammatory reactions, such as T-lymphocyte-mediated granuloma formation.

DIAGNOSIS

Clinical diagnosis depends upon a high index of suspicion, careful evaluation of epidemiological factors, and ultimately, confirmation by serologic testing. Although *C. burnetii* can be isolated by inoculation of animals, embryonated hen eggs, and cell culture, very few laboratories undertake this biohazardous approach.

CONTROL

Antibiotic treatment is more successful in ameliorating acute, self-limited Q fever than in curing life-threatening chronic endocarditis.

1. What are the different types of Rickettsia?
2. How do you differentiate rickettsiae from chlamydiae?
3. Characterize the following diseases
   i. Murine typhus
   ii. Epidemic typhus
   iii. Brill–Zinsser disease
   iv. Q fever

Audy, J.R. (ed.). *Red Mites and Typhus*. University Press, New York. 1968.

Hechemy, K.E., Paretsky, D. Walker, D.H. and Mallavia (eds.). *Rickettsiology: Current Issues and Perspectives*. Vol 590. *NY Acad. Sci*. New York. 1990.

Marrie, T.J. "*Q Fever.*" Vol I. *The Disease*. CRC Press, Boca Raton, FL. 1990.

McDade, J.E. and Newhouse VF. "Natural history of *Rickettsia rickettsii*." *Annu. Rev. Microbiol.* 40:287. 1986.

Moulder, J.W. (ed.). *Intracellular Parasitism*. CRC Press, Boca Raton, FL. 1989.

Walker, D.H. (ed.). *Biology of Rickettsial Disease*. Vols I and II. CRC Press, Boca Raton, FL. 1988.

Wolbach, S.B., Todd, J.L. and Palfrey, F.W. *Etiology and Pathology of Typhus*. The League of Red Cross Societies Harvard Press, Cambridge, Mass. 1922.

Zinsser, H. (ed.). *Rats, Lice, and History*. Little Brown, New York. 1935.

www.cehs.siu.edu/fix/medmicro/ricke.htm microbewiki.kenyon.edu/index.php/Rickettsia

textbookofbacteriology.net/Rickettsia.html

pathmicro.med.sc.edu/mayer/ricketsia.html

www.rickettsia.net/

www.kcom.edu/faculty/chamberlain/Website/Lects/Rickett.html

www.cdc.gov/od/ohs/biosfty

www.wrongdiagnosis.com

rickettsia/intro.html-origin

www.nature.com/genomics

Brucella

Introduction

Bacteria of the genus *Brucella* cause disease primarily in domestic, feral and some wild animals and most are also pathogenic to humans. Human brucellosis is either an acute febrile disease or a persistent disease with a wide variety of symptoms. It is a true zoonosis in that virtually all human infections are acquired from animals. Brucellae are gram-negative coccobacilli (short rods) measuring about 0.6 to 1.5 μm by 0.5–0.7 μm. They are non-sporing, lack capsules and are non-motile. The metabolism of the brucellae is mainly oxidative and they show little action on carbohydrates in conventional media. They are aerobes but some species require an atmosphere with added CO_2 (5–10%). Multiplication is slow at the optimum temperature of 37°C and enriched medium is needed to support adequate growth. Brucella colonies become visible on suitable solid media in 2–3 days. The colonies of smooth strains are small, round and convex but dissociation, with loss of the O chains of the LPS. *B. abortus*, *B. melitensis* and *B. suis* are serious pathogens in humans, *B. canis* causes mild disease and the other two species have not affected humans.

A culture can be identified as belonging to the genus *Brucella* on the basis of colonial morphology, staining and slide agglutination with anti-Brucella serum, smooth or rough colonies.

PATHOGENESIS

Brucellae are facultative intracellular parasites, multiplying mainly in monocyte-macrophage cells. The organisms may gain entry into the body through a variety of portals. Bacteria are transmitted by any of the following mechanisms. They are oral entry, by contact, inhalation of aerosols, percutaneous infection through skin abrasions or by accidental inoculation. After entry, the invading bacteria are rapidly phagocytozed by polymorphonuclear leucocytes. Brucellae are frequently able to survive and multiply in these cells because they inhibit the bactericidal myeloperoxidase-peroxide–halide system by releasing 5′-guanosine and adenine.

After invasion, brucellae are transported to target organs such as lymph nodes, spleen, liver, bone marrow, and (especially in animals) the reproductive organs. The tissue lesions produced by *Brucella* species consist of minute granulomas that are composed of epithelioid cells, polymorphonuclear leucocytes, lymphocytes and some giant cells. In cases of infection with *B. melitensis* these granulomas are particularly small although the toxaemia associated with this organism is great. Necrosis is not common, and abscesses do not form, except in *B. suis* infection. The fact that humans rapidly develop hypersensitivity to brucellar antigens suggests that many of the symptoms of human brucellosis result from the reaction of the host defences.

CLINICAL MANIFESTATION

The presentation of brucellosis is characteristically variable. The incubation period is often difficult to determine but is usually from 2 to 4 weeks. The onset may be insidious or abrupt. Subclinical infection is common. In the simplest case, the onset is influenza-like with fever reaching 38 to 40°C. Limb and back pains are unusually severe. The leucocyte count tends to be normal or reduced, with a relative lymphocytosis. Most affected persons recover entirely within 3 to 12 months but some will develop complications marked by involvement of various organs, and a few may enter an ill-defined chronic syndrome. Complications include arthritis, often sacroiliitis, and spondylitis (in about 10% of cases), central nervous system effects including meningitis (in about 5%), uveitis and epididymoorchitis.

EPIDEMIOLOGY

The reservoirs of brucellosis are various wild, feral and (particularly) domestic animals (Figure 36.1). In humans this organism is spread through milk.

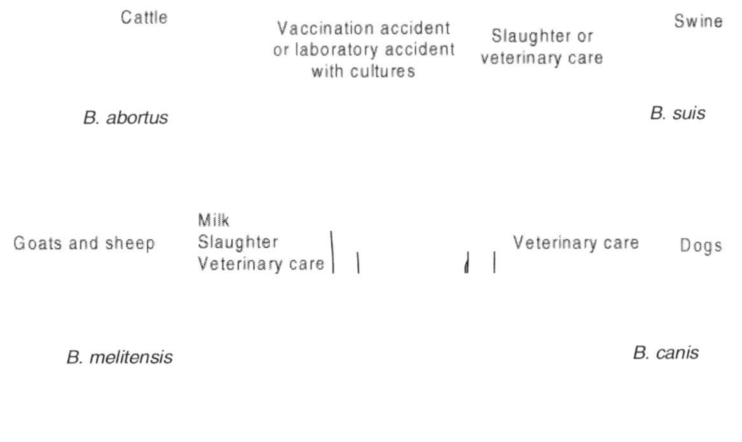

Figure 36.1 Sources of *Brucella* infection

DIAGNOSIS

Blood culture is the method of choice but specimens need to be obtained early in the disease and cultures may need to be incubated for up to four weeks. Modern commercial systems are hampered by the small amount of CO_2 produced during growth. Culture from bone marrow and from presenting foci may be successful. Presumptive identification of cultures from morphology and slide agglutination with specific antiserum should be followed by further work in a reference facility. Molecular techniques for typing are being developed. The standard serum agglutination test (SAT) has been augmented by the modified Coombs' (antiglobulin) technique and the use of 2-mercaptoethanol to separate the actions of specific IgG and IgM (Figure 36.2).

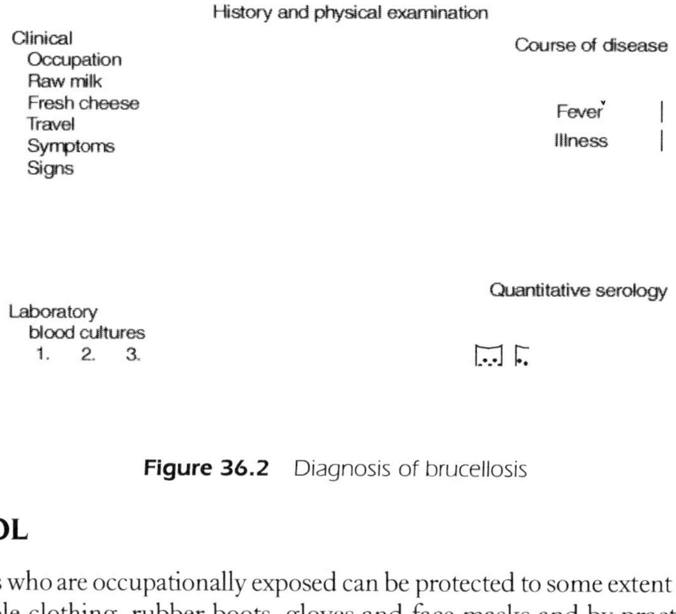

Figure 36.2 Diagnosis of brucellosis

CONTROL

Individuals who are occupationally exposed can be protected to some extent by wearing impermeable clothing, rubber boots, gloves and face masks and by practising good personal hygiene. Pasteurization of milk for drinking and for incorporation into other dairy products is effective in protecting consumers. Humans are treated with combinations of antibiotics for 4 to 6 weeks. Doxycycline and rifampin form the basis, with clotrimoxazole replacing doxycycline in children, but fewer relapses are reported with regimens including two weeks of daily streptomycin. Azithromycin has shown promising results in experimental models.

1. Write short notes on *Brucella*.
2. Discuss the epidemiology and clinical manifestation of *Brucella*.
3. Describe the modes of diagnosis and control of brucellosis.

Allardet-Servent, A., Bourg, G. and Ramuz, M. *et al*. "DNA polymorphism in strains of the genus *Brucella*." *J. Bacteriol.* 170:4603. 1988.

Alton, G.G., Jones, L.M. and Angus, R.D. *et al*. *Techniques for the brucellosis laboratory*. INRA, Paris. 1988.

Bricker, B.J. and Halling, S.M. "Differentiation of *Brucella abortus* bv. 1,2, and 4, *Brucella melitensis*, *Brucella ovis* and *Brucella suis* bv. 1 by PCR." *J. Clin. Microbiol.* 32: 2660. 1994.

Chomel, B.B., DeBess, E.E. and Mangiamele, D.M. *et al*. "Changing trends in the epidemiology of human brucellosis in California from 173 to 1992: a shift towards foodborne transmission." *J. Infect. Dis.* 170:1216. 1994.

Joint FAO/WHO Expert Committee on Brucellosis: Sixth Report, Technical Report Series 740, World Health Organization, Geneva, 1986.

Lang, R. Shasha, B. and Ifrach, N. *et al*. "Therapeutic effects of roxithromycin and azithromycin in experimental murine brucellosis." *Chemotherapy.* 40:252. 1994.

Montejo, J.M., Alberda, I. and Glez-Zarate, P. *et al*. Open, randomized therapeutic trial of six antimicrobial regimens in the treatment of human brucellosis. *Clin. Infect. Dis.* 16: 671. 1993.

Spink, W.W. *The Nature of Brucellosis*. University of Minnesota Press, Minneapolis. 1956.

Young, E.J. and Corbel, M.J. (eds.). *Brucellosis: Clinical and Laboratory Aspects*. CRC Press, Boca Raton. 1989.

www.microbialcellfactories.com/content/

en.wikipedia.org/wiki/Brucella_melitensis

www.medindia.net/bloodtest/microbiology

www.jmm.sgmjournals.org/cgi/content

expasy.org/cgi-bin

www.tigr.org/tigr-scripts/CMR2/GenomePage3.spl?database=

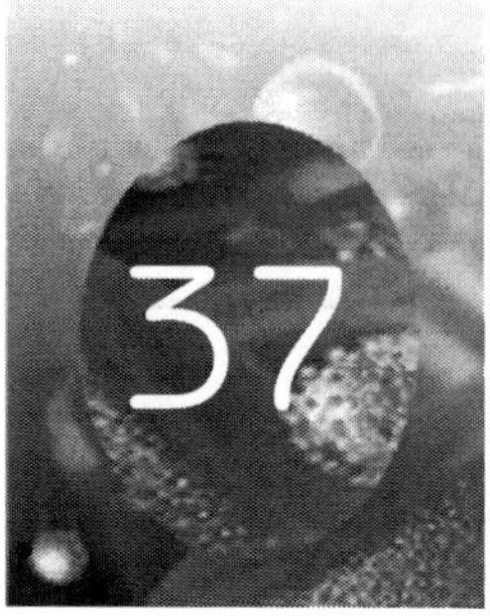

Bacillus anthracis

INTRODUCTION

The anthrax bacillus, *Bacillus anthracis,* was the first bacterium shown to be the cause of a disease. In 1877, Robert Koch grew the organism in pure culture, demonstrated its ability to form endospores and produced experimental anthrax by injecting it into animals. *Bacillus anthracis* is a very large, gram-positive, spore-forming rod, 1–1.2 μm in width and 3–5 μm in length. The bacterium can be cultivated in ordinary nutrient medium under aerobic or anaerobic conditions.

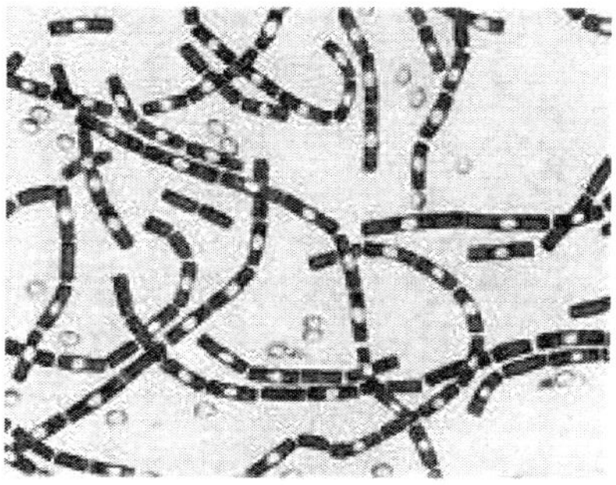

Figure 37.1
Bacillus anthracis—Gram stain

CULTIVATION

Several non-selective and selective media for the detection and isolation of *Bacillus anthracis* have been described, as well as a rapid screening test for the bacterium based on the morphology of microcolonies.

ANTHRAX

Anthrax is primarily a disease of domesticated and wild animals, particularly herbivorous animals such as cattle, sheep, horses, mules, and goats. Humans become infected incidentally when brought into contact with diseased animals, which includes their flesh, bones, hides, hair and excrement. The spores have been found naturally in soil samples from around the world, the organisms cannot be regularly cultivated from soils where there is an absence of endemic anthrax. The most common form of the disease in humans is cutaneous anthrax, which is usually acquired via injured skin or mucous membranes. A minor scratch or abrasion, usually on an exposed area of the face, neck or arms, is inoculated by spores from the soil, a contaminated animal or carcass. The spores germinate, vegetative cells multiply and a characteristic gelatinous oedema develops at the site. This develops into a papule within 12–36 hours after infection. The papule changes rapidly to a vesicle, then a pustule (malignant pustule), and finally into a necrotic ulcer from which infection may disseminate, giving rise to septicaemia. Lymphatic swelling also occurs within seven days. In severe cases, where the bloodstream is eventually invaded, the disease is frequently fatal.

Another form of the disease, inhalation anthrax (Woolsorters' disease), results most commonly from inhalation of spore-containing dust where animal hair or hides are being handled. The disease begins abruptly with high fever and chest pain. It progresses rapidly to a systemic haemorrhagic pathology and is often fatal if treatment cannot stop the invasive aspect of the infection.

Gastrointestinal anthrax is analogous to cutaneous anthrax but occurs on the intestinal mucosa. As in cutaneous anthrax, the organisms probably invade the mucosa through a pre-existing lesion. The bacteria spread from the mucosal lesion to the lymphatic system. Intestinal anthrax results from the ingestion of poorly cooked meat from infected animals. Gastrointestinal anthrax is rare, but may occur as explosive outbreaks associated with ingestion of infected animals. Intestinal anthrax has an extremely high mortality rate.

Meningitis due to *B. anthracis* is a very rare complication that may result from a primary infection elsewhere.

PATHOGENICITY

Bacillus anthracis clearly owes its pathogenicity to two major determinants of virulence: the formation of a poly-D-glutamyl capsule, which mediates the invasive stage of the infection, and the production of the multicomponent anthrax toxin which mediates the toxigenic stage.

The poly-D-glutamyl capsule is itself non-toxic, but functions to protect the organism against the bactericidal components of serum and phagocytes, and against phagocytic engulfment. The capsule plays its most important role during the establishment of the infection, and a less significant role in the terminal phases of the disease, which are mediated by the anthrax toxin.

Anthrax Toxin

One component of the anthrax toxin has a lethal mode of action that is not understood at this time. Death is apparently due to oxygen depletion, secondary shock, increased vascular permeability, respiratory failure and cardiac failure.

The toxin consists of three distinct antigenic components. Each component of the toxin is a thermolabile protein with a MW of approximately 80 kDa. Factor I is the oedema factor (EF) which is necessary for the oedema-producing activity of the toxin. EF is known to be an inherent adenylate cyclase, similar to the *Bordetella pertussis* adenylate cyclase toxin. Factor II is the protective antigen (PA), because it induces protective antitoxic antibodies in guinea pigs. PA is the binding (B) domain of the anthrax toxin which has two active (A) domains, EF (above) and LF (below). Factor III is known as the lethal factor (LF) because it is essential for the lethal effects of the anthrax toxin.

EF+PA has been shown to elevate cyclic AMP to extraordinary levels in susceptible cells. Changes in intracellular cAMP are known to affect changes in membrane permeability and may account for oedema. In macrophages and neutrophils, an additional effect is the depletion of ATP reserves which are needed for the engulfment process. Hence, one effect of the toxin may be to impair the activity of regional phagocytes during the infectious process.

LF+PA have combined lethal activity. The lethal factor is a Zn^{++}-dependent protease that induces cytokine production in macrophages and lymphocytes, and its mechanism of action is slowly becoming understood. The crystal structure of lethal factor, the crucial pathogenic enzyme of anthrax toxin, is known to be a member of the mitogen-activated protein kinase (MAPK) family of enzymes that disrupts cellular signalling. Furthermore, the identity of the human receptor for anthrax PA, named anthrax toxin receptor, has been demonstrated to be a type I membrane protein that binds directly to PA.

CONTROL

The Sterne strain of *Bacillus anthracis* produces sublethal amounts of the toxin that induces formation of protective antibody. The anthrax vaccine for humans, which is used in the U.S., is a preparation of the protective antigen recovered from the culture filtrate of an avirulent, nonencapsulated strain of *Bacillus anthracis* that produces PA during active growth. Anthrax immunization consists of three subcutaneous injections given two weeks apart followed by three additional subcutaneous injections given at 6, 12, and 18 months. Annual booster injections of the vaccine are required to maintain a protective level of immunity.

TREATMENT

Antibiotics should be given to unvaccinated individuals exposed to inhalation anthrax. Penicillin, tetracyclines and fluoroquinolones are effective if administered before the onset of lymphatic spread or septicaemia, estimated to be about 24 hours.

1. Discuss the different types of anthrax.
2. Write short notes on
   i. *Bacillus antracis*
   ii. Anthrax toxin

Claus, D. and Berkeley, R.C.W. "Genus Bacillus Cohn 1872, 174AL. p. 1105." In Sneath PHA, Mair, N.S. Sharpe, M.E. Holt, J.G. (eds.). *Bergey's Manual of Systematic Bacteriology*. Vol. 2. Williams & Wilkins, Baltimore. 1986.

Gordon, R.E., Haynes, W.C. and Pang, CH-N. *The genus Bacillus*. U.S. Department of Agriculture Agricultural Handbook no. 427. U.S. Department of Agriculture, Washington DC. 1973.

Klimpel, K.R., Arora, N. and Leppla, S.H. Anthrax toxin lethal factor contains a zinc metalloprotease consensus sequence which is required for lethal toxin activity. *Molecular Microbiol.* 13:1093. 1994.

Kochi, S.K., Schiavo, G. Mock, M. and Montecucco, C. "Zinc content of the *Bacillus anthracis* lethal factor." *FEMS Microbiol. Lett.* 124:343. 1994.

Kramer, J.M. and Gilbert, R.J. *Bacillus cereus* and other *Bacillus* species. p.21. In Doyle MP (ed.). *Foodborne Bacterial Pathogens*. Marcel Dekker, New York. 1989.

Leppla, S.H. "The Anthrax toxin complex." *In Alouf JE*, Freer JH (eds.). *Sourcebook of Bacterial Proteins*. Academic Press, New York. p. 277.

Norris, J.R., Berkeley, R.C.W., Logan, N.A. *et al.* "The genera *Bacillus* and *Sporolactobacillus*." In Starr, M.P. Stolp, H., Truper, H.G. (eds): *The Prokaryotes: A Handbook on Habitats, Isolation and Identification of Bacteria*. Vol. 2. Springer-Verlag, New York, p. 1711. 1981

Parry, J.M., Turnbull, P.C.B. and Gibson. J.R. *A Colour Atlas of Bacillus species*. Wolfe Medical Atlas no. 19. Wolfe Publishing, London. 1983.

Turnbull, P.C.B. "Studies on the production of enterotoxins by *Bacillus cereus*." *J. Clin. Pathol.* 29:941. 1976.

Turnbull, P.C.B. and Kramer, J.M. "*Bacillus*." In Murray, P.R., *et al.* (eds.). *Manual of Clinical Microbiology*, 6th edn. American Society for Microbiology, Washington, DC. p. 349.

Turnbull, P.C.B., Kramer, J.M. and Melling, J. "*Bacillus*." In Parker, M.T., Duerden, B.I. (eds.). *Systematic Bacteriology. Topley and Wilson's Principles of Bacteriology, Virology and Immunity*. Vol. 2. Edward Arnold, Sevenoaks, England, p. 187. 1990.

www.anthrax.com
www.cdc.gov/ncidod/dbmd/diseaseinfo/anthrax_g.html
en.wikipedia.org/wiki/Anthrax
www.bt.cdc.gov/agent/anthrax/
www.nlm.nih.gov/medlineplus/anthrax.html
www.anthrax.osd.mil
www.fda.gov/cber/vaccine/anthrax.html
www.nature.com/nature/anthrax/
www.health.state.ny.us/nysdoh/consumer/anthrax.html

Actinomyces

Actinomyces spp. are members of a large group of pleomorphic bacteria, many of which have some tendency towards mycelial growth. They are gram-positive filamentous rods that are not acid-fast and are non-motile (Figure 38.1). As in other gram-positive bacteria, the cell wall peptidoglycan contains muramic acid, *N*-acetyl glucosamine, glutamic acid, and one or two additional amino acids. *Actinomyces* species have lysine or lysine plus ornithine in the peptidoglycan. Pili (fimbriae) on *A. viscosus* and *A. naeslundii* are of two types. Type 1 pili are involved in attachment, type 2 pili are involved in coaggregation reactions with other bacteria. They are members of the oral flora of humans or animals. *Actinomyces* species are major components of dental plaque. *A. israelii*, *A. gerencseriae* cause actinomycosis in humans and animals. *Actinomyces* grows well on most rich culture media. They are best described as aerotolerant anaerobes. The species vary in oxygen requirements. For example, *A. viscosus* and *A. naeslundii* grow best in an aerobic environment with carbon dioxide, whereas *A. israelii* requires anaerobic conditions for growth. The major end products of glucose fermentation by *Actinomyces* are acetic, lactic, formic, and succinic acids. All the *Actinomyces* species that have been examined serologically can be separated from the other species and from *P. propionicus*. *A. naeslundii A. viscosus*, *A. odontolyticus*, and *A. bovis* each have at least two serotypes.

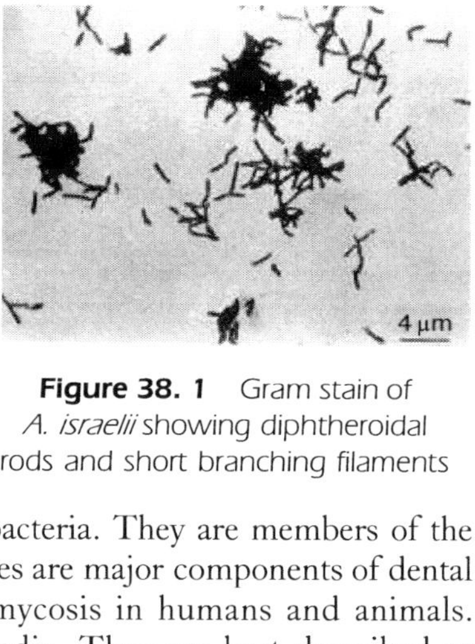

Figure 38. 1 Gram stain of *A. israelii* showing diphtheroidal rods and short branching filaments

Actinomycosis is the disease caused by *Actinomyces.*

PATHOGENESIS

Actinomycosis results when bacteria resident in the mouth are introduced into the tissues. The mechanisms by which *Actinomyces* produce disease are not clear. Pathogenesis may involve the ability of these organisms to suppress some of the immune functions of the host. Studies of oral disease have shown that *Actinomyces* are chemotactic, activate lymphocyte blastogenesis, and stimulate the release of lysosomal enzymes from polymorphonuclear leucocytes and macrophages.

With the exception of *A. pyogenes*, which produces a soluble toxin and a haemolysin that can be neutralized by antiserum, *Actinomyces* and *P. propionicus* do not produce exotoxins or significant amounts of other toxic substances. Factors that would aid in tissue invasion and abscess formation have not been demonstrated.

CLINICAL MANIFESTATION

Actinomycosis is a chronic disease characterized by the production of suppurative abscesses or granulomas that eventually develop draining sinuses. These lesions discharge pus containing the organisms. In long-standing cases, the organisms are found in firm, yellowish granules called sulphur granules.

Actinomycosis is almost always a mixed infection; a variety of other oral bacteria can be found in the lesion with *Actinomyces*. Cervicofacial infections involve the face, neck, jaw, or tongue and usually occur following an injury to the mouth or jaw or a dental manipulation such as extraction. The disease begins with pain and firm swelling along the jaw and slowly progresses until draining sinuses are produced.

Thoracic actinomycosis results from aspiration of pieces of infectious material from the teeth and may involve the chest wall, the lungs, or both. The symptoms are similar to those of other chronic pulmonary diseases, and the disease is often difficult to diagnose. Abdominal actinomycosis is often associated with abdominal surgery, accidental trauma, or acute perforative gastrointestinal disease.

EPIDEMIOLOGY

Actinomyces israelii, *A. gerencseriae*, *A. georgiae*, *A. naeslundii*, *A. odontolyticus*, *A. meyeri*, possibly *A. pyogenes* and *P. propionicus* are normal inhabitants of the human mouth and are found in saliva, on the tongue, in gingival crevice debris, and frequently in tonsils in the absence of clinical disease. Actinomycosis occurs worldwide.

DIAGNOSIS

Gram stain of pus or suppurative exudate shows gram-positive, non-acid-fast rods in diphtheroidal arrangements with or without branching. Specimens are first examined for the presence of granules. If present, the granules are crushed, gram-stained, and examined for gram-positive rods or branching filaments. Washed, crushed granules or well-mixed pus in the absence of granules is cultured on a rich medium, such as brain-heart infusion blood agar and incubated anaerobically and aerobically with added carbon dioxide. Plates are examined after 24 hours and after 5 to 7 days for the characteristic colonies of *A. israelii*, *A. gerencseriae*. Colonies of other *Actinomyces* spp. can take a variety of forms, including smooth domed or flat colonies of various sizes. Isolates morphologically resembling *Actinomyces* are identified by determining the metabolic end products by gas–liquid chromatography and by performing a series of biochemical tests. Immunofluorescence tests are useful for serologic identification of isolates and the direct demonstration of the organisms in clinical samples.

CONTROL

Due to the mixed nature of the infection and the presence of granules, successful treatment of actinomycosis requires long-term antibiotic therapy combined with surgical drainage of the lesions and excision of damaged tissue. *Actinomyces* spp. are susceptible to penicillins, the cephalosporins, tetracycline, chloramphenicol, and a variety of other antibiotics. Penicillin is the drug of choice for infections with all species of *Actinomyces*.

1. Discuss the pathogenesis and clinical manifestation of *Actinomyces*.
2. Write notes on the diagnosis and control of *Actinomyces*.

Barsotti, O., Morrier, J.J. and Decoret, D. *et al.* "An investigation into the use of restriction endonuclease analysis for the study of transmission of *Actinomyces*." *J. Clin.Periodontol*. 20:436. 1993.

Brown, D.A., Fischlschweiger, S. and Birdsell, D.C. "Morphological, chemical and antigenic characterization of cell walls of the oral pathogenic strains *Actinomyces viscosus* T14V and T14AV." *Arch. Oral. Biol.* 25:451. 1980.

Brown, J.R. "Human actinomycosis: a study of 181 subjects." *Hum. Pathol.* 4:319. 1973.

Causey, W.A. "Actinomycosis." p.383. In Vinken, P.V. Bruyn, G.W. (eds.). *Handbook of Clinical Neurology*. Vol. 35. North Holland, Amsterdam. 1978.

Funke, G. Ramos, P.C. and Fernández-Garayzábal, J.F. *et al.* "Description of human-derived Centers for Disease Control Coryneform group 2 bacteria as *Actinomyces bernardiae*." *I.J. Syst. Bacteriol.* 45:57. 1995.

Gerencser, M.A. "Actinomycosis." p. 551. In Balows, A. Hausler, W.J. Jr. and Ohashi, M. (eds.). *Laboratory Diagnosis of Infectious Diseases: Principles and Practice*. Vol. 1. *Bacterial, Mycotic and Parasitic Diseases*. Springer-Verlag, New York. 1988.

Goodfellow, M., Mordarski, M. and William, S.T. (eds.). *The Biology of the Actinomycetes*. Academic Press (London), London. 1984.

Pennington, J.E. (ed.). *Respiratory Infections: Diagnosis and Management*, 2nd edn. Raven Press, New York. 1988.

Mahgoub, E.S. "Mycetoma." p. 633. In Balows, A. Hausler, W.J. Jr. Ohashi, M. *et al.* (eds.). *Laboratory Diagnosis of Infectious Diseases: Principles and Practice*. Vol. 1. Bacterial, Mycotic and Parasitic Diseases. Springer-Verlag, New York. 1988.

Schaal, K.P. "*Genus Arachnia*," p. 1332, and "*Genus Actinomyces*", p. 1383: In Sneath, P.H., Mair, N.S., Sharpe, M.E. and Holt, J.G. (eds.). *Bergey's Manual of Systematic Bacteriology*. Vol. 2. Williams & Wilkins, Baltimore. 1986.

Slack, J.M. and Gerencser, M.A. "*Actinomyces.*" *Filamentous Bacteria: Biology and Pathogenicity*. Burgess, Minneapolis. 1975.

en.wikipedia.org/wiki/Actinomyces

www.life.umd.edu/classroom/bsci424/PathogenDescriptions/Actinomyces

www.gsbs.utmb.edu/microbook/ch034.html

medic.med.uth.tmc.edu/path/00001503.html

www.phac-aspc.gc.ca/msds-ftss/msds2e.html

www.attract.wales.nhs.uk/question_answers

www.thefreedictionary.com/actinomyces

www.emedicine.com/med/topic31

www.ncbi.nlm.nih.gov/entrez/query

PART III

VIROLOGY

39 Characteristic Features of Viruses

Viruses are submicroscopic, unique group of infectious agents and are obligate intracellular parasites. They contain single type of nucleic acid. Viruses exist as two phases. They are intracellular and extracellular viruses. Size ranges from 20–14,000 nm in length. Complete virus particle is called as virion.

Luria and Darnell in 1968 defined viruses as entities whose genome is an element of nucleic acid either DNA or RNA. It reproduces inside of living cells and use their synthetic machinery to direct the synthesis of extracellular infectious viral particles called virions which contain the viral genome and transfer it to the other cell.

Viruses differ from living cells in the following aspects.

1. Simple, acellular organization.
2. Absence of both nucleic acids (RNA and DNA) in the same virion.
3. Inability to reproduce independently.
4. Absence cell division like prokaryotes and eukaryotes.

Viruses are a unique group of infectious agents. The complete viral particle is called virion, consisting of one or more molecules of DNA or RNA enclosed in a coat protein, sometimes also in other layers. Viruses exist in two phases and are intracellular and extracellular. Viruses are smaller than bacteria and larger than cell organelles. Viruses multiply within a host cell.

STRUCTURE OF VIRUS

Virus morphology had been studied over the past few decades. Electron microscopy, X-ray diffraction, biochemical analyses and immunological techniques are adopted to study the detailed morphology of viruses.

Virion Size

Size of the virion ranges from about 10–300 or 400 nm in diameter. The smallest virus is a little larger than ribosomes whereas the poxviruses are about the size of the smallest bacteria (Figure 39.1).

Lateral body
Core
Genome
Parvovirus
Φx174 phage
Poxvirus
Smallest virus
Largest virus

Figure 39.1 *Largest and smallest viruses*

Capsid

The virus consists of a protein coat capsid and a nucleic acid core. Nucleic acid core of the viruses is surrounded by protein coat called capsid. Capsid protects the viral genetic material and aids in its transfer between host cells. The capsid is composed of a number of subunits known as capsomers.

Capsomers are assembled and give rise to viral symmetry. There are three morphological types of capsids. They are

1. Polyhedral/icosahedral/spherical symmetry.
2. Helical/cylindrical/rod-like symmetry.
3. Complex/binal symmetry.

1. *Icosahedral or polyhedral* It is a regular polyhedron with 20 equilateral triangular faces and 12 intersecting points. These capsids appear spherical when viewed at low power in the electron microscope (Figure 39.2). Icosahedral symmetry occurs more frequently. Watson and Crick showed the polyhedral capsids exhibit 3 types of symmetry: e.g. tetrahedral, octahedral and icosahedral. Viruses employ the icosahedral shape because it is the most efficient way to enclose a space.

Capsids are large macromolecular structures constructed from many copies of one or more or few types of protein subunits called protomers. Five or six protomers combine to form a structure, which is represented as a capsomer. A few genes can code for proteins that self assemble to form the capsid.

Capsid is made up of pentamers and hexamers. 5 monomers combine to form a pentamer and is called pentameric capsomer. Each capsid consists of many capsomers. Pentamers are at the vertices, whereas hexamers form its edges and triangular faces. Protomers join to form capsomers through nine covalent bondings. The bonds between proteins within pentamer and hexamers are stronger than those between separate capsomers.

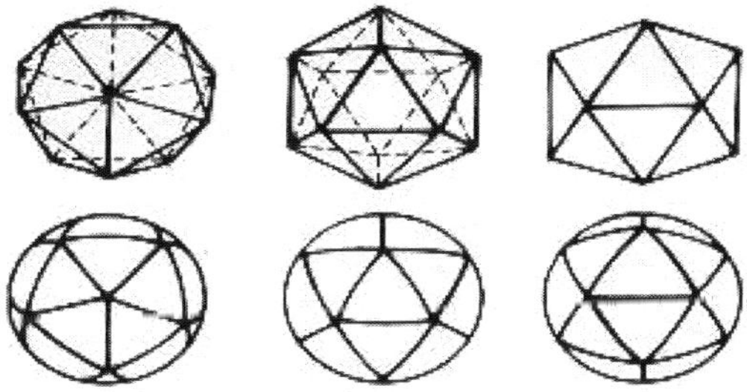

Figure 39.2 Arrangement of capsid proteins in icosahedral and spherical viruses

The minimum number of capsomer is 12 followed by 32, 42, 72, 92, 162, 252, 362, 492, 642 and 812.

The number of capsomers found in a few viruses are listed below.

ϕX 174	12
TYNV	32
Adenovirus	252
Herpesvirus	162
Reovirus	92
Polyoma and papillomaviruses	72

The bond between two capsomers is weaker than between monomers. The capsids break down into capsomers during purification. In most of the viruses, pentamers and hexamers are made up of same type of monomers, an exception is adenovirus.

2. *Helical or cylindrical symmetry* It consists of monomers arranged in a helix around a single rotational axis, e.g. TMV (Figure 39.3).

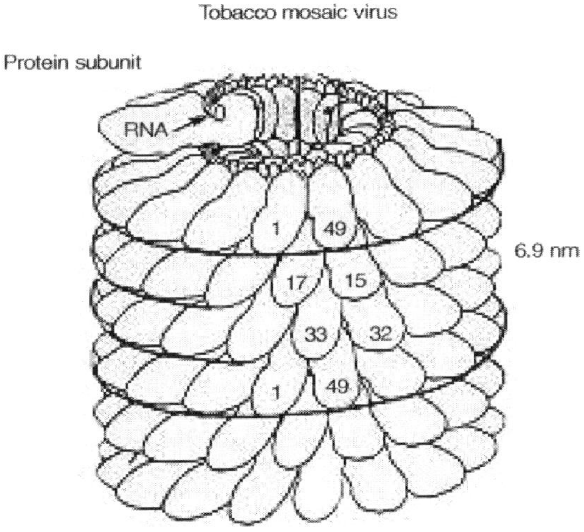

Figure 39.3 Tobacco mosaic virus

Helical capsids are shaped much like hollow tubes with protein wall of the size of the helical capsid that is influenced by its protomers and nucleic acid enclosed within the capsid.

3. *Complex or binal symmetry* Complex viruses have capsid symmetry that is neither purely icosahedral nor helical. They may have tail and other structures or have multilayered wall surrounding nucleic acid.

Complex viruses with both heads and tails are said to have binal symmetry because they possess a combination of icosahedral head and helical tail, e.g. bacteriophage (Figure 39.4).

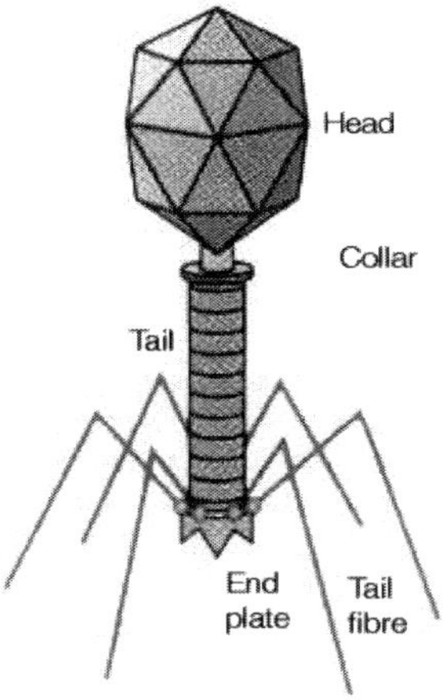

Figure 39.4 Bacteriophage

Viral Envelope

Many animal viruses and some plant viruses are bounded by an outer membranous layer, in addition to capsid and called an envelope. Envelope is about 10–15 nm in thickness. Chemically envelope is made up of carbohydrate, protein and lipids. Lipids and proteins are associated to form lipoproteins. Carbohydrates and proteins associate to form glycoproteins. It is made from host cell plasma membrane, but they vary considerably in size, morphology, complexity and nature of the membrane. Enveloped viruses acquire their membrane, lipid bilayers with associated protein by budding through a cell membrane, the endoplasmic reticulum, Golgi apparatus or nuclear membrane.

Envelope proteins are generally called as glycoproteins that carry covalently linked sugar chains or oligosaccharide. Sugars are added to the proteins post translationally during transport to the cell membrane, from which progeny viruses assemble. Intra or interchain disulphide bonds, another chemical feature of envelope protein. Covalent bonds stabilize the tertiary or quaternary structure of viral glycoproteins.

Viral glycoproteins are embedded in the lipid bilayer by a short membrane-spanning domain. It is a major antigenic determinant and mediate fusion during entry. Internal domains are also available in the envelope and is generally called as surface proteins, which is essential for viral assembly. In some viruses, receptor-destroying enzymes are available to promote viral release, e.g. Influenza "A" contains neuraminidase, coronavirus contains esterase.

In influenza "A" virus, proton channel (M2) proteins are available which may facilitate uncoating and transcriptase activity. Fusion protein is also available in the envelope that mediates fusion. In some viruses, additional viral structural proteins are available, termed the matrix protein. It mediates the interaction of the innermost, genome containing structure with the viral envelope.

Proteins present in some viruses are listed below.

Rhabdovirus	G protein
Orthomyxovirus	HA, NA, M2
Retrovirus	Env
Herpesvirus	gH gL

Envelope is a flexible, membranous structure. Enveloped viruses can have various shapes and are pleomorphic in nature. Envelopes are susceptible to lipid solvents like chloroform, ether and bile salts. Phospholipids of envelope includes phosphatidyl choline, phosphatidyl serine, phosphatidyl ethanolamine. Glycolipids include sphingosine. The functions of the envelope are as follows.

- Gives structure.
- Protects NA and capsid.
- Helps in adherence of virus onto receptors of host cell.
- Confers chemical, biological and antigenic properties of viruses.

Spikes

Some of the proteins project from the envelope and are called spikes or peplomers. It is composed of glycoproteins. The functions of spikes include the following.

They adhere to the receptors of host cell and play an important role in infection.

Spikes mediate host–virus interaction, e.g. influenza virus, rhabdovirus, HIV, etc.

Viral Enzymes

Viral enzymes catalyse a variety of biological reactions and are called as biocatalysts.

1. DNA *polymerase* It is also known as DNA-dependent DNA polymerase. It is used in polymerization of nucleic acid, e.g. involved in T4 DNA replication.

2. RNA *polymerase* RNA viruses code for RNA-dependent RNA polymerase or DNA-dependent RNA polymerase. These enzymes mediate viral specific protein synthesis, e.g. poxvirus.

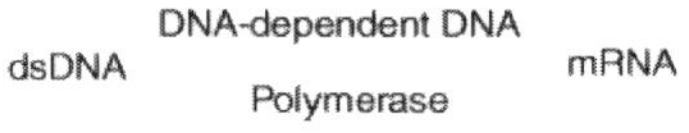

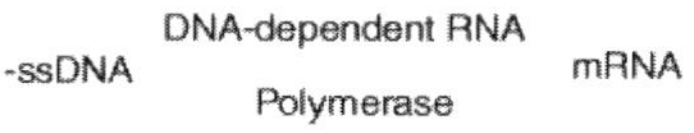

3. RNA-*dependent* DNA *polymerase* This special enzyme is involved in the synthesis of DNA from RNA. This enzyme is also called reverse transcriptase. This enzyme is found in retroviruses.

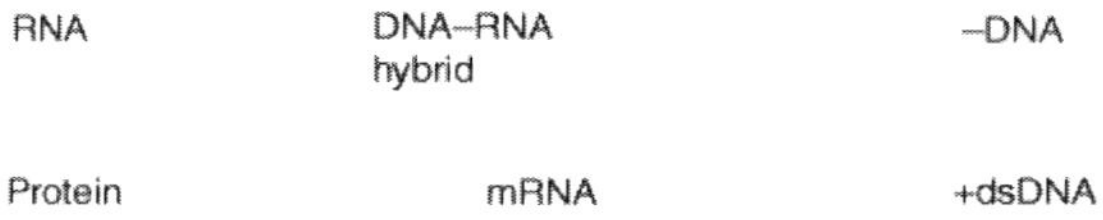

4.*Terminal nucleotidyl transferase* This enzyme adds Poly(A) or Poly(U) sequence at 3′ and 5′ end, e.g. adenovirus, herpesvirus.

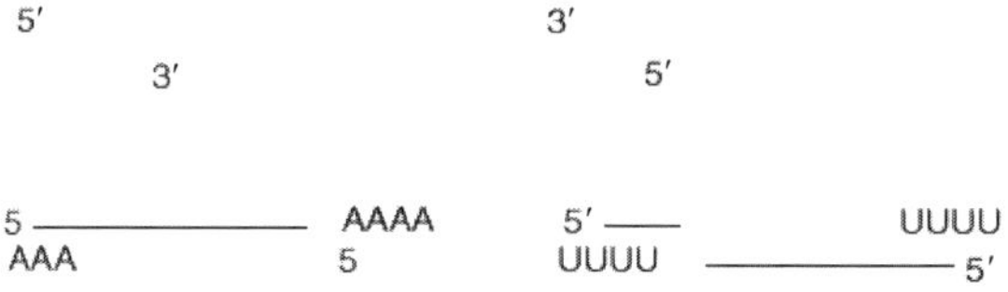

5. *Integrase and protein kinase* Integrase is involved in the integration of host chromosome. Protein kinase is used for the synthesis of protein. Other enzymes like hyaluronidase hydrolyse hyaluronic acid in the respiratory tract; neuraminidase hydrolyses *N*-acetyl muramic acid present in epithelial cells of the respiratory tract.

Viral Nucleic Acid

Viruses are exceptionally flexible with respect to the nature of their genetic material. They have all four possible nucleic acid types—dsDNA, ssDNA, dsRNA and ssRNA.

All the four types are found in animal and plant viruses. Phages contain DNA as genetic material. The size of the genome varies greatly.

The smallest genome is around 1×10^6 D.

1. Define viruses.
2. Define capsid.
3. Define capsomeres.
4. Describe the spikes.
5. Describe about different morphological features of virus.
6. Explain about general characteristics of virus.

1. How do viruses differ from other cellular organelles.

Caspar, D.L.D. "Design principles in virus particle construction." In Horsfall, F.L. and Tamm, I. (eds.): *Viral and Rickettsial Infections in Man*, 4th edn. JB Lippincott, Philadelphia, 1975.

Fields, B.N. (ed.). *Virology*, 3rd edn. Lippincott-Raven Press. 1995.

Gajdusek, D.C. "Unconventional viruses and the origin and disappearance of kuru." *Science*. 197: 943. 1977.

Gelderblom, H.R. "Assembly and morphology of HIV. Potential effect of structure on viral function". *AIDS*. 5, 617–637.1991.

Mattern, C.F.T. "Symmetry in virus architecture." In Nayak, D.P. (ed.). *Molecular Biology of Animal Viruses*. Marcel Dekker, New York. 1977.

Palmer, E.L. and Martin, M.L. *An Atlas of Mammalian Viruses*. CRC Press, Boca Raton. 1988.

Nermut, M.V. and Stevens, A.C. (eds.). *Animal Virus Structure*. Elsevier, Amsterdam. 1989.

micro.magnet.fsu.edu/cells/index.html

biology.about.com/library/weekly/aa110200a.html

web.mit.edu/esgbio/www/cb/virus/virus.html

en.wikipedia.org/wiki/Virus

www.tulane.edu/~dmsander/WWW/335/335Structure.html

web.uct.ac.za/depts/mmi/stannard/virarch.html

www.embl-heidelberg.de/ExternalInfo/fuller/EMBL_Virus_Structure.html

www.uct.ac.za/depts/mmi/stannard/linda.html -

medicine.wustl.edu/~virology/

Classification of Animal Viruses

Many viruses in the environment are shown to be infectious to animals and humans. To distinguish them, these agents are classified. Morphology is probably the most important characteristic feature of virus classification. Modern classifications are primarily based on virus morphology, the physical and chemical nature of virion, constituents and genetic relatedness. Nucleic acid properties such as general type, strandedness, size and segmentation are also included in the classification system. Recent ICTV (International Committee on Taxonomy of Viruses) system of virus classification classifies 28 families of animal viruses. They are summarized in the following section.

THE SINGLE-STRANDED DNA VIRUSES

Family (subfamily)	Genus	Type species
Circoviridae	Circovirus	Porcine circovirus
	Gyrovirus	Chicken anaemia virus
Parvoviridae		
Parvovirinae	Parvovirus	Mice minute virus
	Erythrovirus	B19 virus
	Dependovirus	Adeno-associated virus 2

THE DOUBLE-STRANDED DNA VIRUSES

Order Caudovirales

Family (subfamily)	Genus	Type species
Poxviridae	Orthopoxvirus	Vaccinia virus
Chordopoxvirinae	Parapoxvirus	Orf virus
	Avipoxvirus	Fowlpox virus

Family (subfamily)	Genus	Type species
	Capripoxvirus	Sheeppox virus
	Leporipoxvirus	Myxoma virus
	Suipoxvirus	Swinepox virus
	Molluscipoxvirus	Molluscum contagiosum virus
	Yatapoxvirus	Yaba monkey tumour virus
Asfarviridae	Asfivirus	African swine fever virus
Iridoviridae	Ranavirus	Frog virus 3
	Lymphocystivirus	Lymphocystis disease virus 1
Herpesviridae		
	Simplexvirus	Human herpesvirus 1 and 2
Alphaherpesvirinae	Varicellovirus	Human herpesvirus 3
	Marek's disease-like viruses	Gallid herpesvirus 2
	ILTV-like viruses	Gallid herpesvirus 1
Betaherpesvirinae	Cytomegalovirus	Human herpesvirus 5
	Muromegalovirus	Murid herpesvirus 1
	Roseolovirus	Human herpesvirus 6
Gammaherpesvirinae	Lymphocryptovirus	Human herpesvirus 4
	Rhadinovirus	Simian herpesvirus 2
	Ictalurid herpes-like viruses	Ictalurid herpesvirus 1
Adenoviridae	Mastadenovirus	Human adenovirus 2
	Aviadenovirus	Fowl adenovirus 1
Polyomaviridae	Polyomavirus	Simian virus 40
Papillomaviridae	Papillomavirus	Cotton-tail rabbit papilloma virus

Figure 40.1 shows examples of DNA viruses of animals.

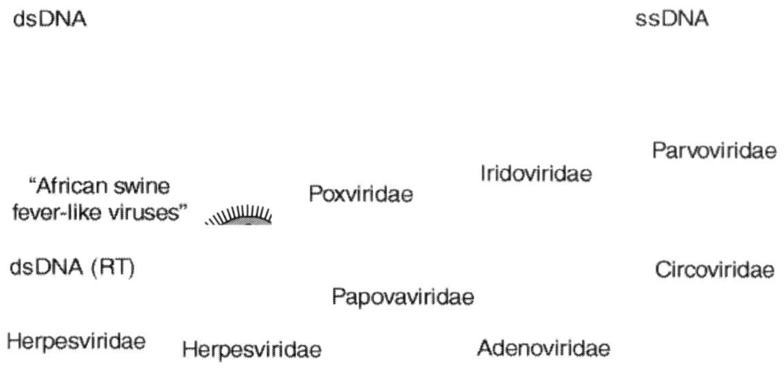

Figure 40.1 *DNA viruses of animals*

THE DNA AND RNA REVERSE TRANSCRIBING VIRUSES

Family (Subfamily)	Genus	Type species
Hepadnaviridae	Orthohepadnavirus	Hepatitis B virus
	Avihepadnavirus	Duck hepatitis B virus
Retroviridae	Alpharetrovirus	Avian leukosis virus
	Betaretrovirus	Mouse mammary tumour virus
	Gammaretrovirus	Murine leukemia virus
	Deltaretrovirus	Bovine leukemia virus
	Epsilonretrovirus	Walley dermal sarcoma virus
	Lentivirus	Human immunodeficiency virus
	Spumavirus	Chimpanzee foamy spumavirus

The double-stranded RNA viruses

Reoviridae		
	Orthoreovirus	Mammalian orthoreovirus
	Orbivirus	Bluetongue virus
	Rotavirus	Rotavirus A
	Coltivirus	Colorado tick fever virus
	Aquareovirus	Aquareovirus A
Birnaviridae	Aquabirnavirus	Infectious pancreatic necrosis virus
	Avibirnavirus	Infectious bursal disease virus

THE NEGATIVE-SENSE SINGLE-STRANDED RNA VIRUSES

Order Mononegavirales

Family (Subfamily)	Genus	Type species
Bornaviridae	Bornavirus	Borna disease virus
Filoviridae	Ebola-like viruses	Zaire Ebola virus
	Marburg-like viruses	Marburg virus
Paramyxoviridae		
Paramyxovirinae	Respirovirus	Sendai virus
	Morbillivirus	Measles virus
	Rubulavirus	Mumps virus
Pneumovirinae	Pneumovirus	Human respiratory syncytial virus
	Metapneumovirus	Turkey rhinotracheitis virus
Rhabdoviridae	Vesiculovirus	Vesicular stomatitis Indiana virus
	Lyssavirus	Rabies virus
	Ephemerovirus	Bovine ephemeral fever virus
	Novirhabdovirus	Infectious haematopoietic necrosis virus
Orthomyxoviridae	Influenzavirus A	Influenza A virus
	Influenzavirus B	Influenza B virus
	Influenzavirus C	Influenza C virus
	Thogotovirus	Thogoto virus
Bunyaviridae	Bunyavirus	Bunyamwera virus
	Hantavirus	Hantaan virus
	Nairovirus	Dugbe virus
	Phlebovirus	Rift Valley fever virus
	Deltavirus	Hepatitis delta virus
Arenaviridae	Arenavirus	Lymphocytic choriomeningitis virus

THE POSITIVE-SENSE SINGLE-STRANDED RNA VIRUSES

Family (Subfamily)	Genus	Type species
Picornaviridae	Enterovirus	Poliovirus
	Rhinovirus	Human rhinovirus A
	Hepatovirus	Hepatitis A virus
	Cardiovirus	Encephalomyocarditis virus
	Aphthovirus	Foot-and-mouth disease virus
	Parechovirus	Human parechovirus

Family (Subfamily)	Genus	Type species
Caliciviridae	Vesivirus	Swine vesicular exanthema virus
	Lagovirus	Rabbit haemorrhagic disease virus
	Norwalk-like viruses	Norwalk virus
	Sapporo-like viruses	Sapporo virus
	Hepatitis E-like viruses	Hepatitis E virus
Astroviridae	Astrovirus	Human astrovirus 1
Nodaviridae		
Nidovirales	Betanodavirus	Striped jack nervous necrosis virus
Coronaviridae	Coronavirus	Infectious bronchitis virus
	Torovirus	Equine torovirus
Arteriviridae	Arterivirus	Equine arteritis virus
Flaviviridae	Flavivirus	Yellow fever virus
	Pestivirus	Bovine diarrhoea virus
	Hepacivirus	Hepatitis C virus
Togaviridae	Alphavirus	Sindbis virus
	Rubivirus	Rubella virus

Figure 40.2 shows examples of RNA viruses of animals.

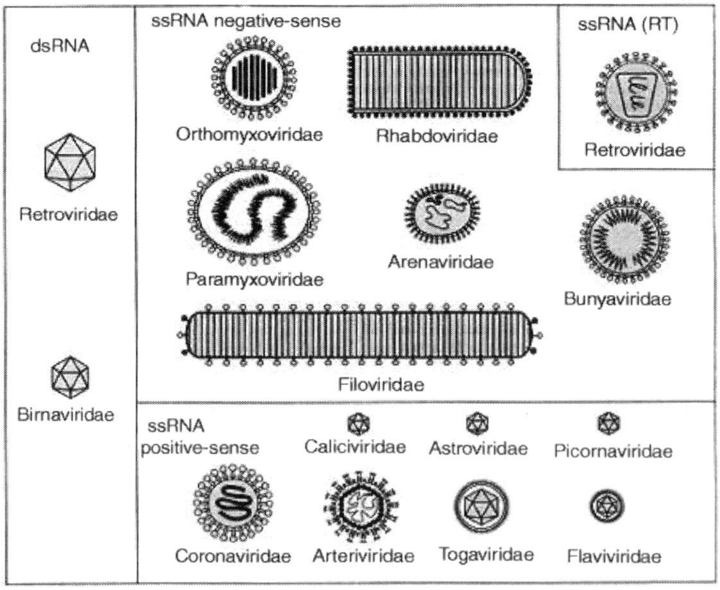

Figure 40.2 RNA viruses of animals

1. How do you classify viruses based on their genome?
2. Describe the different morphological features considered in classification.
3. Is site of replication and assembly one of the criteria for replication?
4. Name any five multi-genome virus.
5. Describe the nature of poxvirus and add a note on its taxonomic position.
6. Define ICTV.

Caspar, D.L.D. "Design principles in virus particle construction." In Horsfall, F.L. and Tamm, I. (eds.). *Viral and Rickettsial Infections in Man*, 4th edn. JB Lippincott, Philadelphia. 1975.

Fields, B.N. (ed.). *Virology*, 3rd edn. Lippincott-Raven Press. 1995.

Mattern, C.F.T. "Symmetry in virus architecture." In Nayak, D.P. (ed.). *Molecular Biology of Animal Viruses*. Marcel Dekker, New York. 1977.

Morse, S.S. (ed.). *The Evolutionary Biology of Viruses*. Raven Press, New York. 1994.

Murphy, F.A., Fauquet, C.M. and Bishop, D.H.L. *et al.* (eds.). *Virus Taxonomy: Sixth Report of the International Committee on Taxonomy of Viruses*. Springer-Verlag, New York. 1995.

Singh, I.P., Chopra, A.K. and Coppenhaver, D.H. *et al.* "Vertebrate brains contain a broadly active antiviral substance." *Antiviral Research.* 27:375. 1995.

Strayer, D.S., Laybourne, K.A. and Heard, H.K. "Determinants of the ability of malignant fibroma virus to induce immune dysfunction and tumor dissemination in vivo." *Microb Pathol.* 9:173. 1990.

en.wikipedia.org/wiki/Virus_classification

www.uct.ac.za/microbiology/tutorial/classif.htm

www.ncbi.nlm.nih.gov/books/bv.fcgi?rid=mmed.chapter.2252

www.gsbs.utmb.edu/microbook/ch041.htm

www.web-books.com/MoBio/Free/Ch1E2.htm

www.answers.com/topic/virus-classification

www.ias.ac.in/meetings/myrmeet/15mym_talks/dchattopadhyay/img7.html

www.sparknotes.com/biology/microorganisms/viruses/section2.rhtml

www.bact.wisc.edu/Microtextbook/index.php

Diagnosis of Viral Infections

INTRODUCTION

Viruses are obligate intracellular parasites, which cause a variety of human infections. These infections range from mild common cold infection to life-threatening AIDS infections. Better treatment of infections needs proper diagnosis. Diagnosis is broadly classified into clinical diagnosis and laboratory diagnosis. Clinical diagnosis is performed by the physician but laboratory diagnosis is performed by medical microbiologists who are specialists in virology. This chapter describes about the diagnosis process.

Viral laboratory diagnosis is broadly classified into four sections. They are:

1. Direct examination
2. Indirect examination
3. Serology
4. Molecular diagnosis

Samples or specimen materials are the prime to be used for the diagnosis of an infection.

COLLECTION OF SPECIMEN FOR VIRAL DIAGNOSIS

Specimen collection and transport play an important role in viral diagnosis. Based on the site of infection, sample collection procedures will vary (Table 41.1).

Table 41.1 Samples used for identifying human viruses

Type of specimen	Viruses associated with the specimen	Quantity of specimen and collection method
Blood	Cytomegalovirus(CMV) Enterovirus Herpes simplex virus Varicella-zoster virus Rubeola virus Arbovirus Hepatitis virus Retrovirus Human parvovirus Epstein–Barr virus Human herpesvirus 6	5–10 ml of anticoagulated blood is usually collected by needle aspiration procedure. Serum is separated by centrifugation process.
CSF	Herpes simplex virus Varicella-zoster virus Cytomegalovirus Epstein–Barr virus Human herpes virus 6 Arbovirus JC virus	2–5 ml of CSF is collected through lumbar puncture. It is collected mainly during meningitis and encephalitis.
Stool or rectal swab	Enterovirus Rotavirus Adenovirus Coxsackie virus Echo virus Calicivirus Astrovirus Poliovirus Hepatitis A virus	2–4 g of formed or liquid stool is collected in a container. Rectal swab is used for neonates.
Urine	Cytomegalovirus Adenovirus Mumps virus Rubeola virus Rubella virus Enteroviruses	5–10 ml of freshly voided urine is collected in wide-mouthed container. Other techniques are suprapubic aspiration and catheter collection.

(Contd.)

Table 41.1 (Continued)

Type of specimen	Viruses associated with the specimen	Quantity of specimen and collection method
Biopsy tissue	Herpesvirus Cytomegalovirus	Tissue specimens obtained should be placed in a sterile container and covered with an adequate volume of viral transport medium.
Ocular specimen	Adenovirus Herpes simplex virus Varicella-zoster virus Enterovirus Influenza virus Mumps virus Measles virus Newcastle disease virus Poxvirus	Conjunctival swab–cotton swab is performed Corneal scraping is performed with the help of fine scalpels. Ocular fluid is collected by draining abscess from the inflamed area.
Skin lesions	Varicella-zoster virus Herpes simplex virus Human papilloma virus Rabies virus Poxvirus	Vesicular lesions, fluid-filled lesions are recovered from crusted vesicles. Dacron or rayon swab is also used.
Amniotic fluid	Cytomegalovirus Rubella virus	2 ml of amniotic fluid is collected.
Respiratory tract specimen	Respiratory syncytial virus Influenza virus Parainfluenza virus Rhinovirus Adenovirus Coronavirus Cytomegalovirus Herpes simplex virus	Nasopharyngeal swab, nasopharyngeal aspirates, throat swab, sputum, bronchial washing, bronchoalveolar lavage fluid

DIRECT EXAMINATION

Clinical specimens are directly examined for the presence of virus particles, viral antigens or viral nucleic acid.

Methods for direct examination include

i. Electron microscopy
ii. Light microscopy for inclusion bodies
iii. Antigen detection by immunofluorescence
iv. Immunoelectron microscopy

Direct examination methods are often called as rapid diagnostic methods because they can give results within the same or next day.

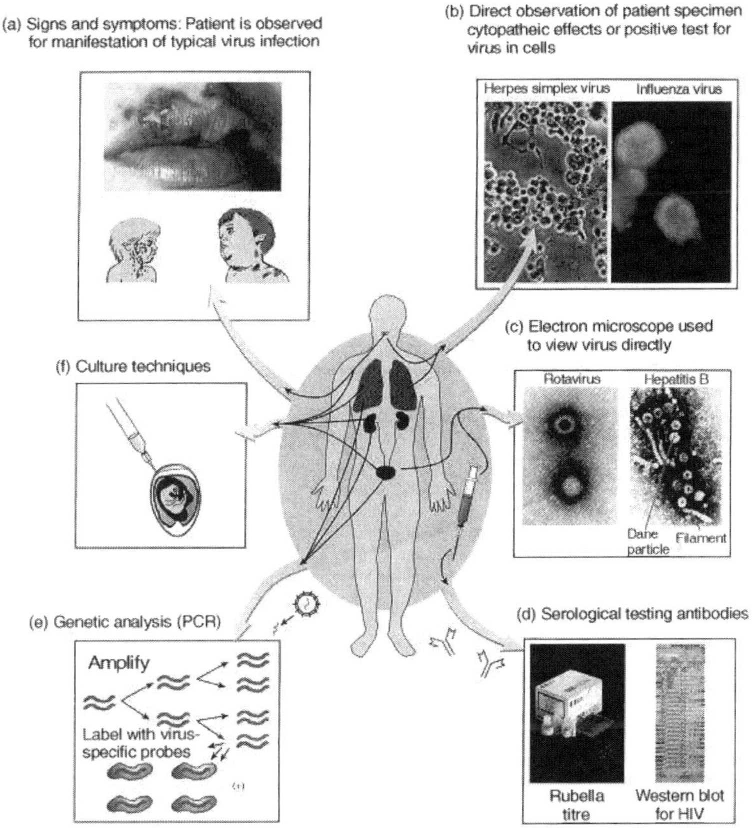

Figure 41.1 Summary of methods used to diagnose virus infections

Antigen Detection

Immunofluorescence (IF) is one of the methods adapted to detect antigens from clinical specimens, e.g. influenza A, B and adenovirus detection from respiratory specimens and detection of rotavirus antigen in faeces.

It is widely used for the rapid diagnosis of viral antigens as well as for the detection of viral specific antibodies. This test makes use of a fluorescence-labelled antibody to stain specimens containing specific viral antigens, so that the stained cells fluoresce under UV illumination.

There are two types of IF, direct IF and indirect IF. Direct IF detects antigens whereas indirect IF test detects antibodies. Fluorescein, rhodamine and phycoerythrine are the dyes used for labelling antibodies.

Electron Microscopy

Viral particles are detected and identified on the basis of morphology using electron microscopy. A magnification of around 50,000 is normally used, e.g. detection of rotavirus, calicivirus from faeces.

Electron microscopes use a beam of highly energetic electrons to examine objects on a very fine scale.

There are two types of electron microscopy:

i. *Direct or immunoelectron microscopy (IEM)* Sample is treated with specific antisera before being examined under electron microscope. Viral particles present will be agglutinated and thus are cointegrated together by the antibodies.

ii. *Solid phase immunoelectron microscopy (SPIEM)* The grid is coated with specific antisera. Virus particles present in the sample will be absorbed on to the grid by the antibody.

Uses

- Electron microscopy is useful for the detection of fastidious gastroenteritis viruses such as, rota, adeno, calici and norwalk-like viruses.
- It is also useful for the rapid diagnosis of herpesvirus infections.
- It is occasionally used for the diagnosis of human papillomaviruses.

Light Microscopy

Replicating viruses often produce histological changes in infected cells. These changes may be characteristic or non-specific. Virus inclusion bodies are basically a collection of replicating virus particles either in nucleus or cytoplasm, e.g. negri bodies in rabies virus infection, cowdry type A inclusion bodies of herpesvirus.

Viruses are cultured and their antigens are detected by using cell cultures, embryonated eggs or laboratory animals.

INDIRECT METHOD

Cell Culture

Cell cultures are the sole system for virus isolation. To prepare cell cultures, tissue fragments are first dissociated with the help of trypsin. The cells are placed in a flat-bottomed plastic container containing a suitable medium (Eagles' medium). After some time, cells will attach to the bottom of the container and start dividing, giving rise to primary cultures.

Primary culture is maintained by changing the fluid 2 or 3 times a week. When the cells become crowded, the cells are detached from the vessel wall by trypsin, and portions are used to initiate secondary cultures. Cells tend to adhere to the glass or plastic container and reproduce to form a monolayer (Figure 41.2).

Normal cells | Transformed cells

A tissue is treated with enzymes to separate the cells

Cells are suspended in culture medium

Normal cells or primary cells grow in a monolayer across the glass or plastic container. Transformed cells or continuous cell cultures do not grow in a monolayer.

Figure 41.2 Growth of cells (monolayer)

Media used for mammalian cell culture The basic component of cell culture medium is a balanced salt solution (BSS). It provides essential inorganic ions, correct osmolarity and correct pH. Two common BSS are Hanks and Eagles' balanced salt solutions. Some additional nutrients are also required for the proliferation of cells. They are 13 essential amino acids, 8 vitamins, antibiotics to prevent bacterial contamination, glucose, phenol red and 5% foetal calf serum. These are available in the BSS, e.g. Eagles' medium.

Types of Cell Cultures

Primary cells Prepared directly from animals and can be subcultured only once or twice, e.g. monkey kidney cells.

Semi-continuous cells Derived from human foetal tissue and can be subcultured 20–50 times, e.g. human embryonic kidney and skin fibroblasts.

Continuous cells Derived from tumours of human, e.g. HeLa (derived from Langerhans' cells of cancer patient Henrietta) and Vero cell lines.

Viruses isolated by cell culture are shown in Table 41.2.

Vero, Hep-2, human diploid cells (HEK and HEL), human amnion, human diploid fibroblasts, MK, BGM, LLC-MK2, Rhabdomyosarcoma, MDCK, RK13 cell lines are used for the cultivation of viruses.

Cultured viruses are identified by cytopatheic effect, haemadsorption, interference and immunofluorescence tests.

Table 41.2 Viruses isolated by cell culture

Readily isolated viruses	Less frequently isolated viruses
Herpes simplex virus	Varicella–zoster virus
Cytomegalovirus	Measles virus
Adenovirus	Rubella virus
Poliovirus	Rhinovirus
Coxsackie B virus	Coxsackie A virus
Echo virus	
Influenza virus	
Parainfluenza virus	
Mumps virus	
Respiratory syncytial virus	

Cytopatheic effect Viruses cause morphological changes in the cell culture and are observed under microscopic examination. These morphological changes are called cytopatheic effects.

Haemadsorption If haemagglutinating viruses are multiplying in the cell culture, the erythrocytes will adsorb onto the surface of cells. This is known as haemadsorption.

Interference The growth of the first virus will inhibit the second virus infection. This is called interference.

Haemagglutination It can be used to detect a virus carrying haemagglutinin in its envelope. The virus is mixed with RBCs, the virus cross-links the RBC and cause agglutination.

Embryonated Eggs

Woodruff and Goodpasture used fertilized chicken egg for viral cultivation. Chorioallontoic membrane (CAM) can be used for the inoculation of vaccinia and rabies viruses. Chick embryo (8–12-days old) is used for viral cultivation.

Inoculation of embryonated eggs is used for growing some viruses such as mumps and influenza viruses on allantoic cavity and myxoma on chorioallantoic membrane. During inoculation a hole is drilled in the shell of the egg. Viral suspension or virus-containing tissue is injected into the proper locations in the egg (since viruses are able to reproduce only in certain parts of the embryo). The injection site determines the membrane on which the viruses will grow.

Death of the embryo or formation of typical pocks or lesions on the egg membranes of the eggs, result from viral growth. The viruses are then identified with the help of serological techniques (Figure 41.3).

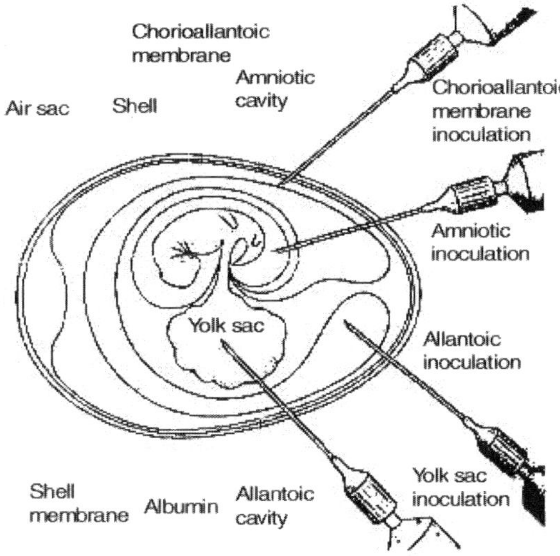

Figure 41.3 Inoculation of an embryonated egg

Laboratory Animals

Some animal viruses are cultured on living animals such as mice, rabbits and guinea pigs. Animal inoculation may be used as a diagnostic procedure for isolation and identification of virus from a clinical specimen. After inoculation, the morphological changes are observed (signs of disease or death of infected tissues) and the virus is identified.

SEROLOGY

Detection of rising titres of antibody between acute and convalescent stages of infection or the detection of IgM in primary infection is called serology. Serum is used in serological diagnosis processes.

Serology forms the mainstay of viral diagnosis. During viral exposure, the first antibody to appear is IgM, which is followed by a much higher titre of IgG. Many different serological tests are available to diagnose viral infections.

Classical techniques of serology are

- Complement fixation test
- Haemagglutination inhibition test
- Immunofluorescence technique
- Neutralization technique
- Single radial haemolysis

New techniques include

- Radioimmunoassay
- ELISA
- Particle haemagglutination
- Western blot
- Recombinant immunoblot assay

Neutralization Test

Neutralization of a virus is defined as the loss of infectivity through reaction of the virus with specific antibody. Virus and the serum are mixed under appropriate conditions and then inoculated into cell culture, eggs or animals. The presence of virus not neutralized may be detected by reactions such as cytopatheic effect, haemadsorption, interference and immunofluorescence tests (Figure 41.4).

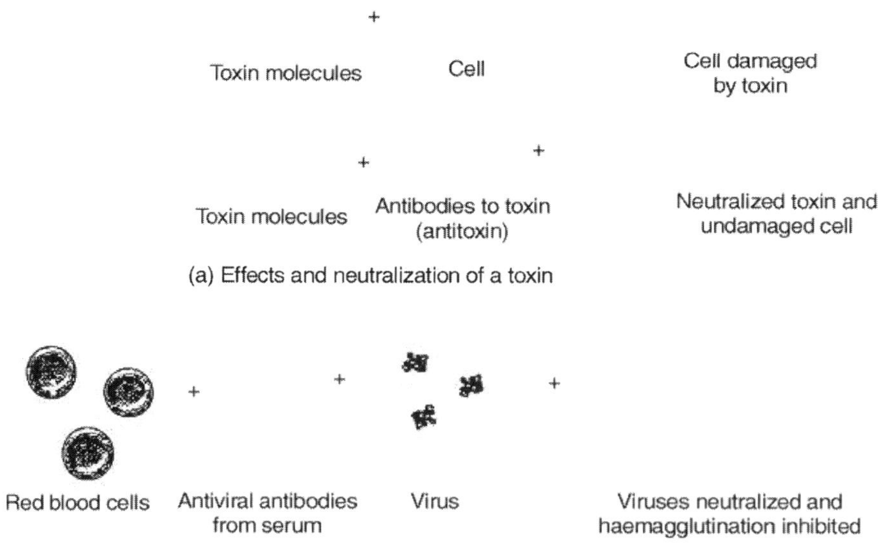

Figure 41.4 Neutralization of virus in positive haemagglutination inhibition test

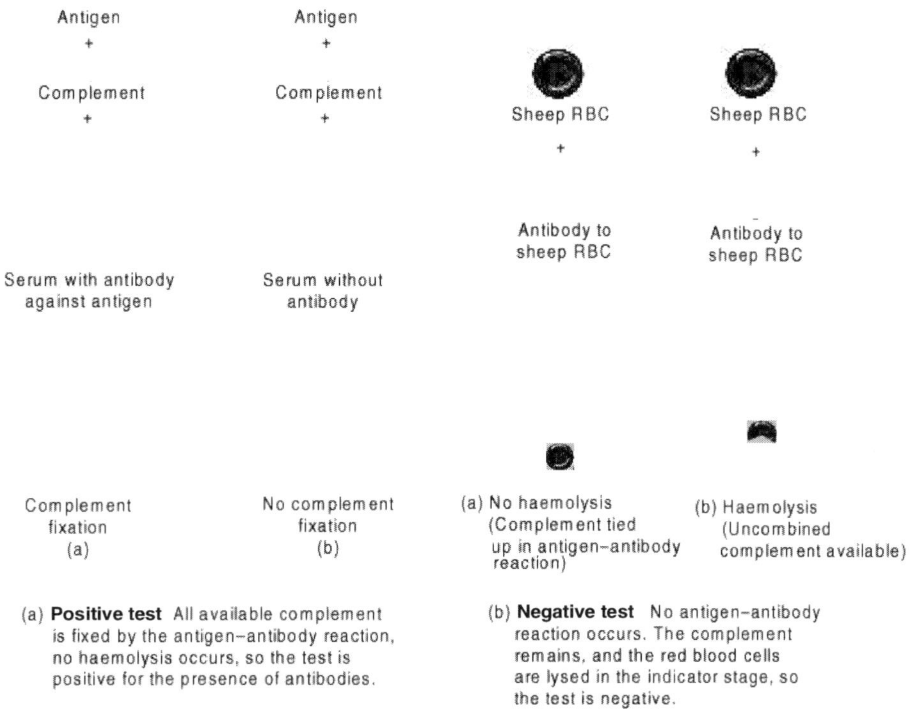

Figure 41.5 Complement fixation test

Complement Fixation Test

Complement fixation test (CFT) was first introduced by Wasserman in 1909 for the diagnosis of syphilis. Now this technique is also used for the diagnosis of viral infections. CFT is convenient and rapid to perform, demand for equipment and reagent is small and varieties of test antigens are readily available.

It is a simple test which has two antigen–antibody reactions, one of which is the indicator system (Figure 41.5).

The test procedure consists of the following steps

1. Two test tubes are taken.
2. Antigen is added to both the tubes.
3. Test serum is added to one tube.
4. If antibody is present, Ag–Ab complex will form and the complement will be fixed when the complement is added.
5. When complement is added, if antibody is present it fixes complement and consumes it.

6. Indicator cells and anti-erythrocytic antibodies are added.
7. If the complement is not fixed and is present in the tube, the indicator cell will lyse (negative result).
8. If the complement has been fixed by the Ag–Ab complex, there is no lysis of cells (positive result).

First reaction takes place between a known virus and a specific antibody in the presence of complement. Complement is fixed by antigen–antibody complex.

The second antigen–antibody reaction consists of reacting sheep RBCs with haemolysin. When this indicator system is added to the reactants, the sensitized RBCs will lyse only in the presence of free complement.

Haemagglutinin Inhibition Test (HAI)

Some viruses have the ability to agglutinate the erythrocytes of mammalian or avian species, e.g. influenza virus, parainfluenza virus, adenovirus, rubella virus, alphaviruses, bunyaviruses, flaviviruses, and some strains of picornaviruses.

Antibodies against haemagglutinin prevent haemagglutination, which is the main principle of this test. HAI test is simple to perform and requires inexpensive equipments and reagents.

Procedure

- The patient's serum is diluted from 1 : 8 to 1 : 1024 with the help of bovine albumin vernol buffer (BAVB).
- Non-specific inhibitors of viral haemagglutination are removed by kaolin, potassium periodate or by heat treatment.
- The serum is added to the fixed dose of viral agglutinin-containing microtitre plate.
- Agglutinable erythrocytes are added and incubated. Only the control erythrocytes show button at the bottom of the well.
- Absence of agglutination indicates infection.

Single Radial Haemolysis

It is routinely used for the detection of rubella with its specific IgG and also for the diagnosis of mumps. Test sera are placed in wells on a plate containing rubella antigen-coated RBC and complement.

The presence of rubella-specific IgG is detected by the lysis of rubella antigen-coated RBC. The zone of lysis around the well is dependent on the level of specific antibody present.

Radioimmunoassay

One of the most sensitive techniques for detecting antigen or antibody is Radioimmunoassay (RIA). S.A.Berson and Rosalyn Yalow developed this technique in 1960. Microtitre wells are coated with a constant amount of antibody specific for an antigen. Serum and specific radiolabelled antigen are then added. After incubation, the supernatant is removed and the amount of radioactivity bound to the antibody is determined. If the sample is infected, the amount of label bound will be less than in the control with uninfected serum.

Gamma-emitting isotope ^{125}I and beta-emitting isotope ^{3}H are commonly used for radioimmunoassay.

Enzyme-linked Immunosorbent Assay (ELISA)

Engval and Perlmann developed the technique of ELISA in 1970. There are three important types of ELISA.

- Competitive ELISA
- Sandwich ELISA
- Indirect ELISA

Competitive ELISA   It is used for the detection of antigens. Antibody is first incubated with a sample containing antigen. The antigen and antibody complex is added to the

Figure 41.6 Enzyme-linked immunosorbent assay (ELISA)

antigen-coated microtitre well. If more antigen is present in the sample, then less free antibody will be available to bind to the antigen-coated well. Addition of an enzyme-conjugated secondary antibody specific to the primary antibody can be used to determine the amount of primary antibody bound to the well. It is a quantitative test for antigen detection.

Sandwich ELISA Antigen can be detected by this method. Antibody is immobilized on a microtitre plate. A sample containing antigen is added and allowed to react with antibody. After washing, enzyme-linked second antibody is added and allowed to react with bound antigen. Substrate is then added to measure colour reaction (Figure 41.6).

Indirect ELISA Antibody can be detected by indirect ELISA. Serum containing antibody can be detected by adding antigen-coated microtitre well and allowed to react with antigen. After washing the non-bound primary antibody, enzyme-conjugated secondary antibody is added, which binds to the primary antibody. Free secondary antibodies are washed and a substrate for the enzyme is added. The amount of colour is directly proportional to the quantity of antibody present in the serum.

VIRAL GENOME DETECTION OR MOLECULAR METHODS

To overcome drawbacks of serological diagnosis genomic or molecular methods are developed. Drawbacks of serological methods are:

- Absence of antibodies during early stages of infection and also in the later stages of some infections.
- Immunosuppressed patients may fail to produce virus-specific antibodies.
- Passively acquired antibody may confuse serological results.

Molecular methods detect viruses, which cannot be cultured *in vitro*. Detection of viral nucleic acid is the main base for molecular diagnosis.

Analysis of nucleic acid involves the separation of nucleic acid and annealing of nucleic acid by base-pairing with known sequence. Specificity and sensitivity are achieved in diagnosis by genome detection methods. Nucleic acid probes are useful for the genomic diagnosis.

Nucleic Acid Probes

Nucleic acid probes are segments of DNA or RNA that have been labelled with enzymes, antigenic substrates, chemiluminescent moieties or radioisotopes. They can bind with high specificity to complementary sequences of nucleic acids. Probes can be directed to either DNA or RNA.

Probes are synthesized chemically, by PCR or by cloning.

- Oligonucleotide probes that are 50 bases long are synthesized chemically.

- PCR produces probes that are several hundred bases long.
- Cloning produces probes of several thousand bases long.

Molecular methods like PCR and nucleic acid-based amplification are used for the detection of viral genome. Classical techniques include dot blot and Southern blot, northern blot and *in situ* hybridization. Newer molecular techniques are polymerase chain reaction, ligase chain reaction, nucleic acid based amplification and branched DNA.

DOT Blots

DNA dot blotting involves spotting of multiple samples of DNA at different dilutions on a nylon or nitrocellulose membrane. Labelled probe is hybridized to the DNA on the membrane. Picogram quantities of DNA can be detected by dot blotting. Applying a series of known quantities of DNA in the same filter helps in quantitation of DNA. It is used for the detection of parvovirus infection. This technique needs larger volume of nucleic acid than other techniques.

Southern Blotting

This method is used to distinguish between closely related viruses, e.g. HSV1 and HSV2, HPV6 and HPV16. Two methods are employed. They are electrophoretic separation of DNA and hybridization. The sample can be a PCR product or restriction enzyme digest.

Northern Blotting

This is similar to Southern blotting but the difference is that the sample is RNA.

In situ Hybridization

This method is used for the localization of specific nucleic acid sequences in cells or tissue sections. Paraffin-embedded tissue sections are dewaxed and are treated with specific proteolytic enzymes to enhance permeability. The sample is denatured and a labelled DNA or RNA probe is used to detect the target sequence of interest. This method is used to localize virus-specific nucleic acid in diseased areas of tissue.

Polymerase Chain Reaction (PCR)

PCR is an extremely powerful research tool, which has become a part of the routine clinical laboratory tests, especially in the field of microbiology, virology and diagnostic genetics.

PCR is used to detect the presence of a specific segment or sequence of nucleic acid, which could be diagnostic for infectious agents such as bacteria, virus and fungus. This technique is based on nucleic acid amplification.

In PCR, the following steps are involved in each cycle of replication (Figure 41.7).

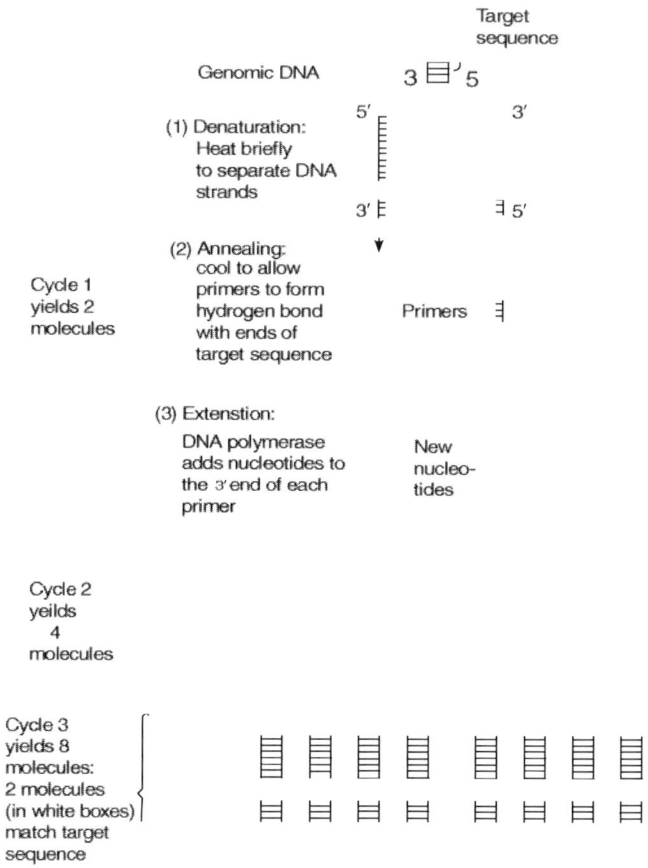

Figure 41.7 Polymerase chain reaction

1. Denaturation of DNA duplex at high temperature.
2. Annealing of oligonucleotide primers to the separated strands.
3. After the primer attachment they "kick start" the polymerization which then elongates the developing nucleic acid.
4. One cycle produces two DNA, one strand is the template and the other is the new strand.
5. The cyclical variation of temperature of the reaction mixture, i.e., 95°C (denaturation), 45°C (annealing) and 72°C (polymerization) is carried out by an automated thermal cycler.

6. Cycle is repeated a number of times, each cycle resulting in a logarithmic increase in the amount of DNA which is amplified.
7. Within 20 cycles a million fold amplification of the starting amount of nucleic acid can be achieved.

DNA quantity is analysed by nucleic acid hybridization using a labelled probe or other known techniques.

Uses

- PCR is useful for strain characterization, e.g. human papilloma virus differentiation.
- HIV detection in infants is possible by using RTPCR or PCR.
- Non-cultivable viruses are detected by this technique.

New quantitative amplification techniques are

1. *RT-PCR* It is a reverse transcriptase PCR, used specifically to detect RNA viruses.
2. *Branched DNA* In this assay, the amplification is achieved by a magnification of the signalled detection molecule rather than the target.
3. *Nucleic acid sequence-based amplification (NASBA)* It is a novel method for RNA amplification using an isothermal process. It employs three enzymes, viz. reverse transcriptase, RNase, T7 RNA polymerase.

A primer defining the 3′ end of the target sequence is annealed to an ssRNA. The primer includes a T7 polymerase promoter sequence, which is essential for RNA polymerase to initiate transcription. Reverse transcriptase completes RNA–DNA hybrid, which is converted into a negative ssDNA by RNase H. The second primer is attached and a double-stranded DNA is synthesized by reverse transcriptase incorporating T7 promoter. T7 RNA polymerase transcribes up to 1000 copies of negative-sense RNA to the target. In the cycle phase primer 2 anneals to the negative-sense RNA. Reverse transcriptase completes a DNA strand by using RNA as a template, resulting in the DNA : RNA hybrid. The RNase H enzyme destroys the RNA strand leaving a positive-sense single-stranded DNA. Primer 1 completes an ssDNA, consisting of target sequence with the help of reverse transcriptase.

NASBA is used for the detection and quantification of RNA viruses and mRNAs.

In this method, signal is amplified. Target nucleic acid is captured to the solid phase via multiple probes. Extended probes are hybridized with adjacent target sequences and contain additional sequences homologous to the branched amplification multimer. Enzyme-labelled oligonucleotides bind to the branched DNA via homologous base pairing and enzyme probe complex is measured by detection of chemiluminesence.

Uses

- Identification and quantification of HIV in peripheral blood.
- Confirmation of HCV infection.
- Detection of herpesvirus sequence in CSF.
- Detection of enterovirus and polyomavirus sequence in CSF.
- Analysis of viral drug resistance.

Western Blotting

It involves separation of viral proteins by SDS-PAGE (Sodium dodecylsulphate-polyacrylamide gel electrophoresis), transferring the resolved antigen to nitrocellulose membrane and identifying specific antigen through antigen–antibody reactions (immunoblotting). It is used for the confirmation of results from ELISA.

DISEASES ASSOCIATED WITH VIRUSES

1. Wart	Papilloma virus
2. Chickenpox	Varicella-zoster virus
3. Smallpox	Poxvirus
4. Influenza	Influenza virus
5. Parainfluenza	Parainfluenza virus
6. Measles	Measles virus
7. Mumps	Mumps virus
8. Hepatitis	HAV, HBV, HCV, HDV, HEV
9. Polio	Poliovirus
10. Gastroenteritis	Rotavirus, calicivirus, adenovirus, astrovirus
11. Common cold	Rhinovirus
12. Myocarditis	Enteroviruses
13. Pericarditis	Adenoviruses
14. Conjunctivitis	Adenoviruses, herpesviruses
15. Retinitis	Herpesviruses
16. Cystitis	Adenoviruses
17. SARS	Coronavirus
18. AIDS	HIV
19. Infectious mononucleosis	Epstein–Barr virus
20. Meningitis	Enteroviruses, herpesviruses, mumps virus
21. Encephalitis	Herpesviruses, arboviruses, JC viruses, mumps virus, rabies virus

22. Rash	Rubella virus, measles virus, HHV-6, parvovirus
23. Congenital infection	Herpesvirus, rubella virus, parvovirus
24. Rabies	Rabies virus

1. Who is the person responsible for viral diagnosis?
2. Name the four stages of viral diagnosis.
3. Explain the role of inclusion bodies on viral diagnosis.
4. Explain the various cell culture systems used for the culture of viruses.
5. How is complement fixation test used for viral diagnosis?
6. Explain the role of blotting technique in viral diagnosis?
7. List few viruses and their associated diseases in human.
8. Explain how sample collection is important in diagnosis.
9. List some of the sample collection procedures for viral diagnosis.
10. Write short notes on
    i. Immunofluorescence
    ii. Embryonated egg
    iii. Neutralization test
    iv. Haemagglutination test
    v. Radioimmunoassay
    vi. ELISA
    vii. Role of PCR on diagnosis

Smallpox

INTRODUCTION

Smallpox is caused by variola virus. Variola virus was first microscopically demonstrated by Buist in 1887. Paschen in 1906 developed a staining technique to observe elementary bodies in smears from smallpox lesions. The last endemic case of smallpox occurred in 1977 and total eradication was confirmed in 1980. Poxviruses are the largest viruses that infect vertebrates.

The family Poxviridae is classified into two subfamilies: Chordopoxvirinae, the poxviruses of vertebrates and Entomopoxvirinae, the poxviruses of insects.

Chordopoxivirinae are classified into six genera:

Orthopoxvirus	Viruses of mammals
Parapoxvirus	Viruses of ungulates
Capripoxvirus	Viruses of goats and sheep
Leporipoxvirus	Viruses of leporids
Avipoxvirus	Viruses of birds
Suipoxvirus	Viruses of pigs

CAUSATIVE AGENT

Variola virus is the causative agent of smallpox while vaccinia virus is used for the vaccination for smallpox. Poxvirus virions are large and brick-shaped. Orthopoxviruses are approximately 240 nm by 300 nm, with short surface tubules, 10 nm wide. Parapoxviruses are narrower (160 nm) and have one long tubule that winds around the virion (Figure 42.1).

Lateral body
Core
Genome

Figure 42.1 Structure of smallpox virus

Virions have a dumb bell-shaped core and two lateral bodies. The genome consists of one molecule of double-stranded DNA, from 130 kb (parapox) to 260 kb (fowlpox), and the core contains enzymes for virus uncoating and genome replication.

Chemical Composition

Virion contains more than 100 polypeptides and is made up of 3% DNA, 90%, protein and 5% lipid. Many enzymes are available in the core.

Resistance

Poxviruses are susceptible to UV and other irradiations, resistant to 50% glycerol and 1% phenol.

REPLICATION

Poxvirus replication takes place in the cytoplasm. The mechanism by which vaccinia or smallpox virus attaches to and enters susceptible host cells is not well understood. The result is release of core into the cytoplasm (Figure 42.2).

Upon the release of core into the host cytoplasm, the core synthesizes viral early mRNAs.

About half of the viral genes are expressed during the early phase of infection. Some early proteins are homologous to cellular growth factors, which can induce proliferation of neighbouring cells. Synthesis of early proteins also induces a secondary uncoating reaction in which a nucleoprotein complex containing the genome is released from the core. One of the early proteins mediate replication of viral genome.

Newly synthesized viral DNA molecules can serve as template for additional cycles of genome replication. Some of the DNA serves as a template for transcription of viral intermediate gene. Activation of transcription of intermediate genes also requires viral initiation protein, the product of early genes, that confers specificity for intermediate promoters on the viral RNA polymerase (vit f 2), which relocates from infected cell nucleus to the cytoplasm. Translation of intermediate gene leads to transcription of late genes. Late genes synthesize structural and functional proteins such as early initiation protein, which must be incorporated into virion during assembly.

Assembly of progeny virus particles begins, probably in association with internal membranes of the infected cells. Initial assembly leads to the formation of immature virions. Spherical-shaped immature virions acquire double membranes from Golgi network and are converted to brick-shaped intracellular mature virions (IMV). These may be released upon lysis.

Some of the virions acquire one more layer from Golgi and are converted into intracellular enveloped viruses (IEV). IEV infects fresh cells.

Several poxvirus genes resemble mammalian genes.

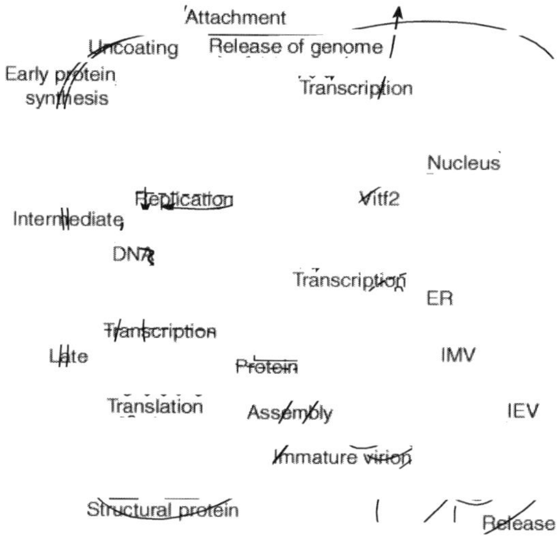

Figure 42.2 Replication of poxvirus

CLINICAL MANIFESTATION

Incubation period is around 12 days.

Poxvirus infections are characterized by the production of skin lesions. Lesion develops at the site of inoculation (usually the hand), and infection may be spread to other sites such as the face and/or genitals by scratching. The lesions of molluscum, usually multiple, are firm, pearly, flesh-coloured nodules.

PATHOGENESIS

The pathogenesis of localized poxvirus infections is simple. Virus invades through broken skin, replicates at the site of inoculation, and causes dermal hyperplasia and leucocyte infiltration. This causes lymphadenopathy and elicits an immune response. The lesion of molluscum is circumscribed by a connective tissue capsule and the dermis.

Human monkeypox is usually acquired via the respiratory tract, and during the 12th day of incubation period viraemia distributes infection to internal organs, which are damaged by virus infection. Spread to the skin initiates the clinical phase, and the lesions progress through the classic stages of macule to papule to vesicle to pustule to crust. Lymphadenopathy, usually involving the cervical and inguinal areas, is often marked.

LABORATORY DIAGNOSIS

Electron microscopy of vesicle or scab material is an effective means of rapid diagnosis.

Immunofluorescence of infected cell cultures will differentiate morphologically similar poxviruses from different genera (e.g. orthopoxvirus and yatapoxvirus).

Poxviruses are easily isolated in tissue culture and/or chicken embryos. Cultivation allows identification by biological and serum neutralization tests.

CONTROL

Control of the common human poxvirus infections depends on knowledge of their epidemiology. Control of infections such as cowpox which has an unknown reservoir, is virtually impossible. Person-to-person transmission is reduced by improving hygiene. The World Health Organization considers that the benefits of vaccination do not outweigh the risks and expense. Control of this disease depends on health education and on breaking the link with the animal reservoir; this last should be achieved by the use of forest land near villages for agriculture.

1. Explain the structure, replication and pathogenesis of smallpox virus.
2. Explain the role of vaccination in controlling viruses.
3. Discuss replication and pathogenesis of smallpox virus.
4. Write short notes on:
   i. Vaccinia virus
   ii. Variola virus

Baxby, D. "Poxvirus infections in domestic animals." p.17. In Darai, G.M. (ed.). *Virus Diseases in Laboratory and Captive Animals*. Martinus Nijhoff, Boston. 1988.

Esposito, J.J. and Knight, J.C. "Orthopoxvirus DNA: A comparison of restriction profiles and maps." *Virology*. 143: 230. 1985.

Fenner, F., Henderson, D.A., Arita, I. *et al*. *Smallpox and its Eradication*. World Health Organization, Geneva. 1988.

Fenner, F., Wittek, R. and Dumbell, K.R. *The Orthopoxviruses*. Academic Press, San Diego. 1989.

Jezek, Z. and Fenner, F. *Human Monkeypox*. Karger, Basel. 1988.

Johanneson, J.V., Krogh, H.K., Solberg, I. *et al*. Human *orf*. *J. Cutaneous Pathol.* 2: 265. 1975.

Scholz, J., Rosen-Wolff, A. and Bugert, J. *et al*. "Epidemiology of molluscum contagiosum using genetic analysis of the viral DNA." *J. Med. Virol.* 27: 87. 1989.

www.bt.cdc.gov/agent/smallpox/

en.wikipedia.org/wiki/Smallpox

www.nlm.nih.gov/medlineplus/ency/article

www.hhs.gov/smallpox/

www.emedicine.com/emerg/topic885.

www.who.int/csr/disease/smallpox

www.kidshealth.org/kid/health_problems/infection/smallpox.html

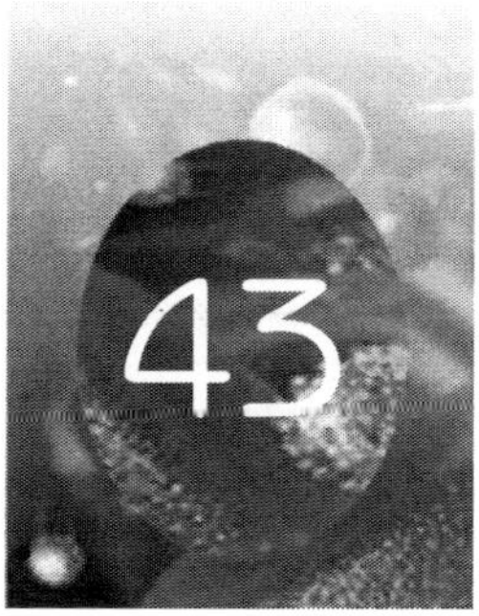

Common Cold

INTRODUCTION

Common cold is probably the commonest infectious disease in man. Bacteria-free filtrates of nasal secretions from patients had been shown to transmit colds to human volunteers as early as 1914. Tyrell and his colleagues isolated the virus in 1960 by inoculating specimens monkey tissue culture. Until their naming as rhinovirus, common cold viruses were initially called as JH, 2060, Salisbury virus or murvirus.

Common cold results from upper respiratory tract infections by any of the 200 known viruses and a few bacteria. Major common cold viruses are orthomyxoviruses, paramyxoviruses, coronaviruses, adenoviruses, enteroviruses and more than 100 strains of rhinoviruses.

Rhinoviruses are the members of Picorna group ("rhino" means "nose"). *Pico* means small and *rna* indicates RNA (small RNA viruses). Many people suffer from 2–4 cold attacks every year. These viruses are infectious only for human and chimpanzees. They have been grown in cultures of human embryonic lung fibroblast and organ cultures of human tracheal epithelium. They grow best at 33°C and at pH 7, the same temperature and pH, that exist in the nasopharynx of humans.

Common cold viruses are divided into two, based on their tissue tropism. M strains grow and produce a cytopatheic effect in monkey kidney cells. H strains are mostly isolated in human embryonic kidney cultures.

STRUCTURE AND PROPERTIES OF PICORNAVIRUS

The picornaviruses are small (22 to 30 nm), non-enveloped, single-stranded RNA viruses with icosahedral symmetry. The virus capsid is composed of four viral proteins VP1 to VP4, which form an icosahedral shell. The genome is a single-stranded RNA (molecular weight, approximately 2×10^6 to 3×10^6). The RNA strand consists of approximately 7,500 nucleotides and is covalently bonded to a non-capsid viral protein (VPg).

Resistance

Picornaviruses can survive for long periods in organic matter and are resistant to the low pH in the stomach (pH 3.0 to 5.0).

Picornaviruses are inactivated by pasteurization, boiling, formalin and chlorine. They are ether-resistant due to absence of essential lipids.

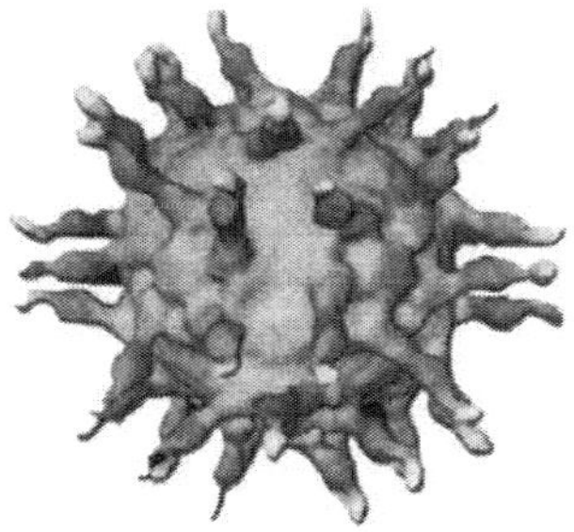

Figure 43.1 Picornaviruses

REPLICATION

Replication occurs in the cytoplasm.

Replication is similar in all picornaviruses (Figure 43.2).

- The virus binds to the cellular receptor.
- The mechanism of uncoating of RNA genome is unknown.
- Translation is initiated at an internal site, 741 nucleotides from the 5′ end of the viral RNA and polyprotein precursors are synthesized.
- Polyproteins are cleaved and produce individual proteins (P1, P2 and P3).
- P1 proteins contain viral structural proteins.
- P2 and P3 are responsible for proteases and RNA synthesis proteins.
- The proteins that are involved in RNA synthesis are transported into the membranous vesicles.
- Positive-sense RNA is also transported into the vesicles.
- It is copied into negative-sense RNA that act as the template for the synthesis of positive-sense RNA.
- Structural proteins are formed by partial cleavage of P1 precursor proteins.
- These proteins are transported into vesicles.
- Assembly takes place within the vesicles.
- Mature virions are released after cell lysis.

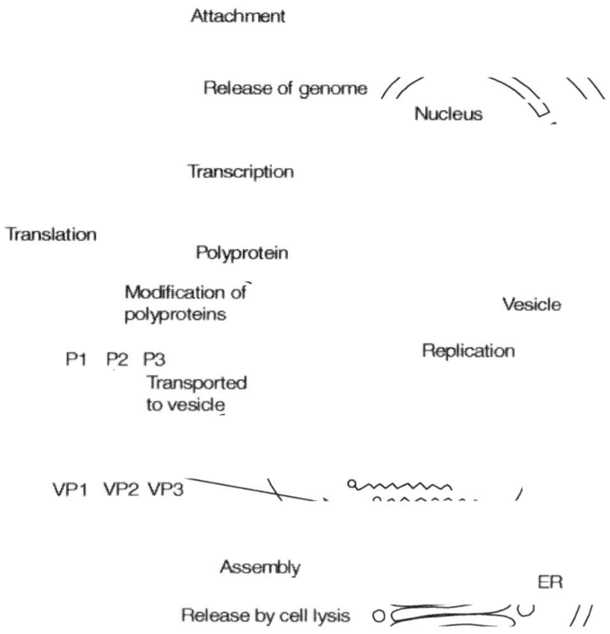

Figure 43.2 Replication of picornavirus/rhinovirus

CLINICAL MANIFESTATION

Incubation period is about 1–2 days. Usual symptoms in adults include irritation in upper respiratory tract, nasal discharge, headache, chill sensation, mild cold, malaise and cough. The nasal and nasopharyngeal mucosa become red and swollen. Secondary bacterial infection may produce otitis media, bronchitis, etc.

PATHOGENESIS

The virus enters the human body via respiratory tract through droplet nuclei. It lodges in the respiratory epithelial cells and infect adjacent cells. Replication takes place inside the cytoplasm. After the incubation period, it produces toxicity which causes cessation of the ciliary action and cells get sloughed off. Due to non-specific immune response, mucous secretion increases and inflammatory reaction also occurs. Infection stops due to interferon release and antibody production.

LABORATORY DIAGNOSIS

Diagnostic method includes culture and serology.

Culture

Nasal washings are used as specimen.

Human diploid fibroblast cells are used for cultivation of rhinovirus and incubated at 3°C. Recovery of the virus is done between 1–7 days.

The viruses are identified by cytopatheic effect and differentiation is done by neutralizing sera.

PREVENTION AND CONTROL

Prevention and control is possible by the following methods:

- Vaccination
- Antiviral agents
- Interruption of transmission

Multiple serotypes and antigenic drift create major problems in rhinoviral vaccine development. Formalin-inactivated, parenterally administered vaccines induce antibody in serum but not in nasal secretions.

A new live attenuated rhinovirus vaccine has been developed but not tested.

TREATMENT

There are no antiviral drugs available to treat common cold. Experimental drugs including such avidone, rhodanine, disoxaril and their analogs that block uncoating of virus are available. Exviroxine inhibits the polymerase.

For the reduction of transmission, hand washing and disinfection of inanimate objects are recommended.

1. List the important properties of rhinovirus.
2. Write a short note on common cold.
3. Describe prophylaxis of common cold.
4. Is there any method available to cultivate rhinovirus? Explain.

McCay, J. and Werner, G. "Different rhinovirus serotypes neutralized by antipeptide antibodies." *Nature.* 329:736. 1987.

Minor, P., Brown, F. and King, A. *et al*. "Classification and nomenclature of viruses family. Picornaviridae." In Fifth Report of International Committee on Taxonomy of Viruses. *Archives of Virology, Supplementum*. Francki, R.I.B. Fauquet, C.A. and Knudson, D.L. (eds.) Springer-Verlag Wien. New York. 1992. pp. 320–326.

World Health Organization: Manual for Managers of Immunization Programmes. WHO/EPI/Polio/89.1, 1989.

Yin-Murphy, M. "Acute hemorrhagic conjunctivitis." In Melnick JL (ed.). *Progress in Medical Virology*. Karger, Basel. 1984. pp. 23–44.

Yin-Murphy, M. "Enteroviruses. Human enteroviruses (Serotypes 68–71)." In: *Encyclopedia of Virology*. Webster, R.G. and Granoff, A. (eds.). United Kingdom Academic Press Ltd., London. 1994. pp. 378–384.

www.commoncold.org/

www.niaid.nih.gov/factsheets/cold.htm

en.wikipedia.org/wiki/Common_cold

www.nlm.nih.gov/medlineplus/commoncold.html

www.nlm.nih.gov/medlineplus/ency/article/000678.htm

www.kidshealth.org/parent/infections/bacterial_viral/cold.html

www.mayoclinic.com/health/common-cold/DS00056

www.commoncold.co.uk/

www.medicinenet.com/common_cold/article.htm

www.giantmicrobes.com/us/products/commoncold.html

Influenza

INTRODUCTION

It is a viral infection of the respiratory tract resulting in fever, headache, muscle ache and weakness. Influenza virus, a member of the orthomyxo family, causes this disease. The name Myxovirus was proposed originally for a group of enveloped RNA viruses characterized by their ability to get absorbed into mucoprotein receptors on erythrocytes, causing haemagglutination.

HISTORY

- Influenza is an acute infectious disease of the respiratory tract, which occurs in sporadic, epidemic and pandemic form.
- Italians gave the name influenza during the year 1358.
- In 1933, Smith isolated the causative agent (influenza A).
- Burner (1935) developed chick embryo technique for propagation of virus.
- Francis and Magill (1940) isolated a serotype of the influenza and named it as influenza B.
- Taylor (1949) isolated the third serotype of influenza virus, type C.

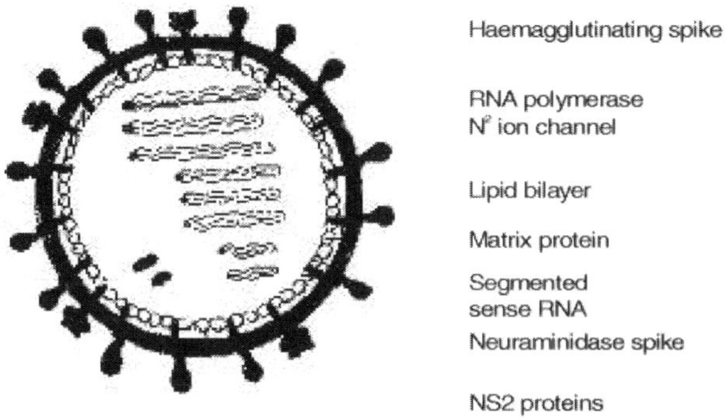

Figure 44.1 Structure of influenza virus

CAUSATIVE AGENT

Influenza virus causes influenza. It belongs to Orthomyxovirus family and consists of three species, namely A, B and C. Type A is usually responsible for the large outbreaks and is a constantly changing virus. New strains of type A virus develop regularly and result in new epidemics every few years. Type B and C are fairly stable viruses. Type B causes smaller outbreaks. Type C usually causes mild illness like common cold. Influenza virus is spherical and is 80–120 nm in diameter. Antisense RNA genome occurs in 8 separate segments containing 10 genes. The segments are complexed with nucleoprotein to form nucleocapsid with helical symmetry. Nucleocapsid is enclosed in an envelope consisting of a lipid bilayer and two surface glycoproteins, a haemagglutinin and neuraminidases. Influenza C virus has 7 segments of RNA and only one surface protein. Influenza viruses are readily inactivated by non-polar solvents and by surface-active agents.

Resistance

The virus is inactivated by heating at 50°C for 30 minutes. It remains viable at 0–4°C for about a week. Infectivity is lost rapidly at 20°C. It is preserved at –70°C or by freeze-drying.

The membrane protein is known as matrix protein or M protein. M2 protein projects through the envelope to form an ion channel. It is responsible for gene transfer. H gene is responsible for haemagglutinin spike and N gene is responsible for neuraminidase.

Spikes measure about 10 nm in length and have a molecular weight of 225000 dalton. 10 genes from 8 segments of antisense RNA are responsible for synthesis of viral proteins.

REPLICATION

Virus replication takes about 6 hours and kills the host cell. The virus enters permissive cells via the haemagglutinin subunit which binds to cell membrane glycolipids or glycoprotein containing sialic acid or *N*-acetylneuraminic acid, the receptor for virus adsorption. The virus is then engulfed by pinocytosis into endosomes. The acid environment of the endosome leads to uncoating of the nucleocapsid thus releasing it into the cytoplasm. A transmembrane protein derived from the matrix gene forms an ion channel for protons to enter the virion and destabilize protein binding allowing the nucleocapsid to be transported to the nucleus, where the genome is transcribed by viral enzymes to viral mRNA. Unlike replication of other RNA viruses, orthomyxovirus replication depends on the presence of active host cell DNA. The synthesized viral mRNA are transported to the cytoplasm, where it is translated by the host ribosome (Figure 44.2).

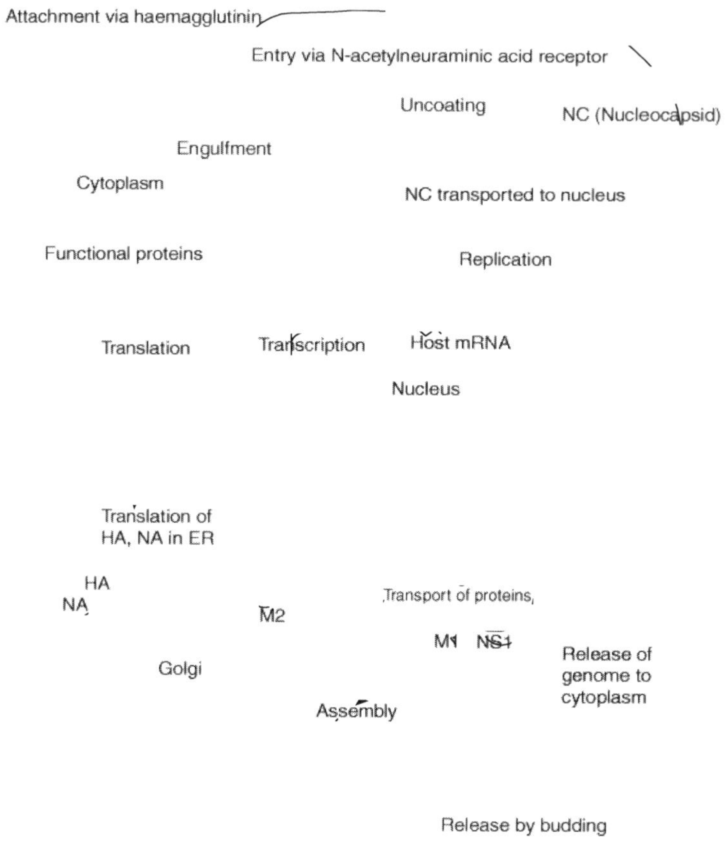

Figure 44.2 *Replication of orthomyxovirus—influenza virus*

mRNAs specifying viral membrane proteins (HA, NA, M) are translated by ribosome bound to endoplasmic reticulum and they undergo glycosylation. The nucleocapsid is assembled in the nucleus. After the attachment of M1 protein to newly synthesized RNA, viral RNA synthesis is stopped and nucleocapsids are transported out. HA and NA proteins are transported to the cell surface and are incorporated into the plasma membrane. Virion nucleocapsids along with NS2 associate with regions of plasma membrane containing HA and NA proteins. After acquiring the envelope, they undergo maturation as they bud through the host cell membrane.

During budding, the viral envelope haemagglutinin is subjected to proteolytic cleavage by host enzymes. It has 3 basic amino acids, which are specifically attacked by host enzymes. These are called taubenberger amino acids. This process is necessary for the released particles to be infectious.

CLINICAL MANIFESTATION

The incubation period is 1–4 days. The symptoms of influenza virus infection include the following.

Fever, malaise, cough with or without mucus, stuffy and congested nose, nasal discharge, sore throat, headache, clammy skin, muscle ache and stiffness, nosebleed, shortness of breath, vomiting, chillness, joint stiffness, sweating, elbow pain and fatigue, loss of appetite and abnormal taste.

PATHOGENESIS

Influenza virus is transmitted from person to person primarily through droplets released by sneezing and coughing. It is a highly contagious disease. Alveoli are the primary target of the virus. Infected cells will slough off, allowing extravagation of fluid and secondary submucosal inflammation. During the initial stage, liquefaction of mucus occurs. The infection in mucosal cell results in cellular destruction and desquamation of the superficial mucosa. The resulting oedema and mononuclear cell infiltration of the involved areas are accompanied by symptoms like cough, sore throat and nasal discharge. Most of the symptoms are because of interferons. Current evidence indicates that the extent of virus-induced cellular destruction is the prime factor determining the occurrence. In an uncomplicated case, virus can be recovered from respiratory secretions for 3–8 days. The disease may extensively involve the alveoli, resulting in interstitial pneumonia, sometimes with marked accumulation of lung haemorrhage and oedema.

Gene Reassortment

Because the influenza virus genome is segmented, genetic reassortment can occur when host cells are infected simultaneously with viruses of two different parent strains. If a cell is infected with two strains of type A virus, for example, some of the progeny virions will contain a mixture of genome segments from the two strains. This process may lead to influenza pandemics. This process is also called antigenic shift. It is the major antigenic change. Smaller antigenic change is called antigenic drift.

EPIDEMIOLOGY

Influenza viruses are classified as types A, B, C on the basis of antigenicity of their nucleoproteins and matrix protein. Rainy season is a peak time for influenza. Influenza epidemic is of two types. Both type A and type B viruses cause yearly epidemics. Type A causes influenza pandemics. Two different mechanisms of antigenic changes are responsible for producing the strains that cause these two types of epidemics. Some of the influenza strains are transmitted from animal to humans. Mostly it is transmitted through person to person contact and by droplets.

LABORATORY DIAGNOSIS

The most commonly employed method for laboratory diagnosis is recovery of the virus from specimens containing respiratory secretions such as nasal wash and throat swab or sputum.

Methods of diagnosis include isolation of virus, cold agglutination, influenza complement fixation test and immunofluorescent technique.

Isolation of Virus

Respiratory secretions are treated with antibiotics and inoculated into amniotic cavity of 10–11-days-old egg or monkey kidney cells and incubated at 35°C for 3 days. Then eggs are chilled and amniotic fluid is harvested. The presence of viral antigens are demonstrated by using haemadsorption test at 4°C. Type B agglutinates both guinea and fowl cells. Type C agglutinates only fowl cells.

ELISA is also useful for demonstration of antigens (Refer chapter 41).

Serological examination is by specific antigen and antibody reaction. They are haemagglutination, inhibition test, neutralization test (Refer chapter 41).

COMPLICATIONS

Reye's Syndrome

It is an acute encephalopathy of children of age 2–16 years. Fatty degeneration of liver is associated with this syndrome. Mortality rate is 10–40%. Otitis media, sinusitis, asthma, bacterial pneumonia and aspergillosis are the other symptoms.

PREVENTION

Inactivated influenza virus vaccines have been used for old people. The virus for the vaccine are grown in chick embryo, inactivated by formalin, purified to some extent and adjusted to a dosage known to elicit an antibody response in most individuals.

Vaccine contains the strains of type A and B. Early October to mid-November is the best time to get vaccinated. Due to delay of vaccine supply, only the following people are recommended to get vaccinated.

- Persons who are above 65 years of age
- Persons who have chronic disorders
- Persons requiring regular medical check-up
- Persons who receive aspirin therapy regularly

TREATMENT

The synthetic drugs amantadine and rimantadine hydrochloride are effectively used to prevent infection and illness caused by type A but not by type B viruses. The drugs interfere with virus uncoating and transport by blocking the transmembrane M2 ion channel. Drugs prevent about 50–67% of infection. Drug resistance also occurs.

CDC. Prevention and Control of Influenza. MMWR, 44:No. RR 3:1–22, 1995.

Couch, R.B. "Influenza: its control in persons and populations." *J. Infect. Dis*. 153:431. 1986.

International Conference on Asian Influenza. *Am. Rev. Respir. Dis.* 83:1. 1961.

International Conference on Hong Kong Influenza. *Bull. WHO* 41 :335. 1969.

Kilbourne, E.D. *Influenza*. Plenum. 1987.

Krug, R.M. (ed.). *The Influenza Virus*. Plenum, New York. 1989.

"Options for the Control of Influenza II." *Excerpta Medica*, Amsterdam, 1993.

www.who.int/csr/disease/influenza/

www.who.int/topics/influenza/en/

www.cdc.gov/flu/

users.rcn.com/jkimball.ma.ultranet/biologypages/influenza.html

www.influenzareport.com/

www.home-remedies-for-you.com/remedy/Influenza.html

www.in.gov/isdh/bioterrorism/pandemicflu/index.html

www.in.gov/isdh/pdfs/pandemicinfluenzaplan.

www.influenza.com/

www.fluinfo.in.gov/

www.pbs.org/wgbh/amex/influenza/

www.uct.ac.za/depts/mmi/jmoodie/influen2.html

www.kolkatabirds.com/birdflu.html

hongkong.usconsulate.gov/acs_avian_influenza.html

www.niaid.nih.gov/factsheets/flu.html

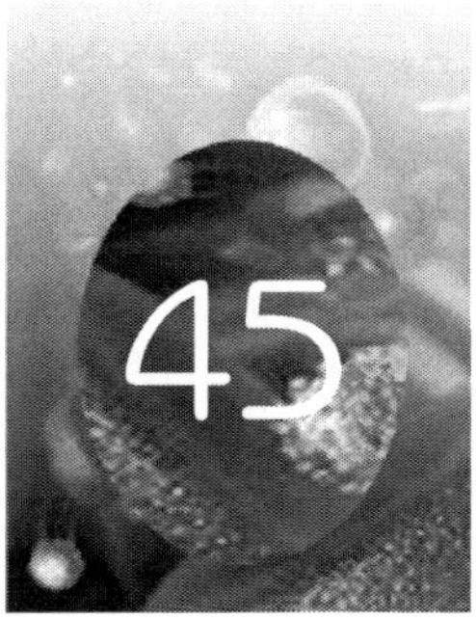

Measles

INTRODUCTION

Measles is a highly contagious skin disease that is epidemic throughout the world. Thomas Sydenham in 1690 gave the first clear and accurate description about measles. In 1846 an outbreak of measles occurred in remote areas of islands. Gold Berger and Anderson established the viral aetiology of measles in 1911 by transmitting the disease to monkey through the inoculation of filtrates of blood and nasopharyngeal secretions.

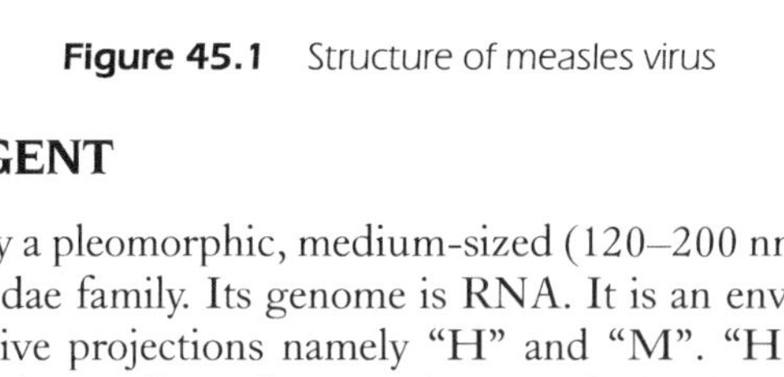

Figure 45.1 Structure of measles virus

CAUSATIVE AGENT

Measles is caused by a pleomorphic, medium-sized (120–200 nm in diameter) virus of the Paramyxoviridae family. Its genome is RNA. It is an enveloped virus. It has two biologically active projections namely "H" and "M". "H" is responsible for viral attachment to host cells and causes haemagglutination. The outer "M" is responsible for fusion of the viral outer membrane with the host cell. The M antigen is also responsible for producing multinucleated giant cells. It has tightly coiled nucleic acid surrounded by the lipoprotein envelope (Figure 45.1).

The virus grows well on human or monkey kidney and human amnion culture, which are the preferred cells for primary isolation. The other name for multinucleated giant cells is Warthin Finkeldey cells.

Resistance

The virus is heat-labile and is readily inactivated by heat, UV rays, ether and formaldehyde.

REPLICATION

Parainfluenza viruses attach to the host cell by haemagglutinin. It binds to the host cell neuraminic acid receptor (Figure 45.2).

Entry of virus into the cell by fusion with the cell membrane is mediated by the F1 and F2 glycopeptides.

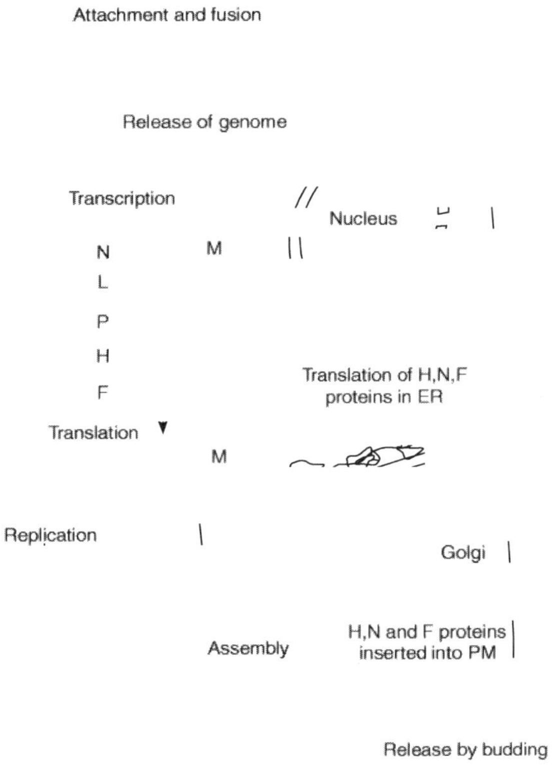

Figure 45.2 Replication of paramyxoviridae

The viral particles contain single-stranded negative-sense RNA. The virion transcriptase initiates transcription into 5–8 complementary positive-sense messenger RNA strands.

They direct the viral protein synthesis and are copied into negative-sense RNA strands which are integrated in the new virions.

After assembly, progeny viruses are released by budding (Figure 45.2).

CLINICAL MANIFESTATION

Incubation period is about 10–12 days. It begins with fever, running nose, cough and swollen weepy eyes. Within a few days, a fine red rash appears on the forehead and spread outward over the rest of the body. Unless complications occur, symptoms disappear within one week.

Unfortunately many cases are complicated by secondary infections caused by bacterial pathogens, mainly *Staphylococcus aureus*, *Streptococcus pneumoniae*, *Streptococcus pyogenes* and *Haemophilus influenzae.*

Very rarely, measles reactivation is observed after two to ten years and produces a disease called Subacute sclerosing pan encephalitis (SSPE) , which is marked by slow progressive degeneration of the brain, resulting in death within two years.

Measles occurring during pregnancy results in an increased risk of miscarriage, premature labour and low birth weight. Protracted diarrhoea is often seen as a complication in children in poor countries.

PATHOGENESIS

The respiratory route and conjunctiva acquire rubeola (measles) virus. It primarily replicates in the upper respiratory epithelium, then spreads to lymphoid tissues and following further replication eventually spreads throughout the body. Mucous membrane involvement is responsible for an important diagnostic sign Koplik's spot (small bluish white ulceration on the buccal mucosa). Damage to the respiratory mucous membrane partly explains the markedly increased susceptibility of measles patients to secondary bacterial infections, especially infection of the middle ear and lung.

The skin rash of measles results from the cytopatheic effect of rubeola virus replication in skin vascular endothelial cells and cellular immune response against the viral antigen in the skin. It is not known why the rash is often healed on the face before it reaches the lower parts of the body.

The measles virus temporarily suppresses the cellular immunity, which can cause reactivation of herpes simplex virus.

LABORATORY DIAGNOSIS

Primary diagnosis is with the help of Koplik's spot formation.

Specimen

Throat/nasopharyngeal swab, urine and whole blood are the specimens used for diagnosis. Turn around time is 14 days.

Cytologic Diagnosis

Specimens should be fixed with formalin and stained with haematoxylin and eosin. Characteristic giant cells containing eosinophilic intranuclear and intracytoplasmic inclusions are observed for the first 2 or 3 days of the specimen.

Antigen Detection

Antigen detection is done by immunofluorescence technique. Immunoenzyme staining increases the sensitivity of the test.

Virus Isolation

The virus is isolated by cell culture technique from respiratory secretions and other samples. Primary cultures of human embryonic kidney cell and monkey kidney cells are more sensitive for viral isolation.

Nucleic Acid Detection

This technique was done in immunocompromised patients who may not be capable of eliciting antibody response. Viral nucleic acid is detected by using reverse transcriptase, PCR *in situ* hybridization or reverse transcriptase, PCR and amplification of RNA extracted from specimens.

EPIDEMIOLOGY

Humans are the only natural hosts for rubeola virus. It has been eradicated from the US but occasionally epidemics were observed.

CONTROL

Young children are vaccinated with MMR vaccine. Preschool children are also vaccinated.

1. Discuss the various features of measles.
2. Describe the features of paramyxovirus.
3. Write short notes on
   i. SSPE
   ii. Koplik's spot
   iii. Warthin Finkeldey cells
   iv. Fusion protein

Anderson, L.J. and Heilman, C.A. "Protective and disease-enhancing immune responses to respiratory syncytial virus." *J. Infect. Dis*. 171:1. 1995.

CDC: Measles-United States, 1994. MMWR 44(26):486. 1995.

Groothuis, J.R. "Treatment and prevention of severe respiratory syncytial virus infection in young children." *Current. Opin. Infect. Dis.* 8: 206. 1995.

Karron, R.A., Wright, P.F. *et al*. "A live attenuated bovine parainfluenza virus type 3 vaccine is safe, infectious, immunogenic, and phenotypically stable in infants and children." *J. Inf. Dis*. 171: 1107. 1995.

McLean, D.M. "Parainfluenza viruses." In Zuckerman, A.J., Banatvala, J.E., Pattison, J.R. (eds.). *Principles and Practice of Clinical Virology,* 3rd edn. John Wiley & Sons Ltd. Chichester. 1994.

Toms, G.L. "Respiratory syncytial virus: virology, diagnosis, and vaccination." *Lung.* 168S: 388.1990.

Welliver, R.C., Won, D.T., Middleton, Jr. E, *et al*. "Role of parainfluenza virus-specific IgE in pathogenesis of croup and wheezing subsequent to infection." *J. Pediatrics.* 101: 889. 1982.

www.kidshealth.org/parent/infections/lung/measles.html

www.cdc.gov/node.do/id/0900f3ec8000750e

www.nlm.nih.gov/medlineplus/ency/article/001569.html

www.measlesinitiative.org

www.home-remedies-for-you.com/remedy/Measles.html

www.who.int/topics/measles/

www.netdoctor.co.uk/diseases/facts/measles.html

46

Mumps

INTRODUCTION

Mumps it results from an acute viral infection. Target of mumps is the parotid gland, which is located just below and in front of the ear. Mumps means "mumble". Mumps begins with painful swelling of one or both the parotid glands.

CAUSATIVE AGENT

Mumps virus causes mumps. It is an enveloped virus included under the family Paramyxoviridae. It is an ssRNA-containing negative-sense virus. It is a helical-shaped virus. Viral aetiology was demonstrated by Johnson and Goodpasture in 1934. Hebel cultivated it in embryonated eggs. In 1955, Henle and Deinhardt grew it in tissue culture. Virus possesses haemagglutinin, neuraminidase and fusion protein.

Resistance

It is a heat-labile and chemically sensitive virus.

CLINICAL MANIFESTATION

Incubation period is 16 to 18 days. Parotid swelling with pain is the first sign. Fever and extreme pain during swallowing occurs.

PATHOGENESIS

Virus is transmitted in saliva and respiratory secretions, and its portal of entry is respiratory tract. This virus multiplies in the respiratory tract and local lymph node in the neck. Virus spreads throughout the body via the bloodstream and produces symptoms only after infecting other tissues such as the parotid glands and meninges. In the salivary gland, the virus multiplies in the epithelium of ducts that convey saliva to the mouth. This destroys the epithelium. The body inflammatory response to the infection is responsible for the severe swelling and pain. In adults, it infects the tubule and causes death of testicular tissue. The immune system of the host eliminates the infection.

EPIDEMIOLOGY

Humans are the only natural hosts of mumps, and natural infection confers lifelong immunity.

LAB DIAGNOSIS

Generally serological diagnosis is not necessary. This virus can be identified with haemagglutination inhibition test. Embryonated eggs and cell culture techniques are used for culturing.

CONTROL

An effective vaccine is available and is often administered as part of the trivalent measles, mumps and rubella (MMR) vaccine. It provides protection for at least 10 years.

1. Is there any cultivation procedure for mumps virus?
2. Discuss the various features of mumps.

Anderson, L.J. and Heilman, C.A. "Protective and disease-enhancing immune responses to respiratory syncytial virus." *J. Infect. Dis*. 171:1. 1995.

CDC: Measles-United States, 1994. MMWR 44(26):486. 1995.

Groothuis, J.R. "Treatment and prevention of severe respiratory syncytial virus infection in young children." *Current Opin. Infect. Dis.* 8: 206. 1995.

Karron, R.A., Wright, P.F., *et al*.: A live attenuated bovine parainfluenza virus type 3 vaccine is safe, infectious, immunogenic, and phenotypically stable in infants and children. *J. Inf. Dis.* 171:1107. 1995.

McLean, D.M. "Parainfluenza viruses." In Zuckerman, A.J., Banatvala, J.E., Pattison, J.R. (eds.). *Principles and Practice of Clinical Virology*, 3rd edn. John Wiley & Sons Ltd. Chichester. 1994.

Toms, G.L. "Respiratory syncytial virus: virology, diagnosis, and vaccination." *Lung*. 168S:388. 1990.

Welliver, R.C., Won, D.T. and Middleton, Jr. E., *et al*. "Role of parainfluenza virus-specific IgE in pathogenesis of croup and wheezing subsequent to infection. *J. Pediatrics*. 101:889. 1982.

www.kidshealth.org/parent/infections/bacterial_viral/mumps.html

en.wikipedia.org/wiki/Mumps

www.cdc.gov/nip/diseases/mumps/

www.nlm.nih.gov/medlineplus/mumps.html

www.mayoclinic.com/health/mumps/DS00125

www.medinfo.co.uk/conditions/mumps.html

www.health.state.ny.us/diseases/communicable/mumps/fact_sheet.html

mumps.enseeiht.fr/

Rubella

INTRODUCTION

Rubella is a milder disease and often unrecognized because it is difficult to diagnose. It is also called three day measles or German measles. It is a mild exanthematous fever characterized by transient macular rash and lymphadenopathy. The rubella virus was isolated in tissue culture in 1962.

CAUSATIVE AGENT

The rubella virus, a member of the Togaviridae family, causes rubella. It is a small, pleomorphic and spherical-shaped, enveloped virus. Genome is positive-sense ssRNA. It is included in the genus *Rubivirus* (non-arthropod-borne virus).

There are four types of antigens:

- Haemagglutinating antigens
- CF antigens
- Precipitating antigens
- Platelet aggregating antigens

Resistance

The virus is inactivated by ether, chloroform, formaldehyde, β-propiolactone and destroyed at 56°C.

Chemical Composition

Chemically the virion is composed of 75% protein, 18.8% lipid, 4% carbohydrate and 2.4% RNA. UV inactivates the virus within 40 seconds.

CLINICAL MANIFESTATION

The incubation period is 14 to 21 days.

Malaise, headache, fever, mild conjunctivitis, rash beginning on forehead and face and enlarged lymph nodes are the symptoms.

PATHOGENESIS

The rubella virus is acquired by inhalation. It multiplies in the nasopharynx and enters the bloodstream, causing viraemia. The blood transports virus to various body tissues, including skin and joints.

Humoral and CMI develop against the virus, and the resulting antigen–antibody reaction leads to rash and joint symptoms and then for the rapid disappearance of the disease.

During pregnancy, the virus crosses the placenta and infects the foetus. All foetus cells are susceptible to infection. Some cells are killed. Resultant foetal abnormalities are referred to as the congenital rubella syndrome . It includes eye, brain damage, deafness, heart defects and low birth weight.

EPIDEMIOLOGY

Rubella is worldwide in distribution and occurs during the winter and spring months.

LABORATORY DIAGNOSIS

Haemagglutination inhibition test, EIA (enzyme immunoassay), latex agglutination, CFT and neutralization test are used to diagnose rubella.

Inoculation of tissue culture media with throat, blood or urine specimens is used for cultivation.

PREVENTION AND TREATMENT

Live attenuated rubella vaccine is administered to children at 15 months. It was approved during the year 1969. The vaccine produces longlasting immunity. MMR is also given.

Gammaglobulin is used for treatment, but it is not a specific treatment. Rubella cannot be treated with antibiotics. Rubella is a mild infection in kids. Infected children can be treated at home. To relieve minor discomfort, ibuprofen may be given.

CONTROL

Rubella can be prevented by routine childhood immunization and by immunization of susceptible adolescents and adult populations with live attenuated rubella vaccine. Immunoglobulin is not very effective in prophylaxis of rubella in pregnant women, and its routine use is not generally recommended. It should be considered only if termination of pregnancy is not acceptable to mother under any circumstances.

1. Explain the pathogenesis of rubella.
2. Discuss the congenital rubella syndrome.

Bakshi, S.S. and Cooper, L.Z. "Rubella (review)." *Clin. Dermatol*. 7:8. 1989.

Centers for Disease Control: Rubella and Congenital Rubella Syndrome United States, January 1, 1991-May 7, 1994. MMWR 43:391. 1994.

Centers for Disease Control: Rubella Prevention. MMWR, Recommendations and Reports 39:1–18. 1990.

Cochi, S.L., Edmonds, L.E., Dyer, K. *et al*. "Congenital rubella syndrome in the United States, 1970–1985: on the verge of elimination." *Am. J. Epidemiol.* 129:349. 1989.

Frey, T.K. "Molecular Biology of Rubella Virus." *Advances in Virus Research.* 44:69.1994.

Frey, T.K. and Abernathy, E.S. "Identification of Strain-Specific Nucleotide Sequences in the RA 27/3 Rubella Virus Vaccine." *J. Inf. Dis.* 168: 854. 1993.

Green, R.H., Balsamo, M.R., Giles, J.P. *et al*. "Studies of the natural history and prevention of rubella." *Am. J. Dis. Child.* 110:348. 1965.

Heggie, A.D. and Robbins, F.C. "Natural rubella acquired after birth." *Am. J. Dis. Child.* 118:12. 1969.

Horstmann, D., Schluederberg, A., Emmons, J.E. *et al*. "Persistence of vaccine-induced immune responses to rubella: comparison with natural infection." *Rev. Infect. Dis.* S7:S80. 1985.

Orenstein, W.A., Bart, K.J., Hinman, A.R., *et al*: "The opportunity and obligation to eliminate rubella from the United States." *J. Am. Med. Assoc.* 251:1988. 1984.

Parkman, P.D., Hopps, H.E. and Meyer, H.M. "Rubella virus: isolation, characterization and laboratory diagnosis." *Am. J. Dis. Child.* 118: 68. 1969.

www.kidshealth.org/parent/infections/skin/german_measles.html
www.mayoclinic.com/health/rubella/DS00332

www.cdc.gov/nip/publications/pink/rubella.

www.nlm.nih.gov/medlineplus/rubella.html

www.who.int/vaccines/en/rubella.shtml

www.medinfo.co.uk/conditions/rubella.html

www.rubella.net/

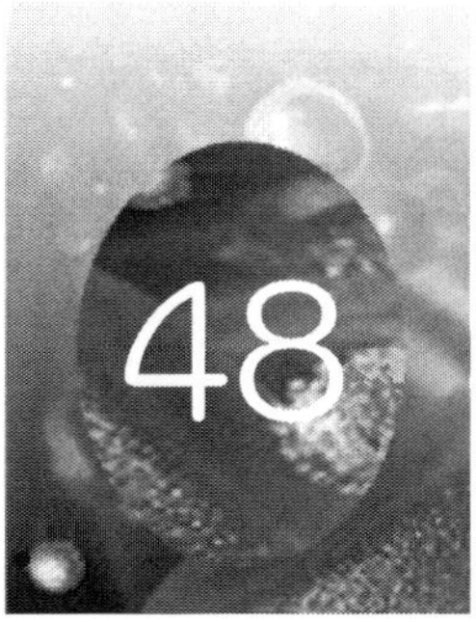

Arbovirus Infections

INTRODUCTION

Arboviruses (Arthropod-borne viruses) are viruses of vertebrates biologically transmitted by insect vectors. Mosquitoes, ticks, flies and other insects transmit the virus. Arboviruses and rodent-borne viruses are placed in the toga, flavi, bunya, rhabdo, arena and filovirus groups.

There are more than 450 arboviruses and rodent-borne viruses. Of these about 100 are known pathogens for humans.

CLASSIFICATION

Table 48.1 provides viruses and its families transmitted through arthropods which is a common feature. Among all viruses, alpha flavivirus prevalance is more severe than others.

Table 48.1 Arthropod-borne virus

Family	Genus	Important species
Togaviridae	Alphavirus	Chikungunya virus, Eastern, Venezuelan and Western equine encephalitis virus, Sindbis virus
Flaviviridae	Flavivirus	Brazilian encephalitis, dengue, Japanese B encephalitis, Murray valley encephalitis, West Nile fever, Yellow fever, St. Louis encephalitis, Russian spring summer encephalitis viruses
Bunyaviridae	Bunyavirus	California encephalitis virus, phlebovirus, Sandfly fever virus, Rift valley fever virus, Nairovirus, Oropaiche, Turlock Crimean Congo haemorrhagic fever virus, Nairobi sheep disease virus, Ganjam virus
Reoviridae	Hantavirus	Hantan, Seoul, Puemala, Prospect Hill, Sin Nombre, virus
	Orbivirus	Colorado tick fever, African horse sickness, Blue tongue viruses
Rhabdoviridae	Vesiculovirus	Vesicular stomatitis virus, chandipura virus

CHARACTERISTIC FEATURES OF ARBOVIRUSES

Various features of arboviruses are presented in Table 48.2.

Table 48.2 Characteristic features of different arboviruses

Character	Togaviridae	Flaviviridae	Bunyaviridae	Reoviridae
Size	70 nm	40–60 nm	80–120 nm	60–80 nm
Shape	Spherical	Spherical	Spherical	Spherical
Availability of envelope	Present	Present	Present	Absent
Genome	+ssRNA	+ss RNA	–ssRNA	dsRNA
Size of genome	9.7–11.8 kbp	10.7 kbp	11–21 kbp	16–27 kbp
Number of genome segments	Single	Single	Three	10–12 segments
Site of replication	Cytoplasm	Cytoplasm	Cytoplasm	Cytoplasm
Site of assembly	Plasma membrane	Endoplasmic reticulum	Plasma membrane	Plasma membrane
Release	Budding	Cell lysis/budding	Budding	Cell lysis
Examples	Alphavirus, chikungunya, Eastern equine encephalitis, Venezuelan and western equine encephalitis virus, Sindbis virus	Flavivirus Brazilian encephalitis Dengue Japanese B encephalitis Murray valley encephalitis West Nile fever Yellow fever virus St. Louis encephalitis virus	Bunyavirus California encephalitis virus Guama virus Lacrosse virus Simbu virus Phlebovirus Sandfly fever virus Riftvalley fever virus Nairovirus	Orbivirus Coltivirus

Other arboviruses are vesiculovirus (Rhabdoviridae), arenavirus (Arenaviridae) and filovirus (Filoviridae).

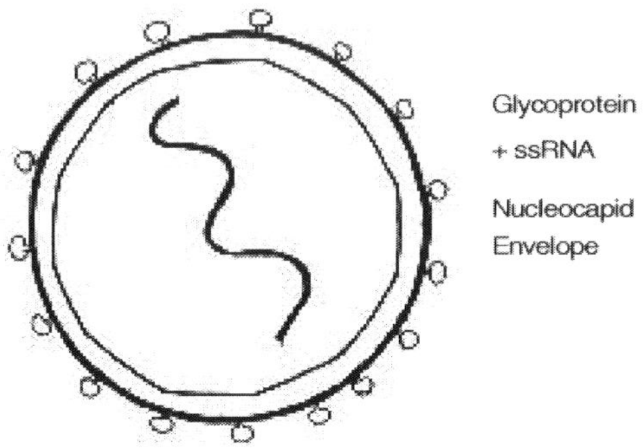

Figure 48.1 Common structure of alpha and flavivirus

ENCEPHALITIS

Encephalitis means inflammation of the brain. Alphaviruses, flaviviruses, bunyaviruses may cause encephalitis in human. Based on ecological distribution, the disease is named accordingly. Table 48.3 describes transmission of different encephalitis and distribution.

Table 48.3 Mosquito-borne encephalitis and its distribution

Diseases type	Distribution	Vectors
Western equine encephalitis (Alphavirus)	Pacific, Mountain	*Culex tarsalis*
Eastern equine encephalitis (Alphavirus)	Atlantic	*Aedes*
Venezuelan equine encephalitis (Alphavirus)	South America	*Aedes* and *Culex*
St.Louis encephalitis (Flavivirus)	North and Central America	*Culex*
West Nile virus causing encephalitis (Flavivirus)	India	*Culex*
Murray valley encephalitis (Flavivirus)	Australia	*Culex*
Japanese encephalitis (Flavivirus)	Japan, India	*Culex*
California encephalitis (Bunyavirus)	California	*Aedes*

CHARACTERS OF CAUSATIVE AGENT

Togavirus, flaviviruses, bunyaviruses, reoviruses, rhabdoviruses, arenaviruses are able to cause encephalitis. Its characters are in Table 48.4.

Table 48.4 Tick-borne encephalitis and its distribution

Disease type	Distribution	Vector
Russian spring encephalitis (Flavivirus)	Russia	Ixodes
Pawassan virus	North America	Ixodes
Louping ill	UK	Ixodes
Kyasanur forest disease	Asia	Haemaphysalis

CLINICAL MANIFESTATION

Incubation period is 4–21 days. Headache, chillness, fever, nausea, vomiting, generalized pain is observed within 24–48 hours. Other symptoms are marked drowsiness, mental confusion, tremors, convolutions and coma. Mortality rate of encephalitis is about 80%.

PATHOGENESIS

Pathogenesis of encephalitis is not well studied. The virus enters the human body through mosquito or tick bite. After entry, the virus multiplies in non-neural tissue and is present in the blood for 3 days till the first sign of involvement of the CNS is seen. Then the virus multiplies in the brain cells, destroys the cell and encephalitis becomes apparent.

LABORATORY DIAGNOSIS

Recovery of Virus

Recovery of the virus is by the inoculation of serum with intracerebral inoculation of suckling mice.

Serology

Neutralizing haemagglutination-inhibiting antibodies are detected within few days. CF antibodies appear later.

TREATMENT

There is no specific treatment.

DENGUE

Dengue is one of the most important human infections caused by the dengue virus, which is included under the family Flaviviridae.

MORPHOLOGY AND REPLICATION

Refer Table 48.2.

CLINICAL MANIFESTATION

Symptoms are sudden fever with headache, bodyache, anorexia, vomiting, retrobulbar pain, pain in the back and limbs, lymphadenopathy and maculopapular rash.

In its serious form, dengue causes two important manifestation. Dengue haemorrhagic fever and dengue shock syndrome. Dengue virus is transmitted from person to person by *Aedes aegypti* mosquitoes. Incubation period is usually 5–6 days. Man is the only definitive host.

LABORATORY DIAGNOSIS

Serological tests are used to diagnose this disease. ELISA is used to detect IgG antibodies.

CONTROL

Control of dengue is possible by controlling vectors. There is no vaccine available.

Monath, T.P. (ed.). *The Arboviruses: Epidemiology and Ecology*, 5 Vols. CRC Press, Boca Raton, FL. 1988.

Murphy, F.A. *et al.* (eds.). "Virus Taxonomy: Classification and Nomenclature of Viruses." *Arch Virol., Suppl.* 10. 1995.

Rey, F.A. *et al.* "The envelope glycoprotein from tick-borne encephalitis virus at 2 Å resolution." *Nature.* 375: 291–298. 1995.

www.dhpe.org/infect/Arbovirus.html
en.wikipedia.org/wiki/Arbovirus
www.cdc.gov/node.do/id/0900f3ec8000d069
www.arbovirus.health.nsw.gov.au/
www.mass.gov/dph/wnv/wnv1.htm
www.britannica.com/eb/article-9009235/arbovirus
arbonet.caeph.Tulane.edu/
www.ndwnv.com/
www.acrcorp.com/ArboInfoCr/

Polio

INTRODUCTION

Polio is an ancient disease. Various Egyptian hieroglyphics dated approximately 2000 BC depict individuals with wasting, withered legs and arms. In 1840, the German orthopaedist Jocob Von Heine described the clinical features of poliomyelitis and identified spinal cord as the problem area. Poliovirus has tropism for epithelial cells of the alimentary tract and cells of the central nervous system. Infection is asymptomatic or causes a mild, undifferentiated febrile illness. Spinal and bulbar poliomyelitis occasionally occurs. Paralytic poliomyelitis is not always preceded by minor illness. Paralysis is usually irreversible, and there is residual paralysis for life.

CAUSATIVE AGENT

All three poliovirus serotypes (1 to 3) can give rise to paralytic poliomyelitis. It is a non-enveloped RNA (Figure 49.1) virus. It is included under the family Picornaviridae. The Picornaviruses are small (22 to 30 nm), icosahedral virions resistant to lipid solvents. The virus capsid is composed of 60 copies each of four viral proteins VP1 to VP4, which form an icosahedral shell. The genome is a single-stranded RNA (molecular weight, approximately 2×10^6 to 3×10^6). The RNA strand consists of approximately 7,500 nucleotides and is covalently bonded to a non-capsid viral protein (VPg) at its 5′ end and to a polyadenylated tail at its 3′ end.

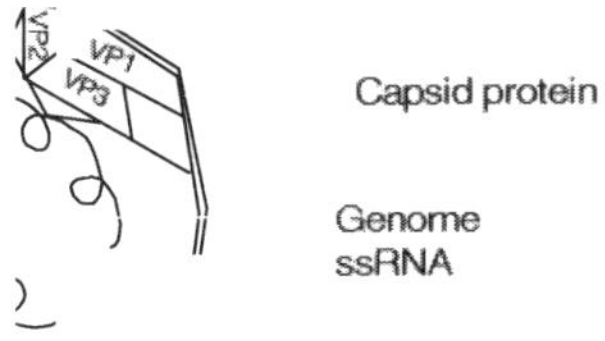

Figure 49.1 Picornavirus

Resistance

Enteroviruses can survive for long periods in organic matter and are resistant to the low pH in the stomach (pH 3.0 to 5.0). Picornaviruses are inactivated by pasteurization, boiling, formalin and chlorine.

REPLICATION

The virus binds to a cellular receptor. The mechanism of uncoating of RNA genome is unknown. VPg protein at the 5´ end is removed from RNA, and the resulting RNA associates with ribosomes.

Translation is initiated at an internal site, 741 nucleotides from the 5´ end of the viral RNA and polyprotein precursors are synthesized.

Polyproteins are cleaved and individual proteins (P1, P2 and P3) are produced. P1 proteins contain viral structural proteins. P2 and P3 are responsible for proteases and RNA synthesis proteins.

The proteins that are involved in RNA synthesis are transported into the membranous vesicles. Positive-sense RNA is also transported into the vesicles. It is copied into negative-sense RNA that act as the template for the synthesis of positive-sense RNA.

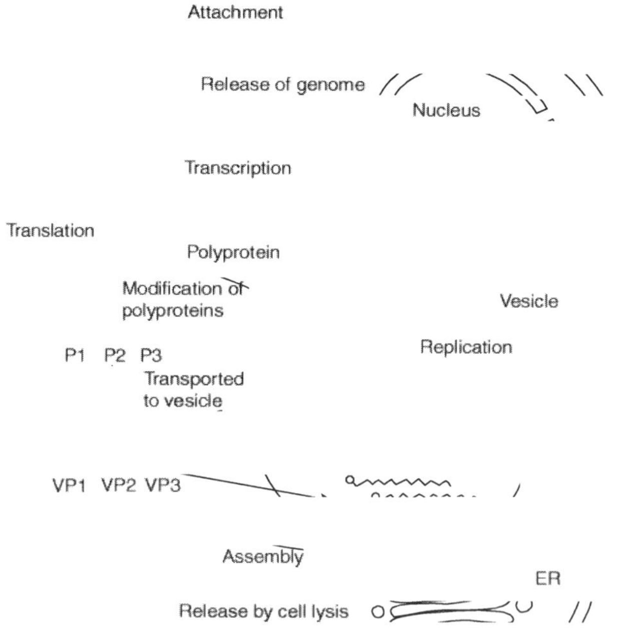

Figure 49.2 Replication of picornavirus

Structural proteins are formed by partial cleavage of P1 precursor proteins. These proteins are transported to vesicles. Assembly takes place within the vesicles. mature virions are released after cell lysis (Figure 49.2).

CLINICAL MANIFESTATION

Incubation period is about 7–14 days. Paralytic poliomyelitis can occur without antecedent minor illnesses. A patient may suffer aseptic meningitis with pains in the back and neck muscles for several days.

PATHOGENESIS

Humans are the only natural hosts of poliovirus. They attach to a specific receptor on these cells, which in humans is encoded by a gene on chromosome 19. Poliovirus infection is quite common in non-immunized individuals, but only about 1 per cent of these cases progress to the paralytic form of the disease. Primary replication of poliovirus takes place in the oropharyngeal and intestinal mucosa (the alimentary phase). From here, the virus spreads to the tonsils and Peyer's patches of the ileum and to deep cervical and mesenteric nodes, where it multiplies abundantly (the lymphatic phase). Subsequently, the virus is carried by the bloodstream to various internal organs and regional lymph nodes (the viraemic phase). More concentrated damage results in flaccid paralysis of the muscles innervated by the affected motor nerves. Muscle involvement peaks a few days after the paralytic phase begins. Paralysis is usually irreversible, and residual paralysis remains for life. The paralytic disease is called spinal poliomyelitis if the weakness is limited to muscles innervated by the motor neurons in the spinal cord and bulbar poliomyelitis, if the cranial nerve nuclei or medullary centres are involved.

Host Defences

Shortly after infection of the respiratory or alimentary tract, increasing amounts of interferon and subsequently virus-specific IgA-antibody are detected in the saliva and the respiratory and gut secretions. Interferon inhibits virus multiplication, and IgA complexes with extracellular virus. This complex not only inhibits the spread of virus to susceptible epithelial cells but also reduces the oral and faecal shedding of infectious virus.

EPIDEMIOLOGY

Poliomyelitis affects all age groups. In areas with poor hygiene and poor sanitation, most infants are infected relatively early in their life and acquire active immunity while still protected by maternal antibodies. In countries which have high poliomyelitis incidence, older age group are also covered in the vaccination programs.

LABORATORY DIAGNOSIS

Enteroviruses and rhinoviruses may be isolated from faeces, pharyngeal swabs, saliva, and nasal aspirates, and some enteroviruses may be isolated from skin lesions, conjunctiva, cerebrospinal fluid, spinal cord, brain, heart, and blood.

Poliovirus cultivation is performed with the help of tissue-culture technique. The most specific of the conventional laboratory tests used to identify picornavirus serotypes is the neutralization test. Serodiagnosis for the whole range of picornaviruses is impractical because of the multiplicity of serotypes. The neutralization test is also used to determine the immune status of a person.

CONTROL

The Salk-type inactivated poliovirus vaccine (IPV) consists of a mixture of three poliovirus serotypes grown in monkey kidney cell cultures and is made non-infectious by formalin treatment. It is given in two intramuscular injections spaced a month apart and requires periodic boosters to maintain an adequate serum-neutralizing antibody level.

The Sabin-type live attenuated oral poliovirus vaccine (OPV) is commercially available as trivalent antigen. The viruses are attenuated by multiple passages in monkey kidney or human diploid cell cultures, and the vaccine potency is stabilized with one molar magnesium chloride or sucrose. This vaccine mimics wild poliovirus infections by inducing serum-neutralizing antibody, as well as interferon and virus-specific IgA antibody in the pharynx and gut. This combined IPV/OPV approach deserves consideration as an additional tool for control and eradication of poliomyelitis in countries where polio is endemic and where there is a danger of poliovirus. In 1988, the World Health Assembly, the governing body of the World Health Organization, set the goal of global eradication of poliomyelitis by the year 2000 but the programme is extended for some more years.

1. Describe structure of poliovirus.
2. Describe the replication of poliovirus.
3. Explain about various features of poliomyelitis.
4. Write short notes on
   i. IPV
   ii. OPV
   iii. Salk
   iv. VP

Goldblum, N., Gerichter, C.B., Tulchinsky, T.H. and Melnick, J.L. "Poliomyelitis control in Israel, the West Bank and Gaza Strip: Changing strategies with the goal of eradication in an endemic area." *Bull. WHO*. 72(5): 783. 1994.

Minor, P., Brown, F., King, A. *et al*. "Classification and Nomenclature of viruses family. Picornaviridae." In Fifth Report of International Committee on Taxonomy of Viruses. *Archives of Virology, Supplementum*. Francki, R.I.B., Fauquet, C.A. and Knudson, D.L. (eds.) Springer-Verlag Wien, New York. 1992. pp. 320–326.

World Health Organization: Global Poliomyelitis Eradication by the Year 2000. Manual for Managers of Immunization Programmes. WHO/EPI/Polio/89.1, 1989.

Yin-Murphy, M. "Acute hemorrhagic conjunctivitis." In Melnick, J.L. (ed.). *Progress in Medical Virology*. Karger, Basel. 1984. pp. 23–44.

www.polioeradication.org/

www.kidshealth.org/parent/infections/bacterial_viral/polio.html -

www.npspindia.org/

www.cdc.gov/nip/publications/VIS/vis-IPV.

www.dhpe.org/infect/polio.html

www.polionet.org/

www.who.int/mediacentre/factsheets/fs114/en/

www.polio.com/

Rabies

INTRODUCTION

Rabies is an acute, fulminant, fatal encephalitis. It is called the "madman disease" which has instilled terror in human society. The reason is that, with rare exceptions, all of the people who were bitten by a rabid animal, got rabies.

Rabies is an important zoonotic infection in which man is the dead end of infection and hence doesn't play any role in its spread to a new host. In most of the developing countries, dogs are the principal reservoir of rabies (canine rabies) whereas sylvatic rabies involve animals such as foxes, raccoons, cats and coyotes. Rabies has been recognized from very ancient times. The word "rabies" is derived from the Latin word *Rabidus*, which means "mad". It is an epidemic disease.

HISTORY

1885 Development of human rabies vaccine

1903 Diagnosis of Negri bodies

1940s Use of rabies vaccine for dogs

1954 Addition of rabies immunoglobulin

1958 Rabies virus grown in cultured cells

1959 Development of fluorescent antibody test

CAUSATIVE AGENT

Rabies virus causes rabies infection.

- It belongs to the family Rhabdoviridae and genus *Lyssavirus.*
- In Greek "Lyssa" means "rabies".

- Rabies virus is bullet shaped, size about 180 × 75 nm with one end rounded or conical and the other planar or concave (Figure 50.1).

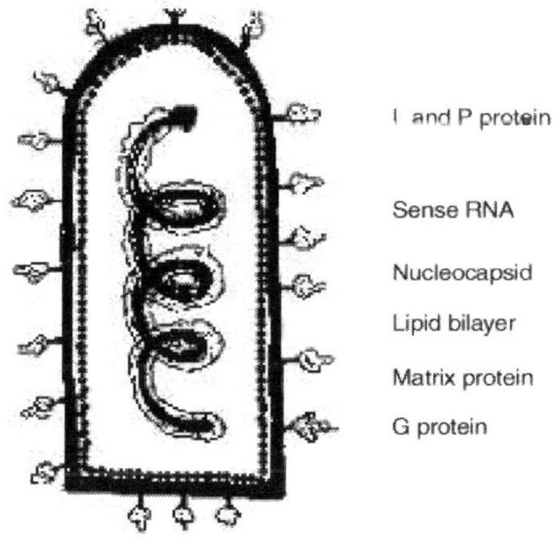

Figure 50.1 Structure of rabies virus

- Genome is negative-sense single-stranded RNA. It is nucleocapsid in nature. Two layers, the matrix layer and the outer envelope cover the genome. Matrix is made up of M protein. Outer envelope is made up of lipid bilayer similar to the plasma membrane. External envelope has spike-like projections. It is made up of glycoproteins. Spikes are responsible for the pathogenic property of the virus. RNA-dependent RNA polymerase is responsible for genome replication. L and P proteins control its activity.
- Rabies viruses of man and animals all over the world appears to be of a single antigenic type. Antigens of rabies viruses are
    - G protein
    - M protein
    - N protein
    - Haemagglutinin

Chemical Composition

Chemically the rabies virus is composed of

- 4% RNA
- 67% protein
- 26% lipid
- 3% carbohydrate

Resistance

- Rabies virus is highly resistant to cold, dryness and decay. The virus is highly thermolabile with a half-life of approximately 4 hours at 40°C and 35 seconds at 60°C. The virus cannot withstand pH less than 4 or more than 10. It is also susceptible to oxidizing agents, most organic solvents, surface-acting agents and quaternary ammonium compounds. Proteolytic enzymes, ultraviolet rays and X-rays rapidly inactivate the virus. Soaps and detergents are effective against the rabies virus because of their lipid-eliminating property, which destroys the outer envelope of the virus.

REPLICATION

1. The rabies virus binds to the cellular receptor and enters the cell via receptor-mediated endocytosis.
2. The viral membrane fuses with the membrane of the vesicle releasing the viral nucleocapsid.
3. This structure comprises of negative-sense RNA coated with nucleocapsid and a small number of L and P proteins, which catalyse RNA replication. Negative-sense RNA is copied into 5 subgenomic mRNA by L and P proteins.
4. The N, P, M and L mRNAs are translated by free cytoplasmic mRNA.
5. G mRNA is translated by ribosomes bound to the endoplasmic reticulum.
6. Newly synthesized P, N and L proteins are involved in RNA replication. This process begins with positive-sense RNA synthesis.
7. Positive-sense RNA of the host serves as the template.
8. Some of the negative-sense RNA enters into viral protein synthesis.
9. G mRNA transcribes and synthesizes glycoproteins.
10. G proteins travel to the plasma membrane.
11. Progeny nucleocapsid and M proteins are transported to the adjacent area of plasma membrane.
12. Assembly takes place and new viruses are released through budding process and infect new cells (Figure 50.2).

CLINICAL MANIFESTATION

Incubation period is about 30–60 days.

Clinical Features in Man

The clinical features in man include hydrophobia, lock jaw, encephalitis, hysteria, acute polyneuritis and poliomyelitis.

Clinical Features in Dog

The clinical features in dog are change in behaviour of dog, change in tone of bark, change in feeding habit, off feed, consumption of abnormal objects, fever, vomiting, excessive salivation, paralysis of lower jaw, restlessness, convalescence and paralysis leading to death.

Clinical Features in Cat

Clinical features in cat are aggressiveness, great sensitivity to touch, profuse salivation, attempt to attack dog or even man.

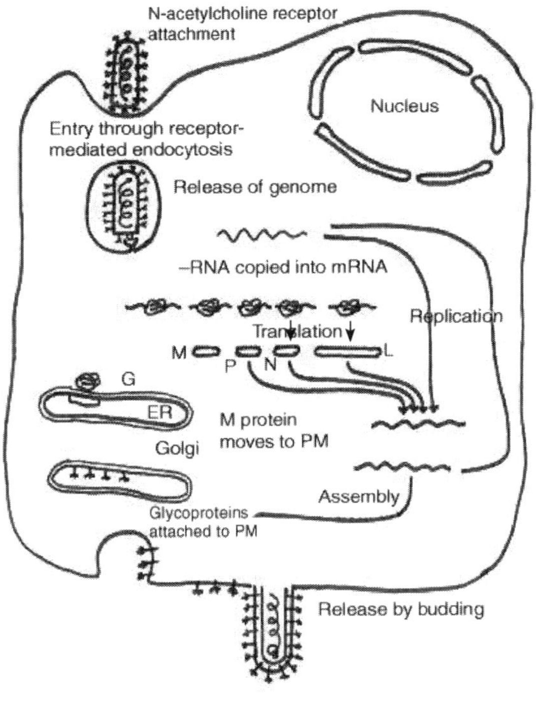

Figure 50.2 Multiplication of rabies virus

PATHOGENESIS

The virus present in the saliva of a rabid animal is deposited at the biting site. The virus attaches via spikes to the nicotinic acetylcholine receptors of host tissue cells. The virus appears to multiply in the muscles, connective tissues or nerves at the site of deposition. It penetrates the nerve endings either immediately or after a varying interval and travels in the axoplasm towards the spinal cord and brain. The movement of the virus in the axons is passive at a speed of about 3 mm per hour. The virus then multiplies extensively in the brain tissue, causing symptoms of encephalitis. Characteristic inclusion bodies called Negri bodies, form at the site of viral replication in the brain but the cells are not lysed. The virus spreads outward from the brain via the nerves to various body tissues, notably the salivary glands, eye and fatty tissue under the skin as well as heart and other vital organs.

The immune response of the host probably plays an important role in pathogenesis since viral antigens are expressed on the surface of the infected cells. Freshly isolated viruses are called *street virus*. If this virus is passed though various cell lines it is converted into *fixed virus*. It is used for vaccination.

LABORATORY DIAGNOSIS

Specimen

Saliva/sputum, skin biopsy, hair follicle, cerebrospinal fluid, blood, corneal swab, urine and brain post-mortem form the specimens for diagnosis of rabies.

Laboratory Tests

- Negri body examination
- Mouse inoculation
- Complement fixation test
- Polymerase chain reaction
- Haemagglutination test
- Passive haemagglutination test
- Electron microscopy
- Enzyme linked immunosorbent assay
- Fluorescent antibody test
- Serum virus neutralization test
- Counterimmunoelectrophoresis
- Immunoperoxidase test
- Haemagglutination inhibition test
- Passive diffusion
- Tissue culture techniques

Negri Body Examination

- Negri bodies are intracytoplasmic inclusion bodies. Sellers staining procedure is followed to demonstrate the presence of Negri bodies.
- First the brain is cut opened to expose the hippocampus region. A small piece is cut and placed on a filter paper with cut surface facing upwards. The filter paper is then placed on a glass slide.
- The cut surface is slightly sponged with the edge of the filter paper to remove blood. A clean microscopic slide is pressed on the tissue piece.
- While the smear is wet, it is flooded with the working stain and allowed to stain for 2–3 seconds and then washed with water, air-dried and examined under 100×.

Observation

Nerve cells	Blue cytoplasm and dark blue nucleus
Stroma	Pink

Erythrocytes	Copper coloured
Negri bodies	Magenta to dark red with dark blue or black inner granules

PREVENTION AND TREATMENT

This section was categorized into three, all three carry equal importance and one should not be given undue importance or neglected, at the cost of other two components. These components are:

- Management of wound
- Post-exposure immunization
- Pre-exposure immunization

Management of Wounds

Since the rabies virus enters the human body through a bite or scratch, it is imperative to remove as much as saliva with soap. After the removal with soap, any quaternary ammonium compound (1% cetrimonium bromide) may be applied as an antiseptic along with antirabies serum.

Post-exposure Immunization

Because of the long incubation period it is possible to institute prophylactic post-exposure immunization. Immunization must be started at the earliest to ensure that the individual will be protected before the rabies virus invades the central nervous system.

Two types of agents are employed to confer immunity to an individual who has been exposed to the rabies virus: antirabies serum/rabies immunoglobulin and antirabies vaccine. Antirabies serum provides passive immunity in the form of ready made antirabies antibody to tide over the initial phase of infection.

Antirabies vaccines that are available in India are

Semple's sheep brain vaccine

Human diploid cell vaccine (HDCV)

Primary chick embryo cell vaccine (PCECV)

Purified vero cell rabies vaccine (PVRV)

Site of vaccination The ideal site for vaccination is the anterior abdominal wall. This area offers enough space to accommodate 10 injections at 10 different sites and cause least discomfort to the patient.

Dose schedule A 6-dose schedule spread over a period of three months is recommended. It is also called Essen schedule (proposed by International Conference on Rabies, held at Essen, Germany). Days of vaccination are 1st, 3rd, 7th, 14th, 30th and 90th days.

Pre-exposure Immunization

There are no vaccines available for mass pre-exposure vaccination. The pre-exposure prophylaxis, hence, is recommended for definite group of individuals who because of their profession or hobby are at a higher risk of getting exposed to rabies virus.

High risk group for rabies

- Laboratory staff handling virus and infected materials
- Veterinarians
- Animal handlers and catchers
- Wildlife officers
- Quarantine officers
- Naturalists in epizootic areas
- Travellers from rabies-free areas to rabies-endemic areas

CONTROL OF RABIES

Any strategy for control of rabies in developing countries shall have the following components:

- Epidemiological surveillance
- Mass vaccination
- Dog population management
- Community participation

1. Draw the structure of rabies virus and explain its various properties.
2. List different tests used to diagnose rabies.
3. Discuss various features of rabies.
4. Explain prophylaxis of rabies.
5. Write short notes on:
   i. Fixed virus
   ii. Street virus
   iii. Negri bodies
   iv. Seller's staining
   v. Madman disease

1. Why is the rabies vaccine given in the abdominal region?
2. Why is rabies called madman disease?

Baer, G.M. (ed.). *The Natural History of Rabies*. CRC Press, Boca Raton, 1991.

Campbell, J.B. and Charlton, K.M. (eds.). *Rabies*. Kluwer Acad. Publ., Boston. 1988.

Centers for Disease Control: Rabies prevention United States, 1991. Recommendations of the Immunization Practices Advisory Committee (ACIP). MMWR 40 (RR-3): 1. 1991.

Krebs, J.W., Strine, T.W. Smith, J.S. *et al*. "Rabies surveillance in the United States during 1993." *J. Am. Vet. Med. Assoc.* 205: 1695. 1994.

Smith, J.S., Orciari, L.A., Yager, P.A. *et al*. "Epidemiologic and historical relationships among 87 rabies virus isolates as determined by limited sequence analysis." *J. Infect. Dis.* 166: 296. 1992.

en.wikipedia.org/wiki/Rabies

www.cdc.gov/ncidod/dvrd/kidsrabies/

www.rabies.com/

www.who.int/mediacentre/factsheets/fs099/en/

www.cfainc.org/articles/rabies.html

www.nlm.nih.gov/medlineplus/rabies.html

www.in.gov/isdh/programs/rabies/rabies.html

www.in.gov/boah/rabies/

www.canismajor.com/dog/rabies.html

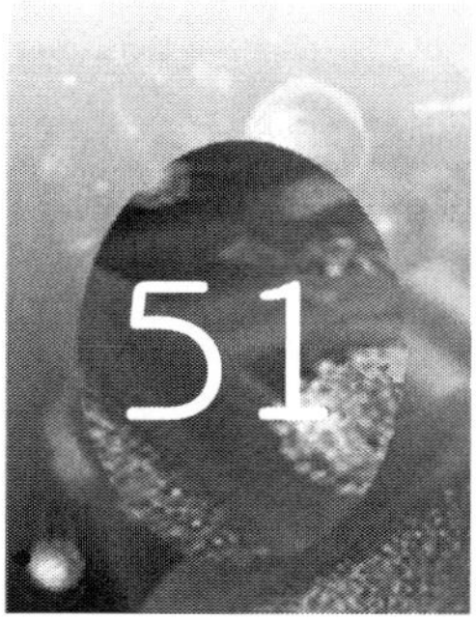

Hepatitis

INTRODUCTION

Hepatitis is the term used for any condition where there is inflammation or necrosis of liver cells. Necrosis is the term used for death of some or all cells in an organ or a tissue. Viral hepatitis has emerged as a major public health problem throughout the world affecting several hundreds or millions of people. Hepatitis and jaundice are not the same. Jaundice is the term used for yellowish discoloration of sclera, the white part of the eyes. Jaundice occurs due to various reasons, one of which is hepatitis.

A heterogeneous group of viruses called hepatotrophic viruses causes viral hepatitis and damage to liver. This group includes

Hepatitis A virus (HAV)

Hepatitis B virus (HBV)

Hepatitis C virus (HCV)

Hepatitis D virus (HDV)

Hepatitis E virus (HEV)

The only common feature of these hepatitis viruses is their primary hepatotrophism.

On the basis of epidemiology and clinical criteria, hepatitis was classified into two types. It is also differentiated on the basis of serology and molecular markers. One type occurred sporadically or as epidemics, affecting mainly children and young adults and transmitted by faecal-oral route. This was called infective or infectious hepatitis, later termed type A hepatitis.

A second type of viral hepatitis is transmitted by serum inoculation or blood transfusion. This was called by various names such as homologous serum jaundice, serum hepatitis and transfusion hepatitis. It was later called type B hepatitis.

All infectious hepatitis is caused by type A viruses and all serum hepatitis is caused by type B viruses.

About 98% of hepatitis is caused by hepatitis A, B, C, D and E viruses. Other viruses cause the remaining 2%. It includes

- Non-A, non-B, non-C, non-E hepatitis viruses
- Cytomegaloviruses
- Epstein–Barr viruses
- Herpes simplex viruses
- Yellow fever viruses

Exact mechanism for liver cell damage is not well understood. But it may be through

- Complex process of body's defence mechanism
- Direct attack and damage to liver cell

HEPATITIS A VIRUS (HAV)

Hepatitis type A is a subacute disease of global distribution, occurring mainly in children and young adults. The term "infectious hepatitis" was coined in 1912 to describe the epidemic form of the disease.

CHARACTERISTIC FEATURES

- HAV was first demonstrated by Feinstone and co-workers during the year 1973 from faeces by immunoelectron microscopy.
- It is a 27-nm non-enveloped, symmetrical (+) RNA virus, with icosahedral symmetry.
- Most of its characters mimic *Picornavirus* family. Only one serotype of the virus is known.
- It was originally designated as enterovirus type 72 but now recognized as the prototype of a new genus *Hepatovirus*.
- It is non-cytopatheic when grown in cell culture.
- It causes necrosis of parenchymal cells and histiocytic periportal inflammation *in vivo*.
- The HAV genome comprises about 7500 nucleotides.

Resistance

- The virus is relatively resistant to inactivation.
- It can withstand 60°C for 1 hour and 100°C for 1 minute. The virus is inactivated by formaldehyde 1 : 4000 at 37°C for 72 hours and chlorine 1 ppm in 30 seconds.

It is not affected by non-ionic detergents. It survives at 4°C or even much cooler temperatures.

CLINICAL MANIFESTATION

The following symptoms are observed during HAV infection:

- Fever
- Malaise
- Anorexia
- Nausea
- Vomiting
- Liver tenderness
- Mortality is very low ranging from 0.1–1%. General symptoms subside with the onset of jaundice.

PATHOGENESIS

Clinical expression of HAV varies considerably. HAV enters the body via ingestion of contaminated food, water and faecal-oral route.

The steps involved in pathogenesis are:

- The virus enters through faecal-oral route, food and water.
- The virus withstands stomach acidity because it is non-enveloped.
- Replication occurs in the oropharynx and epithelial lining of the intestine.
- It crosses the intestinal barrier and reaches liver through the bloodstream and binds to the receptors of liver parenchyma cells and produces the symptoms.
- Virus is then released and it enters the intestine through bile.

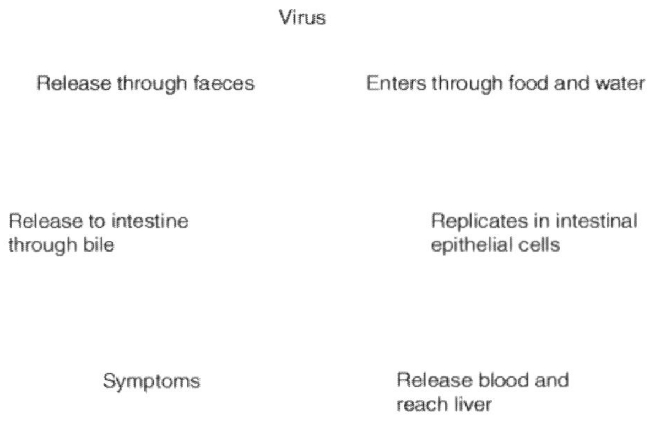

- The virus is then released to the outer environment through faeces, 10 days before the onset of symptoms.
- Incubation period for HAV is about 3–5 weeks.

LABORATORY DIAGNOSIS

Isolation of virus in tissue culture requires prolonged adaptation and it is therefore not suitable for diagnosis.

Serological techniqu es are available that include:

- Immunoelectron microscopy
- Complement fixation
- Immune adherence haemagglutination
- Radioimmunoassay
- Enzyme immunoassay

CONTROL AND TREATMENT

The control methods include

- Improvement of sanitation
- Prevention of faecal contamination and direct contact with infected individuals.

The people at risk of HAV infection are staff and residents of hospitals, day care centre workers, sexually active male homosexuals intravenous drug abusers, sewage workers, medical students, military personnels, low socioeconomic groups and patients with chronic liver disease.

Available vaccines include

- Inactivated vaccines (formaldehyde)—administered through intramuscular route.
- Attenuated vaccine—administrated orally.

Post-exposure treatment include human immunoglobulin and anti-HAV.

PREVENTION

- Consumption of water from safe water sources.
- Use of boiled water for brushing teeth when travelling in an area where there is a high risk of getting HAV infection.
- Avoidance of fruits, salads or uncooked vegetables that have not been washed or treated in boiled water.
- Avoidance of uncovered food or beverages from street vendors.
- Taking HAV vaccine if at high risk of getting infection.

HEPATITIS B VIRUS (HBV)

It is the most important type among hepatitis-causing viruses. HBV causes a type of hepatitis called serum hepatitis. The disease occurs throughout the world. It is an important cause of acute and chronic infection of liver. HBV infection causes more than million deaths per year worldwide.

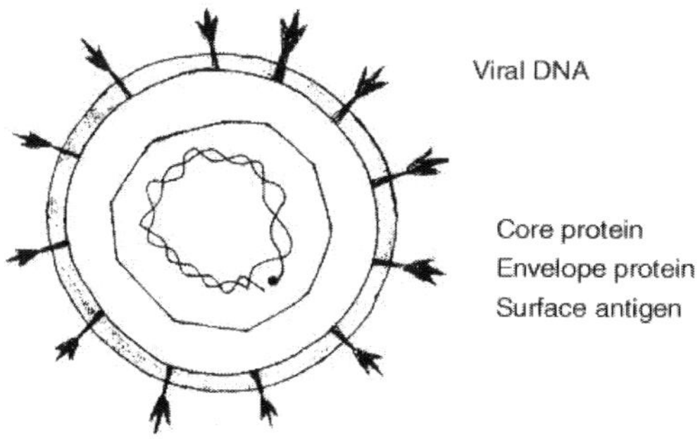

Figure 51.1 Structure of hepatitis B virus

In 1965, Blumberg reported protein antigen in the serum of an Australian patient. This antigen was called the Australian antigen. By 1968, the Australian antigen was shown to be associated with serum hepatitis. This was then found to be hepatitis surface antigens (HBsAg).

CHARACTERISTIC FEATURES

- HBV is a 42-nm spherical virus that possesses several antigens (Figure 51.1).
- There are three envelope polypeptides that are called as HBsAg, HBcAg and HBeAg.
- HBV belongs to the family Hepadnaviridae.
- The nucleocapsid (27 nm) of the virion consists of the viral genome surrounded by the core antigen.
- The genome, which is approximately 3.2 kb in length, has an unusual structure and is composed of two linear strands of DNA held in circular configuration. Negative strand is complete but positive strand is incomplete. 3′ end of the genome is associated with a DNA polymerase molecule.
- In genome there are 4 major open reading frame (ORF)
  - ORF-S consists of Pre-S1 and Pre-S2 regions that code for structural proteins of surface and core respectively.
  - ORF-P encodes polymerase that contains DNA polymerase and RNase H.
  - ORF-X encodes transcriptional activator (Figure 51.2).

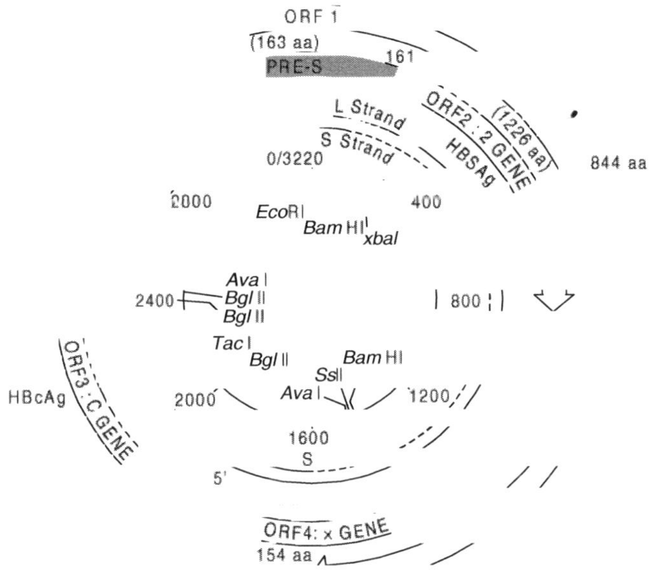

Figure 51.2 Genome organization of HBV

- ORF-C consists of Pre-C region and C region.
- 22 nm Dane particle and tubular structure constitute the HBsAg.

The virus is stable at 37°C for 60 minutes. Its antigen can be destroyed by 0.5% sodium hypochlorite for 3 minutes. HBsAg is stable at pH 2.4 for 6 hours. 2% glutaraldehyde destroys the antigenicity within 3 minutes. HBsAg is resistant to UV.

- The virus does not grow in tissue culture medium.

CLINICAL MANIFESTATION

Incubation period varies widely from 40 days to 6 months, but is often about 2–3 months.

The symptoms include discomfort, tiredness, fever, chills, loss of appetite, nausea, vomiting, headache, pain in abdomen, diarrhoea, jaundice, itching sensation on skin, darkening of urine, light coloured stool and inflammation of joints.

Mortality rate is about 0.5–2% and most people develop carrier state.

The complications include pain in liver, liver cancer, hepatocellular carcinoma, arthralgia, polyarteritis and glomerulonephritis.

PATHOGENESIS

Pathogenesis involves 3 steps.

Entry The virus gains entry by the following routes

1. Blood transfusion
2. Sexual transmission
3. From infected mothers to neonates
4. Through contaminated syringes and needles
5. In rare cases, by arthropods

2. *Multiplication and spread* The target of HBV is the hepatocyte. HBV proteins and genomes were identified in extrahepatic sites also (bone marrow, spleen, lymph nodes and circulating lymphocytes) but they do not produce any damage in these locations.

The steps involved in the viral multiplication and spread are

- Virus attaches to the hepatocyte with the help of its surface antigen (HBsAg).
- HBV enters inside the cytoplasm through receptor-mediated endocytosis.
- DNA of the virion is transported into the nucleus.
- DNA transcription follows resulting in the formation of mRNA.
- Short mRNA transcripts undergo translation in the ER.
- Translation products are Pre-S and S proteins.
- P protein translation occurs in the cytoplasm and it is coupled to capsid protein formation.
- P and capsid proteins are assembled and they enclose the DNA.
- Replicated DNA undergoes nick formation resulting in dsDNA with staggered strand.
- Morphological changes occur on the surface of plasma membrane.

Complete virions are released through exocytosis (Figure 51.3).

Liver cell damage Replication results in injury of hepatocytes and release of progeny virions into the bloodstream. Cell injury is not only caused by cytopatheic effect of virus

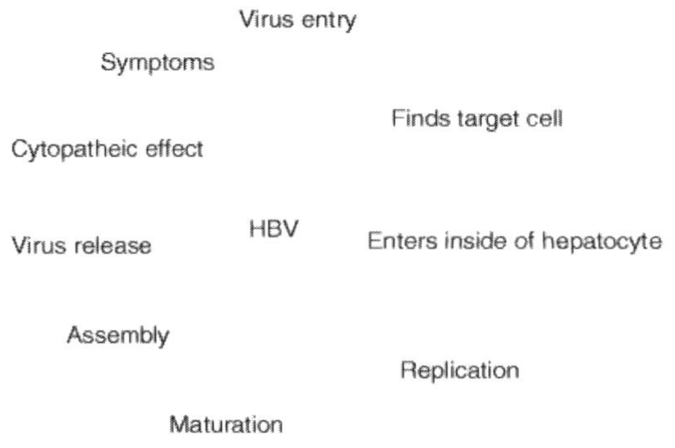

but also by the activation of cytotoxic immune mechanisms. This results in liver-tissue degeneration and the release of liver-associated enzymes into the bloodstream.

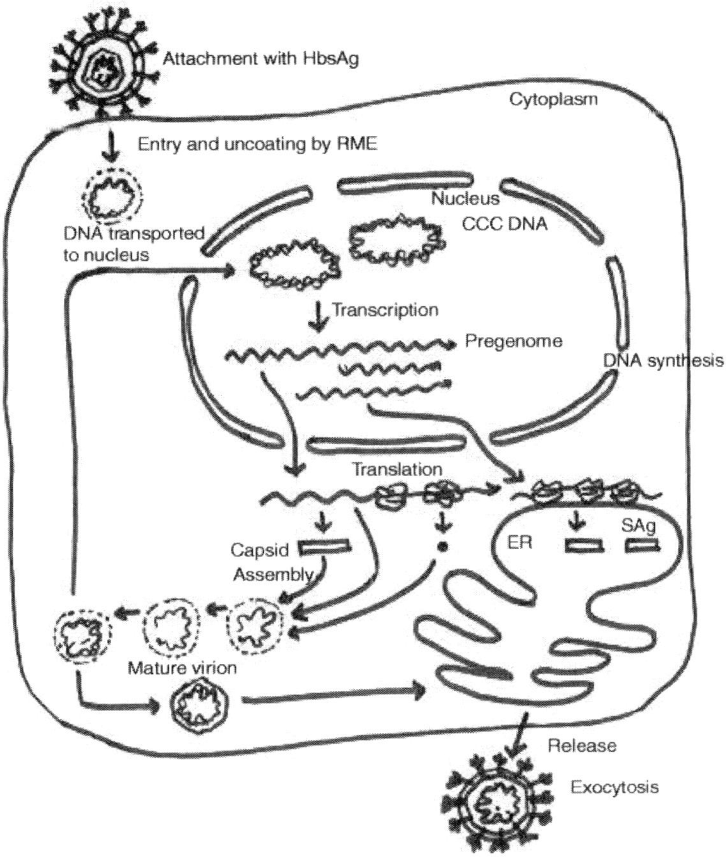

Figure 51.3 Multiplication of hepatitis B virus

This is followed by jaundice, the accumulation of bilirubin in the skin and other tissues with a resulting yellow appearance.

CMI may be important in terminating the infection and in some instances, it will cause immune-mediated liver damage. Exogenous interferon may be effective in treating some patients with chronic hepatitis.

Chronic hepatitis leads to persistent hepatitis, with mild periportal inflammation. This will lead to chronic active hepatitis (CAH) with more widespread inflammation and necrosis. CAH leads to cirrhosis and hepatocellular carcinoma.

LABORATORY DIAGNOSIS

Several types of blood tests are recommended for diagnosis of hepatitis.

Liver Enzyme Test

Blood levels of liver enzymes such as alanine aminotransferase and aspartate aminotransferase are elevated in the early stages of viral hepatitis.

Detection of Hepatitis B Antigens

ELISA test is used for the detection of HBsAg. ELISA can also be used for the detection of hepatitis antibodies.

EPIDEMIOLOGY

- About 140,000–320,000 infections occur per year.
- 70,000–1,60,000 are symptomatic infections.
- Out of symptomatic infections about 8400–19,000 hospitalizations/year and 140–320(0.2%) deaths occur per year.
- Of all infections, 8,000–32000 are chronic and 5000–6000 deaths occur per year.
- Intermediate endemicity is observed in India.
- 5–10% adults, 30% children and 90% neonates act as carriers.
- There are 350 million carriers worldwide, out of these, 45 millions are in India.

  The following are high risk groups for HBV infection:

  - Health care workers who come in contact with contaminated blood or other body fluids.
  - Male homosexuals.
  - People having sexual contact with those having HBV infection.
  - People with kidney diseases who require dialysis.
  - People who receive organ transplants.
  - People undergoing treatment for leukemia.
  - Babies born to infected mothers.
  - Intravenous drug users.

PREVENTION

Prevention is possible by active and passive immunization.

Two types of vaccines are currently available:

1. Recombinant HB vaccine synthesized from yeast cells are safe and effective and provides 90% protection.
2. Plasma-derived vaccine.

Vaccine injection is given in muscles of the upper and outer parts of the arm at birth, at the age of one month and at 6 months. A booster dose is recommended at 5 years of age.

TREATMENT

People with chronic and active inflammation of liver cells due to HBV are treated with interferon. It blocks the entry of virus into the cells. In recent years interferon is prescribed in combination with ribavirin.

Interferon is not recommended for people who have developed cirrhosis or scarring of liver tissue. For those, lamuvidine is recommended. Medicine decreases the multiplication of virus.

Interferon stimulates the body's natural defence.

During IFN treatment, the following side effects are observed.

- Discomfort
- Bodyache
- Fatigue
- Irritability
- Loss of appetite
- Nausea
- Diarrhoea

To cope with the side effects of IFN the following measures are followed.

- Intake of large volumes of water everyday.
- Intake of painkillers with the recommendation of physician.
- Resting as much as possible.
- Practising relaxation techniques such as breathing exercise.
- Regular food habits
- Frequent brushing
- In case of severe nausea, medicines are taken as recommended by physician.

HEPATITIS C VIRUS

- HCV is the leading cause of post-transfusion hepatitis (Figure 51.4).
- Both chronic, asymptomatic carriers and chronic hepatitis have been documented with HCV.

Figure 51.4 Hepatitis C virus

- Frequency of HCV is greater than HBV. Post-transfusion chronic hepatitis may occur in up to 54% of cases.
- Parenterally transmitted non-A, non-B hepatitis, now known as HCV, can be identified by a specific serologic test for anti-HCV antibodies.
- This virus is a 30–60-nm, spherical-shaped, positive-sense RNA virus.
- Its genome has 10,000 nucleotides.
- HCV is considered as the major risk factor because 80% of infections lead to chronicity.
- 70% leads to chronic active hepatitis or cirrhosis.
- Infection is also associated with progression to primary liver cancer and hepatocellular carcinoma.
- HCV rarely seems to cause fulminant hepatitis.
- The genome of HCV resembles those of the pestivirus and flavivirus.
- All genomes contain a single large open-reading frame, which is translated to yield polyproteins from which the viral proteins are derived by post-translational cleavage and other modifications.
- Helicase, polyproteins and proteases are involved in RNA replication.

HEPATITIS DELTA VIRUS (HDV)

- HDV occurs in only those who have HBV infection (Figure 51.5).
- In 1977, Rizzetto and colleagues in Italy identified a new viral antigen in the liver cell nuclei of patients infected with HBV. Later it was called HDV.
- HDV is coated with HBsAg, which is needed for release from the host hepatocyte and for entry in the next round of infection.

HBsAg

Coat

ssRNA

Figure 51.5 Hepatitis delta virus

- HDV is a spherical, 36-nm particle with an outer coat composed of HBV, surface antigen surrounding the circular ssRNA with 1.7 kbp. The internal protein has a molecular weight of 68,000 daltons. Closest relative of HDV is a satellite virus of plants.
- Two types of infections are recognized:
  - Co-infection—HDV and HBV are transmitted together at the same time.
  - Superinfection—Delta infection occurs in a person already harbouring HBV.
  - HDV has 1679 nucleotides.

- RNA replicates with the help of host RNA polymerase II.
- About 5% HBsAg carriers worldwide are infected with HDV.
- Diagnosis is performed with immunofluorescence and ELISA.
- An IgM antibody appears 2–3 weeks after infection and is soon replaced by IgG antibody.

Capsid

ssRNA
+ sense

Figure 51.6
Hepatitis E virus

HEPATITIS E VIRUS

- It is a single-stranded linear RNA virus. It is included under the family Caliciviridae. Largest epidemic occurred in Delhi during the winter of 1955–56, affecting over 30,000 people within 6 weeks.
- Incubation period ranges from 2–9 weeks with an average of 6 weeks.
- Most cases occur in the young to middle-aged adults (15–40 years).
- A unique feature is the clinical severity and high case fatality rate of 20–40% in pregnant women, especially in the last trimester of pregnancy.
- HEV is a spherical non-enveloped virus, 30–32 nm in diameter.
- It enters liver through intestine and blood.
- It can be identified by means of ELISA.
- There is no vaccine currently available for preventing HEV.
- Relapse of HEV infection is common.
- Personal hygiene and sanitation are the only effective ways of prevention.

DIFFERENTIATING PROPERTIES OF HEPATITIS VIRUSES

Characters	HAV	HBV	HCV	HDV	HEV
Family	Picornaviridae	Hepadnaviridae	Flaviviridae	Defective Virus	Calciviridae
Genus	Enterovirus	Hepadnavirus	Hepaci virus	Deltavirus	Herpesvirus
Size	25–29 nm	40–50 nm	30–60 nm	35 nm	30–32 nm
Nucleic acid	ssRNA(+)	dsDNA	ssRNA(+)	ssRNA(–)	ssRNA(+)
Capsid	Icosahedral	Spherical	Spherical	Spherical	Icosahedral
Virion	Non-enveloped	Enveloped	Enveloped	Enveloped	Non-enveloped

1. What are the parameters used to differentiate different hepatitis viruses?
2. Write short notes on
   i. Serum hepatitis
   ii. ORF
   iii. Infectious hepatitis
   iv. Post-transfusion hepatitis
   v. HAV, HBV, HCV, HDV and HEV
   vi. Superinfection
   vii. Co-infection

Beasley, R.P. and Hwang, L.Y. "Overview on the epidemiology of hepatocellularcarcinoma." 532 In: *Viral Hepatitis and Liver Disease*. Hollinger F.B., Lemon, S.M. and Margolis, H.S. (eds.). Williams and Wilkins, Baltimore. 1991.

Gerin, J.L., Purcell, R.J. and Rizzetto, M. (eds.). *The Hepatitis Delta Virus*. Wiley-Liss, New York. 1991.

Houghton, M., Han, J., Kuo, G., Choo, Q.L. *et al*. "Hepatitis C virus: structure and molecular virology." 229. In: *Viral Hepatitis: Scientific Basis and Clinical Management*. Zuckerman, A.J. and Thomas, H.C. (eds.). Churchill Livingstone, Edinburgh. 1993.

Margolis, H.S. (eds.). *Viral Hepatitis and Liver Disease*, Williams and Wilkins, Baltimore. 1991.

Zuckerman, A.J. and Thomas, H.C. (eds.). *Viral Hepatitis: Scientific Basis and Clinical Management*, Churchill Livingstone, Edinburgh. 229. 1993.

en.wikipedia.org/wiki/Hepatitis_A

www.cdc.gov/hepatitis/

www.hepatitis.org/

www.who.int/topics/hepatitis/en/

www.nlm.nih.gov/medlineplus/hepatitis.html

www.hepb.org/

www.hepnet.com/

www.hepfi.org/

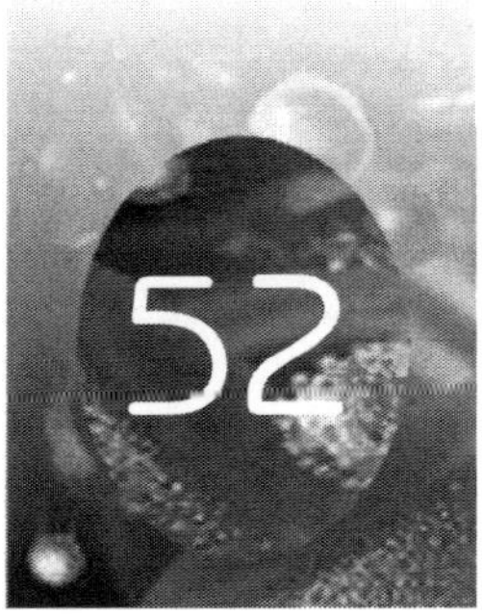

AIDS

INTRODUCTION

As the name implies AIDS (Acquired Immunodeficiency syndrome) is a condition where there is a deficiency in the body's natural defence mechanisms or the immune system. It is not acquired due to hereditary or long-term use of some medicines such as those for the treatment of cancer, but due to certain behavioural patterns. Syndrome is a group of symptoms. When one gets AIDS, there can be a wide range of symptoms all due to the body's diminished ability to fight diseases. AIDS is one of the sexually transmitted diseases, and an epidemic disease with worldwide distribution. It was first described in the United States in 1981. The disease appears to have begun in Central Africa as early as 1950. Montagnier and his colleagues first reported isolation of an etiological agent in 1983 from the Pasteur Institute, Paris.

PREVALENCE

WHO estimates that 8–10 million adults and 1 million children worldwide are infected with AIDS virus. On the basis of report 73.3% of AIDS is acquired by sexual contact, 7.4% by transfusion process, 0.7% by homosexual contact, 9% by intravenous drug abuse. Up to 1995, about 295,473 deaths have been reported in the US. Mortality rate is extremely high. The case fatality rate average was about 92% for adults diagnosed with AIDS before 1987. By the year 2010, AIDS will become the major killer of children with an estimated 250 million infections worldwide.

The first AIDS case in India was reported in 1986 from Chennai. Since then, there has been a rapid spread of HIV infection all over the country. By March 1998, the National AIDS Control Organization had reported that a total of 71,400 people were having HIV infection from among 3.2 million people who were tested for it. Of these, about 10% of AIDS cases were from India. About 80% were males and 20% females. Almost 89% people with AIDS were in the age group of 15–44 years, most economically productive years for any individual. Maharashtra has reported the maximum number of HIV-infected individuals followed by Tamil Nadu and Manipur (Chennai first in Tamil Nadu).

CAUSATIVE AGENT

Primarily the HIV-1 virus causes AIDS. This virus is a retrovirus and is closely related to human T-cell leukemia virus. Sometimes HIV-2 virus also causes it. HIV is an enveloped virus with a cylindrical core inside. The core contains two copies of ssRNA and several enzymes. Ten virus-specific proteins have been discovered. One of them is the gp120 envelope protein, which participates in the attachment to $CD4^+$ cells. Virus is spherical in shape, about 90–120 nm in size. The name HIV was given by International Committee on Virus Nomenclature in 1986.

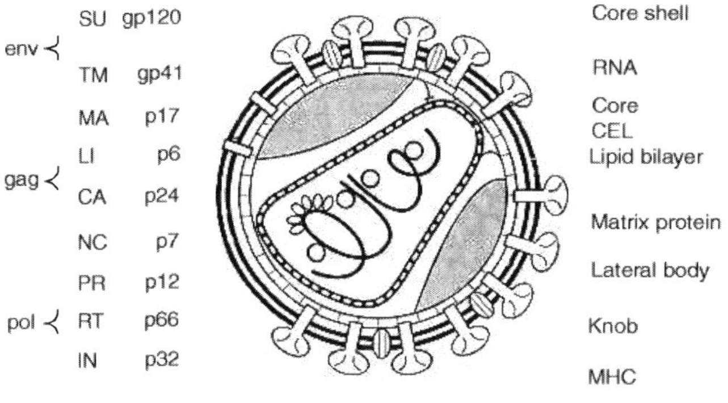

Figure 52.1 Structure of HIV

HIV-1 virion is a pleomorphic structure containing 72 external spikes. The two major viral envelope proteins, gp120 and gp41 form these spikes. The core of HIV-1 contains 4 nucleocapsid proteins. The phosphorylated p25 polypeptide forms the chief component of the inner shell of the nucleocapsid, whereas the p17 contains 2 copies of single-stranded RNA that is associated with the various preformed virus enzymes, including reverse transcriptase, integrase, ribonuclease and protease (Figure 52.1).

Resistance

HIV is thermolabile, being inactivated in 10 minutes at 50°C and in seconds at 100°C. At room temperature, in dried blood it may survive for up to 7 days. HIV is inactivated within 10 minutes by treatment with 50% ethanol, 3.5% isopropanol, 0.5% lysol, 0.5% paraformaldehyde, 0.3% H_2O_2 and 10% household bleach. For treatment of contaminated medical instruments, 2% solution of glutaraldehyde is useful.

TRANSMISSION

HIV is believed to have originated in Central Africa. From here it spread to the rest of the world. HIV is primarily transmitted by:

- Sexual contact homosexual and heterosexual
- Direct exposure of a person's bloodstream to body fluids

- Mother to child through placenta
- Skin erosion
- Intravenous drug abuse
- Transfusion process
- Invasive medical procedure
- Drug abuse with needle sharing
- Breast feeding to newborn

REPLICATION

Once the virus enters inside the body, the viral gp120 envelope protein binds to the CD4 glycoprotein plasma membrane receptor on $CD4^+$ T cells, macrophages, dendritic cells and monocytes. After the viral envelope has fused with the plasma membrane, the virus releases its core protein and the two RNA strands in the cytoplasm. Inside the infected cell, the core protein remains associated with the RNA as it is copied into single-stranded DNA by the RNA/DNA-dependent DNA polymerase activity of the reverse transcriptase enzyme, ribonuclease H, and the DNA is duplicated to form a dsDNA copy of the original RNA genome.

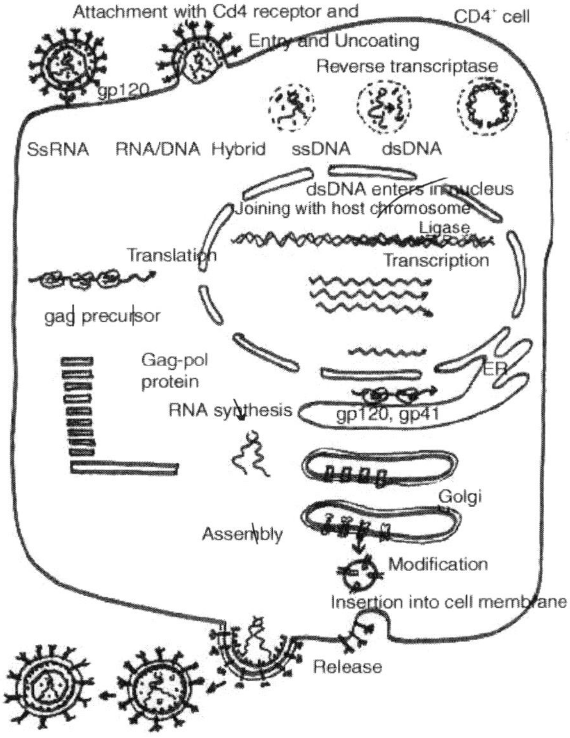

Figure 52.2 Multiplication of HIV virus

The viral dsDNA is then translocated to the nucleus and integrated into the host chromosomal DNA by the viral integrase enzyme. This integrated viral DNA and chromosomal DNA is called provirus. Then transcriptional factors stimulate transcription of proviral DNA into genomic ssRNA and after processing, several mRNAs are formed. Then the viral RNA is exported to cytoplasm. After completion of this process, host cell enzymes catalyse the synthesis of viral protein. HIV ssRNA and proteins assemble beneath the host cell membrane, into which gp41 and gp120 are inserted. The cell enlarges, forms a bud which then forms a new virus. Eventually the host cell lyses.

PATHOGENESIS

The precise mechanism of AIDS pathogenesis is still not known and many hypotheses exist. Many believe that destruction of CD4 function causes AIDS. Once a human's $CD4^+$ cells are infected with HIV, four types of pathological changes may ensue. First, a mild form of AIDS may develop with symptoms, which include fever, lymph node enlargement, oral candidiasis, presence of antibodies to HIV, weight loss, malaise and headache.

These symptoms occur in the first few months after infection, last for 1–3 weeks and recur. This is known as AIDS-related complex (ARC).

The second set of symptoms appears after 2–8 years of HIV infection, although it varies considerably with each individual. These include

- Candidiasis of bronchi, trachea
- Cervical cancer
- Coccidioidomycosis
- Diarrhoeal disease
- Cytomegalovirus disease
- Encephalopathy
- Toxoplasmosis
- Herpes simplex infection
- Histoplasmosis
- Lymphoma
- Tuberculosis
- Pneumonia
- Septicaemia

The third type involves the CNS, since virus-infected macrophages can cross the blood-brain barrier. The classical symptoms are:

- Headache
- Fever
- Abnormal refluxes
- Ataxia
- Autoimmune neuropathies
- Cerebrovascular disease
- Brain tumour
- Inflammation of neurons
- Nodule formation
- Demyelination

The fourth result of HIV infection is cancer. Other diseases are:

- Kaposi's sarcoma

- Carcinoma of the mouth and rectum
- B-cell lymphoma

LABORATORY DIAGNOSIS

HIV infection can be detected by:

Specific Tests

1. Virus isolation
2. Detection of HIV-specific antibodies (ELISA)
3. Western blot
4. Polymerase chain reaction

Non-specific Tests

1. Total and differential WBC count
2. Assay of T cell
3. Platelet count
4. Estimation of IgG and IgA level
5. Skin test for CMI(s1)

PREVENTION

- Avoiding sexual contact with HIV-infected individuals
- Avoiding sharing of shaving materials
- Avoiding drug abuse
- Screening of blood before transfusion
- Using condoms during sexual contact

CONTROL

Therapeutic categories for HIV include antiviral therapies, immune modulators and therapies to treat and prevent opportunistic diseases. Intensive educational efforts are needed to prevent transmission of HIV by sexual intercourse, intravenous drug use, and exposure to blood or blood products. Transmission of HIV can be prevented in health care workers and other medical settings by the application of universal precautions. Current treatment for HIV infection consists of Highly Active AntiRetroviral Therapy (HAART). Azidothymidine (AZT) and other antiviral agents are used for prophylaxis against progression of disease and for treatment. Zidovudine, lamivudine, tenofovir, emtricitabine, lopinavir boosted with ritonavir, are the effective drugs used for treatment under HAART programme.

1. Draw the structure of HIV virus and label its parts.
2. Explain the pathology associated with AIDS.
3. Explain how HIV is transmitted?
4. Describe the syndrome associated with AIDS.
5. Write short notes on
   i. HIV
   ii. ELISA
   iii. AIDS
   iv. ARC

Blattner, W.A. "Retroviruses." In Evan, A.S. (ed.). *Viral Infections of Humans*. Plenum, New York. 1989.

Centers for Disease Control: Revision of the CDC surveillance case definition for acquired immunodeficiency syndrome. MMWR 36:3S. 1987.

Nerurkar, L.S., Wong-Staal, F. and Callo, R.C. "Human retroviruses, leukemia, and AIDS." In: Henderson, E.S. and Lister, T.A. (eds.). *Leukemia*, 5th edn. WB Saunders, Philadelphia. 1990.

Rosenblatt, J.D., Chen, I.S.Y. and Wachsman, W. "Infection with HTLV-I and HTLV-II: evolving concepts." *Semin. Hematol.* 25: 230. 1988.

Williams, L.M. and Cloyd, M.W. "Polymorphic human gene(s) determines differential susceptibility of CD4 lymphocytes to infection by certain HIV-1 isolates." *Virology.* 184: 723–728. 1991.

www.rulabinsky.com/cavd/

www.youandaids.org/About%20HIVAIDS/Symptoms/index.asp

www.unaids.org.in

www.aids2006.org/

www.nlm.nih.gov/medlineplus/aids.html

topics.nytimes.com/top/news/health/diseasesconditionsandhealthtopics/aids/index.html

www.unaids.org/

www.unfpa.org/aids_clock/

www.nari-icmr.res.in/

Herpesvirus Infections

INTRODUCTION

The Herpesvirus family contains most important human pathogens. More than 50 herpesviruses are known to cause infections of which only 8 are associated with human infections. The outstanding property of herpesviruses is their ability to establish lifelong persistent infections in their host and to undergo periodic reactivation. Herpesviruses are, large enveloped DNA viruses. Herpes infections are of particular importance in immunocompromised patients. The term “herpes” means “to creep” (easily spreading nature). Genital herpes was first described in the 18th century.

CAUSATIVE AGENT

- Herpesviruses are large enveloped viruses.
- Virion is spherical, 150–200 nm in diameter.
- It has icosahedral capsid.
- Its genome is linear double-stranded DNA, 124–235 kbp.
- More than 35 proteins are present in the virion.
- Replication occurs in the nucleus of the host cell.
- Genome is large enough to code for at least 100 proteins.

CLASSIFICATION

Classification of herpesvirus family is complicated (Table 53.1).

Table 53.1 Classification of herpesviruses

Order	α-herpesvirinae	β-herpesvirinae	γ-herpesvirinae
Family	Herpesviridae	Herpesviridae	Herpesviridae
Genus I	Simplex virus	Cytomegalovirus	Lymphocryptovirus
Species I	Human herpesvirus 1	Cytomegalovirus	Epstein–Barr virus
Species II	Human herpesvirus 2		
Genus II	Varicella virus	Roseolovirus	Rhadionovirus
Species I	Varicella-zoster virus	Human herpes virus 6	Kaposis-Sarcoma virus
Species II		Human herpes virus 7	

HERPES SIMPLEX VIRUS

These viruses are extremely widespread in the human population. The virus is spread by direct contact with infected secretions. There are two distinct herpes simplex viruses: type 1 and type 2 (HSV-1 and HSV-2). Both types share many common antigens. The structure of herpes simplex virus is shown in Figure 53.1.

Figure 53.1 Structure of herpes simplex virus

Herpes simplex virus type 1 and 2 infect epithelial cells and establishes latent infection in neurons. Type 1 is associated with oropharyngeal lesions. Type 2 primarily infects the genital mucosa and is mainly responsible for genital herpes. Both viruses can cause neurological complications.

Type 1 HSV is spread by contact with infected saliva, whereas HSV-2 is transmitted sexually.

REPLICATION

- The virion binds to the extracellular protein through gB and gC receptor.
- Another viral protein gD interacts with a second cellular receptor.
- This interaction mediates fusion of the virus with the host plasma membrane.
- The virus is uncoated, liberating tegument proteins and nucleocapsid into the cytoplasm.
- Viral nucleocapsid docks at the nuclear pore and releases viral DNA into the nucleus, where the DNA circularizes.
- VP16 enhances transcription of viral genome and stimulate transcription of immediate early genes by host cell RNA polymerase II.
- Immediate early mRNAs are spliced and transported to the cytoplasm, where they are translated.
- Immediate early proteins (α-proteins) are imported into the nucleus, where they activate the transcription of early genes.
- β-protein genes are transported to the cytoplasm after transcription and are translated.
- β-proteins are imported to the nucleus where they induce DNA replication and synthesize substrate for DNA synthesis.
- DNA replication produces long concatemeric DNA molecules, the templates for late gene expression.
- Late mRNAs are transported to the cytoplasm and synthesize gamma protein. These proteins are structural proteins and are needed for viral assembly.
- Some late proteins are inserted into ER and are transported into Golgi apparatus for glycosylation.
- Mature glycoproteins are transported to plasma membrane of the infected cell.
- Some gamma proteins are transported to the nucleus for assembly of nucleocapsid and DNA packaging.
- Newly replicated viral DNA is packaged into preformed capsids.
- These capsids, together with some tegument proteins bud from the inner nuclear membrane into the lumen of ER and acquire envelope.
- Enveloped virus is then transported to the plasma membrane for release by exocytosis.
- Latent infection occurs primarily in neurons found in sensory and autonomic ganglia. During this infection, latency-associated transcript (LTT) promoter is synthesized and are involved in protein synthesis.

CLINICAL MANIFESTATION

Primary infections show the following symptoms:

- Gingivostomatitis
- Fever
- Sore throat
- Lesions in oral cavity
- Malaise
- Submandibular lymphadenopathy
- Anorexia
- Pharyngotonsilitis
- Keratoconjunctivitis

Infection of the Genitalia

Vesiculoulcerative lesions occur in the penis of male. Lesions also occur in cervix, vulva, vagina and perineum of female. They are painful and systemic and are associated with fever, malaise, dysuria and lymphadenopathy.

Infection of other Organs

Skin abrasions and Eczema herpeticum (Kaposi's varicelli form eruption) occurs.

Infection of the brain show the following symptoms:

Encephalitis

Fever

Headache

Stiff neck

Neonates

HSV-2 causes 75% of neonatal infections.

Three types of infections are observed.

Lesions localized to the skin, eyes and mouth

Encephalitis

Disseminated infections

PATHOGENESIS

HSV is transmitted through contact of a susceptible person with an individual excreting virus. When the virus encounter the mucous membrane or a break in the skin, HSV undergoes replication in the parabasal and intermediate epithelial cells, which lyse and invoke inflammatory response. HSV causes cytolytic infection. Pathological changes during primary replication are due to necrosis and inflammation.

Characteristic histopathologic changes include ballooning of infected cells, production of Cowdry type A intranuclear inclusion bodies, margination of chromatin and the formation of multinucleated giant cells. The early inclusions virtually fill the nucleus but later condense and are separated by a halo from the chromation at the nuclear margin. Cell fusion provides an efficient method for cell to cell spread. Oedema fluid accumulates between the epidermis and dermal layer. This fluid contains cell free virus, cell debris and inflammatory cells. As white cell responds, the lesions form scab and heal without scarring. In mucous membranes, the vesicle rupture rapidly and forms ulcers.

Primary infections are usually mild and most of them are symptomatic. After initial replication, the virus migrates to the dorsal root ganglion, where, after further replication latency is established.

LABORATORY DIAGNOSIS

Diagnosis may be made on clinical grounds. Patients with fever and vesicles may be considered to have HSV infection.

Specimen

Vesicular and hepatic lesions of skin, cornea or brain, throat washings, CSF and stool.

Isolation of Virus

Virus isolation is one of the most reliable method for confirmation of the clinical diagnosis.

The specimens are transported through viral transport medium and inoculated into tissue culture. HSV has a wide host range, so many cell culture systems are susceptible. The appearance of typical cytopatheic effects in cell culture in 2–3 days suggests the presence of HSV.

Serology

Several methods have been developed for rapid diagnosis. Antibodies appear in 4–7 days after infection. Neutralization test, CF, ELISA, RIA and immunofluorescence can quantitate antibodies.

EPIDEMIOLOGY

Herpes simplex viruses are worldwide in distribution. No animal reservoirs are involved in human infection transmission. Transmission is by infected secretions.

TREATMENT

- Most of the drugs used for the treatment of HSV, inhibit the DNA synthesis.
- Vidarabine triphosphate inhibits DNA polymerase.
- Acyclovir targeting HSV-infected cells and viral DNA polymerase.
- Topically applied idoxuridine, trifluridine, vidarabine and acyclovir have been used for herpetic keratitis.

PREVENTION

Currently there is no effective therapy for the prevention of HSV infection. Avoidance of direct contact with lesion or infected secretions is the primary means of prophylaxis. Vaccination for herpes would be the ideal preventive measure.

VARICELLA–ZOSTER VIRUS

VZV infection is a common childhood infection of the world. Varicella (chickenpox) is a mild, highly contagious disease, chiefly occurs in children, characterized by a generalized vesicular eruption of the skin and mucous membrane. Zoster (shingles) is a sporadic, disease of adults or immunocompromised individuals that is characterized by a rash limited in distribution to the skin.

Primary infection causes chickenpox and reactivation of the virus in the later life results in shingles.

CLINICAL MANIFESTATION

Incubation period is 10–23 days. Malaise and fever are earliest symptoms, followed by rash that characteristically begins on the scalp and trunk and then spreads. Macules evolve in 2–3-mm vesicles, spread throughout the body. Lesions on mucous membranes are easily transmitted and may appear as ulcers. Lesions appear on the mouth, rectum and vagina. Other symptoms include headache, sore throat, loss of appetite and irritability.

Zoster infection usually starts with severe pain in the area of skin or mucosa supplied by one or more groups of sensory nerves and ganglia. The most common complication of Zoster is Postherpetic neuralgia (PHN). Pain may be characterized by burning, itching or tingling sensations.

PATHOGENESIS

The route of infection is the mucosa of the upper respiratory tract or the conjunctiva. The virus circulates in the blood and undergoes multiple cycles of replication and eventually localizes in the skin. Lesions of varicella infection are associated with cutaneous and mucosal endothelial cells. Swelling of epithelial cells, ballooning degeneration and the accumulation of tissue fluids results in vesicle formation.

Eosinophilic inclusion bodies are found in the nuclei of infected cells. Multinucleated giant cells are common.

Zoster lesions are histopathologically similar to varicella. There is also an acute inflammation of the sensory nerves and ganglia. It is not clear what triggers reactivation of latent Varicella–zoster virus infection in ganglia. It is believed that waning immunity allows viral replication to occur in a ganglion, causing intense inflammation and pain.

LABORATORY DIAGNOSIS

Demonstration of multinucleated giant cells and type A intranuclear inclusion bodies. CF and neutralization test are used for serological diagnosis.

TREATMENT

Vidarabine and Acyclovir are useful for treatment. Zoster immunoglobulin (ZIG) is also useful.

CYTOMEGALOVIRUS (CMV)

It is ubiquitous in nature. They cause a variety of human infections, which range from asymptomatic infection to severe life-threatening diseases. CMV produces cell enlargement and was first described in 1904. In 1956, Smith first isolated human salivary gland virus and were later described as cytomegalovirus by Weller.

CMV has the largest genetic content among all the humanherpes viruses. Its DNA genome (240 kbp) is significantly larger. CMV is very species-specific and cell type specific. It produces characteristic cytopatheic effect and produces perinuclear cytoplasmic inclusions.

CMV infection is an endemic infection. Virus is transmitted through person-to-person contact. Prevalence is related to socioeconomic condition. Infection rates increase in early childhood and peak at 1–2 years of age.

Potential source of virus include saliva, urine, semen, breast milk, cervical and vaginal secretions.

CLINICAL MANIFESTATION

It causes infectious mononucleosis-like syndrome and it is a mild infection. The following symptoms appear.

- Malaise
- Myalgia
- Fever
- Liver function

- Abnormalities
- Lymphocytosis
- Hepatosplenomegaly is frequently observed in children
- Pneumonia occur as a complication

Congenital infection may result in death of the foetus in the uterus. Cytomegalic inclusion disease of newborns is characterized by involvement of CNS and reticuloendothelial cells.

PATHOGENESIS

Incubation period is about 4–8 weeks. This virus causes infectious mononucleosis. Most CMV infections are subclinical in nature and also establish lifelong latent infections. Salivary gland involvement is common and probably chronic. CMI is depressed with primary CMV infection. It may take several months for cellular responses to recover.

LABORATORY DIAGNOSIS

Specimen

Throat washings, urine, cervical swab and blood.

Methods

- Viral isolation
- Serological study
- Electron microscopy
- Histologic evaluation
- Immunohistochemical staining of tissues
- Nucleic acid hybridization technique

Virus is isolated with the help of human fibroblast cultures. Culture may be positive within 1–2 weeks.

Antibodies may be detected by neutralization test, CF, RIA or immunofluorescence test.

TREATMENT AND CONTROL

Ganciclovir is used. Acyclovir is used to treat life-threatening infection.

Vidarabine and interferons are also used.

Active and passive immunization as well as antiviral agents are used for prevention.

Live attenuated CMV vaccines AD 169 and Towne 125 have been used and is safe and highly immunogenic.

EPSTEIN–BARR VIRUS

Epstein–Barr virus is a ubiquitous herpes virus, which causes infectious mononucleosis, and also induces nasopharyngeal carcinoma, Burkitt's lymphoma and other lympho-proliferative disorders.

In 1964, Epstein, Achong and Barr first discovered the virus. Henles developed serological methods for diagnosis of EBV.

EBV is a distinct human herpes virus and its genome consists of about 172 kbp. Incubation period is 30–50 days. B lymphocyte is one of the target cells for EB virus.

EBV requires close person-to-person contact for transmission. Saliva is the main source of infection. Viral replication occurs in epithelial cells of the pharynx and salivary glands. Following replication in epithelial cell, the virus infects B lymphoid cells, where it persists in a latent state. Infection of B lymphocytes induces polyclonal proliferation and EBV antigen expression and a subset of these cells becomes permanently infected.

CLINICAL MANIFESTATION

- Burkitt's lymphoma (a tumour of the jaw)
- Nasopharyngeal carcinoma (cancer of epithelial cells)
- Infectious mononucleosis
- Fever
- Headache
- Malaise
- Fatigue
- Sore throat
- Pneumonitis
- Hepatitis
- Haematologic abnormalities
- Oral hairy leuco-wart like growth on tongue

LABORATORY DIAGNOSIS

Specimens

Saliva
Peripheral blood
Lymphoid tissue

Methods

- Viral isolation
- Nucleic acid hybridization. It is the most sensitive means of detecting EB virus.
- Indirect immunofluorescence
- ELISA

TREATMENT AND CONTROL

Acyclovir reduces EB virus shedding.

There is no vaccine available.

1. How do you classify herpes viruses?
2. Explain different pathological conditions associated with herpesvirus.
3. Differentiate cytomegalovirus and Epstein–Barr virus.
4. Explain about diagonsis of herpesviruses.
5. Write short notes on
   i. Posthepatic neuralgia
   ii. Shingles
   iii. HSV
   iv. Cowdry type A inclusions

Bloom, J.N. and Palestine, A.G. "The Diagnosis of Cytomegalovirus Retinitis." *Ann. Intern. Med.*

Chang, Y. Cesarman, E. and Pessin, M.S. *et al.* "Identification of herpesvirus-like DNA sequences in AIDS-associated Kaposi's sarcoma." *Science*. 266:1865. 1994.

Drew, W.L. "Cytomegalovirus infection in patients with AIDS." *J. Infect. Dis.* 158:449. 1988.

Ho, M. "Cytomegalovirus." p.1351. In: Mandell, G.L., Bennett, J.E. and Dolin, R. (eds.). *Principles and Practice of Infectious Diseases*, 4th edn. Churchill Livingstone, New York. 1995.

Roizman, B. "Herpesviridae: A brief introduction." In press. In: Fields, B.N., Knipe, D.M., Chanock, E., Hirsch, M., Melnick, J., Monath, T. and Roizman, B. (eds.). *Virology*, 3rd edn. Raven Press, New York. 1995.

Roizman, B.R. "New Viral Footprints in Kaposi's Sarcoma." *N. Eng. J. Med.* 1995; 332: 1227.

Schooley, R.T. "Epstein–Barr virus (Infectious Mononucleosis)." p. 1364. In: Mandell, G.L., Bennett, J.E. and Dolin, R. (eds.). *Principles and Practice of Infectious Diseases*, 4th edn. Churchill Livingstone, New York. 1995.

Straus, S. "Introduction to herpesviridae." p. 1330. In: Mandell, G.L., Bennett, J.E. and Dolin, R. (eds.). *Principles and Practice of Infectious Diseases*, 4th edn. Churchill Livingstone, New York. 1995.

Whitley, R.J. "Cercopithecine Herpesvirus 1 (B Virus)." In press. In: Fields, B.N., Knipe, D.M., Chanock, E., Hirsch, M., Melnick, J., Monath, T. and Roizman, B. (eds). *Virology*, 3rd edn. Raven Press, New York. 1995.

Whitley, R.J., Lopez, and Schlitt, M. "Encephalitis caused by herpesviruses, including B virus." p. 41. In: Scheld, W.M., Whitley, R.J. and Durack, D.T. (eds.). *Infections of the Central Nervous System.* Raven Press, New York. 1991.

web.uct.ac.za/depts/mmi/stannard/herpes.html

www.dbc.uci.edu/~faculty/wagner/movieindex.html

www.virology.net/Big_Virology/BVDNAherpes.html

www.herpes.org.uk/

herpesvirus.tripod.com/

www.medterms.com/script/main/art.asp?articlekey=3733

www.cdc.gov/ncidod/eid/vol5no3/campadelli.htm

en.wikipedia.org/wiki/Herpesviridae

www.wrongdiagnosis.com/h/herpesvirus /intro.htm

www.ncbi.nlm.nih.gov/entrez/query

Treatment of Viral Infections

VIRAL VACCINES

Vaccine is defined as any microbial preparation that induces the immune response of animal/human and leads to the formation of memory cells. Vaccines may produce lifelong immunity. Viral vaccines are live or killed preparations of viruses or their components that are used in immunization and offer immunity to the individual viral agents. Vaccination is a process by which vaccine is introduced into human or animals. Vaccines protect the individual from infectious diseases.

ORIGIN OF VACCINATION

In 1796, Edward Jenner extracted the contents of the pustules from the cowpox infected milkmaid, Sarah Nelmes, and injected it into the arm of an 8-year-old James Phipps. He developed immunity against smallpox virus. Jenner practised this to his patients and also French soldiers. Further work on immunization was carried out by Louis Pasteur (1822–1895) and vaccines were produced for chicken cholera, rabies, etc. He gave the name "vaccine" to honour Jenner. In Latin *Vacca* means "Cow".

IMMUNIZATION

The process of inducing immunity in an individual is called immunization. There are two types of immunization, namely, active immunization and passive immunization.

Active Immunization

It refers to the stimulation of immune response with the help of specific antigen, e.g. poliovirus vaccination for the eradication of polio.

Active immunization is of two types, namely, natural active immunization and artificial active immunization. Natural active immunization refers to the development of resistance due to natural infection. Artificial type refers to the development of resistance by the action of vaccines.

Passive Immunization

It is defined as the development of resistance to an infection in a non-immune individual by the administration of sensitized antibodies or lymphocytes. It provides immediate protection.

If the immunity is transferred from the mother to child naturally, it is called natural passive immunization. If the antibodies and immune cells are transferred from an immunized individual to a non-immune individual, it is called artificial passive immunization.

TYPES OF VACCINES

A vaccine should satisfy the following requirements in order to be effective.

Safety
Induction of protective immune response
Biological stability
Ease of administration
Public acceptance
Low cost

The following are the important types of vaccines available.

Live Attenuated Vaccines

Most live vaccines contain viruses that have been attenuated by laboratory manipulation. Attenuated viruses lose their pathogenic property but they retain the property of multiplication within the host cell. It produces lifelong immunity, e.g. Sabin polio vaccine, MMR vaccine.

Advantages

- Produces humoral and cell-mediated immunity
- Produces lifelong immunity
- Requires only a single booster dose
- Ease of administration

Disadvantages

- Less stable
- Virus may revert to virulent form

Inactivated or Killed Virus Vaccines

Killed vaccine contains either inactivated or killed whole virus particles, e.g. Salk polio vaccine, rabies vaccine, hepatitis vaccine. It is inactivated by chemical or physical agents. Formalin, β-propiolactone are the chemical agents used to inactivate the viruses. UV radiation is a physical agent which may inactivate the viruses. Killed vaccines are used along with adjuvants. Adjuvants are immunostimulatory particles that help in antigen presentation, e.g. Alum is one of the good adjuvant.

Advantage

- No mutation or reversion

Disadvantages

- Booster doses are required
- High cost.

Subunit Vaccine

Vaccines that employ components of a pathogenic virus are called subunit vaccines, e.g. gp120 gene of HIV virus was cloned in a cell and gp120 protein was produced. It is purified and then used as a vaccine.

Recombinant Antigen Vaccine or Recombinant Vector Vaccine

Gene of one virus is cloned into suitable cells and used as vaccines, e.g. recombinant HBV vaccine. DNA sequence of HBsAg is cloned with yeast plasmid vector and transferred into yeast cells. Recombinant yeast cells are selected with a suitable medium. Recombinant proteins are then collected and used as vaccine.

Advantages

- Stable
- Stimulates both cellular and humoral immunity
- Small quantity is required

Disadvantage

- Cells may get converted into virulent form

Anti-idiotype Vaccine

This vaccine is made up of antibodies and stimulates specific antibody production against antigens since antibody of one animal acts as antigen for another animal, e.g. Anti-idiotype vaccine targeted to gp120 antibodies.

Synthetic Peptide Vaccine

HBV surface antigen is made up of 48 amino acids. Synthetic peptides are synthesized based on the amino acid configuration of HBV surface antigens. Peptide molecules are encapsulated into microspheres and are directly introduced into the cells. This type of vaccine results in high titres of antibody and T-cell response against HBV surface antigen, e.g. HBV vaccine.

Advantages

- Less toxic
- Viral cultivation not required

Disadvantages

- Requires adjuvant
- Less immunogenic than whole virus vaccine

DNA Vaccine

In DNA vaccines, DNA of the virus is directly used instead of viral compounds.

DNA coding for viral antigen is coated with gold particle and injected into the epidermis of an individual by gene gun method. Antigen is expressed in human system and stimulates MHC-I and $CD8^+$ cytotoxic T cells. $CD8^+$ cells kill the infected cell and release viral antigen. Viral antigen expresses MHC-II, which will reach the immune cell and induce antibody production, e.g. HBV and influenza DNA vaccines are used for immunization in infants.

Homologous Vaccine

These vaccines are prepared from single pathogen and are used against the infection of same pathogen, e.g. Polio vaccine.

Heterologous Vaccine

Vaccine is prepared from one virus but used against infection of a different virus, e.g. cowpox virus is used to prevent smallpox.

ANTIVIRAL AGENTS

Antiviral agents include compounds that are used either as prophylactic agents or as inhibitors of virus multiplication. These antiviral agents have become the most important tool in the cure of viral diseases. There are two types of antiviral agents. They are natural antiviral compounds (e.g. interferons) and synthetic antiviral agents (e.g. zidovudine).

Ideal antimicrobial agents must block viral production without causing lethal damage to uninfected cells.

Ideal antiviral agents should have the following properties:

- The compound should block viral spread.
- It must block viral replication.

- It should be effective against the resistant forms.
- Molecular mechanism of the drug should be known.
- It should be safe.
- It should be inexpensive.
- It should be easy to formulate.
- It should be better than any competitive drug.

Stages in virus replication which are possible targets for chemotherapeutic agents are:

- Attachment to host cell
- Uncoating
- Synthesis of viral mRNA
- Translation of mRNA
- Replication of viral RNA or DNA
- Maturation of new virus proteins
- Budding and release

SYNTHETIC ANTIVIRAL AGENTS

Antiviral drugs are categorized according to their point of action in the viral replication cycle.

Agents that Block Attachment, Uncoating or Both

These compounds inhibit fusion of viral envelope with endosome membrane. They prevent release of nucleocapsid into the cytoplasm. They are used in the treatment of influenza infections. They act by binding to M2 protein, e.g. amantadine and rimantadine (Figure 54.1).

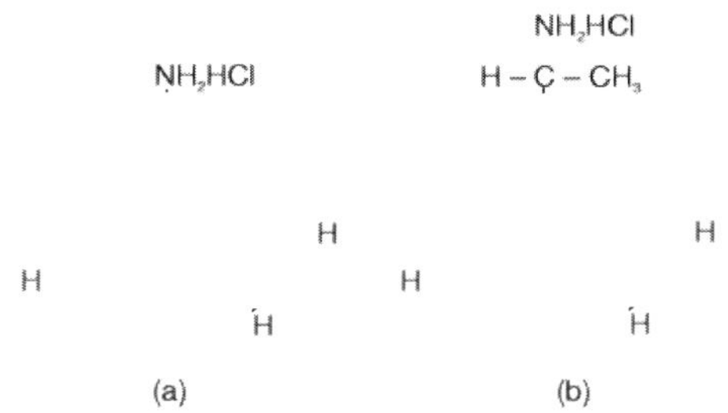

Figure 54.1 *(a) Amantadine (b) Rimantadine*

Agents that Inhibit DNA Polymerase

They are nucleoside analogues. They lack proper radicals for cross-linking to other nucleotides. When they act on a normal DNA chain, termination of DNA synthesis results, e.g. vidarabine, idoxuridine (Figure 54.2).

Vidarabine (*Deoxyadenosine analogue*)

- It is a nucleoside purine analogue.
- It is phosphorylated by kinases.
- Vidarabine phosphate acts as a competitive inhibitor for DNA polymerase and as a chain terminator.
- The tertiary structure has an extra hydroxyl group that prevents attachment of further residues. Thus when vidarabine acts on a nascent DNA chain, synthesis is inhibited.
- It is used to treat herpes simplex virus infection and encephalitis.
- It is available as a topical ophthalmic formulation.
- Ara-A, Vira-A are the trade names.

Idoxuridine (IDU)

- It is a pyrimidine analogue.
- It is phosphorylated by cellular kinases.
- Idoxuridine triphosphate acts as a competitive inhibitor of viral polymerase.
- It is used as a topical agent for herpes infection of cornea.

Stoxil, Herpex are the trade names.

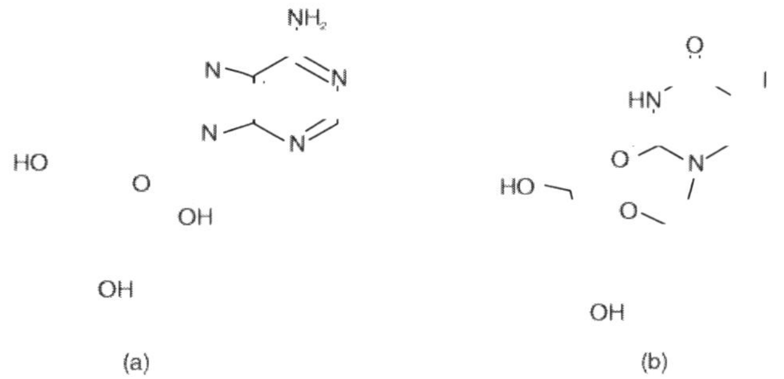

Figure 54.2 (a) Vidarabine (b) Idoxuridine

Target-activated Nucleoside Analogues

Acyclovir

- It resembles guanosine but it is acyclic, i.e., sugar moiety is incomplete.
- It prevents cross-linking.
- It is irreversibly bound to DNA transcriptase and inactivates the enzyme leading to inhibition of viral mRNA synthesis.
- It can be administered orally and intravenously.
- It is used in the treatment of herpes encephalitis.

Ganciclovir

- It is highly reactive against cytomegalovirus (CMV) DNA polymerase.
- It is used in the treatment of infections like retinitis and encephalitis.
- It can be administered orally and intravenously.

The structures of acycloir and ganciclovir are depicted in Figure 54.3.

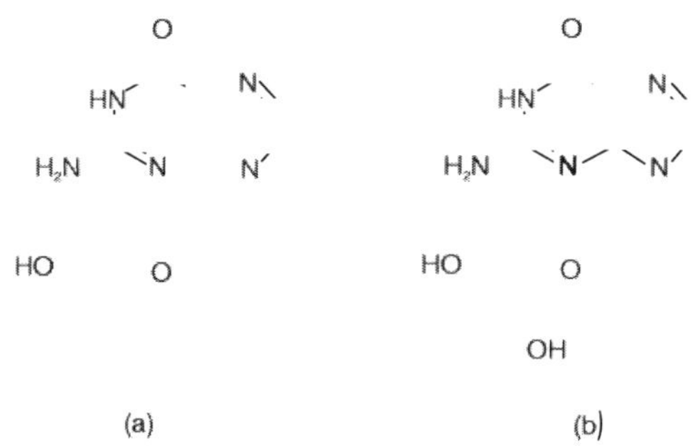

Figure 54.3 (a) Acyclovir (b) Ganciclovir

Agents that Inhibit Reverse Transcriptase

Zidovudine (AZT)

- It is also known as azidothymidine (AZT) (Figure 54.4).
- It resembles thymidine except for substitution of hydroxyl group by nitrogen in the 3´ position of pentose ring.
- It competitively binds to reverse transcriptase and stops its action.
- It also results in chain termination because it lacks the 3´ hydroxyl group essential for formation of phosphodiester linkage between two nucleotides.

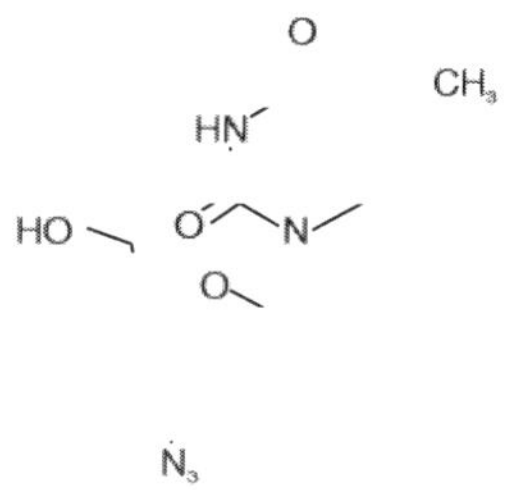

Figure 54.4 Zidovudine

- It is a potent drug for treating HIV infection.

Side effects are anaemia and neutropenia.

Protease Inhibitors

- They are non-hydrolysable synthetic peptides.
- They inhibit HIV protease, by altering amino acid sequences.
- Approved drug for HIV.
- Commercial names are Crixivan, Norvir.

e.g. Saquinavir, Ritonavir, Indinavir, Nelfinavir

Broad-spectrum Nucleoside Analogues

Ribavirin

- It is a nucleoside analogue of guanosine.
- It inhibits enzymes needed for mRNA capping.
- It binds to RNA polymerase and blocks its activity.
- It is a broad-spectrum drug.
- It is used to treat respiratory syncytial virus, hepatitis C virus, herpes simplex virus, measles, mumps and Lassa fever.

Foscarnet

- It inhibits DNA chain elongation by blocking the pyrophosphate binding on the viral DNA polymerase.
- It inhibits reverse transcriptase by binding on to a specific affinity site.
- It is used to treat herpes simplex infections in HIV-positive patients.

The structure of ribavirin and foscarnet are shown in Figure 54.5.

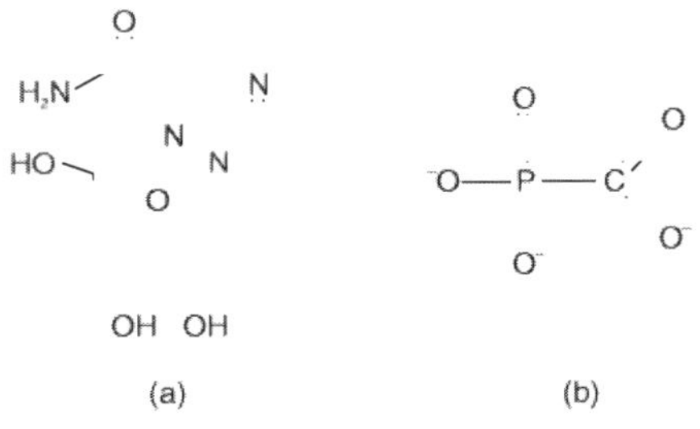

Figure 54.5 (a) Ribavirin (b) Foscarnet

INTERFERONS

Interferons are low molecular weight antiviral glycoproteins synthesized by virally sensitized cells and also in response to intracellular parasites.

Interferons were discovered by Isaacs and Lindermann. They were first observed from the fluid of chorioallantoic membrane infected with influenza virus. The natural antiviral substance was named as interferon because of its interference with viruses.

Interferons are abbreviated as IFN. They are named according to their origin, e.g. Mu-IFN (Murine interferon), Hu-IFN (Human interferon).

CLASSIFICATION OF INTERFERONS

Five major classes of interferons are recognized based on the cells of origin. They are:

- IFNα—synthesized by virus-infected leucocytes. It is made up of 20 different molecules.
- IFNβ—synthesized by virus-infected fibroblasts.
- IFNγ—produced by antigen-stimulated T cells.
- IFNω and IFNτ — produced by placenta.

CHARACTERISTICS OF INTERFERONS

It is a soluble, non-antigenic glycoprotein. Molecular weight ranges from 17–25 kDa. Genes coding for IFNα and IFNβ are found in chromosome 16 and the gene for gamma interferon is in the 6th chromosome.

Biological Effects

Interferons are multipotential regulatory factors. They stimulate NK cell activity and induce antibody production. They possess antiviral properties and antitumour activity. Interferons induce the phagocytic process and inhibit non-viral agents like *Toxoplasma* and *Shigella*-like intracellular parasites.

Immunological Properties

Interferons induce the action of B cells. Macrophages exhibit increased MHC expression and anti-tumour activity. T cells effect increased IL-2 production and also decrease viral replication (Figure 54.6).

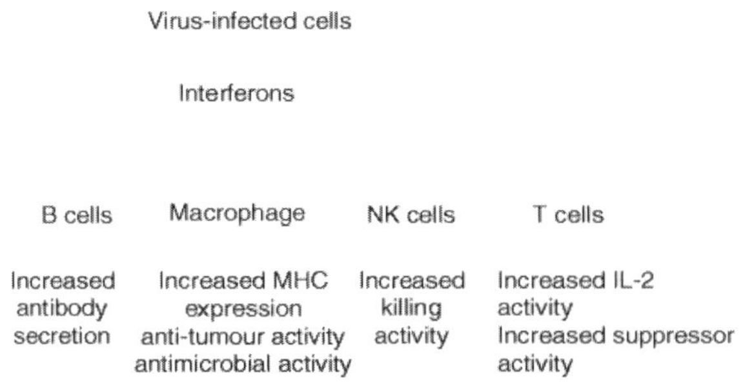

Figure 54.6 Immunological property of interferons

Inducers of Interferons

Both DNA and RNA viruses can induce IFN synthesis. Pure DNA, mitogenic agents, endotoxin, intracellular parasites like *Listeria monocytogenes*, chlamydiae, *rickettsiae* and protozoa are also inducers of interferons (Figure 54.7).

Inducer virus

Attachment

DNA

Transcription

IFN mRNA

Interferon synthesis

Interferon release

Figure 54.7 Synthesis of interferons

INTERFERONS AS THERAPEUTIC AGENTS

Interferons can be used for the treatment of viral diseases and certain cancers. The following are some of the diseases and the interferons used for treating them.

AIDS	IFNα-n3
Genital warts	IFNα-nB
Hepatitis B and C	IFNα-2b
Multiple myeloma	IFNα-2b
Papillomavirus infection	IFNα-n3

But use of IFN for therapeutic purposes is not devoid of side effect. Some of the common side effects are nausea, fatigue, fever, chills, lose of appetite and muscular dysfunction. To reduce the burden of IFN, patients should take plenty of water.

INTERFERONS AND BIOTECHNOLOGY

- Artificially some of the interferons are synthesized by making use of biotechnological process.
- Alpha interferons are available in the market as Intron A, Roferon, Wellferon.
- Human IFN gene is cloned into Chinese hamster ovary cells and the produced IFNβ-1a is sold under the trade name Avonex. It is administered intramuscularly at 30 mcg concentrations.
- IFNβ-1b is produced by cloning of TNF-β gene in *E. coli* and marketed as Betaseron.

MODE OF ACTION OF INTERFERONS

Pathway I

Interferon binding on to cellular receptors leads to the production of an unusual intracellular polymerase which links to ATP by 2´ to 5´ phosphodiester linkages. This enzyme is also known as 2–5 linked oligoadenylate synthetase. Upon activation, this enzyme produces oligonucleotides from ATP. These oligonucleotides activate the enzyme ribonuclease which acts on the viral RNA or mRNA, and cleaves it into free nucleotides.

Pathway II

- IFN also activates the enzyme protein kinase. Protein kinases are the important components of growth and cell cycle.
- Active kinase phosphorylates eukaryotic initiation factor of protein synthesis (eIF-2α). Upon phosphorylation, eIF-2α becomes inactive. This leads to inhibition of protein synthesis in the virus and also in the host.

  The mode of action of interferons is shown in Figure 54.8.

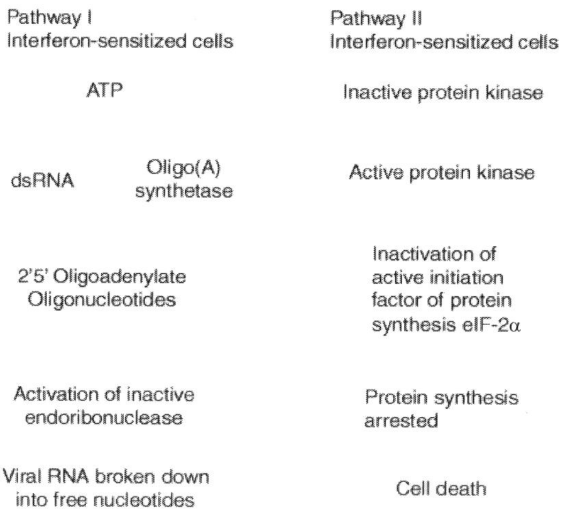

Figure 54.8 Mode of action of interferons

1. Explain the different classes of interferons.
2. Explain the characters of interferons.
3. Explain the clinical applications of interferons.
4. Describe the mode of action of interferons.

5. Explain the advantages and disadvantages of live attenuated vaccines.
6. Describe the properties of antiviral agents.
7. Write short notes on

i. IFN	v. Amantidine
ii. Immunization	vi. Acyclovir
iii. Vaccine	vii. Zidovudine
iv. Vaccination	viii. Interferons

Centers for Disease Control. General recommendation on immunization: Recommendations of the Advisory Committee on Immunization Practices (ACIP). MMWR 43 (no. RR-1):1, 1994.

Hayden, F.C. "Antiviral Agents." In: Mandell, G.L., Bennett, J.E. and Dolin, R. (eds.). *Principles and Practice of Infectious Diseases*, 4th edn. Churchill Livingstone Inc. N.Y., 1995. p. 41.

Kovacs, J.A., Baseler, M., Dewar, R.J., Vogel, S., Davey, R.T., Falloon, J., Polis, M.A., Walker, R.E., Stevens, R., Salzman, N.P., Metcalf, J.A., Masur, H. and Lane, H.C. "Increases in CD4 T lymphocytes with intermittent courses of interleukin-2 in patients with human immunodeficiency virus infections." *NEJM.* 332: 567. 1995.

Newton, A.A. "Tissue culture methods for assessing antivirals and their harmful effects." In: Field, H.J. (ed.). *Antiviral Agents: The Development and Assessment of Antiviral Chemotherapy*. Vol 1. CRC Press, Boca Raton, FL. 1988.

Plotkin, S.A. and Mortimer, E.A. Jr. *Vaccines*, 2nd edn. W.B. Saunders Co., Philadelphia. 1994.

Zoon, K.C. "Human Interferons: Structure and Function." In: *Interferon 8*. Academic Press, London. 1987.

Crumpacker, C.S. "Molecular targets of antiviral therapy." *New Engl. J. Med.* 321: 163. 1989.

De Clercq, E. "Antiviral therapy of HIV infections." *Clin. Microbiol. Rev.* 8: 200. 1995.

Jeffries, D.J. and De Clercq, E. (eds.). *Antiviral Chemotherapy*. John Wiley & Sons, Ltd., Chichester, Sussex, England. 1995. p. 580.

en.wikipedia.org/wiki/Antiviral_drug

www.bioinfo.com/antiviral.html

ww.cdc.gov/flu/professionals/treatment/

www.umm.edu/altmed/ConsDrugs/DrugCats/AntiviralAgents.html

www.roche-australia.com/anti-viral-agents.cfm

virus.stanford.edu/antivirals.html

aje.oxfordjournals.org/cgi/content/full/159/7/623

www.biology-online.org/dictionary/Antiviral_agents

www.thefreedictionary.com/antiviral+agent

www.wrongdiagnosis.com/treat/antivirals.htm

www-micro.msb.le.ac.uk/3035/Vaccines.html - 14k

pathmicro.med.sc.edu/lecture/vaccines.htm

www.nature.com/nbt/journal/v23/n11/full/nbt1105-1370.html

bmb.oxfordjournals.org/cgi/content/abstract/41/1/56

www.ingentaconnect.com/content/adis/rea

www.innovations-report.de/html/berichte/medizin_gesundheit/bericht-25068.html
www.iupac.org/publications/pac/1990/pdf/6207x1433.Polio

PART IV
MYCOLOGY

55 Introduction to Fungi

GENERAL CHARACTERISTICS

- Fungi are eukaryotic organisms. Each fungal cell has a nucleus, nuclear membrane, endoplasmic reticulum and mitochondria.
- Mycology is the study of fungi.
- Fungi are aerobic or facultatively anaerobic chemoheterotrophs.
- Fungi synthesize lysine from α-aminoadipic acid.
- Many fungal species produce flagellated, motile cells.
- Fungi lack the property of photosynthesis.
- They are found as saprophytes in the soil, and have the capability to degrade organic matter.

MORPHOLOGY

Fungi are cosmopolitan in distribution. Some are aquatic, some are terrestrial and few are air-borne. Fungal cell wall is rigid and thick, 15–30% of the dry weight of fungal cell wall contains some essential components such as polysaccharides and proteins. It provides rigidity and strength and protects the cell membrane from osmotic shock. The cell wall thickness of yeasts is about 200–300 nm in diameter and of moulds is about 200 nm in diameter. There are two morphological forms of fungi: yeasts and moulds.

Yeasts

A yeast is a unicellular fungus that has single nucleus and reproduces either asexually by budding and transverse division or sexually through spore formation. Yeasts are round or oval shaped, 4–5 μm in diameter. Some yeasts are 24 μm in diameter. On Sabouraud's dextrose agar (SDA), yeasts produce creamy opaque colonies, e.g. *Cryptococcus neoformans.* Yeast cells are longer than bacteria (Figure 55.1).

Bud

Nucleus

Figure 55.1 Yeast and budding yeast

Yeast-like Fungi

Some yeasts produce false hyphae called pseudohyphae. They are called yeast-like fungi, e.g. *Candida albicans.*

Moulds

Moulds are multicellular fungi composed of filamentous or tubular structures called hyphae. The length of a single hypha is about 5–50 μm. Some hyphae have cross-walls called septa. Hyphae are branched and form a mat-like structure called mycelium. On SDA, moulds produce cottony colonies. In some fungi, protoplasm streams through hyphae, uninterrupted by cross walls. These hyphae are called coenocytic. The hyphae of other fungi have cross walls called septa with single pore or multiple pores (Figure 55.2). These hyphae are termed septate. Many fungi have dimorphic forms. Dimorphic fungi can change from yeast (y) form in the animal to mould form (m) in the external environment. This shift is called the y–m shift.

Nucleus

(a)

Septum

Nucleus

(b)

Figure 55.2 (a) Coenocytic hyphae b) Septate hyphae

REPRODUCTION

Fungi reproduce asexually as well as sexually. Asexual reproduction includes fragmentation, fission, budding and spore formation (Figure 55.3). Sexual reproduction includes oospore, zygospore, ascospore and basidiospore formation. The spores derived from sexual means are called as telomorphic spores and those that are produced by asexual means are called anamorphic spores. Based on spores and conidia, fungi are categorized. Some important conidia are:

- Arthroconidia (Arthrospores)
- Blastoconidia (Blastospores)

- Chlamydoconidia (Chlamydospores)
- Phialoconidia (Phialospores)
- Sporangioconidia (Sporangiospores)

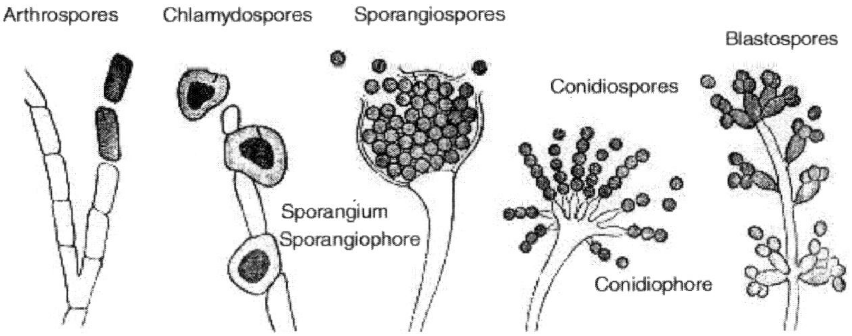

Figure 55.3 Types of spores

MEDICALLY IMPORTANT DIVISIONS OF FUNGI

The medically important divisions of fungi are:

- Deuteromycota
- Zygomycota
- Ascomycota
- Basidiomycota

FUNGAL DISEASES

The fungal diseases can be classified into:

- Superficial mycoses (e.g. Tinea versicolor)
- Subcutaneous mycoses (e.g. Sporotrichosis)
- Cutaneous mycoses (e.g. Tinea capitis)
- Systemic mycoses (e.g. Histoplasmosis)
- Opportunistic mycoses (e.g. Candidiasis)

ECONOMIC IMPORTANCE OF FUNGI

- *Saccharomyces* and *Trichoderma* are used in the production of foods.
- They are used for the biological control of pests.
- Many fungal members cause spoilage of fruits, grains and vegetables and also cause plant diseases.
- Some of the fungi are used as food (e.g. mushroom).

1. How can a fungus be defined?
2. Where are fungi found?
3. What are the morphological features of fungi?
4. Describe about yeast and mould.
5. Mention any two sexual and asexual methods of reproduction of fungi.
6. How will you differentiate yeast from mould?
7. Explain the economic importance of fungi.
8. Write short notes on
   i. y–m shift
   ii. Mycelium
   iii. Hyphae
   iv. Coenocytic form of hyphae
   v. Septate hyphae

Mycoses

INTRODUCTION

Infections caused by fungi are called mycoses. Based on the area of infection, mycoses are classified into three types:

1. Superficial mycoses
2. Subcutaneous mycoses
3. Systemic mycoses

SUPERFICIAL MYCOSES

Superficial mycoses affect the skin, hair or nail. These infections are mild but sometimes chronic in nature. Those fungi that cause superficial infection have the capability to digest keratin and are saprophytes. It induces mild inflammatory responses. There are two types of superficial infections: surface mycoses and cutaneous mycoses.

In surface mycotic infections, fungi survive only on dead layers of the skin. All fungi involved in the superficial mycoses are included under the group mould, e.g. pityriasis versicolor (tinea versicolor), piedra (trichosporosis).

Cutaneous mycoses are observed slightly in the deeper portions of epidermis. In this case, the fungi grow in the cornified layer of the skin and cause inflammatory responses, e.g. Dermatophytosis caused by *Microsporum* spp., *Trichophyton* spp., *Epidermophyton* spp.

SUBCUTANEOUS MYCOSES

These are a group of fungal infections involving the dermis and subcutaneous tissue. These are referred to as mycoses of implantation because they are acquired when the pathogen is inoculated through the skin by minor cuts or scratches or by thorns or splinter wound. Aetiological agents are ubiquitous in nature and usually found in soil or on decaying vegetation. The infections occur on the parts of the body that are most prone to trauma, e.g. feet, legs, and buttocks. Subcutaneous infections are slow in onset and lesions evolve over many months. Persistence may be due to non-invasive properties of this group of organisms.

Subcutaneous infections include

1. Chromomycosis which is caused by *Phialophora verrucosa*, *Cladosporium carrionii*, *Fonsecaea pedrosoi*, *Fonsecaea compacta*, and *Rhinocladiella aquaspersa.*
2. Mycetoma which is caused by *Madurella grisea*, *Phialophora jeanselmei Leptosphaeria senegalensis*, *Cephalosporium* spp., *Acremonium* spp. and *Petriellidium boydii.*
3. Actinomycetoma which is caused by *Actinomadura madurae (Streptomyces) Actinomadura pellettieri (Streptomyces) Nocardia brasiliensis* and *Streptomyces somaliensis.*
4. Sporotrichosis caused by *Sporothrix schenckii.*
5. Rhinosporidiosis caused by *Rhinosporidium* spp.

SYSTEMIC MYCOSES

These are also called deep mycoses. They are acquired by inhalation. There are two types of systemic mycoses, namely primary systemic mycoses and opportunistic systemic mycoses. Those fungi that cause primary systemic mycoses are called primary pathogens. Examples are blastomycosis, paracoccidioidomycosis, coccidioidomycosis and histoplasmosis.

Some of the saprophytic fungi of the environment cause opportunistic systemic mycoses and are called opportunistic fungi. Examples are cryptococcosis, candidiosis, aspergillosis, penicillosis, mucor mycosis, otomycosis and oculomycosis.

1. How are human fungal diseases categorized?
2. List out the human subcutaneous infections.
3. Explain different systemic mycoses.

Laboratory Diagnosis of Fungal Infections

Fungi are significant, sometimes overlooked, human pathogens. For the diagnosis of pathogenic fungi, a knowledge of their morphological and clinical classification is essential.

MORPHOLOGICAL CLASSIFICATION

Filamentous Fungi

Filamentous fungi include:

- dermatophytes
- *Malassezia furfur*
- *Madurella* spp.
- *Aspergillus* spp.
- *Mucor* spp.
- *Candida* spp.
- zygomycetes
- *Phialophora* spp.
- *Pseudallescheria boydii*
- *Penicillium* spp.
- yeast
- *Cryptococcus* spp.

Dimorphic Fungus

Dimorphic fungi include:

- Blastomyces dermatitidis
- *Histoplasma capsulatum*
- *Coccidioides immitis*
- *Paracoccidioides brasiliensis*
- *Sporothrix schenckii.*

The morphological and clinical classification of fungi is shown in the following table.

Fungi

- Mould
  - Aseptate hyphae
    - Subcutaneous
      - *Basidiolobus*
      - *Conidiolobus*
      - *Rhinosporidium*
  - Septate hyphae
    - Monomorphic
      - Superficial
        - *Microsporum*
        - *Trichophyton*
        - *Epidermophyton*
        - *Malassezia furfur*
        - *Piedraia hortae*
      - Subcutaneous
        - *Phialophora*
        - *Cladosporium*
        - *Madurella*
        - *Acremonium*
        - *Petriellodium*
        - *Leptosphaeria*
    - Dimorphic
      - Systemic
        - *Histoplasma*
        - *Rhizopus*
        - *Blastomyces*
        - *Absidia*
        - *Paracoccidioides*
        - *Mucor*
- Yeast
  - With pseudohyphae
    - *Candida*
  - Without pseudohyphae
    - *Cryptococous*

SOME IMPORTANT MORPHOLOGICAL FEATURES OF PATHOGENIC FUNGI

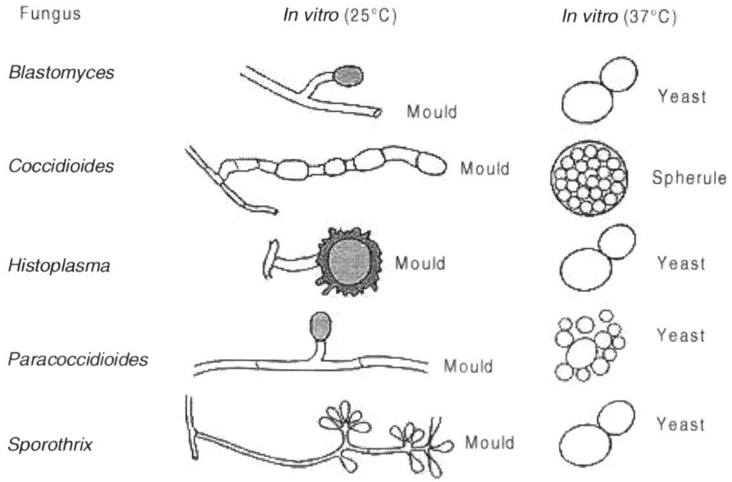

Fungal infections range from mild to life-threatening ones. Clinical laboratories must be prepared to identify the range of new fungal species that causes diseases.

LABORATORY DIAGNOSIS

Diagnosis is based on a combination of clinical and laboratory investigations.

Laboratory procedure includes:

- Demonstration of fungi by microscopy
- Identification by culture
- Detection of specific humoral response
- Detection of fungal antigens and metabolites in body fluids

Success of the laboratory diagnosis methods depends on the following:

- Specimen selection
- Specimen collection
- Specimen transport
- Processing
- Microscopic examination
- Culturing

Specimen Selection

Direct specimens are the best choice for fungal infection, but peripheral specimens may also be useful. For detection, material from active site is required.

Specimen Collection

Specimen collection for the detection of aetiologic agent of mycoses is very similar to specimen collection for bacteria. But fungal detection requires large quantity of specimen than bacterial identification. Finger nails and toe nails suspected of dermatophytosis should be cleaned extensively with 70% ethanol.

Swabs are not optimal specimens that can be tested for fungal identification.

Specimen Transport

Fungal cell wall provides an effective barrier against various toxic agents. For transport of fungal specimens, aerobic transport media are used. Transport containers should be sterile, humidified and leakproof. The specimen is kept moist by adding a minimum volume of sterile saline.

Desiccation, elevated temperatures, low temperatures and starvation affect the viability of some fungi. Bacterial overgrowth is another factor.

Transport time should be kept to a minimum (2 hours) because some fungal agents are sensitive to environmental stress. Specimens from sites with possible contaminating bacteria or leucocytes should be transported rapidly because bacterial overgrowth and leucocytes could cause decreased fungal viability.

Specimens that cannot be transported to the clinical laboratory within 2 hours should be stored at 4°C. Sterile body fluids (30–37°C) and dermatological specimens should not be refrigerated because some dermatophytes are sensitive to cold. Stools should not be kept at or above room temperature for more than one hour.

Specimen Examination

Specimen should be evaluated microscopically and macroscopically. Microscopic examination reveals vital information about the relative concentration and type of fungus present in the specimen.

Once a smear of a specimen is made, several stains can be used to detect fungi. The common staining methods are:

- Calcofluor white
- KOH mount
- KOH with quick ink
- India ink
- Hanks' modified acid-fast stain

After visualization of fungal elements, the following distinguishing characteristics are reported:

- Hyphae
- Hyphae and arthroconidia
- Hyphae, arthroconidia and yeast cells
- Hyphae and yeast cells
- Yeast cells only
- Yeast cells with capsules
- Spherules
- Sclerotic cells
- Sulphur granules with fungal hyphae

Specimen Processing

In general, procedures used for bacterial cultures are also used for fungal cultures. Sample distribution may affect the likelihood of fungal isolation. Also, the number of fungal organisms per volume of specimen is less than that of bacteria. To compensate these problems, the volume of sample is increased or the sample is concentrated.

Liquid specimens may have low concentration of fungus and will require centrifugation to increase fungal cell concentration. Hard specimens should be minced before inoculation. For fungal growth, media should be incubated at 28–30°C. Only *Malassezia furfur* requires 35°C for adequate growth.

To prevent bacterial pathogens, antibiotics should be used or the pH of the medium is reduced. Vaginal cultures may be incubated for only 7 days; other cultures should be incubated for 4 weeks. The media is observed once in two days; most fungi will produce colonies by the end of the 3rd week. *Histoplasma capsulatum* and *Blastomyces dermatitidis* may require 4–8 weeks.

Media Selection

Media selection depends on the body site, nature of infection and nature of contamination.

- Brain-heart infusion agar with antibiotics
- Potato dextrose agar
- Mould agar
- Sabouraud's glucose agar
- Niger seed agar
- Yeast extract phosphate medium
- Buffered charcoal yeast extract agar
- Rose bengal chloramphenicol agar

From the above list of mycotic media, any one can be used for cultivation.

Inoculation

Direct inoculation Many specimens can be directly inoculated. 3–5 drops of fluid should be inoculated to each tube of medium. The plates are inoculated with 2 drops of specimen and then the specimen is streaked over the medium surface.

Examples

- Bone marrow aspirates
- CSF
- Vaginal swab–vortex swab in 5 ml of distilled water and then cultured
- Hairs and skin scrapings are pressed firmly on to the medium surface
- Nail should be pulverized using scalpels and the fragments are pressed firmly on to the medium surface
- Bronchial washings

Concentration Large volume of fluid should be concentrated by centrifugation (2000 rpm for 10 minutes). The resulting pellet is used for culture and KOH examination.

Examples

Body fluids, urine, CSF, sputum (digested with *N*-acetyl L-cysteine)

Mincing or homogenization Biopsy samples, tissues and nails must be processed to increase the recovery. With sterile scalpel, specimen is minced into small pieces in a petri plate with a few drops of sterile distilled water. If *Histoplasma* is suspected, specimen is homogenized by using a tissue grinder after mincing. The media is then inoculated and the plates sealed properly.

Incubation All fungal media are incubated at 28 to 30°C for 4 weeks. Vaginal cultures should be incubated only for 7 days.

EXAMINATION OF FUNGAL GROWTH ON PRIMARY MEDIA

The primary plates are examined daily for the first week, every other day for the second week and twice weekly for the remaining two weeks. When growth appears, yeasts are differentiated from moulds by microscopic examination.

Once colonies are formed on the media, the organism should be viewed microscopically. Zygomycetes are observed under dissecting microscope.

Yeasts are observed by:

- Wet preparation
- Lactophenol cotton blue staining
- India ink preparation

Moulds are observed by:

- Wet preparation
- Scotch tape
- Tease preparation
- Slide cultures
- Lactophenol cotton blue staining
- Ascospore

IDENTIFICATION OF MOULDS ON PRIMARY CULTURE

If an organism is not identifiable on primary culture, subculture should be performed. Yeasts are subcultured on to yeast morphology agar and inoculated by means of streaking. Moulds are subcultured by several methods.

- Block inoculation
- Scarification
- Slide culture

All subcultures are performed in a class II biological safety cabinet.

Block Inoculation

- Approximately 0.5 by 0.5 cm block of the fungal colony is cut with a sterile scalpel.
- The block is placed upside down on to the fresh medium.
- The plates are incubated at 28 ± 2°C.

Scarification

- The surface of the mould is scraped with a blunt needle.
- The surface of the fresh medium is scratched in the centre of the plate with colony material. The material should be inserted into the scratch but should also be exposed at the surface of the medium.
- The plates are incubated at 28 ± 2°C.

Slide Culture

- Agar blocks (1 cm) are cut from an agar plate and they are lifted on to the surface of the agar.
- Each side of the block is inoculated with a small amount of hyphae.
- The block is placed on the surface of the slide.
- A coverslip is placed over the surface of agar.
- The plates are incubated at 28 ± 2°C for 24 hours.
- The coverslip is removed and then a small drop of lactophenol cotton blue is placed on a slide and then examined.

Examination

- Fungal spores will develop in the air space between the agar surface and the coverslip.
- The development of mature spores is monitored.
- When sporulation appears, the coverslip is carefully lifted and then stained with LPCB and examined microscopically.
- The nature of the spore and the spore-bearing structure are determined.

Selection of Media

- It is a crucial step for successful and timely identification of moulds.
- Yeasts must be subcultured on to SDA and also on differential media (CHROM agar, Niger seed agar).

- For mould, a relatively rich medium (V-8 juice agar) and nutritionally poor medium (PDA, oatmeal agar) should be used. Brain heart infusion agar (BHI) is not an optimal spore-inducing media but is needed for systemic fungi.
- A third medium is needed to encourage fruiting body formation, e.g. cotton seed agar, hay infusion agar, water agar.

Media lacking antimicrobials should be used for subcultures. For slide culture, clear or moderately translucent media are preferred.

Dimorphic fungi convert to yeast more quickly and abundantly on specific media. When conversion is attempted, two plates of the same medium should be inoculated. One is placed at 28°C and the other at 37°C. The following media should be used for subculturing and cultivation of fungus.

All moulds should be subcultured on SGA, PDA and V-8 juice agar and cultivated. Other groups of fungi subcultered on differnt media. They are

Dermatophyte fungi	Oatmeal agar
Red yeast	Cornmeal agar
Malassezia	SGA with olive oil
White yeast	Morphology agar
Aspergillus, *Penicillium*, *Paecilomyces*	Czapec dox agar
A. fumigatus	SGA at 45°C
Coelomycetes	BHIA at 28°C and 37°C
Cryptococcus	Niger seed agar

YEAST IDENTIFICATION

Presumptive identification test for yeasts includes the following:

- CHROM agar test
- Germ tube test
- *C. albicans* screen test
- Rapid urease test
- Rapid nitrate reductase test
- Caffeic acid disc test
- India ink staining method

CHROM Agar Test

CHROM agar contains enzymatic substrates that are linked to chromogenic compounds. When specific enzymes cleave the substrates, the chromogenic substrates produce colour. The action of different enzymes produced by yeast species results in colour variations, useful for the presumptive identification of some yeasts. The test provides only presumptive identification of yeast, e.g. *C. krusei, C. tropicalis, C. albicans*.

Cultures were streaked on the medium and incubated at 35°C in humidified dark chamber for 48–72 hours. After 72 hours, colour change is observed. On the basis of observation, the following interpretation is given.

C. *albicans*

Medium-sized, green, smooth, matte colony with a very slight green halo on the surrounding medium.

C. *tropicalis*

Smooth, medium sized matte colony, which is blue to bluish grey with a pale pink edge. The colony may have a dark brown to purple halo, which diffuses into the agar.

C. *krusei*

A large, spreading, rough pink colony with a pale pink to white edge.

Germ Tube Test

This test is used for the presumptive identification of *C. albicans.*

- A yeast colony is lightly touched with a wooden applicator stick.
- The yeast cells are suspended in appropriately labelled tube of serum.
- The tubes are incubated at 37 ± 1°C for 2.5 to 3 hours.
- A drop of the suspension is placed on a slide.
- A coverslip is placed over the suspension.
- It is then examined under high power for the presence or absence of germ tube.
- Sabouraud's glucose agar is the best medium to isolate yeast for germ tube production.

KOH Mount

KOH treatment allows rapid observation of fungal elements present in tissues and fluids such as skin, nails, biopsy materials and sputum.

- A drop of KOH is placed in the centre of the slide.
- A fragment of tissue, purulent material or scraping is placed in the KOH.
- The material is teased to prepare a thin film and mounted with a coverslip.

- The preparation is allowed to digest for 10 minutes at room temperature.
- The slide is gently warmed and pressed.

Variations to the KOH *Method*

- The addition of dimethyl sulphoxide to KOH eliminates the need to heat the specimen.
- To 100 ml of a 60% solution of dimethyl sulphoxide (DMSO) in water, 20 g of KOH is added and dissolved completely.
- Store in an air-tight container in dark.

The preparation is followed as in KOH mount.

KOH-DMSO-Ink

A variation of the KOH-DMSO method is to add Parker's quick ink. The ink enhances contrast and is particularly useful for detecting *Malassezia furfur* in skin scrapings.

To prepare the solution, equal volumes of quick ink and KOH-DMSO are mixed.

Screening of Specimens for Fungal Elements by the KOH-calcofluor Fluorescent Stain Method

Calcoflour white is a non-specific fluorochrome stain for the rapid screening of clinical specimens for fungal elements. It binds with chitin and other large polysaccharides. Artificial staining occurs with collagen. It can also be used to visualize the microscopic morphology.

- A slide with a small amount of the specimen should be prepared.
- One drop of KOH is added to the specimen.
- One drop of calcofluor is added.
- The slide is then covered with a coverslip.
- It is then examined under fluorescent excitation light of 300–412 nm after 5–10 minutes of incubation.
- With a 500-nm barrier, fungi will appear yellowish green. Without an emission filter, fungi will fluoresce bright white.

Result

Hyphae, yeast cells and fungal elements will fluoresce.

India Ink

An India ink preparation can be used for the rapid detection of encapsulated yeasts based on negative staining. The capsule repels the carbon particles of the India ink, giving a clear well-demonstrated halo around each encapsulated stain.

- Place a drop of sediment on a slide.
- Then place a coverslip.
- Place a drop of India ink.
- Allow the ink to diffuse under the slide.
- Place a drop of sediment.
- Add a small drop of India ink and mix.
- Preparation should be of a brownish colour.
- If the preparation appears black, add 1–2 drop of sterile distilled water, mix and mount with coverslip.

Giemsa Stain for *Histoplasma Capsulatum*

- Giemsa stain is used for examining intracellular structures and is applied to primary specimens of bone marrow tissue and WBCs in which *H. capsulatum* is suspected.
- A thin smear is prepared as in differential staining procedure.
- The smear is fixed with absolute alcohol for 1 minute.
- The alcohol is immediately drained off and the smear is allowed to dry.
- The slide is flooded with Giemsa stain that has been diluted 1:10 with distilled water for 5 minutes.
- The slide is washed with water and air-dried. It should not be blot-dried.

Result

- Necrotic cells in the specimen will have pink cytoplasm.
- Normal cells are light blue with violet lavender cytoplasm.
- Phagocytozed yeast cells will stain light to dark blue, and each will have a clear halo around it.
- Purple pseudoencapsulated yeast forms of *H. capsulatum* inside PMN cells and monocytes can be observed.

Ascospore Production (Sporulation)

Two of the characteristics examined to identify most fungi are method of sporulation and the arrangement and morphology of the spores produced. In some fungi, sexual reproduction produces spores in an ascoscarp. This contains small sacs called asci, which contain ascospores. Selection of the medium to demonstrate the production of ascospores in yeast depends on which species of yeast is suspected, i.e., *Saccharomyces* spp., *Schizosaccharomyces* spp. and *Candida norvegensis* produce ascospore on acetate agar, but other yeasts produce ascospores on V-8 juice agar.

- The medium is inoculated with yeast to be identified and incubated at 25–30°C for 3–10 days. After 3 days, a wet preparation is performed to check for ascospores. If ascospores are seen, a suspension of the yeast in a drop of water on a slide is prepared.
- The smear is allowed to dry, then heat-fixed and stained by the Kinyoun's acid-fast staining.
- The slide is examined microscopically for acid-fast ascospores.

Lactophenol Cotton Blue Solution

- LPCB stain is an excellent technique for examination of fungal material. Phenol kills the fungi, and the lactic acid increases preservation. Chino (cotton) blue is a stain for chitin and cellulose.

Cellophane Tape Mount

- This method usually maintains the original morphology of fungal structures. Colony should be grown on plated medium.
- The tip of the loop is held securely with forceps, and the lower, sticky side is pressed very firmly to the surface of the fungal colony.
- The tape is pulled away gently. Aerial hyphae will adhere to the tape.
- The tape strip is opened up and a small drop of LPCB is placed on it on a glass slide.
- The slide is then examined under the microscope.

Tease Preparation

- A tease preparation is the quickest and most common technique for mounting fungi for microscopic examination.
- A drop of lactophenol cotton blue (LPCB) is placed on the slide.
- With the inoculating needle, a small portion of growth is removed midway between the centre of the colony and the edge.
- The material is placed in the LPCB.
- With the two dissecting needles, the fungi is gently teased apart. A coverslip is then placed.
- The edges of the coverslip are then sealed with nail polish to preserve the mount.

MOULD IDENTIFICATION BASED ON SPORE OR CONIDIUM PRODUCTION

Conidium is one of the asexual parts of mould. Size and shape of the conidia will vary depending on the group. Mould conidia may be identified on the basis of slide culture technique.

There are seven forms of conidia:

Aleuroconidia

It is usually borne singly, either directly on the hyphae, or on short hyphal projections, e.g. *B. dermatitidis*, *H. capsulatum* and the dermatophytes (*Microsporum*, *Epidermophyton* and *Trichophyton* spp.)

Annelloconidia

It arises from the apex of annelids (flask-shaped or cylindrical) and is produced with the youngest conidium at the base, e.g. *Exophiala werneckii, Exophiala spinifera, Scopulariopsis* spp.

Arthroconidia

Septation and the disjunction or rounding off of simple or branched conidiogenous hyphae produce them, e.g. *Coccidioides immitis, Trichosporon.*

Blastoconidia

They are blown out from another cell, from hyphae. It is produced either by simple budding or by germ tube, e.g. *Fonsecaea pedrosoi*, *Cladosporium*, *Candida* spp.

Phialoconidia

They are produced within a conidiogenous cell called phialide. It is a rounded or elongated, vase-like structure, e.g. *Aspergillus, Acremonium, Phialophora, Penicillum, Phaecilomyces, Exophiala dermatitidis.*

Sporoconidia

They are produced through minute pores in the outer wall of the conidiophore. The conidia may be septate, pigmented and solitary or in chains with the youngest at the tip, e.g. *Alternaria, Curvularia, Bipolaris.*

Sympoduloconidia

It is blown out from a conidiophore that enlarges, as each new conidium is produced. The primary conidium is produced at the tip of the conidiophore. Successive conidia are produced at the side and above each preceding conidium, e.g. *Fonsecaea pedrosoi, Sporothrix schenckii, Fusarium*, *Beauveria*, *Rhinocladiella.*

CONVERSION OF DIMORPHIC, SYSTEMIC MOULDS TO THEIR PARASITIC MORPHOLOGIES

All white, tan, buff or brown moulds should be considered as potential dimorphic pathogens. These organisms should be worked within a class II cabinet. The lid or cap of the culture vessel should never be opened outside the cabinet.

Infection caused by dimorphic fungi may be life-threatening. Morphologic conversion depends on temperature and growth medium.

The conversion medium is inoculated with a suspected colony and incubated at suitable temperature aerobically or in the presence of 5% carbon dioxide.

Radiating growth is checked from the inoculum fragment after 3–7 days. If growth is seen, a slide preparation is made with LPCB and examined for yeasts or spherules.

EXOANTIGEN TEST FOR IDENTIFICATION OF DIMORPHIC MOULD

The exoantigen test is a specific immunodiffusion test developed to provide rapid information on the immunoidentity of an unknown isolate. Reacting a concentrate of the organism's soluble antigen against paired positive control reagents for each of the systemic fungus does this.

Preparation of ID Agar

- To a 125-ml flask, 50 ml of distilled water is added.
- 0.9 ml of sodium chloride and 0.4 g of sodium nitrate is added to the flask.
- To this 0.25 ml of liquid phenol (90%) and 7.5 g of glycine is added and mixed thoroughly.
- Purified agar (1g) is added and made up to 100 ml with distilled water.
- It is autoclaved for 10 minutes at 120°C.
- Final pH is 6.3–6.4.

Preparation of ID plates

- Melted phenol glycine agar (6.5 ml) is added to a petri plate (base agar).
- The base layer gel is let for at least 30 minutes.
- Then 3.5 ml of hot glycine phenol agar is pipetted on to one side of the solidified base.

Extraction of exoantigens

- A mature culture with at least 15 to 30 minutes of growth on SDA is found. Pasteur pipette is used to cover this growth with 8–10 ml of a 1:5000 aqueous solution of thimerosal.

- The solution is allowed to extract with mycelial growth for at least 24 hours at room temperature.
- Cellular extract (5 ml each) is transferred to a concentrator (amicon minicon micro solute B-15). One drop of extract is placed for sterility test.
- The extract is concentrated from cultures.
- Cultures without characteristic conidia are tested for antigens.

Performance of ID *test*

- Antiserum is placed in the centre well.
- The serum is allowed to diffuse at room temperature.
- The control antigens are placed in the upper and lower wells and duplicate unknown supernatant antigens are placed in the lateral wells on the same slide.
- The charged ID plates are placed in a moist chamber at room temperature for 24 hours and the plates are then read.

HAIR PERFORATION TEST

Dermatophytes and keratinolytic fungi will perpendicularly penetrate hair *in vitro*. The hair perforation test involves the following steps:

- Into a sterile petri plate, place several fragments of sterile hair.
- Add 20–25 ml of sterile distilled water. To this add 0.1 ml of 10% sterile yeast extract.
- Several fragments of fungal cultures are placed and incubated at room temperature.
- The plates are examined at weekly intervals for a period of 4 weeks for hair perforation.
- The hair from the culture is placed on a drop of LPCB and mounted with coverslip.
- In a positive test, hair will be penetrated.

1. What are the major criteria followed to succeed in mycology laboratory investigation?
2. Discuss about sample transport for fungal isolation.
3. Name the fungal elements that are used for diagnosis.
4. List any 5 microscopic techniques for the detection of fungal elements.
5. List different fungal isolation media their selection.
6. Explain about block inoculation and scarification.
7. Discuss about slide culture technique.

8. Describe various methods used to identify yeast.
9. How do you demonstrate *Histoplasma capsulatum* from specimens?
10. Discuss about mould identification.
11. Write short notes on
    i. Germ tube technique
    ii. KOH mount
    iii. India ink preparation
    iv. Tease preparation
    v. LPCB
    vi. Hair perforation test
    vii. Exoantigen test
    viii. Sporulation

Superficial Mycoses

Superficial mycoses are rare infections. The fungus grows only on the superficial layer of skin. Tinea versicolor piedra; Tinea and dermatophytosis are examples for these group of infections.

PITYRIASIS VERSICOLOR

INTRODUCTION

- It is also called tinea versicolor.
- It is a chronic and less serious infection of the skin.
- It is a superficial skin infection involving the stratum corneum characterized by discoloration or depigmentation of skin.
- The organism is found worldwide as a commensal on the smooth skin of humans. It infects skin only.
- It is caused by *Malassezia furfur.* It is a dimorphic fungus.
- "Malassezia" is the name of the mycelial form.
- Chest, abdomen, upper limbs and back are susceptible to *M. furfur* (Figure 58.1).

Short hyphae

Yeast
cells

Figure 58.1 *Malassezia furfur*

There are two types of yeasts: oval yeasts included under serovar C *(Pityrosporum ovale)* and round yeasts included under two serovars A and B *(Pityrosporum orbiculariae)*.

PATHOGENICITY

- It affects the external appearance of an individual.
- It produces non-inflammatory macular lesions with fine scaling.
- Lesions are not usually itchy.
- Mycelial phase fungus is associated with tinea versicolor.
- Fungus interferes with melanin production by producing dicarbolic acid.
- Fungus inhibits the activity of tyrosinase, which is responsible for the synthesis of melanin.
- Some individuals develop folliculitis.

LABORATORY DIAGNOSIS

Specimen

Pus, skin scrapings

Microscopy

- Direct microscopic observation of the specimen with 10–20% KOH.
- Short unbranched hyphae and spherical cells are observed.
- Lesions also fluoresce under Wood's lamp test.

Culture

- Modified SDA supplemented with olive oil with antibiotics is used for cultivation. *M. furfur* is a lipophilic yeast.
- Shiny or pasty white to cream-coloured colonies are formed after 1–2 weeks. The phialoconidia are thick-walled, round or oval in shape. They typically occur in clusters. Individual cells look like yeast.

EPIDEMIOLOGY

- Incidence increases where the climate is hot and humid and is highest in tropics.
- Its occurrence has been related to the presence of certain amino acids and hydrophobic compounds on the skin.

TREATMENT

Application of selenium sulphide is used as treatment.

TINEA NIGRA

INTRODUCTION

It is a superficial, chronic and asymptomatic infection of the stratum corneum caused by the dematiaceous yeast *Exophiala werneckii*.

Aetiological agent is one of the saprophytic fungi. Dimorphism is observed in cultures. Fungus is ubiquitous in nature. During ageing mycelial forms are converted into yeast forms (Figure 58.2).

Colonies are shiny, moist and often white to grey in colour initially. Within a few days, the colony darkens and becomes olive to black. Later, mycelium develops and the colony appears dull and fuzzy.

Figure 58.2 *Exophiala werneckii*

PATHOGENICITY

- The aetiological agents are most frequently found in tropical areas. Approximately 95% cases occur in teenagers.
- Lesions usually consist of a solitary, innocuous macule with sharply defined margins. The brownish colour of the lesion is darkest at the advanced periphery, where most of the actively growing organisms are located.
- Many cases involve the palm. Other parts of the skin may also be infected, including fingers and face. The lesions resemble faded silver nitrate stain.

LABORATORY DIAGNOSIS

Specimen

Skin scrapings and lesions

Microscopy

- Specimens are mounted on 20% KOH or calcofluor white stain.
- Observation reveals brown pigmented septate hyphae and budding yeast cells.

Culture

Culture with Sabouraud's dextrose agar with or without antibiotics should recover the organism.

TREATMENT

It responds well to topical keratolytic solutions of sulphur, salicylic acid or tincture of iodine.

PIEDRA

There are two types of piedra: Black piedra and white piedra.

BLACK PIEDRA

- Black piedra is a nodular infection of the hair shaft caused by *Piedraia hortae*.
- The organism is a plant parasite.
- Human infection caused by this parasite is called black piedra.
- Colour of the colony on SDA is greenish black or black.
- Asci and ascospores are rarely seen in culture.
- Black piedra consists of slow-growing brown to reddish black mycelia. It produces spindle-shaped ascospores.

PATHOGENICITY

Hard nodules are found along the infected hair shaft. Nodules have a hard carbonaceous consistency.

LABORATORY DIAGNOSIS

- Hair is used for diagnosis. SDA medium is used for cultivation. Wood's lamp test is also performed.

WHITE PIEDRA

- *Trichosporon beigelii* is the causative agent.
- It presents as larger, softer, yellowish nodules on the hair.
- *Trichosporon beigelii* is sensitive to cycloheximide.
- Culture is dimorphic.
- Yeast cultures are white and have a pasty consistency.
- As the culture ages, colonies develop deep radiative furrows and take on a yellowish colouration with the creamy texture.

- Microscopic examination reveals septate hyphae that fragments rapidly to form arthroconidia.
- The arthroconidia rapidly round up and many cells form blastoconidia.
- It affects hairs of the scalp, mustache and beard.
- It is characterized by the development of cream-coloured soft pasty growth along the infected hair shaft.
- Piedra is endemic in tropical underdeveloped countries.

LABORATORY DIAGNOSIS

Specimen

Infected hair

Microscopy

- KOH mount
- Wood's lamp test

Culture

Growth is dimorphic.

Specimen is inoculated on SDA medium.

TREATMENT

It is accomplished by the topical use of keratolytic agents such as preparation containing selenium disulphide, hyposulphite, thiosulphite of salicylic acid. Miconazole nitrate is also used. Shaving or cropping of infected hairs close to the scalp surface yields good results.

DERMATOPHYTOSIS

INTRODUCTION

Dermatophytosis is a fungal infection that involves the superficial areas of the body. These diseases are caused by the dermatophytes that invade the keratinized portion of the hair, skin and nails.

Dermatophytes are transmitted by close contact and the organisms may spread rapidly within families and enclosed communities. The dermatophytes can be conveniently classified into three groups on the basis of their reservoir and host preference.

- Anthropophilic species are those which have man as their major host.
- The zoophilic pathogens infect animals.
- Geophilic organisms are found in soil but may infect animals and man.

Anthropophilic	Zoophilic	Geophilic
Epidermophyton floccosum	*Microsporum canis*	*Microsporum gypseum*
Microsporum audounii	*Microsporum nanum*	*Microsporum raecemosum*
Trichophyton mentagrophytes	*Trichophyton verrucosum*	*Microsporum fulvum*
Trichophyton rubrum		
Trichophyton violaceum		

CLINICAL MANIFESTATION

- The superficial cutaneous infection caused by dermatophytes is called tinea or ringworm infection.
- The lesions of the Tineas often appear as pink circular lesions and gradually advance to form new borders. Scaling and peeling is common in affected areas.
- The lesions are rarely painful but can be very itchy. Anthropophilic organisms produce more intense inflammatory reactions with pustular lesions or a large inflammatory mass known as kerion.
- Infection of the nail is chronic and produces discoloration and thickening. Scalp infection leads to scaling and inflammation with hair loss which may sometimes be associated with scarring.
- Dermatophyte infections are clinically classified by anatomical location.

Table 58.1 Clinical entities tend to be caused by different species

Clinical manifestation	Site of infection	Causative agent
Tinea capitis	Scalp and hair	*Microsporum audouinii* *Microsporum canis* *Microsporum nanum* *Trichophyton mentagrophytes*
Tinea barbae	Beard	*Microsporum canis* *Trichophyton mentagrophytes* *Trichophyton rubrum* *Trichophyton violaceum* *Trichophyton verrucosum*

(Contd.)

Table 58.1 (Continued)

Clinical manifestation	Site of infection	Causative agent
Tinea corporis	Glabrous skin	*Microsporum canis* *Microsporum nanum* *Microsporum gypseum* *Microsporum audouinii* *Epidermophyton floccosum* *Trichophyton rubrum* *Trichophyton mentagrophytes* *Trichophyton tonsurans* *Trichophyton verrucosum* *Trichophyton violaceum*
Tinea cruris	Groin	*Epidermophyton floccosum* *Microsporum nanum* *Trichophyton mentagrophytes* *Trichophyton raubitschekii* *Trichophyton rubrum*
Tinea manuum	Hand	*Epidermophyton floccosum* *Microsporum canis* *Microsporum gypseum* *Trichophyton mentagrophytes* *Trichophyton verrucosum* *Trichophyton rubrum*
Tinea pedis	Feet	*Epidermophyton floccosum* *Microsporum persicolor* *Trichophyton mentagrophytes* *Trichophyton violaceum* *Trichophyton rubrum*
Tinea unguium	Nails	*Epidermophyton floccosum* *Trichophyton tonsurans* *Trichophyton mentagrophytes* *Trichophyton violaceum* *Trichophyton rubrum*
Tinea faciei	Face	*Microsporum canis* *Microsporum nanum* *Trichophyton mentagrophytes*
Tinea glutealis	Buttocks	*Trichophyton rubrum* *Epidermophyton floccosum* *Tichophyton mentagrophytes*

AETIOLOGY

On the basis of clinical, morphological and microscopic characteristics, three different genera are recognized as dermatophytes:

- *Epidermophyton*
- *Microsporum* and
- *Trichophyton*

PATHOGENESIS

These fungi invade only dead cornified layers of the skin, hair and nails. Only a few organisms are necessary to induce an infection following trauma.

Dermatophytes grow in filamentous form within the stratum corneum. The downward extension of this hyphae is restricted because certain nutrients, principally iron, are not available in the deeper tissues. Consequently, lateral invasions from the focus of inoculation occur, leading to the annular pattern seen on the skin surface.

The dermatophytes do not directly damage the skin, but a delayed hypersensitivity reaction appears at the borders of the growing fungi after the host becomes sensitized to soluble fungal antigens.

The inflammatory response results in a circinate pattern of erythema and oedema followed by exudation. Invasion of hair follicle produces inflamed nodules, deep seated pustules and abscesses. The skin becomes inflamed only after start of cell-mediated immunity to the dermatophytes.

There are two types of dermatophytic infections.

1. The acute or inflammatory type of infection (associated with CMI to the fungus, which generally heals spontaneously or responds readily to treatment)
2. The non-inflammatory type (which is difficult to eradicate)

Hairy regions are particularly favourable for dermatophytic growth, since the hair is composed of non-living keratin.

Dermatophytic infection affecting the hair of the scalp, eyebrows and eyelashes is called tinea capitis. It may be non-inflammatory or markedly inflammatory with scarring.

The terms ectothrix and endothrix refers to the location of the arthroconidia that infect the hair.

In ectothrix (outside hair) invasion, the hyphae fragments into arthroconidia that accumulate around the hair shaft or just beneath the cuticle, destroying it. The arthroconidia that are outside the hair shaft form a mosaic sheath, a pattern of arthroconidia resembling a tile mosaic.

Hair invaded in ectothrix fashion typically becomes greyish, dull and discolored, eventually the hair becomes brittle and break off. When many hairs are lost in this fashion, irregular greyish areas are left on the scalp, resulting in grey patch tinea capitis.

Tinea capitis is not inflammatory, but scaling of the scalp is a prominent feature of the infection. Grey patch tinea capitis occurs primarily in children.

In endothrix (inside) invasion, arthroconidia form by fragmentation of the hypha within the hair shaft. The cuticle is not destroyed. Arthroconidia are not seen. The presence of the conidia weakens the hair so that it loses its lustre, becomes brittle and breaks off above the surface of the scalp. The conidia in the shafts of the hair appear as black dots. So it is called as black dot tinea capitis.

The dermatophytid or 'id' reaction occurs in some people as an allergic response to a dermatophyte infection elsewhere in the body. The 'id' lesions do not contain the fungi causing dermatophytosis but they are itchy and sometimes painful.

LABORATORY IDENTIFICATION OF DERMATOPHYTES

Specimen Collection

- Skin should be scraped from the margin of the lesion on to a folded black paper.
- Hair should be plucked, not cut, from the edge of the lesion.
- Hairs that fluoresce under a Wood's lamp should be chosen or, if none fluoresce, choose broken or scaly ones.
- Nail clipping should be obtained from the nail bed or from infected areas after the outer layers are discarded.

Direct Examination

- A small volume of the specimen is selected for direct microscopic examination and investigated for the presence of fungal elements.
- The specimen is mounted in a small amount of 10% or 20% potassium hydroxide or calcofluor white.
- The KOH slides are gently heated and allowed to clear for 30 to 60 minutes before examining on a light or phase contrast microscope.
- Calcofluor white slides are examined on a fluorescent microscope.

Media and Tests Used

Bromocresol Purple (BCP)

Bromocresol purple (BCP)–milk solid glucose is a differential media useful in the characterization of dermatophyte species. Some species produce an alkaline reaction (changes media to purple), others do not produce a pH change (the media remains in sky blue colour). Hydrolysis of the milk solids results in a zone of clearing around the colony.

Hair perforation

Many dermatophytes have the ability to degrade hair. The hair perforation test determines whether the organism simply erodes the hair shaft or produces an enzyme that will penetrate and invade the shaft resulting in perforating bodies or cones.

Potato Dextrose Agar (PDA)

Potato dextrose agar (PDA) is a media useful for the production of pigment.

Phytone Yeast Extract (PYE)

Phytone yeast extract agar is a nutritionally enriched media that supports luxurious growth of most fungi. It contains antibiotics to inhibit bacterial growth.

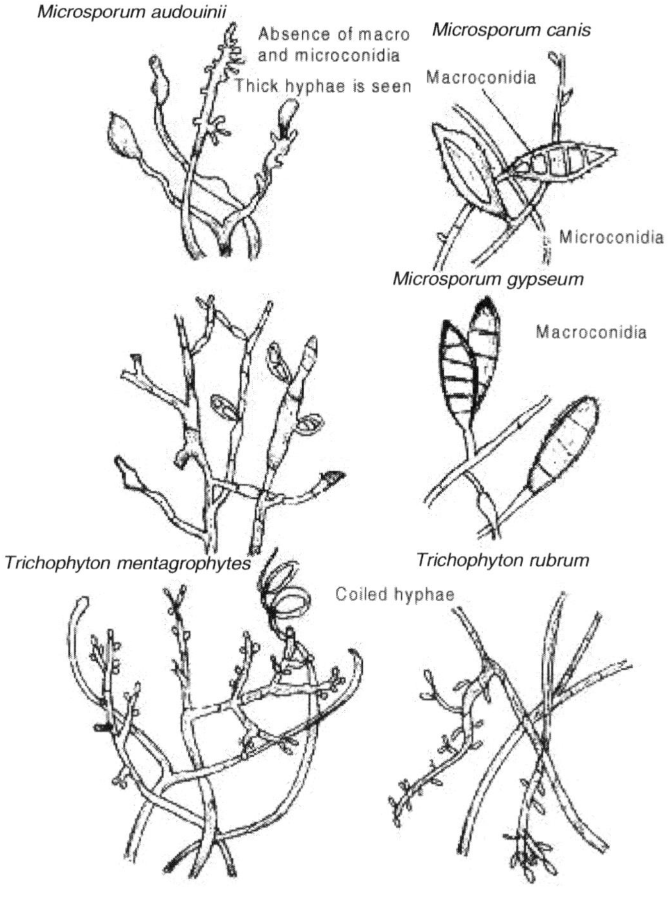

Figure 58.3 Dermatophytes

Rice Grain Test

Microsporum audouinii grows poorly on rice grains and produces a dark discoloration useful in differentiating it from atypical *M. canis* strains. The rice grains also enhance the production of macroconidia in some species.

Scotch Tape Mount

The scotch tape mount is used for examining the microscopic structures of filamentous fungi. With forceps a piece of clear, transparent tape is picked and the surface of the colony is touched with it. The tape is then placed on to a drop of mounting media on a slide; another drop is added and a coverslip is placed. It is then examined under a phase-contrast microscope.

Sabouraud's Dextrose Agar (SDA)

Sabouraud's dextrose agar (Emmon's modification) is a non-selective media which supports the growth of most fungi.

Slide Culture

The slide culture is a method of examining the microscopic structures of a fungus. The organism is grown on a glass coverslip placed on a block of agar. When sufficient growth has occurred, the coverslip is placed on a drop of mounting media on a microscope slide and examined by phase-contrast microscopy.

Urease Test

Christensen's urea broth indicates the presence of the enzyme urease, which splits urea into ammonia, resulting in an alkaline environment. The phenol red indicator turns the media from a straw yellow to fuschia at pH 8.4.

Vitamin Requirements

Certain species of dermatophytes have distinctive nutritional requirements that may be beneficial to differentiate them from other species. The agar base is vitamin-free and various vitamins are added to the basal media. The growth on the vitamin-enriched media is compared to the vitamin-free media to determine enhancement of aerial mycelium.

SOME IMPORTANT CHARACTERISTICS OF DERMATOPHYTIC FUNGI

Species	Microscopy	Culture
E. floccosum: Anthropophilic, worldwide epidemic, infects groin, body, athlete's foot, occasionally nails, does not infect hair.	Smooth, thin-walled, club-shaped macroconidia often in clusters. No microconidia.	Slow-growing, greenish-brown suede like surface, often raised and folded in the centre, deep yellowish-brown reverse.
M. canis: Zoophilic—cats and dogs, less commonly monkeys, guinea pigs, horses, mice, cows, rabbits; worldwide; less common in North America, UK; affects body in adults, scalp in children, rarely the nails.	Macroconidia are typically spindle shaped with 5–15 cells, verrucose, thick-walled and often have a terminal knob.	Colonies are flat, white to cream coloured, with a dense cottony surface and usually have a bright golden reverse pigment but non- pigmented strains may occur.
M. gypseum: Geophilic, worldwide, rare in North America, Europe; common in South America; affect feet, hand, body, scalp, rarely nails; usually acquired from soil but occasionally from animals and flies.	Macroconidia are ellipsoidal, thin-walled, verrucose and 4–6 celled.	Colonies are usually flat, suede-like to granular, with a deep cream to tawny buff to pale cinnamon coloured surface and a yellowish- brown reverse pigment.
M. nanum: Zoophilic—pigs, worldwide affects, scalp, body, ectothrix.	Macroconidia are small, ovoid to pyriform, mostly 2-celled with relatively thin, finely echinulate walls and broad truncate bases.	Colonies are flat, cream to puff in colour with a suede- like to powdery surface texture with a dark reddish-brown reverse.
T. rubrum: Anthropophilic, seldom isolated from animals, never found in soil; worldwide; affects feet, nails, body, groin, rarely scalp; both endothrix and ectothrix; the most frequently isolated dermatophyte.	Most cultures have numerous clavate to pyriform microconidia and moderate numbers of smooth thin-walled macroconidia	Colonies are flat to slightly raised, white to cream, suede-like with a pinkish-red reverse.

(Contd.)

Table (Continued)

Species	Microscopy	Culture
T. mentagrophytes: Anthropophilic and zoophilic—rodents, small and large mammals; worldwide; found in soil; affects feet, body, nails, beard, scalp, hand, groin; zoophilic are ectothrix; anthropophilic do not infect hair.	Numerous single-celled, spherical to subspherical microconidia, often in dense clusters, multicelled macroconidia also seen.	Colonies are flat, white cream in colour, with powdery to granular surface, reverse pigmentation is usually a yellowish brown to reddish brown colour.
T. tonsurans: Anthropophilic; worldwide; affects scalp, body shower sites, occasionally nails; endothrix.	Hyphae are relatively broad, irregular, microconidia are broad and pyriform, occasionally macroconidia are seen.	Flat with raised centre. Colour may vary from pale puff to yellow.
T. verrucosum: Zoophilic—cattle and other domestic and wild animals; worldwide; affects scalp, beard, body, occasionally nails; ectothrix.	It produces chlamydoconidia often referred to as "chains of pearls".	Colonies are slow-growing, small button or disc shaped, white to cream coloured, reverse pigmentation is also observed.
T. violaceum: Anthropophilic, near and Middle East, Eastern Europe, North Africa, occasionally Latin America and Mediterranean, imported to North America and Western Europe; affects scalp, body "shower sites", rarely feet and nails; endothrix; institutional outbreaks.	Hyphae are relatively broad much branched and distorted, no conidia usually seen, microconidia seen occasionally.	Colonies are slow-growing, glabrous or waxy, heaped and folded and a deep violet in colour, colonies often become pleomorphic, forming white sectors.

Some other dermatophytes and their general characteristics are listed below:

1. *M. audouinii* Anthropophilic; worldwide' rare except in Africa and Asia, affects scalp and body;once caused epidemic tinea capitis in prepubescent children; now rare; also spread by guinea pigs and dogs.
2. *M. equinum* Zoophilic—horses; worldwide; affects body; rare in man.
3. *M. persicolor* Zoophilic—mice; not found in soil; sporadic in Europe, UK; affects skin, not hair.
4. *T. equinum* Zoophilic—horses; worldwide; Rare in man.
5. *T. kanei* Anthropophilic; affects body, feet, nails; rare in man.
6. *T. megninii* Anthropophilic; Spain, Portugal, rarely in Africa, Mediterranean; affects body, scalp; beard; ectothrix.
7. *T. raubitschekii* Anthropophilic; Southern Asia, India, Mediterranean; affects body, rarely scalp; most frequently isolated dermatophyte.
8. *T. soudanense* Anthropophilic; Africa, occasionally North America, UK, Brazil; affects scalp and body primarily "shower sites".
9. *T. terrestre* Geophilic; worldwide; non-pathogenic.

TREATMENT

Treatment is with topical ointments such as miconazole, tolnaftate or clotrimazole for 2 to 4 weeks. Griseofulvin and itraconazole are the only oral fungal agents currently approved.

1. What are the two piedras that infect human?
2. Briefly describe the major tineas that affect human.
3. How do you differentiate different dermatophytic fungi?
4. Discuss about clinical manifestation of dermatophytes.
5. Explain about pathogenesis of dermatophytic fungi.
6. Write short notes on
   i. Tinea versicolor
   ii. Wood's lamp test

59 Subcutaneous Mycoses

Mycoses that cause infection in subcutaneous layer are called subcutaneous mycoses. Sporotrichosis, chromoblastomycoses, maduramycoses, phycomycoses, phaeohypomycoses and rhinosporidiosis are the examples for subcutaneous mycoses.

SPOROTRICHOSIS

- Sporotrichosis is one of the subcutaneous infections.
- It is a chronic granulomatous infection of human skin.
- This disease is worldwide in distribution.
- Aetiology of sporotrichosis was first described in the early periods of the last century (1900s).
- It is caused by *Sporothrix schenckii*.
- Aetiological agent is a thermally dimorphic fungus.
- Because of the association of the organism with vegetation, the disease is also called "Rose thorn disease".
- The fungus is saprophytic in nature.
- Organism is found on plant thorns and timber.
- Infection is acquired through thorn pricks or other minor injuries.

The incubation period is highly variable, ranging from few days to months. Average is three weeks.

PATHOGENESIS

The pathogen is temperature-sensitive. Temperature of skin surface and connective tissue are slightly lower than that of deep tissue. This will support the growth of aetiological agent.

Mycelial fragments, conidia or any fraction of the *Sporothrix* is implanted into the deep layers of the skin through trauma.

Initially small hard painless nodules appear at the site of injury and it enlarges into a fluctuant mass that eventually breaks down and ulcerates.

The subcutaneous nodule becomes discolored and the overlaying skin darkens to a reddish colour and eventually blackens.

Primary lesions enlarge and several nodules begin to develop along the lymphatics. Infection rarely extends beyond regional lymphatics. After a few weeks the primary lesions will heal. The entire clinical disease is commonly known as lymphocutaneous sporotrichosis.

75% of human infections are lymphocutaneous sporotrichosis. Immunity of an individual reduces the severity of infection.

Systemic spread is very low but dissemination may occur especially in debilitated patients.

Figure 59.1 Sporothrix

Following the primary nodule, multiple subcutaneous nodules develop along the lymphatic channels and they become hard. It is called chronic sporotrichosis.

Fixed sporotrichosis refers to the presence of only one lesion. It is a non-lymphatic nodule. It is less progressive and common in endemic areas. Primary pulmonary sporotrichosis may result from inhalation of the conidia.

EPIDEMIOLOGY

Incidence is high among agricultural workers. It is an occupational risk disease. Highest prevalence of infection was observed in Mexico. Animals are susceptible to this disease but animal to man transmission is not observed. 75% of infected individuals are males.

LABORATORY DIAGNOSIS

Specimen

- Aspirated fluid
- Pus
- Biopsy tissue
- Exudative material

Microscopy

- Specimens are examined directly by KOH mount.
- Calcofluor white stain is also used.
- In infected tissue, the fungus is seen as cigar-shaped yeast cells without mycelium.
- Asteroid bodies are seen in lesions composed of central fungus cells with eosinophilic materials radiating as extensions. It is an antigen–antibody complex along with complement.

Culture

Clinical specimen is inoculated into the inhibitory agar or antibiotics containing Sabouraud's dextrose agar and incubated at 25–35°C. Colonies are observed within 1–3 days. Young colonies are blackish and shiny, becoming wrinkled and fuzzy on ageing.

If the medium is incubated at 37°C, yeast cells are seen.

Microscopic examination of colonies shows thin, branching mycelium, septate hyphae with small conidia and flower like arrangements formed on delicate sterigmata.

Serology

Specific antibodies are not observed in the early stages of infection. Serological test is of little help in the diagnosis.

Agglutinin to the yeast cell can be used to monitor the course of infection. Sporotrichin will elicit delayed skin reaction in sensitive persons.

TREATMENT

Oral solution of saturated potassium iodide is given. Dosage is increased daily at 0.5–1 ml. Surface lesions are treated with 2% potassium iodide in 0.2% iodine.

Other antifungal agents involved in the treatment are griseofulvin, amphotericin B, flucytosine, dihydroxystilbamide, ketoconazole.

CHROMOBLASTOMYCOSES

It is caused by dematiaceous fungi, which are imperfect fungi that produce varying amount of melanin-like pigments. These pigments are found in the conidia or hyphae or in both.

AETIOLOGY

Phialophora verrucosa

Cladosporium carrionii

Fonsecaea pedrosoi

Fonsecaea compacta

Rhinocladiella aquaspersa

- Melanin production is associated with the virulence of the aetiological agents. It is caused by traumatic implantation of any one or several dematiaceous fungal species.
- The infection is chronic and characterized by the slow development of verrucous, cutaneous vegetation.
- The natural reservoir of these fungi is soil and plant debris.
- Dematiaceous fungi are similar in their pigmentation, in their antigenic structure, morphology and physiological properties.
- The colonies are compact, deep brown to black and develop a velvety, often wrinkled surface. The agents of chromoblastomycosis are identified by their mode of conidiation.
- In tissue, it produces spherical brown cells termed muriform or sclerotic bodies that divide by transverse septation.
- Phialophora conidia are produced from flask-shaped phialides with cup-shaped collarettes.

Phialophora verrucosa It produces branching chains of conidia by distal budding and it also produces elongated conidiophores with long branching chains of oval conidia.

Rhinocladiella aquaspersa It produces lateral or terminal conidia from a lengthening conidiogenous cell.

Fonsecaea pedrosoi It is a polymorphous genus. It produces phialides similar to *Cladosporium*.

Fonsecae compacta It produces spherical conidia. Structures are smaller and compact.

PATHOGENESIS

Fungi are introduced into the skin by trauma, often in the exposed areas of legs or feet. Over months to years the primary lesions become verrucous and wart-like extensions along the draining lymphatics. This is due to mononuclear cellular infiltrate. Cauliflower-like nodules with abscess eventually cover the area. Small ulceration is also formed. Within short duration, these lesions become raised and appear scaly and dull red or greyish in colour. Patients experience minimal discomfort. Systemic invasion is extremely rare.

LABORATORY DIAGNOSIS

- Skin scrapings is the specimen of choice.
- 10% KOH mount is performed.
- Detection of sclerotic body is the main diagnostic tool.

TREATMENT

Surgical excision with wide margin is the therapy. Fluconazole and itaconazole are also recommended.

MYCETOMA

Mycetomais a chronic subcutaneous infection induced by traumatic inoculation with any of the several saprophytic fungi. The disease was first observed in Madurai district of Tamil Nadu. So this disease is also called as "Madura foot" and "Maduramycosis".

The clinical features defining mycetoma are local swelling and interconnecting, often draining sinuses that contain granules, which are microcolonies of the agent embedded in tissue material.

Based on the aetiological agent, mycetoma is classified into three types:

1. Actinomycetoma is a mycetoma caused by actinomycetes.
2. Eumycetoma is a mycetoma caused by a fungus.
3. Botryomycosis is a mycetoma caused by group of bacteria other than actinomycetes.

The natural history and clinical features of all types of mycetoma are similar, but Actinomycetoma may be more invasive, spreading from the subcutaneous tissue to the underlying muscle. It is worldwide in distribution.

The fungal agents of mycetoma are *Madurella grisea, Phialophora jeanselmei, Leptosphaeria senegalensis, Cephalosporium* spp., *Acremonium* spp., *Petriellidium boydii, Pseudallescheria boydii.*

The fungi causing mycetoma are members of fungi imperfecti. Based on the fruiting bodies, these fungi are identified.

Actinomycetes agents of mycetoma are *Actinomadura madurae (Streptomyces), Actinomadura pellettieri (Streptomyces), Nocardia brasiliensis, Streptomyces somaliensis.*

A frequent agent of mycetoma is *Pseudallescheria boydii*. In addition to systemic and pulmonary diseases it may cause sinusitis, keratitis and otomycosis.

In tissues, mycetoma agents produce granules, it may range up to 2 mm in size. The colour of the granule may provide information about the agent.

Mycetoma agents that produce black granules are *Madurella grisea, Petriellidium boydii, Pseudallescheria boydii, Acremonium falciforme, Exophiala jeanselmei, Aspergillus nidulans, Fusarium* spp. and *Curvularia geniculata.*

M. mycetomatis—dark red to black colour granules.

Actinomadura madurae—white, yellow or pink coloured granules.

Actinomadura pellettieri—red granules.

Nocardia brasiliensis—white coloured granules.

Nocardia caviae—white to yellow coloured granules.

Nocardiopsis dassonville—cream-coloured granules.

Streptomyces somaliensis—yellow to brown coloured granules.

Bacteria like *Staphylococcus, Streptococcus, E. coli, Proteus, Pseudomonas aeruginosa* produce white colour granules and *Actinobacillus lignieresi* produce yellow-coloured granules.

Granules are hard and contain intertwined septate hyphae.

PATHOGENESIS

Mycetoma develops after traumatic inoculation of any one of the agents of mycetoma. Subcutaneous tissues of the foot, lower extremities, hands and exposed areas are most often involved in mycetoma formation.

Pathology is characterized by suppuration and abscess formation, granulomas and the formation of draining sinuses containing the granules.

LABORATORY DIAGNOSIS

Specimen

Pus, exudates or biopsy material.

The granule colour, texture, size and the presence of hyaline or pigmented hyphae are helpful in determining the aetiology.

Culture

Specimen is inoculated into SDA with antibiotics.

Madurella mycetomatis Colonies are smooth or folded, glabrous or powdery, leathery and white to yellowish brown in colour. Conidia are small, ovoid to globose. It also produces septate hyphae with many chlamydospores.

Madurella grisea Colonies are rapidly growing, grey or olive coloured. Grey aerial hyphae may be formed. Older colonies become reddish brown. The hyphae are brown walled, septate. Hyphae forms mycelia. Its width is too long and produces chains of arthroconidia.

Nocardia brasiliensis The colonies are slow-growing, wrinkled or folded, yellow to orange in colour, dry and chalky or glabrous with an earthy odour. The smear shows gram-positive branched filament that fragment into bacillary forms. It is an acid-fast bacillus.

Nocardia caviae Growth is rapid and colonies are similar to *Nocardia brasiliensis*. On special media, cream to peach-coloured colonies are formed. Tufts of short aerial mycelia are seen. It is also an acid-fast bacillus.

Actinomadura madurae The colonies rapidly grow on LJ medium and slow on SDA medium. The growth appears white to tan, pink or orange, glabrous, wrinkled, hard and adherent. Short chains of spherical conidia may be formed. It is not an acid-fast bacillus.

Actinomadura pellettieri The colonies are slow-growing, either folded or wrinkled, small, dry and granular. The colour is at first pink to peach, that develops into a deep red with waxy surface. Conidia are rare.

Streptomyces somaliensis Colonies are creamy, wrinkled flaky growth. Tufts of aerial filaments are formed.

TREATMENT

Management of mycetoma is difficult. Amphotericin B is recommended. ketoconazole, nystatin, flucytosine, KI and miconazole are also used.

LOBOMYCOSIS

- It is a chronic subcutaneous infection of humans and dolphin caused by a fungus called *Loboa loboi.*
- It was first observed in 1931.
- Most patients are adults and males.
- Other names of the disease are lobo's disease or keloid blastomycosis.
- Infection is restricted to the dermis.
- Lymph nodes are not involved.

- The infection is chronic progressive slowly.
- Lesions are not painful but ulcerative.
- The size of the lesions depends on the duration of the infection.
- The initial lesions are small, hard subcutaneous nodules usually appearing on the extremities, face or ear, presumably as a result of traumatic inoculation.
- Inflammatory lesions show many foamy macrophages and multinucleated giant cells. Many of the phagocytes contain intact or fragmented cells. The lymphocytes and plasma cells appear as scattered clusters.
- Diagnosis is by direct microscopic examination of the skin scrapings, biopsy and wet preparation of exudative lesions.
- The fungus appears in the tissue as large spherical or oval yeast that exhibit multiple budding. They are multinucleated and thick-walled.
- Occasionally asteroid bodies are observed.
- It cannot be cultured in the laboratory.
- Acid schiff or methenamine silver stain is used for staining.
- For the treatment, infected tissue must be carefully removed. Sulpha drugs are used for this purpose.

PHAEOHYPHOMYCOSIS

It is a term used to denote infections characterized by the presence of dark pigmented septate hyphae in tissue. Both systemic and cutaneous mycosis have been described.

The name "subcutaneous phaeohyphomycosis" is because the causative agents are seen below the dermis.

The disease is caused by dematiaceous fungi like, *Exophiala jeanselmei, Wangiella dermatitidis, Exophiala spinifera, Phialophora hoffmannii, Phialophora repens, Phialophora richardsiae, Tetraploa aristata, Bipolaris spicifera.*

All these fungi are exogenous moulds that normally exist in nature.

In phaeohyphomycosis, cyst usually develops and may enlarge to several centimetres.

Based on the pathogenicity, phaeohyphomycosis is classified into four types:

1. *Cutaneous phaeohyphomycosis* Colonized on skin. May be seen over cracked, fissured areas of sole of foot. Ulceration is observed.

2. *Subcutaneous phaeohyphomycosis* Localized infection following deep inoculation of the fungus into subcutaneous tissues. The lesions may occur on feet, legs, hands, arm, etc. Nodules develop.

3. *Systemic and cerebral phaeohyphomycosis* These are rare and are the most serious form of this disease. Cerebral phaeohyphomycosis often present with brain

abscess. The frontal lobes are mostly affected and encapsulated. Abscesses filled with brown hyphae are present in neurological biopsy.

4. *Paranasal sinus phaeohyphomycosis* It is an indolent disease of the sinus cavity and may spread to the adjacent areas.

LABORATORY DIAGNOSIS

Microscopy

Abscesses are examined in KOH mount.

These fungi are pigmented, dark brown in colour because of melanin pigment.

In tissue, the hyphae are large and often distorted and may be accompanied by yeast cells.

Culture

Clinical specimens are plated on SDA with cycloheximide and are incubated at 25–37°C. Most agents grow slowly and growth is visible only after one or two weeks.

Exophiala jeanselmei and *Exophiala spinifera* produce black and slimy colonies with many yeast like cells. Hyphae are brown septate.

Phialophora hoffmannii, Phialophora repens and Phialophora richardsiae produce phialides with collarettes.

Wangiella dermatitidis produce young colonies are black, soft, moist, shiny and yeast like.

TREATMENT

Itraconazole and amphotericin B are used for treatment.

RHINOSPORIDIOSIS

- Rhinosporidiosis is a chronic infection characterized by the development of polypoid masses of the nasal mucosa.
- It is a granulomatous disease of the mucocutaneous tissue of man.
- It is caused by *Rhinosporidium seeberi.*
- This disease was first reported in 1990 from Argentina. Over 200 cases have been recognized. 90% of the cases are from India and Srilanka of which 90% are males.
- It produces large spherules in lesions and epithelial cells.

- It produces three types of disease in humans:
  1. Cutaneous rhinosporidiosis
  2. Nasal rhinosporidiosis
  3. Ocular rhinosporidiosis

LIFE CYCLE AND PATHOGENESIS

The organism starts its life cycle in the tissue as rounded or oval spore with a cell membrane and clear cytoplasm. It undergoes some developmental processes and is converted into a mature spore. The organism develops into a sporangium containing endospores and having a thick outer chitinous wall. The spores escape from the sporangium and are then carried to the lymphatics.

Lesions are most often found in the mucosa of the nose, nasopharynx and soft palate. Lesions are initially flat but develop into discolored cauliflower type polypoid masses. In nasal area, respiration may be blocked and there is a profuse seropurulent discharge.

During ocular rhinosporidiosis, blood-stained discharge may be noted. Large lesions may prevent occlusion of the eyelids causing exposure conjunctivitis.

LABORATORY DIAGNOSIS

- Histologic tissue examination reveals epithelial hyperplasia and cellular infiltrate of neutrophils, lymphocytes, plasma cells and giant cells.
- Large thick-walled sporangium is also present. It is packed with thousands of endospores. The cell wall of spherule is multilayered. At maturation cell wall becomes thin and endospores are liberated.

TREATMENT

Ethylstilbamide is used for treatment.

PHYCOMYCOSIS

It is a chronic self limiting infection of the subcutaneous tissue caused by *Basidiolobus haptosporus*.

- It was first observed in 1956.
- The colony is colourless or brownish, thin, flat and glabrous.
- Fungus produces aerial mycelium. The hyphae are 8–20 μm wide and produces chlamydospores, forcibly ejected spores and spherical smooth-walled zygospores.
- Highest incidence was observed among children between 5–9 years. 70–80% of infected individuals are males.

- Infection begins on limb with a small firm movable nodule in the subcutaneous tissue. The nodule enlarges and oedema develops and may become massive involving the entire leg or shoulder. The skin becomes rough. Lesions are not painful. They persist for several months, then resolve spontaneously.
- Direct microscopic examination of tissue reveals multiple granular giant cells and eosinophils.
- Broad hyaline branching hyphae with infrequent septa are surrounded by eosinophilic material.
- Culture on SDA media without cycloheximide at 25°C will yield colonies within 2–3 days.
- Potassium iodide is used for treatment.

1. Write short notes on
   i. Sporotrichosis
   ii. Blastomycosis
   iii. Chromoblastomycosis
   iv. Eumycotic mycetoma
   v. Maduromycosis

Systemic Mycoses

Systemic mycoses are deep mycotic infections. It causes infection in inner parts of the body, e.g. lungs, blood, etc. They are classified into two based on aetiology and individuals involved—primary and opportunistic systemic mycoses.

PRIMARY SYSTEMIC MYCOSES

COCCIDIOIDOMYCOSIS

INTRODUCTION

Coccidioidomycosis is an infection caused by the dimorphic fungus called *Coccidioides immitis*. The infection may be inapparent, benign, severe or even fatal. It is an endemic disease in the dry, arid regions of the United States. Infection is acquired through inhalation of dust containing arthrospores of the fungus.

Etiological agent is a dimorphic fungus that normally lives in soil. The organism was discovered in 1892 in tissue from fatal case and was named *Coccidioides* (coccidian like) the species name *immitis* means "not mild". It is a systemic disease.

Smith discovered the natural reservoir for *Coccidioides immitis*.

MORPHOLOGY

It grows in the media as mould. It produces a white, grey or brownish colour, powdery to cottony texture colonies. Microscopically colonies are observed as hyaline branching septate hyphae and as the culture ages, characteristic arthroconidia are produced.

In older cultures, hyphal fragments release unicellular barrel-shaped arthroconidia. Arthroconidia are highly resistant to desiccation, temperature extremes and deprivation of nutrients and may remain viable for years.

In the infected host, *C. immitis* exists as spherules, a spherical thick-walled structure 15–18 μm in diameter, that are filled with a few to several hundred endospores. As the spherule enlarges, the nuclei undergoes mitosis, the cytoplasm condenses around the nuclei and cell wall forms around each developing endospore. At maturation, the spherules rupture to release its endospores. These endospores enlarge to form mature spherule. It is observed in tissues and may appear in sputum of patients with coccidioidal cavities in lungs.

ANTIGENIC STRUCTURE

Coccidioidin is a crude antigen extracted from the filtrate of liquid mycelial culture of *C. immitis*. Spherulin is produced from a filtrate of broth culture of spherules. Both antigens are positive to delayed skin reactions. HS, F and HL are some of the exoantigens of *C. immitis*.

PATHOGENESIS

Inhalation of arthroconidia leads to primary infection. Respiratory tract is a major site for infection. 60% of the infections are asymptomatic.

The cell wall of the infectious particle has several layers. Outer layer represents the original cell wall, middle layer is called thin fibrous rodlets and inner thick wall. Outer layer and rodlet layer contains mannon, protein and lipid and are readily solubilized.

The inner layer contains chitin, 3-aminomethyl mannose. When arthroconidia develops into spherules, it releases antigen.

Arthroconidia and endospores are readily engulfed by alveolar macrophages and kill *C. immitis.* Macrophage activation may enhance killing activity may lead to inflammation.

Primary coccidiodomycosis has a incubation period of 10–16 days. Up to 20% of the patients with pulmonary coccidioidomycosis manifests allergic reactions, usually erythema nodosum.

Some patients may develop a chronic but progressive pulmonary disease with multiplying or enlarged cavities or nodules.

In severe coccidioidomycosis, patients have elevated antibody titres, circulating immune complexes and depressed cellular immunity. Recovery of infection often leads to restoration of immune functions.

The impaired immune functions are due to the documented increase in the population of suppressor cells, blocking factors, immune complexes and impaired lymphocyte circulation.

Immune complexes may contribute to the immunopathology by two mechanisms:

1. Deposition of the complexes may lead to local inflammatory responses.
2. Immunosuppression may result from the binding of complexes to cells bearing Fc receptors.

Chronic coccidioidomycosis develops from initial lesions that appear on the face or neck. Osteomyelitis may also develop.

CLINICAL MANIFESTATION

- Fever
- Chest pain
- Cough
- Weight loss
- Extrapulmonary infection involves the meninges, skin or bone.

LABORATORY DIAGNOSIS

Specimen

- Sputum
- Exudates from cutaneous lesion
- Spinal fluid
- Urine and
- Tissue biopsy

Microscopy

Clinical exudates should be examined directly in 10% or 20% KOH or calcofluor white stains. Tissue specimens are stained with haematoxylin and eosin. Microscopic examination shows spherules and endospores.

Culture

Clinical specimens are inoculated into the inhibitory mould agar or Sabouraud's agar with antibiotics like cycloheximide, chloramphenicol and gentamicin. Media is incubated at room temperature or at 37°C. Colonies of *Coccidioides immitis* may develop within 1 to 2 weeks and are examined microscopically for the production of characteristic arthroconidia.

Spherule formation is observed by the incubation of complex medium at 40°C with 20% CO_2.

Immunodiffusion test demonstrates the presence of specific antigen.

The CF test for antibodies to coccidioidin is a powerful diagnostic and prognostic tool.

Skin Test

The coccidioidin skin test reaches its maximum duration between 12–48 hours after cutaneous injection of 0.1 ml of a standardized dilution.

EPIDEMIOLOGY

Organisms are most prevalent in semi-arid climate, alkaline soil and characteristic indigenous desert plants and rodents. Coccidioidomycosis is considered as an occupational hazard for construction workers, archaeology students.

TREATMENT

Symptomatic primary infections are self-limited and require only supportive treatment. Patients with severe disease require treatment with amphotericin B, which is administered intravenously.

HISTOPLASMOSIS

Histoplasmosis is the most prevalent pulmonary mycosis of humans and animals. It is caused by the dimorphic soil saprophytic fungi *Histoplasma capsulatum*. It is distributed throughout the world. Infection is initiated by inhalation of the fungal conidia.

Dr. Samuel Darling discovered histoplasmosis and also gave a clear-cut description of the disease. This disease is also called the "Darling's disease".

Two colonial forms are produced during cultivation. They are type A or albino type and type B or the brown type. Both phenotypes produce identical yeast and tissue forms. *H. capsulatum* is a thermally dimorphic fungus. At temperature below 35°C, it grows as a mould, often white or brown colour and at 35°C, it grows as yeast with small heaped and pasty colonies. It grows very slowly. Under optimal conditions the mould colony develops after 1 or 2 weeks.

Both microconidia and macroconidia are produced at temperatures below 37°C. The hyaline, septate hyphae produce microconidia (2–5 μm) and large thick-walled macroconidia with peripheral projections of cell wall material (8–16 μm). *H. capsulatum* is a facultative intracellular parasite. In tissue, yeasts are seen within macrophages.

Histoplasmin is an antigen extracted from *H. capsulatum.*

PATHOGENESIS

The aetiological agent enters into the lungs through inhalation. Conidia develop into yeasts after settling of *H. capsulatum* mycelium in the alveoli.

Yeast cells are engulfed by alveolar macrophages.

Within the macrophage, yeast cells are able to multiply and are disseminated to the reticuloendothelial tissues such as liver, spleen, bone marrow and lymph nodes through bloodstream.

Tissue reaction may involve an early infiltration of neutrophils and lymphocytes, which leads to granulomatous inflammatory response and also produces epitheloid cell tubercle.

Acute Pulmonary Histoplasmosis

- Symptoms range from a mild flu-like illness that clears spontaneously to a moderate or severe disease.
- Incubation period varies from one to several weeks.
- Symptoms are fever, cough, chest pain, dyspnea, hoarseness, night sweats and weight loss.

Chronic Pulmonary Histoplasmosis

- It is most often seen in males. It is considered to be an opportunistic complication of lung disease.
- Symptoms are low grade fever, a productive cough, progressive weakness and fatigue.

Disseminated Histoplasmosis

- Dissemination may be completely benign and inapparent except for the calcified lesions, usually in the organs of reticuloendothelial tissues.
- It may be acute and progressive.
- Symptoms are splenomegaly and hepatomegaly, weight loss, anaemia and leucopenia. Granulomatous lesions and macrophages packed with yeast cells can be observed throughout the reticuloendothelial systems.
- Acute progressive histoplasmosis is often fulminant and rapidly fatal.
- Presumed ocular histoplasmosis syndrome (POHS) is also observed in some cases.

LABORATORY DIAGNOSIS

Specimen

- Blood (buffy coat)
- Bone marrow
- Sputum
- Scrapings from the superficial lesions
- Pus from the sinus tract

Microscopy

Smears of infected specimens fixed with methanol and stained with Wright or Giemsa stain will reveal characteristically ellipsoidal yeast cells inside the macrophages.

Culture

Purulent portion of the specimen should be selected for culturing.

In endemic areas, specimens should be inoculated in at least four media which are:

- Sabouraud's agar without antibiotics incubated at 37°C.
- Sabouraud's agar with antibiotics incubated at 25–30°C.
- Brain heart infusion agar with 5% sheep blood without antibiotics incubated at 37°C.
- Brain heart infusion agar with 5% sheep blood with antibiotics cycloheximide incubated at 25–30°C.

pH of the media should be neutral, incubated at least for 4 weeks because the etiological agent grows very slowly.

H. capsulatum is identified by characteristic macroconidia at 25–30°C and observation of yeast cells at 37°C.

Skin Test

Histoplasmin skin test is a valuable tool in epidemiology. Within two weeks after infection, most persons test positive.

Serology

Two tests are widely used for diagnosis. They are complement fixation test and immunodiffusion test.

TREATMENT

Amphotericin B is the drug of choice. A total of 1.5 g is recommended.

BLASTOMYCOSIS

This is a chronic infection characterized by granulomatous and suppurative lesions initiated by inhalation of a thermally dimorphic fungus, *Blastomyces dermatitidis*. This disease is also called North American blastomycosis because initial cases were confined to the United States.

Blastomycosis was first described in its cutaneous form in the 1980s by Gilchrist. Hence blastomycosis is also known as "Gilchrist's or Chicago disease".

Blastomycosis is primarily a pulmonary infection characterized by spread to the skin and other parts of the body. Soil is considered to be the source of infection. The organism is acquired by inhalation. The causative agent *B. dermatitidis* is a dimorphic fungus. On Sabouraud's glucose agar at 25°C, the organism grows as a mould, producing a colony of uniform hyaline, septate hyphae and conidia. Colony development requires at least 2 weeks. Many strains produce a cottony mycelium that becomes tan to brown on ageing.

On enriched media at 37°C, it grows as a yeast with folded pasty and moist colonies.

Microscopically the mycelial form produces abundant conidia from the aerial hyphae and lateral conidiophores. The conidia are spherical, ovoid or pyriform in shape and are 3–5 mm in diameter.

Extracts of culture filtrates of *B. dermatitidis* contain blastomycin.

PATHOGENESIS

- Blastomycosis is acquired by inhalation of exogenous, infectious particles.
- Initial site of infection is the lungs.
- In the alveoli *B. dermatitidis* induces an inflammatory response characterized by the infiltration of both macrophages and neutrophils and subsequent formation of granulomas.
- Both conidia and yeast cells are susceptible to the oxidative killing mechanisms of neutrophils and the fungicidal activity of macrophages. Neutrophils and CMI co-operate to produce effective resistance to blastomycosis.
- Three forms of blastomycosis are recognized: pulmonary blastomycosis, chronic cutaneous blastomycosis and disseminated blastomycosis.

Pulmonary Blastomycosis

It may be asymptomatic or may occur as acute or subacute pneumonia. It may also persist locally or spread to other organs. Symptoms are fever, malaise, night sweats and cough. Pulmonary lesions heal by fibrosis and resorption.

Chronic Cutaneous Blastomycosis

The initial skin lesions appear as one or more subcutaneous nodules that eventually ulcerate. Lesions are most common on exposed skin surfaces such as face, hands and lower legs. If untreated, elevated granulomatous lesions with advancing borders will develop. The yeast cells can be found in microabscesses near the dermis.

Disseminated Blastomycosis

It may be widespread blastomycosis, most commonly involving the extra pulmonary sites. This infection may be chronic. From the lungs, yeasts spread through the bloodstream. Primary cutaneous blastomycosis is initiated by traumatic autoinoculation or contamination of open wound with the infectious material.

LABORATORY DIAGNOSIS

Specimen

- Sputum
- Pus
- Exudates
- Urine

Microscopy

In calcofluor or KOH preparations of pus and sputum, diagnosis can be made by detection of yeast cells. The yeasts are large and they typically have thick cell wall.

In tissue stained with haematoxylin and eosin, the yeast cytoplasm stains dark and the cell wall appear colourless. The cells may be multinucleated.

Culture

Specimens are cultured on inhibitory mould agar or Sabouraud's agar and sheep blood enriched media.

Skin test

Delayed type hypersensitivity can be detected.

Serology

The most useful serological procedure is an immunodiffusion test for specific precipitins.

An enzyme immunoassay for antibodies to antigen A is recently evaluated.

TREATMENT

Amphotericin B is effective against blastomycosis.

PARACOCCIDIOIDOMYCOSIS

Paracoccidioidomycosis or South American blastomycosis is a systemic mycotic infection caused by the dimorphic fungus *Paracoccidioides brasiliensis.* It is a chronic granulomatous infection. Primary site of the infection is the lungs. The infection may spread to other parts and produce ulcerative granulomas in the mucosal surfaces of the nose, mouth and gastrointestinal tract. Internal organs may also become infected.

On SDA at 25–30°C, *P. brasiliensis* grow very slowly, reaching a diameter of 1–2 cm after 2–3 weeks of incubation.

Various conidia are produced by *P. brasiliensis* including chlamydospores, arthroconidia and singly-borne conidia.

By growing on a rich medium at 35–37°C, the yeast forms can be induced. Yeasts are larger and have thinner walls than the yeast forms of *B. dermatitidis*. The buds are attached by a narrow connection. Multiple budding yeasts are also formed.

Yeast cells of *P. brasiliensis* die after approximately 2–3 weeks in broth cultures. This is due to accumulation of toxic phenolic compounds.

PATHOGENESIS

P. brasiliensis is inhaled and initial lesions are observed in the lungs. After inhalation, because of the body temperature of 37°C, the mycelial form is converted into the yeast form. Yeast is considered to be an active pathogenic agent and involved in dissemination. Yeast cell wall polysaccharides such as alpha glucan is associated with virulence.

After a period of dormancy, which may last for decades, the pulmonary granuloma may become active, leading to chronic progressive pulmonary disease.

Organism spread from lungs to other organs like skin, mucosubcutaneous tissue, lymph nodes, spleen, liver and other sites.

Many patients present with painful sores involving oral mucosa.

The disease is common in males than females since a protein of *P. brasiliensis* binds to the oestrogen but not to testosterone or any other hormone. Binding prevents the conversion of mycelial form to yeast form at 37°C. This will explain the resistance of females against paracoccidiodomycosis.

LABORATORY DIAGNOSIS

Specimen

- Sputum
- Tissue
- Scrapings

Microscopy

Specimens are observed by KOH mount or calcofluor method.

Wet preparations are examined for yeasts.

Culture

Culture is performed by SDA and incubated at 25°C and 37°C.

Serology

Both immunodiffusion and complement fixation tests are performed.

TREATMENT

Ketoconazole is the drug of choice. Amphotericin B is also effective. Sulpha drugs such as sulphamethoxypyridazine are also used.

OPPORTUNISTIC SYSTEMIC INFECTIONS

CANDIDIASIS

The members of the genus *Candida* cause candidiasis. These organisms are the members of the normal flora of the skin, mucous membranes and gastrointestinal tract. It occurs worldwide and is the most common systemic mycosis. Of more than 100 species of *Candida*, *Candida albicans* causes most infections followed by *Candida tropicalis.*

Figure 60.1 Germ tubes of *Candida albicans*

MORPHOLOGY AND PHYSIOLOGY

C. albicans is capable of producing yeast cells with pseudohyphae and true hyphae. In serum at 37°C, *C. albicans* produce true hyphae. This is called germ tube production (Figure 60.1).

Cultural Characteristics

On most media *C. albicans* produces raised, cream-coloured opaque colonies within 24–48 hours.

Microscopic Appearance

It produces ellipsoidal or spherical budding yeasts about 3–6 mm in size.

Speciation

There are two serotypes of *C. albicans*, designated as A and B.

Determinants of Pathogenicity

Hyphal production and resistance to phagocytic killing is associated with virulence. High doses of extracts of *C. albicans* exhibit endotoxin activity. Germ tubes are more adhesive than yeast cells. Most strains produce protease enzyme that cleaves immunoglobulins.

EPIDEMIOLOGY

Many factors predispose to opportunistic candida infection. They are

- Cutaneous and mucosal candidiasis
- Old age
- Burns
- AIDS
- Antibiotic treatment
- Systemic candidiasis
- Surgery
- Cytotoxic drugs
- Catheters
- Artificial heart valves
- Aplastic anaemia
- Leukemia
- Trauma
- Peritoneal dialysis
- Cellular immunodeficiency
- Iron metabolic disorders
- Hypovitaminosis A
- Pregnancy
- Infancy
- Cellular immunodeficiency
- Diabetes
- Birth control pills
- Immunosuppression
- Steroid treatment
- Antibiotics
- Hyperalimentation
- Chronic granulomatous disease
- Agranulocytosis
- Malignancy
- Intravenous drug abuse
- Chronic mucocutaneous candidiasis
- Hypoparathyroidism
- Heredity

PATHOGENESIS

Cutaneous Mucosal Candidiasis

Superficial candidiasis is established by an increase in the total census of *Candida* and damage to the skin or epithelium that permits local invasion by the yeast and pseudohyphae.

Oral thrush can occur on tongue, lips, gums or palate. It is a patchy to confluent whitish pseudomembranous lesion composed of epithelial cells, yeast and pseudohyphae.

Yeast invasion of the vaginal mucosa leads to vulvovaginitis characterized by irritation, pruritus and vaginal discharge.

Candidial invasion of the nails and around the nail plate causes onychomycosis, a painful erythematous swelling of the nail fold resembling a pyogenic paronychia, which may eventually destroy the nail.

Systemic Candidiasis

It occurs when *Candida* enters the bloodstream and the phagocytic host defences are inadequate to control the growth and dissemination of the yeasts. Host defences against candidiasis are both specific or non-specific. Cellular and humoral serum components such as opsonins, complement, transferrin may inhibit the survival of *Candida*.

Numerous systemic manifestations of *Candida* may follow the introduction of *Candida* into the bloodstream. They include oesopharyngitis, intestinitis, infant diarrhoea, bronchopulmonary candidiasis, pyelonephritis, cystitis, endocarditis, myocarditis, endophthalmitis, meningitis, arthritis, osteomyelitis, peritonitis, macronodular skin lesions.

Candidemia may result from contamination of indwelling catheters, surgical procedures, trauma to the skin, etc.

Cutaneous Candidiasis

Candidiasis of the mucous membrane is often referred to as thrush. Vaginal thrush occurs more often in pregnant women.

LABORATORY DIAGNOSIS

Specimen

- Scrapings
- Blood
- Spinal fluid
- Tissue biopsies
- Urine
- Exudates

Microscopy

Fluid specimens are examined in gram-stained smears for pseudohyphae or true hyphae along with budding yeast cells.

Skin or nail scrapings are examined in a drop of 10% KOH and calcofluor white for hyphal forms.

Culture

Specimens are inoculated on SDA agar and incubated at 37°C for 24–48 hours. Creamy colonies are observed if the specimen contains Candida. Samples are observed for the presence of pseudohyphae.

TREATMENT

Cutaneous candidiasis is treated with topical antimycotic substances like ketoconazole, nystatin, miconazole. For the treatment of systemic candidiasis amphotericin B, flucytosine are recommended (Figure 60.2).

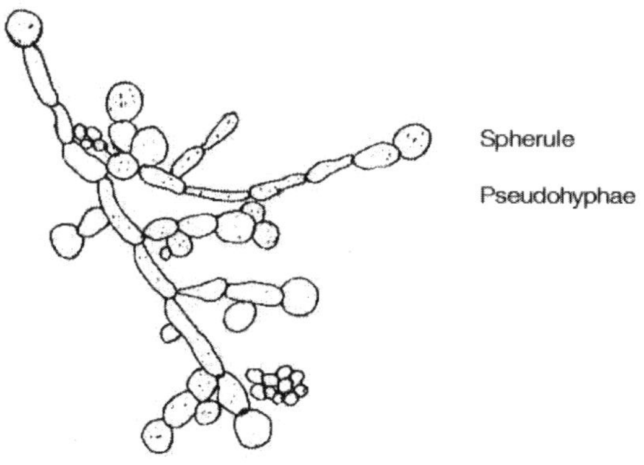

Figure 60.2 Chlamydospore of *Candida*

CRYPTOCOCCOSIS

Cryptococcosis is usually associated with immunosuppression. The disease is worldwide in distribution. It is caused by encapsulated yeast. The natural reservoir for *Cryptococcus neoformans* is the soil and the avian faeces and infection follows airborne exposure and inhalation of the yeast. It is ubiquitous in nature, the incidence of cryptococcosis is relatively low.

CHARACTERISTIC FEATURES

Cultural Characteristics

Visible colonies of *C. neoformans* develop on routine laboratory media within 36–72 hours. They are white to cream coloured, opaque and may be several millimetres in diameter. Colonies are typically mucoid in appearance and the amount of capsule produced can be judged by the degree of colony wetness.

Microscopic Appearance

Most clinical isolates are spherical, budding encapsulated yeast cells in both tissue and cultures. Rarely short hyphal forms are also observed. The hallmark of *C. neoformans* is its capsule, which may be twice the width of the cell. Elevated glucose, CO_2 or temperatures enhances capsule formation.

Properties of Capsule

During infection or immunization with whole yeast cells, antibodies are formed to the capsule. Four types of serotypes are designated based on the capsulated antigens. They are serotypes A, B, C and D. A and B are designated as one variety, C and D designated as another one variety. *C. neoformans* var *neoformans* represents serotype A and B. *C. neoformans* var *gatti* corresponds with serotypes C and D. Serotypes are correlated with prevalence of infection. *C. neoformans* var *neoformans* is globally distributed and *C. neoformans* var *gatti* is associated with eucalyptus trees of tropical countries.

Physiology

Special feature of the pathogenic *C. neoformans* is its best growth at 37 °C and inhibition of growth at 41°C. All *Cryptococcus* species are non-fermentative, hydrolyse starch, assimilate inositol and produce urease.

EPIDEMIOLOGY

Cryptococcosis is a sporadic infection with a worldwide distribution. *C. neoformans* is found in the soil and in avian faecal material, which apparently provide a reservoir of organisms. Cryptococcosis occurs equally in both sexes.

PATHOGENESIS

The high prevalence of *C. neoformans* in nature and relatively low frequency of disease suggests that many persons are probably exposed without any symptoms. Cryptococcosis is initiated in the lungs after inhalation of yeast cells of *C. neoformans*. Based on symptoms there are two types of cryptococcosis. They are pulmonary cryptococcosis and disseminated cryptococcosis.

Pulmonary Cryptococcosis

Primary infection may be symptomatic or may have an influenza like respiratory infection. It is rarely fulminant and hilar lymphadenopathy, calcification and cavitation are seldom observed. Symptoms are cough, sputum production, weight loss and fever.

Disseminated Cryptococcosis

C. neoformans is neurotrophic and disseminates to the central nervous system. Meningitis may be acute or chronic. Symptoms are fever, headache, stiff neck and disorientation, and accompanied by spinal fluid that typically is clear, increased opening pressure and presence of cells.

LABORATORY DIAGNOSIS

Specimen

- Spinal fluid
- Aspirate
- Skin lesions
- Sputum
- Tissue

Microscopy

Specimens should be examined directly in an Indian ink preparation for the presence of yeast cells with capsule. Encapsulated yeasts in tissue sections appear to be surrounded by large empty spaces because of the poor staining of the capsular polysaccharide.

Culture

Culture is performed by SDA and incubated at 37°C.

TREATMENT

Cryptococcosis is treated with both amphotericin B and flucytosine.

ASPERGILLOSIS

Aspergillosis refers to a spectrum of diseases that may be caused by a number of *Aspergillus* species. *Aspergillus* species are ubiquitous in nature and are saprophytes. It occurs worldwide.

About 150 different species and subspecies of *Aspergillus* have been recognized. They can be isolated from vegetation, especially nuts and grains. *A. fumigatus* is the most pathogenic species for humans, although many species like *A. flavus* and *A. terrus* are known to produce infection.

CHARACTERISTIC FEATURES

Cultural Characteristics

Aspergillus species grow very rapidly, producing aerial mycelium that become powdery and pigmented conidia. Based on the colour of the colony and morphology of conidia various species of *Aspergillus* are identified.

Microscopic Appearance

Aspergillus is characterized by conidiophores, which expand into large vesicles at the end and are covered with phialides that produce long chains of conidia. Phialides may arise directly from the vesicle.

PATHOGENESIS

Most cases of aspergillosis develop in individuals who have structural abnormalities within the lungs or have severely impaired resistance to infections.

Inhalation of fungal mycelia leads to aspergillosis. Based on the type of infection aspergillosis is clinically classified into three types:

- Allergic bronchopulmonary aspergillosis
- Aspergilloma and extrapulmonary colonization
- Invasive aspergillosis

In the lungs, alveolar macrophages are able to engulf and destroy the conidia. Conidia swell and germinate to produce hyphae that have a tendency to invade pre-existing cavities or blood vessels.

Allergic Bronchopulmonary Aspergillosis

In some atrophic individuals, development of IgE antibodies to the surface antigens of *Aspergillus* conidia elicits an immediate asthmatic reaction. The conidia germinate and hyphae colonize the bronchial tree without invading the lung parenchyma. This phenomenon is a characteristic of allergic bronchopulmonary aspergillosis.

Symptoms are asthma, eosinophil accumulation, type I and II hypersensitivity reaction.

Aspergilloma

After the inhalation of conidia, it enters into the existing cavity, germinate and produce abundant hyphae in the abnormal pulmonary space. Patients with tuberculosis and sarcoidosis are at risk. Some patients develop cough, dyspnea, weight loss, fatigue and haemoptysis. Cases of aspergilloma rarely become invasive. Many patients with aspergilloma are asymptomatic.

Invasive Aspergillosis

Following inhalation and germination of the conidia, invasive disease develop as an acute pneumonic process with or without dissemination. Symptoms include fever, cough, dyspnea. Hyphae invade the lumen and walls of blood vessels causing thrombosis and necrosis.

RISK FACTORS

Aspergilloma

- Tuberculosis
- Sarcoidosis
- Mycosis cavitary carcinoma
- Burns
- Cellular immunodeficiency

Allergic aspergillosis

- Atopy
- Cystic fibrosis

Invasive aspergillosis

- AIDS
- Antibiotic treatments
- Immunosuppression
- Steroid treatment
- Catheters
- Artificial heart valves
- Aplastic anaemia
- Leukemia
- Trauma
- Cellular immunodeficiency
- Iron metabolic disorders
- Diabetes
- Birth control pills
- Surgery
- Cytotoxic drugs
- Hyperalimentation
- Chronic granulomatous disease
- Agranulocytosis
- Malignancy
- Intravenous drug abuse
- Hypoparathyroidism
- Heredity

LABORATORY DIAGNOSIS

Specimen

- Sputum
- Lung biopsy

Microscopy

On direct microscopic examination of sputum with 10% KOH or calcoflour white, the hyphae of *Aspergillus* are observed as hyaline, septate with uniform width.

Culture

Aspergillus grows within few days on media at room temperature. Cycloheximide containing media should not be used.

TREATMENT

- Aspergilloma is treated with amphotericin B.
- Allergic forms are treated with corticosteroids.
- Itraconazole and flucytosine have also been used.

MUCORMYCOSIS

It is an opportunistic mycotic infection caused by *Mucor*. Mucor may cause zygomycosis, otomycosis and allergies. The mycelium is usually septate. Single or branching sporangiophores support round, spore filled with sporangia. The columella is variable in shape light to pigmented in colours. No rhizoids are present. On Sabouraud dextrose agar, mucor produce white, fluffy mycelium within 12 hours. It becomes grey to brown with age. This infection is found in immunocompromised patients.

1. Describe the symptoms of deep mycoses.
2. How do opportunistic fungal infections differ from other infections?
3. Describe the cultivation infection diagnosis of deep mycosis.
4. Which is the principal organ affected by deep mycosis?
5. Write short notes on
   i. Germ tube
   ii. Darling's disease
   iii. Chicago disease
   iv. Dimorphic fungus
   v. Coccidioidin skin test
   vi. SDA
   vii. Amphotericin B
   viii. Gilchrists' disease
   ix. *Candida albicans*
   x. Aspergilloma

PART V
PARASITOLOGY

61 General Characteristics of Parasites

Parasitic infections are among the most prevalent diseases in the developing and developed countries. Medical parasitology is an area of microbiology that studies invertebrate animals capable of producing diseases in humans and other animals. These invertebrate animal parasites are classified into the Kingdom Animalia and are separated into two subkingdoms Protozoa and Metazoa (Helminths or Worms).

- Protozoa are unicellular, eukaryotic chemoheterotrophs.
- "Protozoa" means "first animal".
- Protozoa are the largest organisms included in the microbial world.
- Except a few, they lack chlorophyll or other photosynthetic pigments.
- They mostly live in soil, fresh water and animals.
- They have complex cells with a pellicle, a cytosome, true nucleus, flagella and cilia.
- Nutrition is by ingestion, i.e., they engulf the food particles by phagocytosis or through special organs.
- Four major classes of protozoa are

Class Sarcodina (Amoebae)	Class Mastigophora (Flagellates)
Class Ciliophora (Ciliates)	Class Sporozoa (Non-motile forms)

- The vegetative form is called trophozoite.
- Asexual reproduction is by fission, budding or schizogony.
- Sexual reproduction is by conjugation.
- Some protozoa can produce a cyst to resist adverse conditions.
- Medically important protozoans are:
  - *Entamoeba*, *Naegleria* and *Acanthamoeba* (amoebae use pseudopods for motility).
  - *Giardia lamblia* causes intestinal infection called Giardiasis (flagellated protozoan).
  - Plasmodium is an apicomplexan that causes malaria.

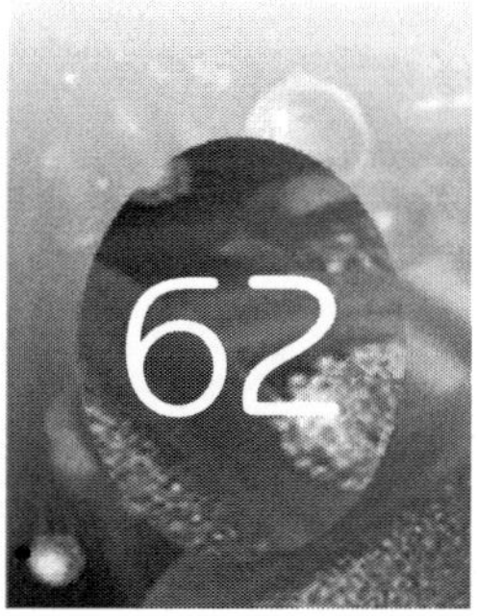

Classification of Pathogenic Protozoa and Helminthes

Protozoa are minute animalcules visible only under the microscope. The name protozoa was given by Gold furs (1871). Protozoa is the first phylum in the animal kingdom. This phylum is classified based on locomotion and mode of nutrition.

Subphylum	Class	Genus	Group
Sarcomastigophora	Mastigophora	*Trypanosoma**	Flagellates
		Giardia *	
		*Trichomonas**	
		*Leishmania**	
		Trichonympha	
	Sarcodina	*Amoeba**	Amoebae
		Elphidium	
		Coccodiscus	
Apicomplexa	Sporozoa	*Labyrinthula*	
		*Plasmodium**	
		*Toxoplasma**	
		Eimerida	
		*Cryptosporidium**	
Microspora	Mycosporea	*Nosema*	
	Microsporea	*Haplosporidium*	
		Myxosoma	

(Contd.)

Table (Continued)

Subphylum	Class	Genus	Group
Ciliophora		*Paramecium*	Ciliates
		*Balantidium**	
		Entodium	
		Vorticella	
		Didinium	
		Stentor	
		Tetrahymena	
		Tokophrya	
		Nyctotherus	

* Important Human Pathogens

Helminths (metazoa) are classified into two phyla: Phylum Nemathelminthes, which consists of roundworms that have cylindrical bodies, and Phylum Platyhelminthes consisting of flatworms which have flattened bodies that are leaf-like or resemble a ribbon. It is again classified into trematodes and cestodes.

Examples of medically important helminthes are:

Nematodes

Ascaris lumbricoides

Enterobius vermicularis

Trichinella spiralis

Trematodes

Fasciola hepatica

Fasciola buski

Paragonimus westermani

Schistosoma spp.

Cestodes

Taenia solium

Hymenolepsis nana

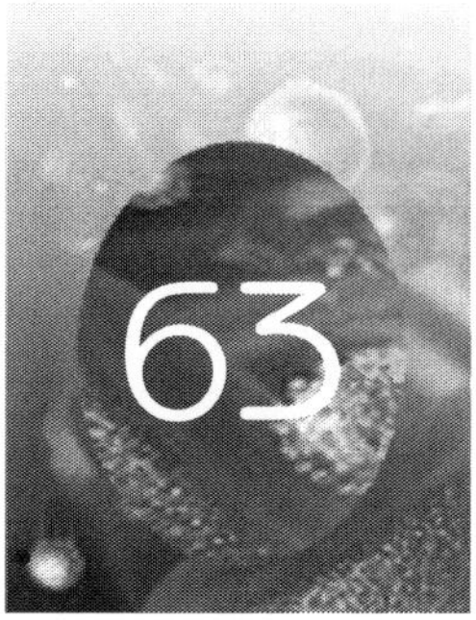

Protozoan Infections

A. AMOEBIASIS

INTRODUCTION

Amoebiasis is one of the intestinal disorders, which affects 10% of the world population. Annual deaths are estimated between 40,000 and 100,000. Amoebiasis is caused by *Entamoeba histolytica.* The taxonomic position of *Entamoeba* is given in Box 63.1.

Phylum	Protozoa
Subphylum	Sarcomastigophora
Superclass	Sarcodina
Class	Rhizopoda
Subclass	Lobosia
Order	Amoebida
Genus	*Entamoeba*
species	*histolytica*

Amoebae are unicellular organisms common in the environment, many are parasites of vertebrates and invertebrates. Relatively few species inhabit the human intestine and only *Entamoeba histolytica* is identified as a human intestinal pathogen.

DISTRIBUTION

- *Entamoeba histolytica* is worldwide in distribution.
- It is commonly available in tropical, subtropical, temperate regions.
- Infection rate is higher in rural and densely populated areas.

HABITAT

- Endoparasite
- Present in man and other mammals
- Lives in mucous layer of colon

Feed involves tissues, bacteria and RBCs, causes fatal and serious disease. The infected individuals discharge mucous and blood in their stool.

STRUCTURE

E. histolytica has a relatively simple life cycle that alternates between trophozoite and cyst stages.

Trophozoite

- The trophozoite is the actively metabolizing, mobile stage.
- Trophozoites vary remarkably in size—from 10 to 60 μm or more in diameter.
- When they are alive, they may be actively motile (unidirectional motility).
- Amoebae are anaerobic organisms and do not have mitochondria.
- The finely granular endoplasm contains the nucleus and food vacuoles, which in turn may contain bacteria or red blood cells.
- The parasite is sheathed by a clear outer ectoplasm.
- Nuclear morphology is best seen in permanent stained preparations.
- The nucleus has a distinctive central karyosome and a rim of finely beaded chromatin lining the nuclear membrane.
- Finger-like pseudopodia are available (Figure 63.1).

Cyst

- The cyst is dormant and environmentally resistant stage.
- The cyst is a spherical structure, 10–20 μm in diameter, with a thin transparent wall.
- Fully mature cysts contain four nuclei with the characteristic amoebic morphology.
- Rod-like structures (chromatoidal bars) are present variably, but are more common in immature cysts.
- Inclusions in the form of glycogen masses may also be present (Figure 63.1).

Other ameobae in the human intestine include:

Entamoeba hartmanni
Entamoeba gingivalis

Entamoeba poleckii
Entamoeba coli
Endolimax nana
Iodamoeba butschlii

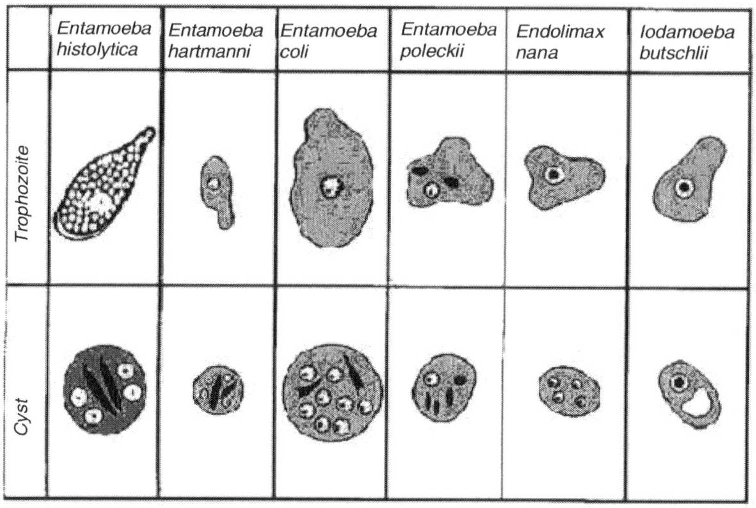

Figure 63.1 Some important amoebae in human

MULTIPLICATION, LIFE CYCLE AND PATHOGENESIS OF *E. HISTOLYTICA*

Encystment occurs apparently in response to desiccation as the amoeba is carried through the colon. After encystment, the nucleus divides twice to produce a quadrinucleate mature cyst. Encysted quadrinucleate cysts of *E. histolytica* are ingested through faecal-oral transmission or through food and water. Ingested cysts of *E. histolytica* excyst in the small intestine. Trophozoites are carried to the colon, where they mature and reproduce. Excystment occurs after ingestion and is followed by rapid cell division to produce four amoebae which undergo a second division. Every cyst thus yields eight tiny amoebae (trophozoites).

Quadrinucleated amoeba adheres to the colonic mucosal cells. The amoeba adherence molecule has been identified as a lectin, which can bind to either of the two common carbohydrate components of cell membrane, galactose and *N*-acetyl galactosamine. Amoeba attacks and kills the host cell. This cytolytic event is a result of incorporation in the host cell membrane of an amoeba-produced, pore-forming protein, amoebapore. This protein forms ion channels in lipid cell membranes and results in cell death within minutes of cell contact with amoeba.

The initial lesion is in the colonic mucosa, most often in the caecum or sigmoid colon. The slow transit of the intestinal contents in these two locations seems to be an important factor in the invasion of the mucosa, because it affords the amoeba greater mucosal contact time and because it permits changes in the intestinal milieu that may facilitate invasion. The initial superficial ulcer may deepen into the submucosa and muscularis to become the characteristic flask-shaped, chronic amoebic ulcer (Amoeboma). Spread may occur by direct extension, by undermining of the surrounding mucosa until it sloughs, or by penetration that can lead to perforation or fistulous communication to other organs or the skin. If the amoebae gain access to the vascular or lymphatic circulation, metastases may occur first to the liver and then by direct extension or further metastasis to other organs, including the brain. If amoebae pass down the colon they encyst under the stimulus of desiccation, and then are evacuated with the stool (Figure 63.2).

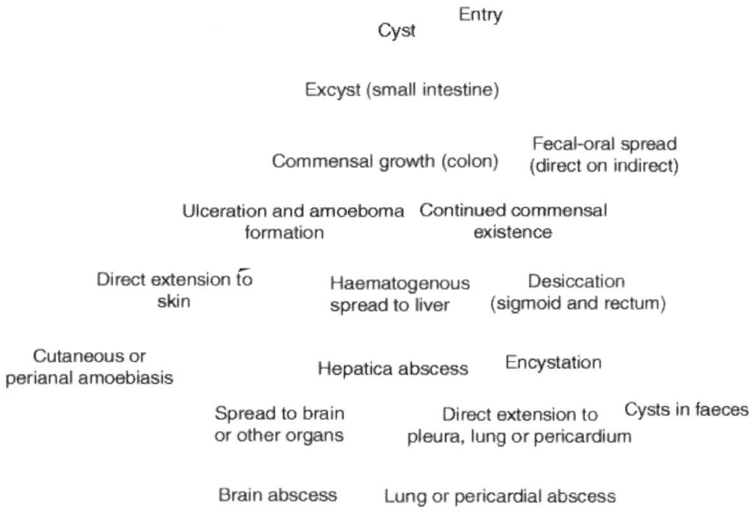

Figure 63.2 Life cycle and pathology of *Entamoeba histolytica*

CLINICAL MANIFESTATION

- Fever
- Amoebic dysentery—fulminant ulceration
- Non-dysenteric gastroenteritis
- Amoeboma formation
- Amoebic colitis
- Hepatomegaly
- Amoebic abscess
- Visceral amoebiasis

DEFENCES

- The gastric acid "barrier" and the steady movement of food through the intestine are non-specific defence mechanisms.
- A role for colonic mucins in protection and depletion of these mucins in infection has been suggested.

EPIDEMIOLOGY

- Faecal-oral transmission occurs when food preparation is not sanitary or when drinking water is contaminated.
- Contamination may come directly from infected food handlers or indirectly from faulty sewage disposal.

LABORATORY DIAGNOSIS

Specimen

- Stool
- Aspirates from intestine and other organs
- Exudates
- Biopsy materials
- Mucous from rectal ulcer

Microscopy

Saline and iodine wet mount

Concentration methods (Refer Practical Section)

The microscopic examination of direct smear has several purposes.

- To assess the worm burden of a patient.
- To provide quick diagnosis of a heavily infected specimen.
- To check motility of organisms.

Serology

Serological methods include:

- Gel diffusion
- Immunoelectrophoresis
- Countercurrent electrophoresis
- Indirect haemagglutination
- Indirect fluorescent antibody

- Skin tests
- Enzyme linked immunosorbent assay (ELISA) and
- Latex agglutination

Cultivation of E. *histolytica*

Fresh stool specimens are used for cultivation.

Egg albumin is the main source for growth. Before preparation of media, the white of four eggs is added to one litre of Ringers solution*.

Entamoeba isolation medium is prepared and poured into the tubes and allowed to settle.

The medium should be kept in an incubator and taken out immediately before inoculation.

Just before inoculation, a little sterile rice starch is added to the medium with the help of inoculation loop.

It will facilitate *Entamoeba* growth. One ml of fresh specimen is inoculated with the help of sterile capillary tube and incubated at 37°C for 2–3 days. The growth is subcultured each and every 2–3 days. All phases of growth are observed in culture media.

Composition of E. *histolytica* Isolation Medium

Ringer's solution

* Sodium chloride	9 g
Potassium chloride	0.2 g
Calcium chloride	0.2 g
Distilled water	1000 ml
pH 7.2–7.8	

Egg white from 4 eggs

CONTROL

Preventive measures are limited to environmental and personal hygiene. Treatment depends on drug therapy. Acute intestinal disease is best treated with metronidazole at a dose of 750 mg three times a day orally for 10 days. In children the dose is 40 mg/kg/day divided into three doses and given orally for 10 days. Iodoquinol at an adult dose of 650 mg orally three times daily for 20 days or diloxanide furoate at an adult dose of 500 mg orally three times daily for 10 days. Amoebic liver abscess is best treated with metronidazole, dihydroemetin, chloroquine or dehydroemetine.

B. GIARDIASIS

Giardiasis is one of the intestinal infections caused by the protozoan *Giardia lamblia*. It was first seen by Leeuwenhoek in 1681 while examining his own stool. Most *Giardia* infections are asymptomatic.

Phylum	Protozoa
Subphylum	Plasmoderma
Class	Mastigophora
Order	Diplomonadida
Genus	*Giardia*

HABITAT

The organism is present in the duodenum and upper part of ileum.

STRUCTURE

The *Giardia* life cycle involves two stages: the trophozoite and the cyst (Figure 63.3). The *G. lamblia* trophozoite is easily recognized under a microscope. It is about 12 to 15 μm long, shaped like tennis racket. The dorsal surface is convex and the ventral surface is concave with a sucking disc, and has two nuclei that resemble eyes, structures called median bodies that resemble a mouth and four pairs of flagella that look like hair; these combine to give the stai ned trophozoite the eerie appearance of a face. The flagella help these organisms to migrate to a given area of the small intestine, where they attach by means of an adhesive disc to the epithelial cells and thus maintain their position despite peristalsis. It is bilaterally symmetrical. The anterior end broad and the posterior end tapers to a sharp point.

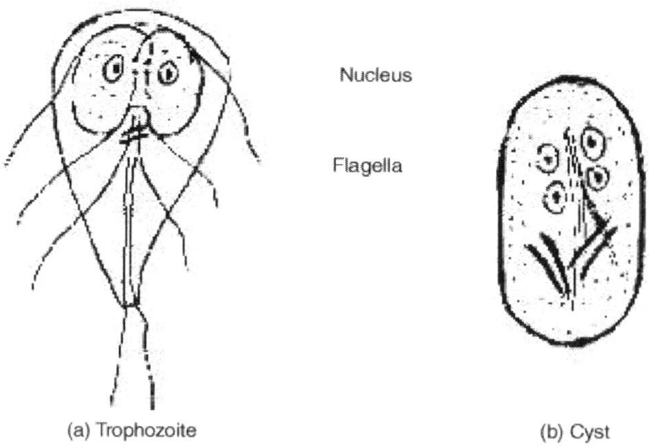

Figure 63.3 *Giardia lamblia*

The *Giardia* cyst—the form usually seen in the faeces—is ovoid, 6 to 12 μm long, and can often be seen to contain two to four nuclei at one end and prominent diagonal fibrils. Flagella and sucking disc are seen inside the cytoplasm.

MULTIPLICATION AND LIFE CYCLE

Giardia infection is acquired by ingesting cysts. The exposure of cysts to host stomach acidity and body temperature triggers excystation, which is completed in the small intestine with the emergence of trophozoites that promptly attach to the host intestinal epithelium.

The trophozoite or the actively metabolizing motile form, lives in the upper two-thirds of the small intestine (duodenum and jejunum) and multiplies by binary fission. Trophozoites that are swept into the faecal stream lose their motility, round up and are excreted as dormant, resistant cysts (Figure 63.4). The mechanisms that cause the signs and symptoms of giardiasis are not known.

Excreted trophozoites disintegrate. The cyst, although not as resistant as many bacterial endospores, is sufficiently hard to survive host to host transfer. For example, some *Giardia* cysts excyst successfully after more than 2 months of storage in water at refrigerator temperatures.

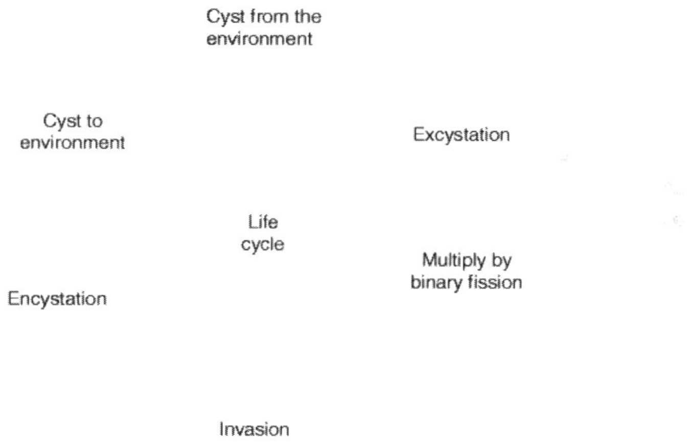

Figure 63.4 Life cycle of *Giardia lamblia*

CLINICAL MANIFESTATION

Diarrhoea or loose, foul-smelling stools, steatorrhoea (fatty diarrhoea), malaise, abdominal cramps, excessive flatulence, fatigue and weight loss.

EPIDEMIOLOGY

Giardia infection occurs worldwide, with an incidence usually ranging from 1.5 to 20 per cent. Higher incidences are likely where sanitary standards are low.

Although people of all ages may harbour these organisms, infants and children are more often infected than are adults.

LABORATORY DIAGNOSIS

In case of giardiasis, cysts are found in the formed stool. Diarrhoeal specimens may also contain trophozoites. Three stool specimens should be obtained at approximately 48-hour intervals.

Examination of these specimens permit detection of the organism in most cases. The chance of finding cysts in a light infection increases if the stool specimen is subjected to a concentration method, such as the zinc sulphate and floatation technique.

CONTROL

Attention to personal hygiene is the key to preventing the spread of giardiasis. Controlling the spread of *Giardia* in drinking water should be possible where community water treatment methods (e.g. disinfection and filtration) are available. For example, iodine and chlorine kill *Giardia* cysts under appropriate conditions.

The drug of choice for treating *Giardia* infections is quinacrine hydrochloride.

C. TRYPANOSOMIASIS

Trypanosomiasis is a common term used to describe the disease caused by the protozoan trypanozoan.

AMERICAN TRYPANOSOMIASIS (CHAGAS' DISEASE)

Chagas' disease is caused by *Trypanosoma cruezi*. It is a parasite of blood plasma. It completes its life cycle in two hosts: man and reduviid bugs.

Phylum	Protozoa
Subphylum	Plasmoderma
Class	Mastigophora
Order	Protomonadida
Genus	*Trypanosoma*
species	*cruezi*

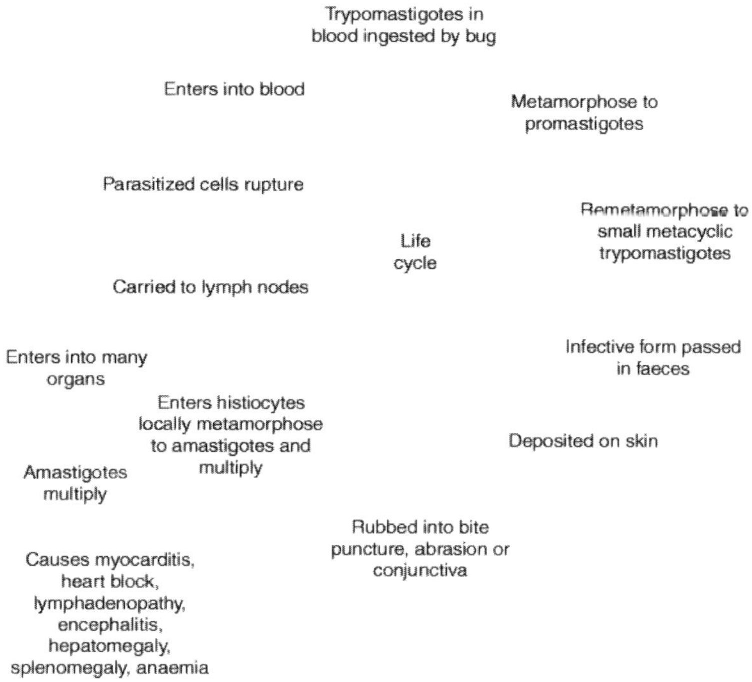

Figure 63.5 Life cycle of *Trypanosoma cruezi*

STRUCTURE

T. cruezi is found in the peripheral blood as a 20-μm trypomastigote. It has a large oval kinetoplast and a poorly developed undulating membrane. In the tissues (mainly heart, skeletal and smooth muscle and reticuloendothelial cells) the parasite occurs as a 3–5-μm amastigote (Figure 63.6). Two morphological forms are found in man.

1. Trypanosomal form—a blood form, appearing as C- or U-shaped structures.
2. Leishmanial form—a tissue form, 2–4 mm in diameter.

MULTIPLICATION AND LIFE CYCLE

In the vertebrate host, multiplication is carried out only by the amastigotes, which divide inside cells or muscle fibres to form groups called pseudocysts.

Trypomastigotes are ingested when the insect takes blood meal from an infected host (Figures 63.5 and 63.6).

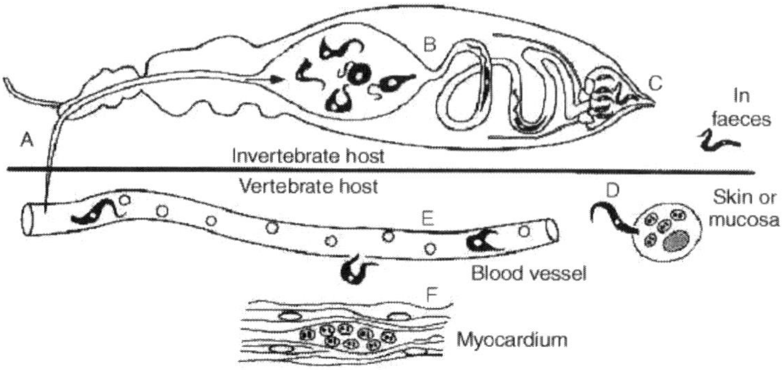

Figure 63.6 Life cycle of *T. cruezi* in the intestine of a triatomine bug and in the vertebrate host. (A) The trypanosomes transform to epimastigotes in the stomach and midgut. (B) Epimastigotes attach to the walls of the rectal sac and produce infective metacyclic trypomastigotes, which are eliminated in faeces (C) and enter the vertebrate host through breaks in the skin. The parasites transform to amastigotes inside local cells (D) and multiply to release blood typanosomes, which invade other tissues (E and F).

PATHOGENESIS

Surface glycoproteins and certain serum factors bound to the parasite may be important in adherence and penetration of cells. Inflammatory reactions at the sites of rupturing pseudocysts can lead to pathological manifestation, such as acute myocarditis and destruction of parasympathetic ganglia of the heart and myenteric plexus, which may cause the changes observed in the chronic phase of illness. Parasite enzymes may also cause cell and tissue damage. Antibodies against endocardial vascular interstitial tissue (EVI), neurons, striated muscle and laminin have been demonstrated. Chronic myocardiopathy may be accompanied by focal myocarditis, extensive fibrosis, myocytolysis, collagenolysis, destruction of nerve fibres and microvascular involvement, which can lead to sudden death.

CLINICAL MANIFESTATION

Chagas' disease begins as a localized infection followed by parasitaemia and colonization of internal organs and tissues. Infection may first be evidenced by a small tumour (chagoma) of the skin. These typical inflammatory reactions are usually accompanied by the swelling of satellite lymph nodes that persist for 1 to 2 months. Symptoms and signs include fever, general oedema, adenopathy, moderate hepatosplenomegaly, myocarditis with or without heart enlargement, and sometimes, meningoencephalitis in children.

LABORATORY DIAGNOSIS

In the early stages of the disease the parasite is demonstrated relatively easily by direct microscopic blood examination (thick and thin blood film examination), by xenodiagnosis (allowing clean, laboratory-reared insects to feed on a suspected victim and later examining the insect faeces), or by culturing the blood. Other specimens include bone marrow and CSF.

Serologic tests like indirect haemagglutination, indirect immunofluorescence, enzyme-linked immunosorbent assay (ELISA) can be used for the diagnosis of Chagas' disease.

CONTROL

Community participation strategies, including education, housing improvements and vector surveillance appear to be more cost-effective than vertical programs for control. Vaccination trials in animals have yielded only partial protection. Live attenuated vaccines apparently are most effective but are too risky for use in humans. In endemic areas, serologic screening in blood banks is important to prevent transmission by transfusion.

AFRICAN TRYPANOSOMIASIS (SLEEPING SICKNESS)

Sleeping sickness (African trypanosomiasis) is caused by *Trypanosoma brucei*. Two morphologically identical strains of *T. brucei* are,

1) *Trypanosoma brucei gambiense* (Gambien trypanosomiasis or Mid- and West-African trypanosomiasis)
2) *Trypanosoma brucei rhodensiense* (Rhodesion trypanosomiasis or East-African trypanosomiasis).

These are parasites of blood plasma.

STRUCTURE

The two subspecies of *T. brucei* are morphologically indistinguishable. They may be pleomorphic, ranging from 12 to 42 μm long (mean 30 μm) and 1.5 to 3.5 µm broad. It has a small kinetoplast and a well-developed undulating membrane. The posterior end is more rounded. It is a curved fusiform organism. The nucleus is large, oval and central in position. When stained with Leishman stain, the cytoplasm and undulating membrane appear blue.

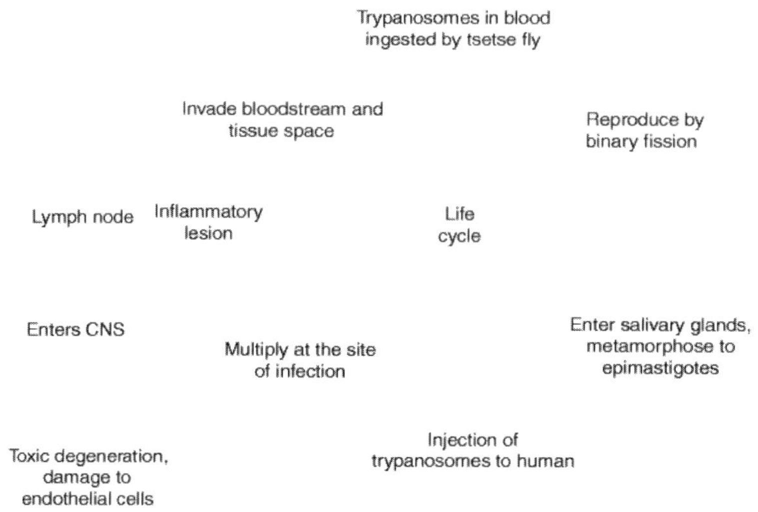

Figure 63.7 Life cycle of *Trypanosoma brucei*

MULTIPLICATION AND LIFE CYCLE

T. brucei multiplies in the blood or cerebrospinal fluid. Trypanosomes ingested by a feeding fly must reach the midgut within a few days, where they reproduce actively as epimastigotes that are attached to the microvilli of the intestine until they transform into metacyclic trypomastigotes, which are found free in the lumen. From the gut through forward motion trypomastigotes accumulate into the ducts of salivary gland of fly. Around 15 to 35 days after infection, the fly becomes infective through its bite (Figure 63.7).

PATHOGENESIS

As the disease progresses, inflammatory changes lead to demyelinating encephalitis. Antibodies against myelin have been detected, suggesting that this condition may have an autoimmune basis. The antigens stimulate high concentrations of IgM antibodies. Periodic changes occur in the surface antigens, thereby circumventing the host's immune responses. On the other hand, the common antigens are liberated in every trypanolytic crisis (episode of trypanosome lysis) and lead to antibody- and cell-mediated hypersensitivity reactions.

CLINICAL MANIFESTATION

After a period of local multiplication, the trypanosomes enter the general circulation via the lymphatics. Recurrent fever, headache, lymphadenopathy and splenomegaly may occur. Later, signs of meningoencephalitis appear, followed by somnolence, cachexia, coma, and death. Enlargement of the posterior cervical chain of lymph nodes (Winterbottom's sign) is more common in *T. b. gambiense* infection.

EPIDEMIOLOGY

Both forms of African trypanosomiasis are transmitted during the day time by the bite of infected tsetse flies (*Glossina* species), which inhabit the open savannah of eastern Africa (*T. b. rhodesiense*) or riverine areas in Western and Central Africa (*T. b. gambiense*). Wild game mammals (bushbuck, hartebeest, lion, hyena) as well as cattle act as reservoirs of *T. b. rhodesiense*.

DIAGNOSIS

In the early stages of the disease, the parasites can be demonstrated in lymph nodes and blood; later, they appear in the cerebrospinal fluid.

Serologic tests such as indirect immunofluorescence, direct card agglutination, and indirect haemagglutination are used successfully for diagnosis.

It can be cultivated in complex synthetic media. The medium most commonly used is NNN (Novy, MacNeal and Niedle) medium, which has a solid phase of rabbit blood agar and a liquid phase of a physiologic salt solution. Liquid media are also available. Only the invertebrate stages appear in such media, and they may or may not be infectious for the vertebrate hosts, depending on the species.

CONTROL

Tsetse fly populations have been reduced successfully by the use of insecticides or traps with an attractant bait plus insecticide. Drugs such as pentamidine and the arsenical suramin are successful in treatment, more recently, eflornithine (difluoromethyl ornithine) are used in advanced disease.

D. LEISHMANIASIS

Leishmaniasis is a general term for diseases caused by species of the genus *Leishmania*, which are transmitted by the bite of infected sandflies. The genus *Leishmania* was created by Ross in 1903.

Phylum	Protozoa
Sub phylum	Plasmoderma
Class	Mastigophora
Order	Protomonadida
Genus	*Leishmania*
species	*L. donovani, L. tropica* and *L. braziliensis*

STRUCTURE, MULTIPLICATION AND LIFE CYCLE

All *Leishmania* species have similar type of morphology and life cycle. The parasite exists in two forms:

1. Aflagellar form—occurs in man and is called leishmanial form.
2. Flagellar form—occurs in the gut of insect (sandfly) and is called leptomonad form.

Leishmanial Form

These forms reside in reticuloendothelial cells. It is a round or oval body measuring 2–4 mm long. Cell membrane is delicate, nucleus is round and is situated in the middle of the cell. Kinetoplasts are rod-shaped.

Leptomonad Form

It is short oval or pear-shaped body measuring 5–10 mm in length. Nucleus is found centrally. Flagellum may be of the same size of the cell.

Multiplication

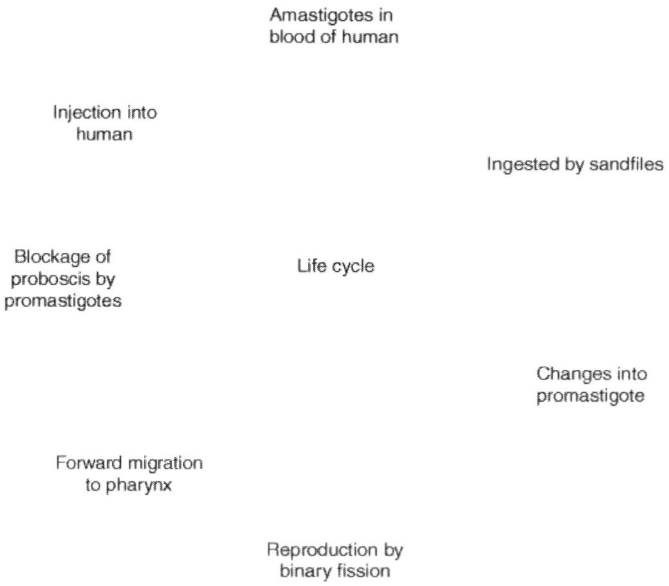

Figure 63.8 Life cycle of *Leishmania* in insects

Parasites ingested by a female sandfly that sucks the blood of an infected person or animal pass into the stomach, transform into promastigotes, and multiply actively. A promastigote form also occurs in sandflies. The parasites finally attach by the flagellum to the walls of the oesophagus, midgut and hindgut of the fly, and some eventually reach the proboscis and are inoculated into a new host. Infective sandflies may become so blocked by parasites that probing alone leads to transmission (Figures 63.8 and 63.9).

Promastigotes (metacyclic forms) from the proboscis of an infected female sandfly are injected into the skin and taken up by local macrophages and metamorphose into amastigotes. Amastigotes are phagocytozed by macrophages but resist digestion. These undergo binary fission and rupture the infected cell. All the cells are sometimes localized as cutaneous form or metastatise in mucosal cell as mucocutaneous form or cells are released into the bloodstream and deposited into reticuloendothelial cells

VISCERAL LEISHMANIASIS (KALA AZAR)

HABITAT

It is an intracellular parasite.

PATHOGENESIS

The parasite can resist the internal body temperature, invade internal organs (liver, spleen, bone marrow, and lymph nodes) where they occupy the reticuloendothelial cells. The pathogenic mechanisms of the disease are not fully understood, but, clearly, in those organs that exhibit marked cellular alteration, hyperplasia of histiocytes leads to hypertrophy.

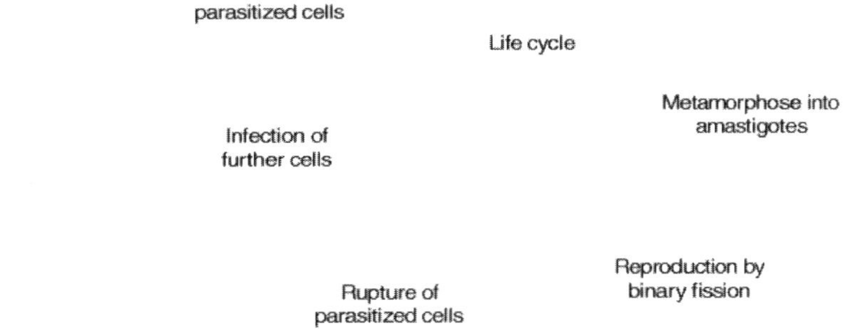

Figure 63.9 Life cycle of *Leishmania* in humans

Leishmania donovani is the important species that causes kala azar.

CLINICAL MANIFESTATION

Visceral leishmaniasis begins with a nodule at the site of inoculation. This lesion rarely ulcerates and usually disappears spontaneously in a few weeks or months. Symptoms and signs are undulating fever, malaise, diarrhoea, splenomegaly, hepatomegaly, lymphadenopathy, emanciation, anaemia and leucopenia. Infiltrative or nodular lesions of the skin may appear after treatment (post-kala-azar dermal leishmaniasis—black colour skin).

LABORATORY DIAGNOSIS

The typical symptoms, particularly hepatosplenomegaly and the pancytopenia, strongly suggest visceral leishmaniasis. The parasite usually can be demonstrated in stained or cultured bone marrow or spleen material. Serologic tests (ELISA, immunofluorescent antibody) are useful.

Direct Evidence

Peripheral Blood Examination (thick and thin film) (*See* Practical Section)

Blood Culture

Leishmania donovani is cultured in a medium composed of two parts of salt agar and one part of rabbit blood. It was first introduced by Novy and MacNeal, later modified by Nicolle and is commonly referred to as the NNN medium. The medium is incubated at 22–24°C.

Salt Agar Composition

Sodium chloride	6 g
Water	900 ml
Agar	14 g

Four tubes are inoculated with a drop of specimen. From the 10th day onwards the tubes are examined for leptomonads.

Other tests include the following.

1. *Spleen puncture examination* This is performed when the spleen is considerably enlarged. The sterilized needle is introduced into the splenic area. Splenic pulp and blood is collected by aspiration. The needle is quickly withdrawn. Few drops of splenic aspirate are placed to identify the slide and a thin film is prepared. It is stained with leishman stain to identify amastigotes. Aspirate is also inoculated on to NNN medium. Medium is looked for promastigote after incubation.

2. *Bone marrow examination* Needle aspiration method is used to collect bone marrow specimen. Sample is looked for protozoa by thin film leishman staining procedure.

3. *Complement fixation test*

4. *Leishmanin test*—0.1 ml of killed leptomonad culture is injected intradermally and observed for induration.

CUTANEOUS AND MUCOCUTANEOUS LEISHMANIASIS

Leishmania tropica and *L. braziliensis* are the important species that can cause cutaneous and mucocutaneous leishmaniasis.

STRUCTURE

All species of *Leishmania,* parasitic in man are morphologically similar and appear as intracellular amastigotes 3 to 6 μm long by 1.5 to 3 μm in diameter. Promastigotes develop in the intestine of the sandfly.

PATHOGENESIS

Promastigotes (metacyclic forms) from the proboscis of an infected female sandfly are injected into the skin and taken up by local macrophages. Lesions of oriental sore and ulcer normally resolve spontaneously after a few months. Nevertheless, in the latter, destructive and chronic lesions of the pinna of the ear are observed in 50 to 60 per cent of the patients. In *L. panamensis* and *L. guyanensis* infections, lesions usually become chronic, sometimes with lymphatic compromise and haematogenous dissemination. In *L. braziliensis* infection, highly destructive spread to the oral or nasal mucosa frequently occurs.

CLINICAL MANIFESTATION

The lesions of cutaneous and mucocutaneous leishmaniasis are limited to the skin and mucous membranes. Cutaneous leishmaniasis appears 2 to 3 weeks after the bite of an infected sandfly as a small cutaneous papule. This lesion slowly grows, becoming indurated and often ulcerated, and develops secondary infection. Secondary or diffuse lesions may develop. The disease is occasionally self-limiting but usually chronic. Leishmaniasis from a primary skin lesion may involve the oral and nasopharyngeal mucosa.

LABORATORY DIAGNOSIS

Peripheral blood examination (thick and thin film) (Refer Practical Section)

Spleen Puncture Examination (Refer page 496)

CONTROL

Leishmaniasis transmitted in or near houses can be prevented with insecticides. Pentavalent antimonials such as sodium antimony gluconate (Pentostam) and meglumine antimoniate (Glucantime) are available for treatment. Amphotericin B has been used in cases with mucosal involvement and in cases of diffuse disease where antimonial therapy fails.

E. MALARIA

INTRODUCTION

Malaria has been a major disease of mankind for thousands of years. Malaria is caused by the protozoan of the genus *Plasmodium.* Four species cause disease in humans: *P. falciparum, P. vivax, P. ovale* and *P. malariae.* Malaria is spread to humans by the bite of female mosquitoes of the genus *Anopheles*. *P. falciparum* and *P. vivax* account for the vast majority of cases. *P. falciparum* causes the most severe disease.

STRUCTURE AND LIFE CYCLE

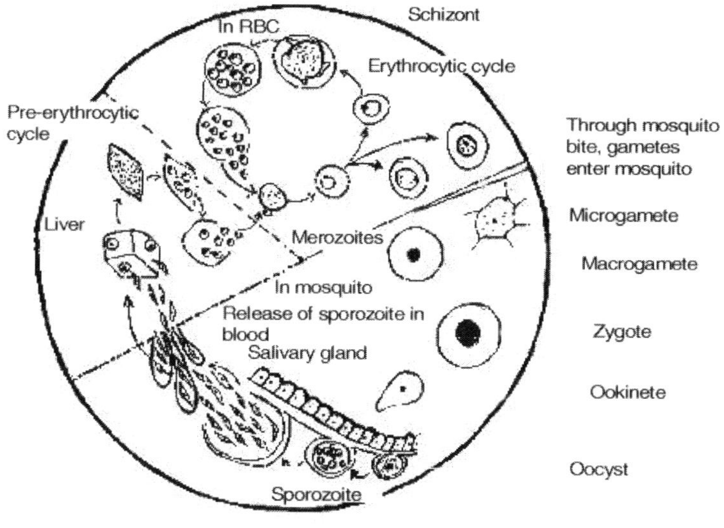

Figure 63.10 Pathophysiology of malarial parasite

Plasmodia pass through a number of stages in the course of their two-host life cycle. The stage infective for humans is the uninucleate, lancet-shaped sporozoite (approximately $1 \times 7\ \mu m$). Sporozoites are produced by sexual reproduction in the midgut of vector anopheles mosquitoes and migrate to the salivary gland. When an infected *Anopheles* mosquito bites a human, she may inject sporozoites along with saliva into small blood vessels. Sporozoites are thought to enter liver parenchymal cells within 30 minutes of inoculation. In the liver cell, the parasite develops into a spherical, multinucleate liver-stage schizont which contains 2,000 to 40,000 uninucleate merozoites. This process of enormous amplification is called exoerythrocytic schizogony. This exoerythrocytic or liver phase of the disease usually takes between 5 and 21 days, depending on the species of *Plasmodium*. However, in

P. vivax and *P. ovale* infections, maturation of liver-stage schizonts may be delayed for as long as 1 to 2 years. These liver-phase parasites are called hypnozoites (Figure 63.10).

The mature schizonts eventually rupture, releasing thousands of uninucleated merozoites into the bloodstream. Each merozoite can infect a red blood cell. Within the red cell, the merozoite develops to form either an erythrocytic stage (blood-stage) schizont (by the process of erythrocytic schizogony) or a spherical or banana-shaped, uninucleate gametocyte. The mature erythrocytic stage schizont contains 8 to 36 merozoites, each 5 to 10 μm long, which are released into the blood when the schizont ruptures. These merozoites proceed to infect another generation of erythrocytes.

The gametocyte, which is the sexual stage of the *Plasmodium*, is infectious for mosquitoes that ingest it while feeding. Within the mosquito, gametocytes develop into female and male gametes (macrogametes and microgametes, respectively), which undergo fertilization and then develop over 2 to 3 weeks into sporozoites that can infect humans. The delay between infection of a mosquito and maturation of sporozoites means that female mosquitoes must live a minimum of 2 to 3 weeks to be able to transmit malaria. This fact is important in malaria control efforts.

PATHOGENESIS

Clinical illness is caused by the erythrocytic stage of the parasite. No disease is associated with sporozoites, the developing liver stage of the parasite, and the merozoites released from the liver, or gametocytes.

The first symptoms and signs of malaria are associated with the rupture of erythrocytes when erythrocytic stage schizonts mature. This release of parasite material presumably triggers a host immune response. The cytokines, reactive oxygen intermediates and other cellular products released during the immune response play a prominent role in pathogenesis, and are probably responsible for fever, chills, sweats, weakness and other systemic symptoms associated with malaria. In the case of falciparum malaria (the form that causes most deaths), infected erythrocytes adhere to the endothelium of capillaries and postcapillary venules, leading to obstruction of the microcirculation and local tissue anoxia. In the brain this causes cerebral malaria; in the kidneys it may cause acute tubular necrosis and renal failure; and in the intestines it can cause ischemia (deficiency of blood) and ulceration, leading to gastrointestinal bleeding and to bacteraemia secondary to the entry of intestinal bacteria into the systemic circulation.

CLINICAL MANIFESTATION

The most characteristic symptom of malaria is fever. Other common symptoms include chills, headache, myalgias, nausea and vomiting. Diarrhoea, abdominal pain and cough are occasionally seen.

The first is a 15 to 60-minute cold stage characterized by shivering and a feeling of cold.

In the second stage there is a 2 to 6-hour hot stage, in which there is fever, sometimes reaching 41°C, flushed, dry skin, and often headache, nausea and vomiting.

Finally, there is a 2 to 4-hour sweating stage during which the fever drops rapidly and the patient sweats.

In all types of malaria, the periodic febrile response is caused by rupture of mature schizonts. In *P. vivax* and *P. ovale* malaria, a brood of schizonts matures every 48 hours, so the periodicity of fever is Tertian (tertian malaria), whereas in *P. malariae* disease, fever occurs every 72 hours (quartan malaria). The fever in falciparum malaria may occur every 48 hours, but is usually irregular, showing no distinct periodicity.

HOST DEFENCES

- Antibodies to sporozoites can prevent invasion of hepatocytes.
- Interferon gamma and other cytokines can inhibit development of liver stage.
- Cytotoxic T lymphocytes can lyse infected hepatocytes.
- Antibodies to merozoites can prevent invasion of erythrocytes.
- Cytokines can destroy infected RBCs.
- Acquired immunity can also protect against malarial infection and the development of malarial disease.

LABORATORY DIAGNOSIS

Diagnosis of malaria generally requires direct observation of malarial parasites in Giemsa-stained thick and thin blood smears. Thick blood smears are more difficult to interpret than thin blood smears but they are much more sensitive, as more blood is examined. Thin blood smears, in which parasites are seen within erythrocytes, are used to determine the species of the infecting parasite.

New diagnostic methods include a rapid antigen-capture dip stick test and a technique for detecting parasites with a fluorescent stain. Both of these tests are fast, easy to perform and are highly sensitive and specific.

Other diagnostic methods include assays to detect malarial antibodies and antigens, and polymerase chain reaction/DNA and RNA probe techniques. These techniques are used primarily in epidemiological studies and immunization trials and rarely in the diagnosis of individual patients.

CULTIVATION

- Modified Bars and Johns method is used for cultivation.
- A flask of 50 ml capacity is taken, in which 20–30 glass beads are inserted and plugged with cotton.

- 5 ml of the patient's blood is added and defibrinated by gentle rotation.
- Into a sterile test tube, a drop of 50% dextrose is added at the bottom by means of capillary pipette. Defibrinated blood is then added so as to make 2.5 cm depth.
- The tube is incubated vertically in an incubator at 37°C.
- The column of blood settles in three layers. (Clear plasma is seen at the top, thin layer of leucocytes in the middle and RBCs settle to the bottom.)
- At the end of 12–24 hours, thin blood smears are prepared from RBC layer and stained with Leishman's stain.
- Above mentioned method was improved by Dr. Block by adding 20–30 ml of blood with 0.1 ml of 50% of dextrose solution.
- Development of gametocytes has been observed after 72 hours.

CONTROL

Drugs used for treatment are primaquine, chloroquine, mefloquine, quinine, quinidine, pyrimethamine-sulphadoxine, doxycycline, halofantrine, artemisinin, proguanin.

Prevention

Individuals should avoid contact with the mosquito by wearing protective clothing.

An insect repellent containing *N,N*-diethyl *m* toluamide (DEET) should be used.

Insecticide-impregnated bednets should be used for sleeping.

For most of the travellers, mefloquine is the drug of choice and doxycycline is an acceptable alternative.

Chloroquine plus proguanil is another possible regimen for chloroquine-resistant areas, but this regimen is much less effective than mefloquine or doxycycline.

Prophylaxis with chloroquine or mefloquine should begin 2 weeks before entering the malarious area.

1. Differentiate between the cyst and trophozoites of *E. histolytica*.
2. How do you cultivate *Entamoeba* and *Plasmodium*?
3. Describe the four major classes of protozoa.
4. Explain the pathogenic protozoa classification.
5. Explain about saline and Iodine wet mount Giemsa staining.
6. Discuss about cultivation of *Leishmania* and *Trypanosoma*.
7. Explain the life cycle and pathology of amoebiasis.

8. Describe about *Giardia lamblia* infection.
9. Discuss various stages of malarial infection.
10. Explain the common and differentiation features of blood and tissue forming *Leishmania* and *Trypanosoma*.
11. Write short notes on
    - i. Amoebiasis
    - ii. Chagas disease
    - iii. Giardiasis
    - iv. Haemoflagellate
    - v. Leishmania
    - vi. Trypanosome
    - vii. Amastigote
    - viii. Guadinucleated cyst
    - ix. Trophozoite
    - x. NNN
    - xi. Tsetse fly
    - xii. Sand fly
    - xiii. Metazoa
    - xiv. Tertian malaria
    - xv. Quartan malaria

Gow, N.A.R. and Gadd, G.M. (eds.). *The Growing Fungus*. Chapman and Hall, London. 1995.

Kwon-Chung, K.J. and Bennett, J.E. *Medical Mycology*. Lea and Febiger, Philadelphia. 1992.

Latge, J.P. and Boucias, D. (eds.). *Fungal Cell Wall and Immune Response*. Springer-Verlag, New York. 1991.

Odds, F.C. *Candida and Candidiasis*, 2nd edn. Bailliére Tindall, Philadelphia. 1988.

Beaver, P.C., Jung, R.C. and Cuppa, W.W. *Clinical Parasitology*. 9th edn. Lea & Febiger, Philadelphia. 1984.

Spice, W., Cruz-Reyes, J. and Ackers, J. *Entamoeba histolytica. Molecular and Cell Biology of Opportunistic Infections in AIDS*. Chapman & Hall, London.1922.

World Health Organization: Amoebiasis: Report of a WHO Expert Committee. Who Tech Set. 1969.

Brener, Z. and Krettli, A.U. "Immunology of Chagas' disease." In: Wyler, D.J. (ed.). *Modern Parasite Biology: Cellular, Immunological and Molecular Aspects*. WH Freeman, New York. 1990.

Pentreath, V.W. "Neurobiology of sleeping sickness." *Parasitol. Today*. 5: 215. 1989.

Petry, K. and Eisen, H. "Chagas's disease: A model for the study of autoimmune diseases." *Parasitol Today*. 5: 111. 1989

World Health Organization: Control of the leishmaniasis. Report of a WHO Expert Comittee. WHO Tech. Rep. Ser. 793. 1990.

Beadle, C., Long, G.W., Weiss, W.R. *et al*. "Diagnosis of malaria by detection of *Plasmodium falciparum* HRP-2 antigen with a rapid dipstick antigen-capture assay." *Lancet.* 343: 564. 1994.

Bruce Chwatt, L.J. *Essential Malariology*, 3rd edn. Edward Arnold, Boston. 1993.

Centers for Disease Control: Health Information for International Travel 1993. HHS Publication (CDC) 938280. Washington, DC. 1993.

Good, M.F. and Saul, A.J. (eds.) *Molecular Immunological Considerations in Malaria Vaccine Development.* CRC Press, Boca Raton. 1993.

www.protozoa.org

www.parasit.org/e.his/

en.wikipedia.com

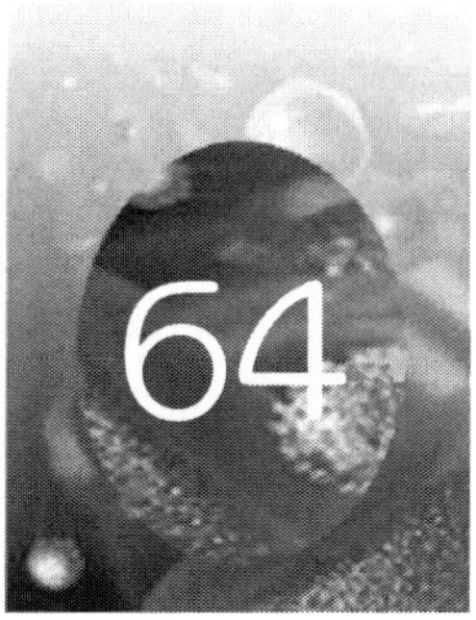

Nematode Infections

INTRODUCTION

Nematodes are helminthic roundworms, the adults of which are characterized by a tapered, cylindrical body, with longitudinally oriented muscles and a triradiate oesophagus. Enteric nematodes are among the most common and widely distributed animal parasites of humans. There are two categories of nematodes:

1. The intestinal nematodes

   They are again classified into two:

   i. Those that infect human in the egg stage, e.g. *Ascaris lumbricoides, Trichuris trichiura, Enterobius vermicularis.*

   ii. Those that infect human in the larval stage, e.g. *Necator americanus, Ancylostoma duodenale, A. braziliense* and *A. caninum.*

2. Tissue nematodes

   Wuchereria bancrofti—Bancrofti filariasis

   Brugia malayi—Malayan filariasis

   Brugia timori—Timorian filariasis

   Onchocerca volvuls—Onchocerciasis

   Loa loa—Loiasis

ASCARIASIS

Ascaris lumbricoides

It is also called large intestinal round worms. WHO has estimated 800 to billion infections worldwide.

STRUCTURE

Ascaris lumbricoides is the largest and most common intestinal nematode of humans. Females are approximately 30 cm long; sexually mature males are smaller. The diameter varies from 2 to 6 mm. Mated females produce fertile eggs that are oval to subspherical, 45 to 75 μm by 35 to 50 μm, and are covered by a thick shell with a light brown, mammillated, albuminous outer coat. Unmated females produce unfertilized eggs that are thin-shelled, ellipsoidal, and measure 78 to l05 μm by 38 to 55 μm. The mammillated coat of unfertilized eggs is irregular and the contents are granular and disorganized (Figure 64.1).

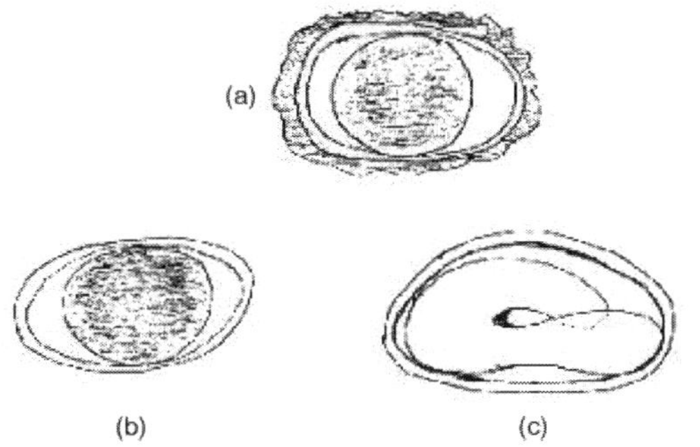

Figure 64.1 *Ascaris lumbricoides* (a) fertilized egg (b) non-fertilized egg (c) embryonated

MULTIPLICATION AND LIFE CYCLE

Infection begins when the embryonated eggs are ingested with food or drink.

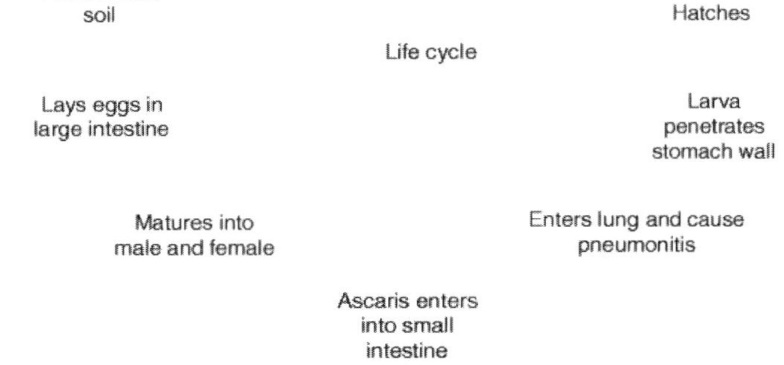

Figure 64.2 Life cycle of *Ascaris lumbricoides*

A. lumbricoides is found in the small intestine, particularly the jejunum. After ingestion by man, the eggs pass to the duodenum where they hatch; the released larvae penetrate the intestinal mucosa, enter the lymphatics and portal system, and are carried to the liver, heart and lungs. This migratory phase requires a few days. The larvae then break out of the capillaries into the alveoli, pass up the respiratory tree, and are swallowed. They reach the intestine and continue their development 8 to 12 weeks after infection. They become sexually mature adults. The adults live for about a year and are subsequently passed in the faeces. Females produce as many as 240,000 eggs per day and as many as 65 million in a lifetime. The eggs are unsegmented and are passed in the faeces. In moist, warm, shady soil, the eggs embryonate and an infective larva develop within the egg in about 3 weeks (Figure 64.2).

Hands carry infective ova from soil contaminated with human excreta, vegetables, etc.

PATHOGENESIS

The initial pathology is associated with migrating larvae; it can cause pneumonitis with eosinophilic infiltration and granuloma formation. The reactions lead to pneumonitis and a condition known as Loeffler's syndrome. Adult worms may cause blockage of the intestines. Acute pancreatitis and biliary stones may occur. The rare fatalities usually result from intestinal obstruction or biliary ascariasis.

CLINICAL MANIFESTATION

Abdominal discomfort, nausea, vomiting, weight loss, fever, and diarrhoea are the symptoms. Allergic manifestation in hypersensitized persons lead to pneumonitis, cough, low-grade fever and eosinophilia.

EPIDEMIOLOGY

A. lumbricoides is distributed widely in tropical and subtropical areas, especially in the developing countries of South America, Africa and Asia. More than one billion infections are estimated to exist at any given time. In rural areas of Asia, it is not unusual to find 85 per cent of the population passing ascaris eggs.

LABORATORY DIAGNOSIS

Definitive diagnosis requires finding characteristic eggs in faeces. Concentration techniques involving floatation or sedimentation of eggs also may be used.

CONTROL

The most effective method to control ascariasis, as well as other soil-transmitted helminthiasis, is sanitary disposal of faeces. Mebendazole, the drug used, is effective against numerous intestinal nematode infections and causes few side effects. Levamisole is also useful, as are pyrantel pamoate, piperazine citrate, thiabendazole and albendazole.

ENTEROBIASIS

Enterobius vermicularis

Enterobius vermicularis causes enterobiasis. It is also called seat worm.

STRUCTURE

It is small and white in colour. It is more or less spindle-shaped with characteristic cephalic swellings and a large muscular oesophagus with a large posterior bulb. Females are approximately 1 cm long and males are half that size. The curved posterior end of male worms has a single copulatory spicule. The males are rarely seen because they die shortly after copulation and are expelled. The eggs are thin-shelled, ovoid, flattened on one side, and measure 50 to 60 μm by 20 to 30 μm.

MULTIPLICATION AND LIFE CYCLE

Ingested eggs hatch in the small intestine, each releasing a larva. These larvae migrate into the large intestine and mature there. Many of the larvae are found at the base of the intestinal crypts. The gravid, female worms migrate out of the anus at night when the anal sphincter is relaxed and lay eggs that adhere to the perianal skin. The female essentially ruptures, releasing as many as 10,000 eggs. The eggs embryonate and

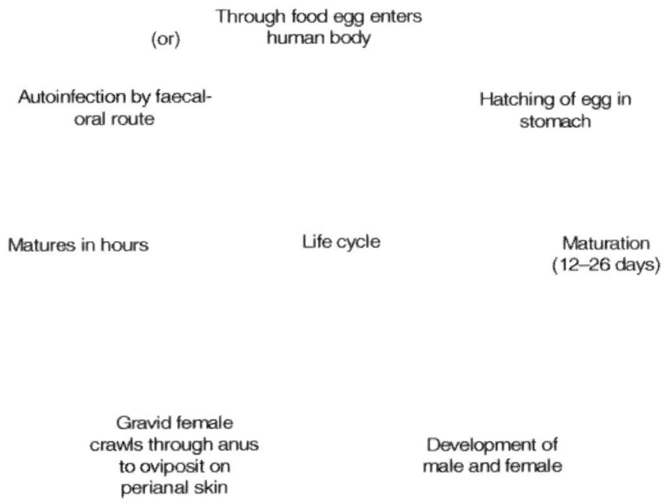

Figure 64.3 Life cycle of *Enterobius vermicularis*

become infective within a few hours after being deposited onto the skin. Infection is transmitted from hand to mouth. The ingested eggs hatch in the small intestine, each releasing an infective stage larva. The parasite moves to the caecum and matures into an adult in 2 to 4 weeks after infecting the host. Infections are self-limited; reinfections can occur (Figure 64.3).

PATHOGENESIS

The significant pathology is the irritation caused by the gravid females around the anus. The migrating females often enter into the female genital tract and female urethra causing inflammation. These worms may enter into the peritoneal cavity through the fallopian tube.

Symptoms may include an eczematous condition around the anus and perineum, salpingitis, nocturnal enuresis and sometimes inflammation.

CLINICAL MANIFESTATION

Enterobiasis or pinworm infection, usually causes mild disease. The most common symptom is perianal itching which disturbs sleep and in children, may be responsible for loss of appetite. Abdominal pain and irritability may also be signs of enterobiasis. The parasite has been suspected as a cause of appendicitis.

LABORATORY DIAGNOSIS

Children suffering sleepless nights because of perianal itching often have pinworms. Eggs are rarely found in the faeces, and the diagnosis is made by finding eggs on perianal swabs made of scotch tape. The tape is pressed first onto the perianal region and then onto a microscope slide, and is examined microscopically.

CONTROL AND TREATMENT

When an infection is recognized, efforts should be made to improve personal hygiene. Fingernails should be cut short, the perianal region washed in the morning, and bedding and sleeping garments washed daily. Other members of a patients' family should be checked; the entire family may need treatment to eliminate infection. Although several antihelmintics are effective in treating enterobiasis, the drugs presently recommended are pyrvinium pamoate, pyrantel pamoate and mebendazole. It is advisable to re-treat the patient one month later.

FILARIASIS

It is also called elephantiasis. It is caused by filarial worm. The filariae are thread-like parasitic nematodes (roundworms) that are transmitted by arthropod vectors. The adult worms inhabit specific tissues where they mate and produce microfilariae, the characteristic tiny, thread-like larvae. Filarial diseases are a major health problem in many tropical and subtropical areas. The disease produced by a filarial worm depends on the tissue locations preferred by adults and microfilariae.

Wuchereria bancrofti and *Brugia malayi* are the lymphatic filarial worms that cause filariasis.

STRUCTURE AND CLASSIFICATION

Adult *Wuchereria* and *Brugia* are elongated and slender (30 to 100 mm by 100 to 300 μm); males are about half the size of females. Microfilariae have the diameter as that of a red blood cell and 250 to 300 μm long. They are enclosed in a characteristic sheath (Figure 64.4).

Figure 64.4 Structure of microfilaria

MULTIPLICATION AND LIFE CYCLE

Definitive host : Human

Intermediate host : Mosquito

Microfilariae ingested by a vector mosquito migrate out of the midgut to the thoracic muscles, where they develop, molt several times, and finally migrate to the mouthparts of the mosquito as infective larvae. As the infective mosquito feeds on another host, the larvae leave the proboscis and enter the puncture wound made by the mosquito. Larvae quickly migrate to the lymphatics, where they mature, mate, and produce microfilariae in the new host. The period from time of infection until detection of microfilariae in the blood varies from 3 to many months. (Figure 64.5).

PATHOGENESIS

Infective larvae from a feeding vector mosquito migrate to the regional lymphatic vessels and by inducing a host inflammatory reaction cause the eventual blockage and oedema characteristic of *W. bancrofti* and *B. malayi* infections. The pathology varies greatly from one individual to another, and the exact mechanisms are not completely understood. The host reaction to the parasite is considerable and worsens when the worms molt, when the females first begin to produce microfilariae and when the worms die and degenerate. Lymphatic vessels are often partially or completely blocked by lymph thrombi, by masses of dead worms, or by endothelial proliferation, fibrin deposition and granulomatous reactions. Lymph stasis favours secondary bacterial and mycotic infection. The initial inflammation of regional lymph nodes and major lymphatic vessels may be followed by a prolonged asymptomatic period and then by recurring attacks of lymphangitis and "filarial fever" over a period of years. Repeated exposure appears to lead to production of abnormally large amounts of collagenous material and fibrosis of the tissue around the affected lymphatics.

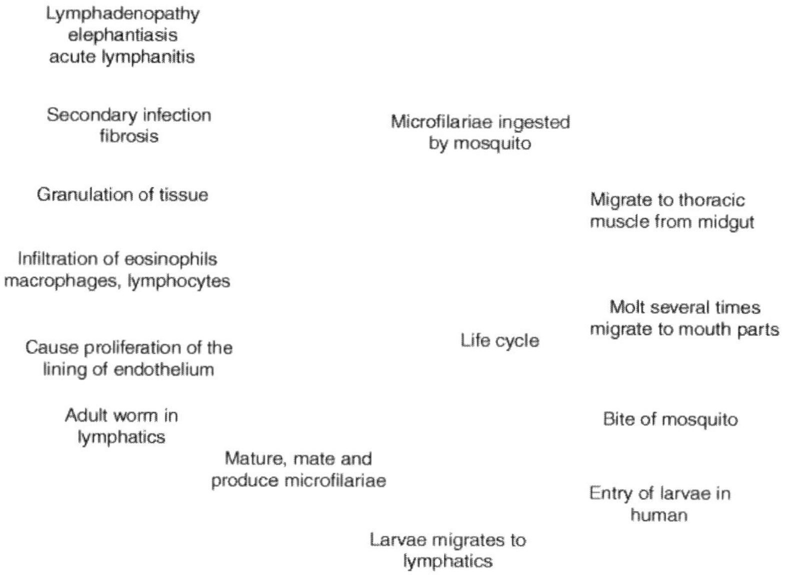

Figure 64.5 Life cycle of *Wuchereria bancrofti*

CLINICAL MANIFESTATION

Early symptoms of lymphatic filariasis consist of intermittent fever and enlarged, tender lymph nodes. The lymphatic vessels that drain into the lymph nodes and those that harbour the developing and adult worms also become inflamed and painful. In more chronic infections, there may be pain also in the epididymis and testes.

Elephantiasis is the end result of *W. bancrofti* infection when most of the mature worms die. The affected part becomes enormously enlarged producing a tumour-like solidity.

EPIDEMIOLOGY

W. bancrofti is prevalent in many parts of the tropics and subtropics. Species of three major genera of mosquitoes that serve as vectors are the common house mosquito

Culex in many urban centres, *Aedes* species in the South Pacific islands and *Anopheles* species in the more isolated rural areas.

LABORATORY DIAGNOSIS

- Enlarged and tender lymph nodes.
- Definitive diagnosis may be accomplished by identifying microfilariae in thick blood smears. Species identification is based on the presence of a sheath and the position of terminal nuclei and the size of the cephalic space. Because of the nocturnal periodicity of microfilariae, blood smears are better made at night when microfilarial levels are usually higher.

CONTROL AND TREATMENT

Attempts to reduce the prevalence of lymphatic filariae include vector control and mass treatment campaigns using diethylcarbamazine citrate. This drug significantly reduces the level of microfilariae in the blood. Ivermectin, a drug that has recently been shown to be effective in the treatment of onchocerciasis, is being evaluated for use in lymphatic filariasis.

1. Describe about the eggs of *Ascaris lumbricoides*.
2. Discuss the types of nematodes with examples.
3. Describe in detail the various nematode infections.
4. Write short notes on
   i. Ascariasis
   ii. Enterobiasis
   iii. Filariasis

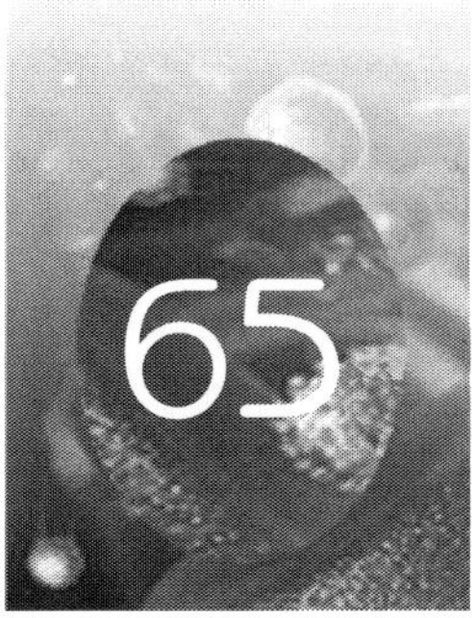

Trematode Infections

INTRODUCTION

Trematodes or flukes, are parasitic flatworms with unique life cycles involving sexual reproduction in mammalian and other vertebrate definitive hosts and asexual reproduction in snail intermediate hosts. These organisms are divided into four groups on the basis of their final habitats in humans:

- The hermaphroditic liver flukes which reside in the bile ducts and infect humans on ingestion of watercress (*Fasciola*) or raw fish (*Clonorchis* and *Opisthorchis*);
- The hermaphroditic intestinal fluke (*Fasciolopsis*), which infects humans on ingestion of water chestnuts;
- The hermaphroditic lung fluke (*Paragonimus*), which infects humans on ingestion of raw crabs or crayfish; and
- The bisexual blood flukes (*Schistosoma*), which live in the intestinal or vesical (urinary bladder) venules and infect humans by direct penetration through the skin.

Flukes do not multiply in humans, so the intensity of infection is related to the degree of exposure to the infective larvae.

Fasciola hepatica or Liver fluke infection

Fasciola hepatica infection has two distinct phases: The first occurs in the initial 6 to 9 weeks of infection when the larvae migrate through the liver; the second begins when they enter the bile ducts. The acute clinical syndrome is characterized by prolonged fever, pain in the right hypochondrium, and sometimes hepatomegaly and urticaria; also marked eosinophilia usually occurs during this period. *F. hepatica* infection is also called liver rot.

LIVER FLUKES

F. hepatica is a hermaphroditic liver fluke that infects mainly ruminants but may incidentally, infect humans. It resides in the bile ducts of liver, where it produces large number of eggs daily for many years. The eggs pass into the lumen of the small intestine and leave the body in the faeces. If deposited in fresh water, the eggs hatch into ciliated form, which upon penetration of the correct species of snail, undergo several developmental stages and produce large numbers when ingested by the definitive host, the larvae excyst, penetrate through the gut wall into the peritoneal cavity, enter the liver capsule, and begin to wander through the hepatic tissues. This migratory phase continues for about 7 weeks. The half-grown flukes then enter the bile ducts, mature and begin to produce eggs (Figure 65.1).

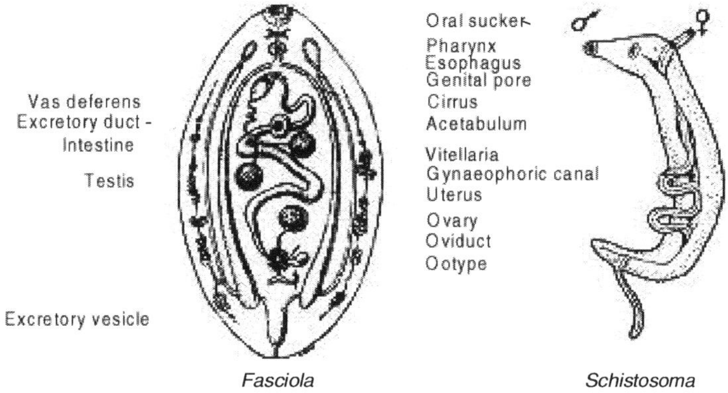

Figure 65.1 Internal structure of *Fasciola* and *Schistosoma*

The acute clinical syndrome of *F. hepatica* infection occurs during migration of the fluke larvae in the liver parenchyma, with resultant inflammation and localized destruction of liver cells. Once in the bile ducts, liver flukes of all species produce inflammation due to mechanical irritation and toxic secretions. Secondary infection occurs due to biliary stasis, resulting in chronic recurrent, suppurative infections; cholangiocarcinoma may also occur.

F. hepatica occurs worldwide in ruminants and causes significant morbidity and mortality in sheep and cattle. For the most part, human infection is sporadic; only a few hundred cases have been reported in the world literature, usually associated with ingestion of wild watercress. Fascioliasis in humans has almost always been identified during the acute migratory stage of infection; occasionally, worms are found in the bile ducts at surgery or autopsy. Foci of chronic human fascioliasis have been found in rural areas of Peru.

LIFE CYCLE

I. Definitive host: Humans, ruminants, dogs, cats and other companion animals
   1. Operculated type-mature trematode egg with miracidium
   2. Free swimming miracidium-stage infective to snail intermediate host

II. First intermediate hosts: Snail of the genera *Bulimus* spp. *Pseudofossarulus* spp. *Lymnaea* spp. *Bithynia* spp. and others.
   3. Sporocyst
   4. Mother/daughter redia
   5. Free swimming cercariae

III. Second intermediate host: (if applicable, i.e., *Opisthorchis* spp.) encysted in the tissue of cyprinid—freshwater fish.
   6. Cercariae encysted in the fish's soft tissue

IV. Infective stage to the definitive host
   7. Encysted metacercariae on aquatic/semi-aquatic vegetation, i.e., watercress, fodder (Figure 65.2).

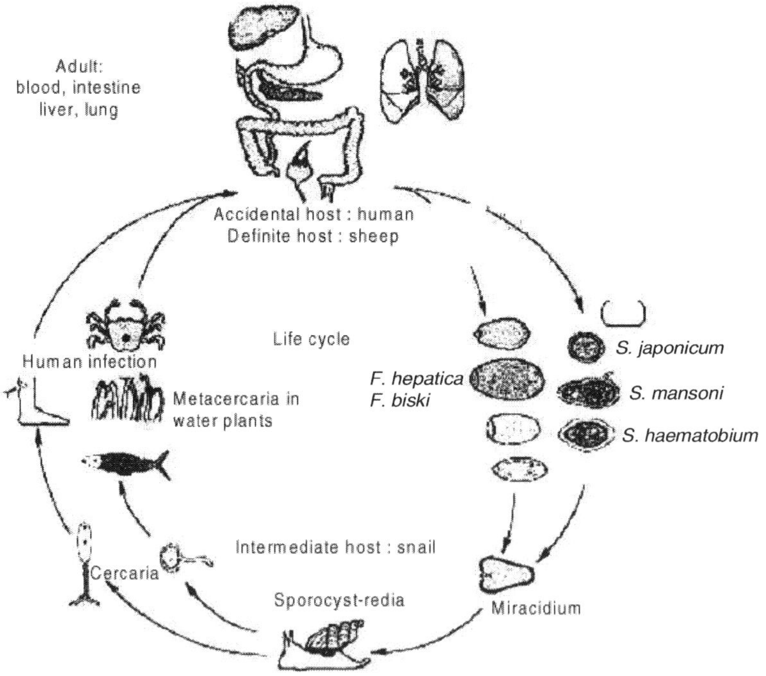

Figure 65.2 Life cycle of *Fasciola* and *Schistosoma*

CONTROL

Infection can be avoided by careful preparation and cooking of freshwater fish, crustacea and vegetables. For *F. hepatica*, the current drug of choice is bithional, given orally for 15 days.

SCHISTOSOMA INFECTION

It is a bisexual blood fluke. It produces non-operculated eggs with spines. Two larval forms are observed in the life cycle. Disease produced by the *Schistosoma* are generally called Schistosomiasis.

Schistosome eggs, which induce granulomatous inflammation in the venules or tissues.

Schistosoma are long-lived worms, having the life span of 20–30 years.

Bilharz (1851) discovered the adult worm of *Schistosoma haematobium.*

Fujii in 1847 mentioned about the Katayama fever occurring in Japan.

Three major species of schistosomes infect humans: *Schistosoma mansoni, S. japonica,* and *S. haematobium*. The adult male and female schistosomes reside in human mesenteric or vesical venules. Fertilized female worms produce large numbers of eggs, which pass out of the blood vessels, through the tissues, and into the lumen of the gut (*S. mansoni* and *S. japonicum*) or urinary bladder (*S. haematobium*), from which they are shed into the environment, where they may infect a snail intermediate host. After a period of asexual multiplication in the snail, the cercariae pass out into the water from which they directly penetrate into the human skin. The young schistosomes migrate from the skin to the lungs and then to the hepatoportal system, where they mature, mate and pass down into the mesenteric or vesical venules (Figure 65.1).

STRUCTURE AND LIFE CYCLE

I. Definitive host: Humans passing eggs into the environment
   1. Mature eggs reach the freshwater environment in excreta.
   2. Miracidium hatches from the egg under optimal conditions of temperature, light and freshwater and seeks an appropriate intermediate host. Asexual reproduction takes place within the intermediate host.

II. Intermediate host: Aquatic snails
   1. Miracidium penetrates the snail's soft tissue and undergoes differentiation into mother sporocysts which produce a second order daughter sporocyst.
   2. Daughter sporocyst migrates to the digestive gland where it produces cercariae by asexual reproduction.
   3. Motile cercaria emerges from the snail and seeks contact with a vertebrate host.

III. Migrating larvae and other life cycle stages resident in human host tissue

1. Cercaria penetrates through the skin of the vertebrate host where it undergoes transformation into schistosomulum.
2. The schistosomulum undergo growth and differentiation while migrating into the vasculature through the heart, lungs and hepatic portal system. Several weeks are required for the maturation of adult worms (4–5 weeks for *S. japonicum*, 6–7 weeks for *S. mansoni* and 10–12 weeks for *S. haematobium*).
3. Juvenile worm resides in the hepatic portal system continuing maturation to male and female adults.
4. Adult worms mate in the hepatic portal vessels and migrate to the mesenteric venules where they continuously lay eggs.
5. Some eggs are swept into the precapillary sinusoids of the liver where they become the nidus of a granulomatous reaction.
6. Chronic hyper-responsiveness to soluble egg antigens leads to the immunopathological consequence of "clay pipe-stem" fibrosis consisting of enlarged, whitish, fibrous surrounding portal vein lumen and resembling clay pipe stems in cross section (Figure 65.2).

CLINICAL MANIFESTATION

Three major disease syndromes occur in schistosomiasis:

Usually those of birds and small mammals. It appears most commonly in *Schistosoma japonicum* infection, is much less common in *S. mansoni* infections, and is seen rarely in patients infected with *S. haematobium*. Acute schistosomiasis is the self-limited febrile illness that occurs on primary exposure to the parasite.

A wide variety of symptoms have been associated with chronic schistosomiasis caused by *S. mansoni* and *S. japonicum*, including fatigue, abdominal pain and intermittent diarrhoea or dysentery.

In schistosomiasis, chronic disease may appear many years later, developing usually in individuals with a heavy worm burden. In schistosomiasis caused by S. *japonicum, S. mansoni*, chronic disease is characterized by hydronephrosis and eventually to uraemia.

EPIDEMIOLOGY

Trematodes do not multiply directly in humans, but instead mate and produce large numbers of eggs that pass out of the body in the faeces, urine or sputum. Thus, the intensity of human infection is related largely to the rate of exposure to infective larvae, to the frequency and extent of contact with water and of consumption of contaminated foods.

LABORATORY DIAGNOSIS

The anatomic locations of the symptoms and signs suggest a diagnosis of schistosomiasis or of disease caused by the liver, intestinal, urinary tract or lung flukes.

Demonstrating the eggs of the parasite in the excreta provides the definitive diagnosis in all cases of fluke infection.

CONTROL

A key control measure for all fluke infections is to prevent egg-containing excreta from contaminating water sources. Another approach is to control snail populations, largely by the use of molluscicides. Contact with infected water should be minimized to prevent schistosomiosis. Treatment of all fluke infections other than fascioliasis is accomplished by a one-day course of oral drug praziquantel.

1. Explain in detail the life cycle of *Schistosoma*.
2 Explain the lab diagnosis of nematode and trematode infections.
3. Discuss the various parasitic control and treatment methods available in the world.
4. Write short notes on
   i. Bisexual worm
   ii. Schistosome
   iii. Cercaria
   iv. Metacercaria
   v. Redia
   vi. Definitive host
   vii. Intermediate host
   viii. Microfilarie
   ix. Hermaphraditic
   x. Elephantiasis
   xi. Loeffler's syndrome
   xii. Filarial fever
   xiii. *Aedes*
   xiv. Liver rot

Newport, G. and Colley, D. *Schistosomiasis*. In Warren KS (ed): "Immunology and Molecular Biology of Parasitic Infections." Blackwell, Scientific Publications, Oxford, 1993.

Pearce, E. *The Immunology of Schistosomiasis*. In Boothroyd J.C. and Komuniecki, R. (eds.). Molecular Approaches to Parasitology. Wiley-Liss, New York. 1995.

Soulsby, E.J.L. *Immunology of Helminthes*. In Good, R.A. and Day, S.B. (eds.). Immunology of Human Infection.

Part II: Virus and Parasites. Plenum, New York. 1982.

Chan, M.S., Medley, G.F., Jamison, D. and Bundy, D.A.P. The evaluation of potential global morbidity attributable to intestinal nematode infections. Parasitology 109:373. 1994.

Grove, D.I. Strongyloidiasis; a major roundworm infection of man: Taylor and Francis, London. 1989.

Guyatt, H.L., Chan, M.S. and Medlet, G.F. *et al*. Control of *Ascaris* infection by chemotherapy: which is the most cost effective option? Tran Roy Soc Trop Med Hyg 89:16. 1995.

Parija, S.C., Malini, G. and Rao, R.S. "Prevalence of hookworm species in Pondicherry, India." *Trop. Geogr. Med.* 44:378. 1992.

Pawlowski, Z.S., Schad, G.A. and Stott, G.J. Hookworm infection and anemia: approaches to prevention and control. World Health Organisaiton Geneva 1991.

Turner, P.F., Rockett, K.A., Ottesen, E.A., Francis, H., Awadzi, K. and Clark, I.A. "Interlukin-6 and tumor necrosis factor in the pathogenesis of adverse reactions after treatment of lymphatic filariasis and onchocerciasis." J. Infect. Dis. 169:1071. 1994.

Wamae, C.N. Advances in the diagnosis of human lymphatic filariasis: a review. East Afr. Med. J. 71:171. 1994.

World Health Organisation: Lymphatic Filariasis: The disease and its control. WHO Tech. Rpt. Ser. 821, Geneva, 1992.

PART VI

MYCOPLASMA AND OTHER INFECTIONS

66 Mycoplasma

INTRODUCTION

Mycoplasma are bacteria without cell wall belonging to the group mollicutes. The causative agent of bovine pleuropneumonia was first isolated in the year 1898. The same type of organisms were isolated from animals in later days, all resembling pleuropneumonia causative agents and were called pleuropneumonia-like organisms (PPLO). These organisms are now called as mycoplasmas. Mycoplasmas are the smallest and simplest self-replicating bacteria. The mycoplasma cell contains a plasma membrane, ribosomes and a genome consisting of a double-stranded circular DNA molecule. They have no cell walls, and so are placed in a separate class mollicutes ("mollis" means "soft"; "cutis" means "skin"). Mycoplasmas have been nicknamed the crabgrass of cell cultures because their infections are persistent, frequently difficult to detect and diagnose and are difficult to cure.

Mycoplasma pneumoniae

The term primary atypical pneumonia was coined in the early 1940s to describe pneumonias caused by mycoplasma. Unidentified filterable agent was isolated by Eaton and his associates and it was called Eaton agent. This agent was identified as *Mycoplasma pneumoniae*.

It causes subclinical infection, upper respiratory tract disease and bronchopneumonia in humans. Symptoms are remittent fever, cough and headache that persist for several weeks. One of the most consistent clinical features is a lung convalescence, which may extend from 4 to 6 weeks. Several unusual complications have been noted, including haemolytic anaemia, polyradiculitis, encephalitis, aseptic meningitis and central nervous system illness such as Guillain–Barré syndrome. In addition, pericarditis and pancreatitis have been observed.

UREAPLASMA AND OTHER MYCOPLASMAS

Ureaplasma urealyticum causes non-gonococcal urethritis in men. Ureaplasmas have also been associated with chorioamnionitis, habitual spontaneous abortion and low-weight infants. *Mycoplasma hominis*, a common inhabitant of the vagina of healthy women, becomes pathogenic once it invades the internal genital organs, where it may cause pelvic inflammatory diseases such as tubo-ovarian abscess or salpingitis. *Mycoplasma genitalium*, isolated in 1981 from the urethral discharge of two homosexual men, may account for the non-gonococcal urethritis. *M. genitalium* is a highly fastidious organism.

DISTINGUISHING PROPERTIES

The coccus is the basic form of all mycoplasmas in culture. The diameter of the smallest coccus capable of reproduction is about 300 nm. In most mycoplasma cultures, elongated or filamentous forms (up to 100 μm long and about 0.4 μm thick) also occur. The filaments tend to produce truly branched mycelioid structures, hence the name mycoplasma (*myces* means "a fungus"; *plasma* means "a form"). Mycoplasmas reproduce by binary fission. Some mycoplasmas possess unique attachment organelles, which are shaped as a tapered tip in *M. pneumoniae* and *M. genitalium. M. pneumoniae* is a pathogen of the respiratory tract, adhering to the respiratory epithelium, primarily through the attachment organelle.

One of the most useful distinguishing features of mycoplasmas is their peculiar fried-egg colony shape, consisting of a central zone of growth embedded in the agar and a peripheral one on the agar surface.

The mycoplasma genome is typically prokaryotic, consisting of a circular, double-stranded DNA molecule. The mycoplasma and ureaplasma genomes are the smallest recorded for any self-reproducing prokaryote.

M. pneumoniae, grow very slowly, particularly on primary isolation. *U. urealyticum*, a pathogen of the human urogenital tract, grows very poorly *in vitro*, reaching maximal titres of 107 organisms/ml of culture. *M. genitalium*, another human pathogen, grows so poorly *in vitro* that only a few successful isolations have been achieved.

PATHOGENESIS

All mycoplasmas are parasites of humans, animals, plants or arthropods. The primary habitats of human and animal mycoplasmas are the mucous surfaces of the respiratory and urogenital tracts. Most mycoplasmas that infect humans and other animals are surface parasites, adhering to the epithelial linings of the respiratory and urogenital tracts (Figure 66.1). Initial interaction between the host cell and the parasite releases toxic metabolites, which will cause tissue damage. Membrane fusion releases hydrolytic enzymes into the host cell leading to host cell damage.

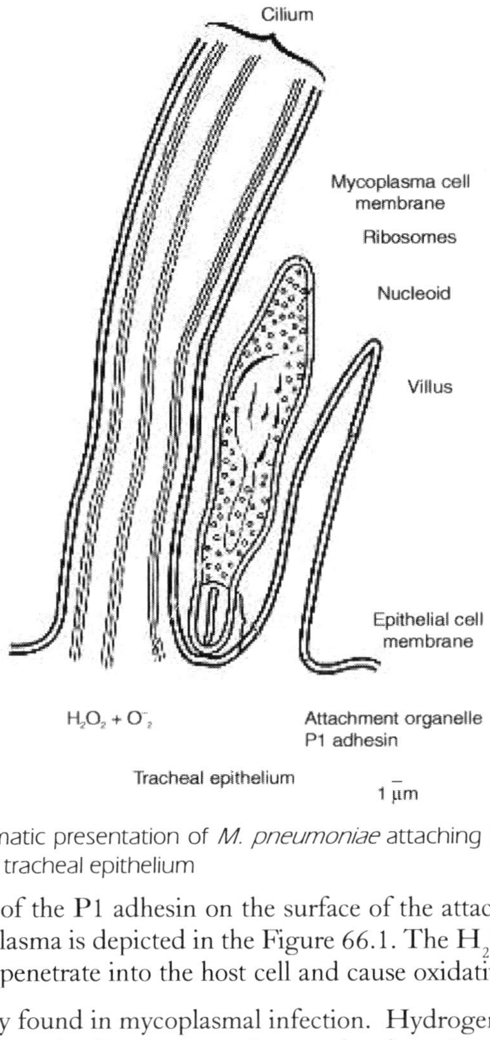

Figure 66.1 Schematic presentation of *M. pneumoniae* attaching to the surface of the ciliary tracheal epithelium

The clustering of the P1 adhesin on the surface of the attachment organelle at the tip of the mycoplasma is depicted in the Figure 66.1. The H_2O_2 and O_2 released by the mycoplasma penetrate into the host cell and cause oxidative damage.

Toxins are rarely found in mycoplasmal infection. Hydrogen peroxide (H_2O_2), the end product of respiration in mycoplasmas, has been implicated as a major pathogenic factor ever since it was shown to be responsible for the lysis of erythrocytes by mycoplasmas *in vitro*; however, the production of H_2O_2 alone does not determine pathogenicity, as the loss of virulence in *M. pneumoniae* is not accompanied by a decrease in H_2O_2 production. For H_2O_2 to exert its toxic effect, the mycoplasmas must adhere close enough to the host cell surface. *M. pneumoniae* inhibits host cell catalase by releasing superoxide radicals (O_2^-) (Figure 66.2).

Mycoplasmas activate macrophages, and induce cytokine production and lymphocyte proliferation. Thus, in the case of *M. pneumoniae*, the host may be largely responsible for pneumonia by mounting a local immune response to the parasite. *M. pneumoniae* infection can be found in most children, 2 to 5 years of age, although the illness occurs with greatest frequency in individuals, 5 to 15 years of age.

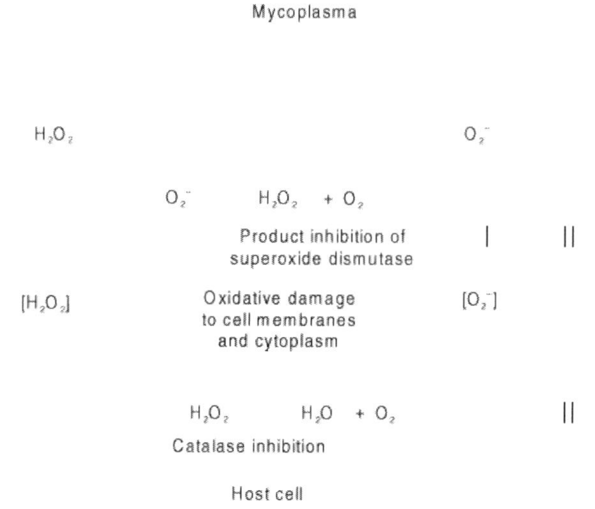

Figure 66.2 Proposed mechanism of oxidative damage to host cells by adhering *M. pneumoniae* by increasing concentrations of H_2O_2 and O_2^-

EPIDEMIOLOGY

M. pneumoniae accounts for 8 to 15 per cent of all pneumonias in young children. In older children and in young adults, the organism is responsible for approximately 15 to 50 percent of all pneumonias. *M. pneumoniae* requires close personal contact for spread; successful spreading usually occurs in families, schools and institutions. The incubation period ranges from 2 to 3 weeks.

U. urealyticum is spread primarily through sexual contact. Colonization is linked to the frequency of sexual intercourse and the number of sexual partners. Women may be asymptomatic reservoirs of infection.

LABORATORY DIAGNOSIS

Culture is essential for definitive diagnosis. A routine mycoplasma medium consists of heart infusion, peptone, yeast extract, salts, glucose or arginine, and horse serum (5 to 20 per cent). Foetal or newborn calf serum is preferable to horse serum. To prevent the overgrowth of the fast-growing bacteria that usually accompany mycoplasmas in clinical materials, penicillin, thallium acetate or both are added as selective agents. For ureaplasma culture, the medium is supplemented with urea and its pH is brought to 6.0. For *M. pneumoniae* isolation, nasopharyngeal secretions are inoculated into a selective diphasic medium (pH 7.8) made of mycoplasma broth and agar and supplemented with glucose and phenol red. When *M. pneumoniae* grows in this medium, it produces acid, causing the colour of the medium to change from purple to yellow. Broth from the diphasic medium is subcultured to mycoplasma agar when a colour change occurs, or at weekly intervals for a minimum of 8 weeks.

One of the most useful distinguishing feature of mycoplasmas is their peculiar fried-egg colony shape, consisting of a central zone of growth embedded in the agar and a peripheral one on the agar surface (Figure 66.3).

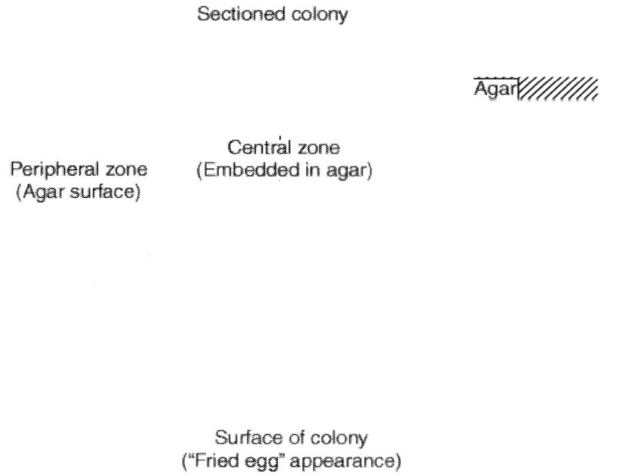

Figure 66.3 Colony morphology of mycoplasma

Serodiagnosis and Molecular Probes

More sophisticated tests, including electrophoretic analysis of cell proteins, DNA–DNA hybridization tests, mycoplasmal DNA cleavage patterns by restriction endonucleases and PCR tests employing species-specific primers for amplification, may be performed in a research laboratory.

PREVENTION AND CONTROL

Chemoprophylaxis is not recommended and no vaccine is available.

TREATMENT

The mycoplasmas are sensitive to tetracyclines, macrolides and the newer quinolones. Tetracycline or erythromycin is recommended for treatment of *M. pneumoniae* pneumonia. To prevent *U. urealyticum* infection, sexual partners should be treated simultaneously with tetracycline.

1. State the characteristic features of *Mycoplasma*.
2. How do you differentiate *Mycoplasma* and *Ureaplasma* based on their pathogenesis.
3. Explain pathogenesis and diagnostic features of mycoplasmas.
4. Write short notes on

   i. PPLO, ii. Mycoplasma, iii. Mollicutes, iv. Eaton agent

Zoonotic Infections

INTRODUCTION

Rudolf Virchow introduced the word "zoonosis" in 1880 to denote the diseases shared by man and animals. In 1959, WHO defined zoonosis as those diseases and infections, which are naturally transmitted between vertebrate animals and man. Zoonotic diseases are a major public health problem in India. Zoonoses occur throughout the world. Over the last two decades, there has been considerable change in the importance of certain zoonotic diseases in many parts of the world, resulting from ecological changes such as urbanization, industrialization, etc.

CLASSIFICATION

Zoonotic infections are classified based on the aetiological agents, mode of transmission and reservoir of host .

Classification Based on Aetiological Agents

Bacterial zoonoses	:	Anthrax, brucellosis, plague, leptospirosis, salmonellosis, Lyme disease, Kyasanur forest disease
Viral zoonoses	:	Rabies, arbovirus infections, Kyasanur forest disease, yellow fever, influenza
Rickettsial zoonoses	:	Murine typhus, tick typhus, scrub typhus, Q-fever
Protozoal zoonoses	:	Toxoplasmosis, trypanosomiasis, leishmaniasis
Helminthic zoonoses	:	Echinococcosis, taeniasis, schistosomiasis
Fungal zoonoses	:	Histoplasmosis, cryptococcosis, superficial dermatophytosis
Ectoparasites	:	Scabies, myasis

Classification Based on Mode of Transmission

Direct zoonoses These are transmitted from an infected vertebrate host to a susceptible host (man) by direct contact, by contact with a fomite or by a mechanical vector, e.g. rabies, anthrax, brucellosis, leptospirosis, toxoplasmosis.

Cyclozoonoses These require more than one vertebrate host species, but no invertebrate host is required for the completion of the life cycle, e.g. echinococcosis, taeniasis.

Metazoonoses These are transmitted biologically by invertebrate vectors, in which the agent multiplies and/or develops in the host, e.g. plague, arbovirus infections, schistosomiasis, leishmaniasis.

Saprozoonoses These require a vertebrate host and a non-animal developmental site like soil, plant material, pigeon droppings, etc. for the development of the infectious agent, e.g. aspergillosis, coccidioidomycosis, histoplasmosis, zygomycosis.

Classification based on Reservior Host

Anthropozoonoses These are infections transmitted to man from lower vertebrate animals, e.g. rabies, leptospirosis, plague, arboviral infections, brucellosis and Q fever.

Zooanthroponoses These are infections transmitted from man to lower vertebrate animals, e.g. streptococcal and staphylococcal infections, diphtheria infections, caused by enterobacteriaceae, human tuberculosis in cattle and parrots.

Amphixenoses These are infections maintained in both man and lower vertebrate animals and transmitted in either direction, e.g. salmonellosis, staphylococcosis.

FACTORS INFLUENCING THE PREVALENCE OF ZOONOSES

- Ecological changes in man's environment
- Handling of animals their products and wastes (occupational hazards)
- Increased movements of man
- Increased trade in animals products
- Increased density of animal population
- Transportation of virus-infected mosquitoes
- Cultural anthropological norms

HIGH RISK GROUPS FOR ZOONOSIS

- Laboratory staff handling infected materials
- Veterinarians
- Animal handlers and catchers

- Wildlife officers
- Quarantine officers
- Naturalists
- Slaughter house workers
- Animal researchers

TRANSMISSION OF ZOONOTIC AGENTS

The organisms enter the human body in varieties of ways:

1. By penetration of the skin
2. By inhalation
3. By ingestion

Humans are the dead-end host.

Penetration

Direct penetration The epidermis of the skin may be breached in a number of ways to permit entry of microbes through abrasions or open wounds.

Disease	Animal reservoir	Causative agents
Anthrax	Domestic animals	*Bacillus anthracis*
Brucellosis	Goats, sheeps	*Brucella melitensis*
Erysipeloid	Swine poultry	*Erysipelsthrix rhusiopathiae*
Leptospirosis	Rodents, foxes	*Leptospira interrogans*
Melioidosis	Rodents, foxes	*Pseudomonas pseudomallei*
Glanders	Domestic animals	*Pseudomonas mallei*
Tularemia	Rabbit	*Francisella tularensis*
Dermatophytosis	Dogs, cats, etc.	*Trichophyton*

Arthropod vectors Some arthropod vectors serve as flying syringes (e.g. mosquitoes).

Disease	Causative agent	Vector
Lyme disease	*Borrelia burgderferi*	Tick
Plaque	*Yersinia pestis*	Flea
Relapsing fever	*Borrelia recurrentis*	Tick
Rocky mountain spotted fever	*Rickettsia rickettsii*	Tick
Yellow fever	Flavivirus	Mosquito
Encephalitis	Flavi and Alphavirus	Mosquito
Leishmaniasis	*Leishmania* spp.	Sandfly
American trypanosomiasis	*Trypanosoma cruezi*	Kissing bug
African trypanosomiasis	*Trypanosoma* spp.	Tsetse fly

Animal Bite

It introduces organisms into the deep tissues.

Disease	Causative agent	Animal reservoir
Pasteurellosis	*Pasteurella multocida*	Dog
Rat bite fever	*Spirillum minus*	Rat
Rabies	Rabies virus	Dog
Blastomycosis	*Blastomyces dermatitidis*	Dog

Inhalation

Some of the animal microbes are inhaled in two ways:

- From droplet nuclei
- From an inanimate reservoir–soil

Disease	Causative agent
Anthrax	*Bacillus anthracis*
Q fever	*Coxiella burnetii*
Histoplasmosis	*Histoplasma capsulatum*

Ingestion

Most zoonoses are acquired by ingestion. Foodstuff are the major source

Disease	Source
Brucellosis	Dairy
Listeriosis	Vegetables
Salmonellosis	Egg
Giardiasis	Water
Cryptosporidiosis	Water

CONTROL OF ZOONOTIC INFECTIONS

Zoonotic controlled by infections are:

- Use of human and veterinary medicines
- Sanitary engineering
- Eradication of infected animal reservoir
- Protection of animals
- Pasteurization of milk

- Proper cooking of food
- Eradication of flies and mosquitoes
- Mass vaccination

1. How zoonotic infections are transmitted to human?
2. Describe the control of zoonotic infections.
3. Explain various transmission methods of zoonotic infections.
4. Write a short notes on
   i. Zoonoses
   ii. Cyclozoonoses
   iii. Metazoonoses

Maniloff, J. McElhaney, R.N. Finch, L.R. and Baseman, J.B. (Eds.). Mycoplasmas, Molecular Biology and Pathogenesis. American Society for Microbiology, Washington. 1992.

Razin, S. and Barile, M.F. (Eds.). *The Mycoplasmas*, Vol. 4, Mycoplasma Pathogenicity Academic Press, Orlando. 1985.

Razin, S. and Tully, J.G. (Eds.). *Molecular and Diagnostic Procedures in Mycoplasmology*, Academic Press. Orlando. 1995.

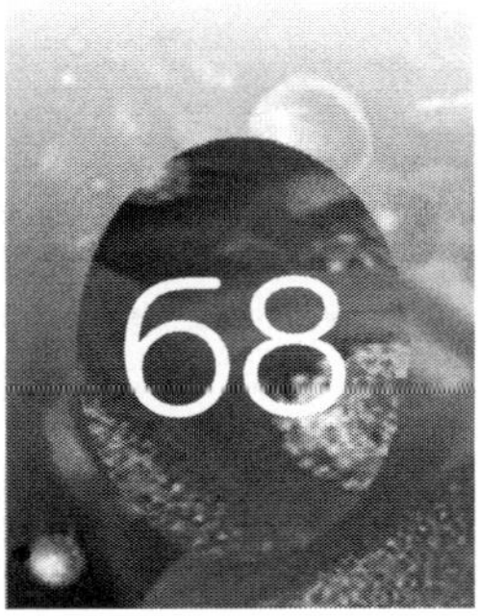

Nosocomial Infections

INTRODUCTION

Nosocomial infections (hospital-acquired infections) are infections acquired in the hospital. The term "nosocomial" comes from the greek word *nosos* meaning disease and *komeion* meaning hospital. Nightingale was the first person to improve medical care in military and civilian hospitals. Semmelweis and Holmes followed Nightingale's medial care procedures to reduce the transmission of pathogenic microorganisms within the hospital.

Pasteur demonstrated that microorganisms could be transmitted through air. During this time, Lister realized the significance of Pasteur's study and was convinced that microorganisms present in the air contaminate surgical wounds and he followed antiseptic surgery. Subsequently he also introduced the technique of disinfecting operating rooms of hospitals and greatly reduced the mortality of surgical patients.

PREVALENCE OF NOSOCOMIAL INFECTIONS

Nosocomial infections occur in approximately 5% (1 in 20 patients) of all patients admitted. Infection rate will vary depending upon the type of hospital. Nosocomial infections in acute care institutions are as follows: 40% of hospital infections are observed in urinary tract, 20% in surgical wounds, 10% in respiratory route, 5–10% are caused by primary bacteria and others are due to opportunistic fungi and viruses (20–25%). Opportunistic infections are observed in immunocompromised patients.

Approximately 1% of all nosocomial infections directly cause death and 3% contribute to death.

IMPORTANT PATHOGENS OF NOSOCOMIAL INFECTIONS

Staphylococcus aureus

Enterococcus spp.

Escherichia coli

Group B Streptococci

Coagulase-negative *Staphylococcus*

Proteus spp.

Pseudomonas aeruginosa

Antibiotic-resistant gram-negative rods

Candida albicans

Torulopsis spp.

Aspergillus spp.

Anaerobic bacteria

Sources of Microbes

There are two categories of infection:

1. Exogenous—caused by microbes from an external source.
2. Endogenous—caused by microbes that are part of a person's own normal flora.

The following are exogenous sources:

- Hospital personnel
- Surgical procedures
- Food
- Air
- Visitors
- Medication
- Drains, implants, catheter
- Other patients
- Water
- Fomites
- Insects

RISK FACTORS

- Invasive device
- Tissue transplant
- Extensive skin burn
- Sickle cell anaemia
- Bone marrow failure

- Malignant disorder
- HIV infection
- Malfunction of spleen
- Implants

METHODS TO AVOID NOSOCOMIAL INFECTIONS

1. Strict isolation for prevention of diseases such as diphtheria, smallpox, conjunctivitis, Lassa fever, pneumonic plague and chickenpox.
2. Contact isolation for prevention of wound infections, gonococcal conjunctivitis, herpes simplex infections, influenza and pneumonia.
3. Respiratory isolation for the prevention of measles, meningitis, mumps, whooping cough and pneumonia.
4. Enteric precautions for prevention of Amoebic dysentery, gastroenteritis, typhoid fever and cholera.
5. Universal blood and body fluid precautions to be taken in the case of AIDS, HBV, malaria, syphilis and gonorrhoea.

CONTROL OF NOSOCOMIAL INFECTIONS

Because of the seriousness of nosocomial infections, the American Hospital Association (AHA) and the Center for Disease Control (CDC) recommend that each hospital develop an infection control programme. One of the most important activities of the control program is surveillance, the systematic observation and recording of cases of transmissible diseases. To accomplish this, the National Nosocomial Infection Surveillance System (NNIS) was established in 1970 by the CDC.

In 1999, CDC updated hospital control program to reduce the spread of infection.

Methodologies of hospital disease control programme are

1. Hand washing
2. Aseptic technique
3. Isolation of an infected patient
4. Proper sanitation
5. Disinfection
6. Sterilization

COMPONENTS OF CONTROL PROGRAMME

1. Hospital policy-making committee—develops policies for control of nosocomial infection.
2. Infection control committee—develops guidelines for patient care, monitors effectiveness of control program
3. Microbiology laboratory—isolates and identifies organisms and source of infection, monitors disinfection and sterilization processes.
4. Infection control officer—implements infection control practices and surveillance and investigation of infection in patients and in personnel, and maintains a continuing education program.

By combining importance of infection and control program Florence Nightingale said that "the very first requirement in a hospital is that it should do the sick no harm."

1. Explain the risk factors and sources of hospital-borne infection.
2. Write a short note on nosocomial infection.
3. List organisms responsible for hospital infection.
4. Describe about the committees involved in hospital infection control programme.
5. How do you control hospital-borne infection?
6. Describe methods used to control nosocomial infection.

PRACTICALS

Aim

To estimate the total erythrocyte count using haemocytometer.

Clinical Significance

Increase in total erythrocyte count is seen in haemo concentration due to burns, cholera, etc. Decreased count of erythrocyte is observed in polycythemia, old age, pregnancy and anaemia.

Normal Value

Male	4.56 million cells/cu.mm.
Female	4 to 4.5 million cells/cu.mm.

Specimen

Capillary blood is recommended.

Requirements

Microscope with low-power and high-power objective.

Neubauer counting chamber with coverslip

Shali pipette

Test tube

Cotton

RBC diluting fluid

Trisodium citrate	3 g
Formalin	1 ml
Distilled water	100 ml

Procedure

1. Fill the red blood cell pipette with capillary blood exactly up to 0.5 mark by holding the pipette horizontally.
2. Now draw the diluting fluid up to the mark 101.
3. The pipette should be gently rotated to obtain good mixing.
4. The coverslip is placed over the Neubauer chamber so as to cover the ruled platform evenly.
5. Now load the chamber. This is done in 3 steps.
    i. Mix the contents of pipette for 3 minutes.

   ii. Expel 6 drops from the pipette to remove the fluid in the stem which has not been mixed with the blood.
   iii. By holding the pipette at an angle of 45° and touching the space between the coverslip and the chamber by the tip of the pipette, an appropriate drop of the mixture is allowed to run under the cover glass by capillary action.
6. Allow 2 minutes for settling of the cells and count.
7. Place the counting chamber on the stage of the microscope. Switch to low-power objective. Adjust the light and locate the large square in the centre with 25 small squares.
8. The RBCs in the four corner squares and in the centre squares are counted.
9. Total the cells counted in the 5 squares and calculate the RBC/cu.mm by the following equation.

Calculation

$$\text{Number of RBC/ml} = \frac{\text{No. of cells counted} \times \text{Dilution}}{\text{Area counted} \times \text{Depth of fluid}}$$

i.e.,

No. of RBC/m/ = No. of cells × Factor

Dilution = 200

Area counted = 5 × 0.04 = 0.2

Depth of fluid = 0.1 mm

$$\text{Factor} = \frac{\text{Dilution}}{\text{Volume}} = \frac{200}{0.2 \times 0.1}$$

$$= 10{,}000$$

So RBC/cu.mm of the blood = Number of cells counted × 10,000

Result

The given blood sample contains million cells/cu.mm of red blood cells.

Aim

To estimate the amount of white blood cells (leucocytes) in human blood.

Principle

The blood is diluted to 1 : 20 with the WBC diluting fluid. Glacial acetic acid lyses RBC's while the Gentian violet stains the nuclei of leucocytes. The cells are counted under low-power objective by using the counting chamber. The number of white blood cells in undiluted blood are calculated and reported as the number of WBC/cu.mm of whole blood.

Clinical Significance

Increase in total leucocyte count above 10,000 cells/cu.mm is known as leucocytosis and a decrease of less than 4,000 cells/cu.mm is known as leucopenia. Pathological leucocytosis is observed in infections such as pneumonia, tonsilitis, meningitis, abscess, rheumatic fever, diphtheria, smallpox, chickenpox, erythroblastosis foetalis, uraemia, ulcer, pregnancy, menstruation, high temperature, severe pain and muscular exercise.

Leucopenia is observed in influenza, typhoid, tuberculosis, measles, brucellosis, agranulocytosis, hepatitis B infection, dengue, sandfly fever, rheumatoid arthritis, primary bone marrow depressions, megaloblast conditions or when exposed to radiation.

Normal Value

At birth	10,000 to 25,000 cells/cu.mm
1 to 3 years	6,000 to 18,000 cells/cu.mm
4 to 7 years	6,000 to 15,000 cells/cu.mm
8 to 12 years	4,500 to 14,000 cells/cu.mm
Adult	4,000 to 10,000 cells/cu.mm

Specimen

Capillary blood is recommended for total leucocyte count.

Requirements

Microscope

Neubauer counting chamber with coverslip

Shali pipette

WBC diluting fluid

Glacial acetic acid	2 ml
Gentian violet	1 g
Distilled water	97 ml

Mix all with a stirrer, and the solution is ready to use.

Procedure

1. Draw blood in a clean dry pipette up to the mark 0.5.
2. Wipe off the pipette with cotton.
3. Now draw the diluting fluid up to the mark 11(dilution 1 in 20).
4. Hold pipette horizontally and mix the contents of the pipette for 5 minutes for complete haemolysis of RBCs.
5. Dispel the first 4 drops of the content.
6. Place the haemocytometer on a flat surface with the coverslip on the counting chamber.
7. Load the haemocytometer with the mixture. By holding the pipette at 45° angle and touching the space between the cover glass and the chamber by the tip of the pipette, an appropriate drop of the mixture is allowed to run under the cover glass by capillary action. It must be sufficiently large to cover the whole ruled platform, yet not large enough to fill the moats. Also there must be no air bubbles.

8. Leave the counting chamber undisturbed for 3 to 5 minutes to allow the cells to settle.
9. Place the counting chamber on the stage of the microscope. Switch to the low-power objective, adjust the light and locate the large square on one of the corners.
10. Scan all large corner squares and count the cells which are identified by their blue colour with a large nucleus. Make a total of all the cells counted in the 4 squares and calculate the WBC/cu.mm using the following calculation.

Calculation

$$\text{Number of WBC/cu.mm} = \frac{\text{No of WBC counted} \times \text{Dilution}}{\text{Area counted} \times \text{Depth of the fluid}}$$

Dilution = 20

Area counted = 4

Depth of the fluid = 0.1 mm

i.e., WBC counted in whole blood = No of cells counted × factor

$$\text{Factor} = \frac{20}{4 \times 0.1} = 50$$

So WBC/cu.mm. of the blood = Number of cells counted × 50

Result

The given blood sample contains cells/cu.mm

Aim

To differentiate leucocytes.

Principle

Three major steps involved in differential count are preparation of smear, staining of smear and microscopic observation.

The smear is made with a drop of blood taken directly from the skin puncture, which gives the true picture of blood morphology. Staining is done with polychromatic stain, which includes methylene blue and eosin (Leishmann's stain).

The polychromatic stain induces multiple colours when applied to the cells. The stain is dissolved in methanol and a buffer. Methanol acts as a fixing agent and also as a solvent. Following staining, the basic components of white cells are stained by the acidic eosin dye and are described as eosinophilic or acidophilic, where the acidic components of blood (nucleus and nucleic acids) are stained blue to purple by the basic dyes and are called basophilic. The neutral components present in the blood cells are probably stained by both of the dyes.

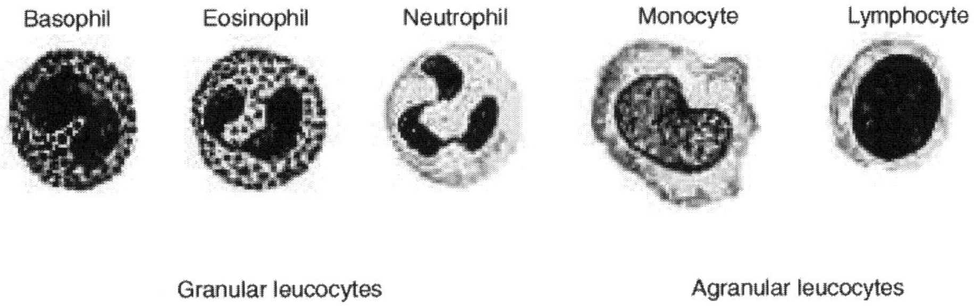

Normal Value

Neutrophils 50–70 %
Lymphocyte 20–40 %
Eosinophils 1–3 %
Monocyte 1–5 %
Basophils 1 %

Specimen

Blood sample

Materials Required

Reagents

Leishmann's stain

Leishmann's stain powder	2 g
Methanol	100 ml

Mix these two components thoroughly and warm up to 50°C. Filter the solution, and it is ready to use.

Equipment

Compound microscope
Slides

Procedure

Preparation of Smear

1. A drop of skin-puncture blood is transfered to a clean slide without grease.
2. The drop is placed approximately 1 cm from the end. With the help of a spreader slide, a blood smear is made.
3. The blood smear is dried.

Fixing and Staining

Methanol present in the stain fixes the smear. The slide is kept on the staining rack with the blood smear facing up. Apply undiluted Leishmann's stain over the smear and leave it undisturbed for 2 minutes. Add buffered water over the stained smear for 5–7 minutes. Later wash the slide with a stream of buffered water until it acquires a pinkish tinge.

Examination of Blood Smear

Examine the stained blood smear under a low-power microscope. In an ideal smear, 3 zones are identified, the thick area or head following which is the body and the tail at the thin end of smear. Choose the portion of blood smear which appears slightly before the tail end of the smear. Place a drop of immersion oil on the slide; various types of WBC's are identified and enumerated based on the staining and morphology of the nucleus.

Granulocytes These are granulated cells, which are stained pink and include eosinophils, basophils and neutrophils.

Neutrophils Pale pink cytoplasm with fine coloured granules. Nucleus is banded and segmented.

Eosinophils Cytoplasm stains pink and contains red and orange granules. Nucleus is bilobed.

Basophils Cytoplasmic granules are large, dark and blue-black coloured.

Lymphocytes Large sized lymphocytes have clear blue cytoplasm on the margin of the nucleus. In smaller sized lymphocytes, a dark violet colour almost fills the entire cell and has rim of clear cytoplasm.

Result and Discussion

Aim

Determination of haemoglobin by Shali haematin method.

Clinical Significance

A decrease in haemoglobin below the normal range is an indication of anaemia. Haemoglobin level is also decreased in pregnancy, blood loss, malnutrition, polycythemia, congenital heart disease due to reduced oxygen supply and in haemo concentration due to loss of body fluids in severe diarrhoea and vomiting.

Normal Value

Male	14–18 g/dl
Female	11.5–16.5 g/dl
Children	11–14.5 g/dl

Specimen

Anticoagulated blood

Materials Required

Shali haemoglobinometer
Calibrated tube for haemoglobin measurement
Shali pipette
0.1N HCl
Distilled water
Pasteur pipette

Principle

When blood is added to 0.1N HCl, haemoglobin is converted to brown coloured acid haematin. The resulting colour after dilution is compared with the standard reference in the haemoglobinometer.

Procedure

1. Add 0.1N HCl up to the lowest mark of the calibrated tube by using Pasteur pipette.
2. Draw 20 μl of the blood using Shali pipette, wiping the excess blood with the help of dry cotton.
3. Transfer the blood to the acid in the calibrated tube.
4. Mix well and allow the tube to stand for at least 10 minutes at room temperature.
5. Dilute the solution with few drops of water and mix well.
6. Repeat the dilution until the colour matches with the standard glass in the haemoglobinometer.
7. Then the colour is matched against the natural light.
8. The level of the fluid is noted at its lowest meniscus, and the reading corresponding to the level on the scale is recorded in g/dl.

Observation and Result

Aim

To estimate the erythrocyte sedimentation rate (ESR) by Westergren method.

Clinical Significance

ESR is a non-specific test that reflects changes in plasma protein, which accompanies most of the acute chronic infections. These pathological conditions accelerate routleaux formation of red cells. As a result of routleaux formation, ESR increases, which suggests the possible pathological condition. Normalization of ESR indicates possible recovery from the disease.

Normal Value

Men	5–15 mm/hour
Women	5–20 mm/hour

Specimen

Citrated blood 1.6 ml, i.e., mixed with 0.4 ml of 3.8% sodium citrate solution.

Requirements

3.8% sodium citrate solution
Westergren tube calibrated in millimetres
Rack for holding Westerngren tubes (ESR stand)
Timer
Vials or test tubes

Procedure

1. Deliver 0.4 ml of 3.8% sodium citrate into a plain test tube.
2. Add 1.6 ml of blood into the test tube containing sodium citrate solution, mix gently and thoroughly.
3. Place the Westergren rack in a levelled plain surface, away from direct sunlight and air-drift.
4. Fill the Westergren tube exactly up to the mark.
5. Set the timer exactly for 30 minutes.
6. Note the level to which the red cell column has fallen at the end of 30 minutes.
7. Repeat the reading up to 60 minutes.
8. Report the result in mm/30 minutes and mm/60 minutes.

Observation and Result

Aim

To estimate the amount of cholesterol present in the given sample.

Clinical Significance

Cholesterol occurs in appreciable amounts in the body. Most of the cholesterol is synthesized by the liver from acetyl-CoA. Cholesterol is concerned with the metabolism of lipids and is an important precursor for steroid hormones. Serum cholesterol level is increased in diabetes mellitus, nephrosis, biliary cirrhosis and hypothyroidism. The level decreases in severe infection, severe anaemia, hyperthyroidism and malnutrition.

Principle

This test is performed by means of oxidation reaction. The two reaction centres present in the cholesterol molecules are the double bond and the OH group. Cholesterol reacts with strong acid reagents to produce coloured substances chiefly chestadiene sulphonic acid. Acetic acid and acetic anhydride are used as the solvent and the dehydrating agent respectively while sulphuric acid is used as the dehydrating and oxidizing agent.

In this method cholesterol in the serum is extracted in the presence of acetic acid. Addition of sulphuric acid gives a blue coloured chromatophore.

Reagents

Stock Ferric Chloride Reagent

840 mg ferric chloride is weighed and dissolved in 100 ml glacial acetic acid.

Ferric Chloride Precipitating Reagent

10 ml stock ferric chloride is diluted into 100 ml glacial acetic acid.

Ferric Chloride Diluting Reagent

8.5 ml of stock ferric chloride is diluted into 100 ml of glacial acetic acid.

Standard Cholesterol

200 mg of cholesterol is dissolved in 10 ml of ferric chloride precipitating reagent and made up to 100 ml with glacial acetic acid.

Procedure

1. To 0.1 ml of serum, add 4.9 ml of ferric chloride reagent.
2. Mix well using glass rod and centrifuge for 15 minutes.
3. From this, take 2.5 ml of filtrate and add 2.5 ml of ferric chloride diluting reagent followed by 4 ml of concentrated sulphuric acid with thorough mixing.
4. Various concentrations of cholesterol (10 μg–100 μg) is diluted to 5 ml with ferric chloride diluting reagent. Then add 4 ml of concentrated sulphuric acid to all tubes. Mix well and read the colour developments at 560 nm. Plot a standard graph and calculate the amount of cholesterol present in the given sample.

Normal Value

150–250 mg of cholesterol/100ml serum

Calculation

2.5 ml of filtrate contains 100 μg of cholesterol

$$\text{2.5 ml came from 5 ml of diluted sample} = \frac{100 \times 5}{2.5}$$

$$= 200\ \mu g$$

This 5 ml was made from 0.1 ml of the sample.

$$\text{0.1 ml sample contains } 200\ \mu g$$

$$\text{100 ml sample contains} = \frac{200 \times 100}{0.1}$$

$$= 200\ \text{mg/dl}$$

Result

Aim

To estimate the amount of sugar present in the given sample.

Principle

The aldehyde group of glucose condenses with *o*-toluidine in glacial acetic acid (colourless). On heating, a blue-green colour is developed which is due to the formation of *N*-glucosamine.

Reagents

o-Toluidine Reagent

Mix *o*-toluidine, acetic acid and water in the ratio of 15:75:10. Add 2.5 g of boric acid and 2.5 g of thiourea, mix well and keep it in a brown bottle.

10% Trichloroacetic Acid

Stock Standard

Dissolve 100 mg of glucose in 100 ml of distilled water.

Working Standard

100 ml of standard solution is made up to 100 ml using distilled water. 100 ml of working standard contains 10 mg of glucose.

Procedure

1. Take various concentrations of working standard solution ranging from 20 mg in a series of test tubes and make up to 1 ml with distilled water.
2. Take 1 ml of distilled water as blank.
3. To 0.1 ml of serum, add 2 ml of 10% trichloroacetic acid to precipitate proteins. Allow it to stand, and centrifuge.
4. Take 1 ml of protein-free filtrate in a test tube and mark it as test.
5. To all the above tubes add 3 ml of *o*-toluidine reagent and keep in a boiling water bath for 15 minutes.
6. A greenish colour is developed and is read at 620 nm.

Normal Value

80 to 120 mg/dl

Result and Discussion

Aim

To detect the level of SGPT and SGOT in the given serum sample.

Principle

SGOT

GOT catalyses the following reaction

L-Ketoglutamate + L-Aspartate L-Glutamate + Oxaloacetate

Oxaloacetate is coupled with 2,4,dinitrophenylhydrazine which gives known colour in alkaline medium and is measured calorimetrically at 505 nm.

SGPT

α-ketoglutarate + L-alanine L-glutamine + pyruvate

Pyruvate couples with 2,4, dinitrophenylhydrazine (2,4 DNPH) to give the corresponding hydrazone, which gives known colour in alkaline medium, and is measured calorimetrically at 505 nm.

Clinical Significance

SGOT

Coronary infarction	20–200 IU
Hepatic damage	250–1500 IU
Jaundice	50–100 IU

SGPT

Coronary infarction	17–75 IU
Hepatic damage	50–1000 IU
Jaundice	100–200 IU

Normal Value

SGOT up to12 IU

SGPT up to12 IU

Advantages

Highly popular and simple method.

Gives reproducible results.

Substrate and standard are in stabilized state.

Very economical.

Requirements

Phosphate buffer Dissolve 5.965 g of disodium hydrogen phosphate and 1.09 g potassium dihydrogen phosphate in water and dilute to 500 ml. Adjust the pH to 7.4.

Standard solution Dissolve 0.022 g sodium pyruvate in 100 ml of phosphate buffer. Make fresh when needed.

GOT substrate Transfer 0.146 g of α-ketoglutarate and 13.3 g of DL-aspartic acid to a beaker. Add 1N sodium hydroxide until the solution is complete. Adjust the pH to 7.4 with sodium hydroxide and dilute to 500 ml with phosphate buffer.

GPT substrate Take 0.146 g of α-ketoglutarate and 8.9 g of DL-alanine taken in a beaker. Add 1N sodium hydroxide until the solution is complete. Adjust the pH to 7.4 with sodium hydroxide and dilute to 500 ml with phosphate buffer.

DNPH colour developer Dissolve 0.099 g of 2,4 dinitrophenyl hydrazine in 500 ml of 1N hydrochloric acid. Store in dark bottle in the refrigerator.

NaOH solution 0.4N Dissolve 8 g of NaOH in distilled water and dilute to 500 ml with distilled water.

Procedure

1. Pipette 0.5 ml of GOT or GPT substrate into a test tube and place it in a water bath at 37°C for 5 minutes.
2. Add 0.1 ml of serum and mix.
3. Incubate exactly for 30 minutes for SGPT and for 60 minutes for SGOT.
4. Add 0.5 ml of DNPH and mix.
5. Let it stand at room temperature for 20 minutes.
6. Add 5 ml of 0.4N NaOH, mix vigorously and let it stand for 5 minutes.
7. A colour change developed and is read calorimetrically at 505 nm. Follow the same procedure for the standards.

Precautions

Serum sample must be free from haemolysis, since RBCs are rich in this enzyme.

Calibration Curve/Standard Graph

The standard graph is prepared by using the following table.

Reagent	Blank	1	2	3	4	5	Incubation
Distilled water	0.2 ml	0.2 ml	0.2 ml	0.2 ml	0.2 ml	0.2 ml	
Substrate	1 ml	0.9 ml	0.8 ml	0.7 ml	0.6 ml	0.5 ml	5 minutes/water bath
Standard	—	0.1 ml	0.2 ml	0.3 ml	0.4 ml	0.5 ml	30 PT/60OT/water bath
DNPH	1 ml	1 ml	1 ml	1 ml	1 ml	1 ml	20 minutes/room temp
NaOH	10 ml	10 ml	10 ml	10 ml	10 ml	10 ml	5 minutes/room temp

Result

Tube no.	True OD	SGOT/IU	True OD	SGPT/IU
1		22		25
2		55		50
3		95		85
4		150		135
5		215		200
TEST				

Aim

To estimate the amount of protein present in the urine.

Background Information

Most plasma proteins are too large to pass through the glomeruli of the kidney. The small amount of protein, which does filter through, is normally reabsorbed back into the blood by the kidney tubules. Only trace amounts of protein (less than 0.15 g per 24 hours) are found in normal urine. These amounts are insufficient for detection by routine laboratory tests.

When more than trace amounts of proteins are found in urine, it is termed as proteinuria. This condition is often referred to as albuminuria because when there is glomerular damage, most of the protein which passes through the glomerular filtrate is albumin because this protein molecule is smaller than most of the globulins.

Causes of Proteinuria

Tubular urinary disease
Pyogenic/tuberculus pyelonephritis
Severe lower urinary tract infections
Nephrotic syndrome
Urinary schistosomiasis
Severe febrile illness
Hypertension accompanying haematuria

Specimen

Urine

Materials Required

Test tubes
pH paper
Sulphosalicylic acid reagent

Sulphosalicylic acid	5 g
Distilled water	25 ml

Procedure

1. Take two test tubes and label one as 'C' for control and another as 'T' for test.
2. Pour about 2 ml of clear urine into each tube (if the urine is turbid or cloudy, filter or centrifuge it to obtain a clear sample).
3. Using the pH paper, test the pH of the urine. If neutral or alkaline, add a drop of glacial acetic acid to each tube and mix.
4. Add 2–3 drops of sulphosalicylic acid reagent to tube 'T'.
5. Holding the tubes against a dark background examine for cloudiness in tube 'T'. The appearance in tube 'T' was reported as follows:

No cloudiness	negative
Slight cloudiness	+
Moderate cloudiness	++
Marked cloudiness	+++
Cloudiness with precipitate	++++

Result

Aim

To detect the presence determination of bile salt in the urine.

Clinical Significance

Bile salt is not present in normal urine.

It is detected in urine, when it is unable to leave the blood circulation through hepatic route into the intestine. The yellow to green colouration of the urine shows the presence of bile salt in urine (hepatitis).

Specimen

Urine

Materials Required

Test tubes

Sulphur powder

Principle

Bile salts lower the surface tension of the urine. When sulphur powder is added to the surface of the urine, it sinks to the bottom of the tube. In normal urine, it floats on the surface.

Procedure

1. Take about 10 ml of urine in a test tube.
2. Sprinkle a little dry sulphur powder on the surface of the urine.
3. Observe the sulphur particles and record the results.

Result

Aim

To determine the urine sugar quantitatively.

Clinical Significance

Presence of glucose in urine (glucosuria) after 12 hours of meals is an indicator of diabetes mellitus.

Principle

The aldehyde group present in the glucose changes the blue-coloured alkaline cupric sulphate to yellow-red cuprous oxide. Several other sugars such as lactose, fructose, galactose, pentose, and non-sugars such as uric acid, creatinine, ascorbic acid and homogentisic acid having a similar reducing property.

Specimen

Urine

Materials Required

Test tubes
Test tube stand
Dropper
Pipette
Bunsen burner
Benedict's reagent

Sodium citrate	173 g
Sodium carbonate	100 g

Dissolve the salt in 900 ml of distilled water, boil for 2–3 minutes, cool and add 17.3 g cupric sulphate.

Dissolve the contents and make the final volume to one litre. The reagent is stable at room temperature.

Procedure

1. Label the required test tubes and arrange them in a stand.
2. Add 5 ml of Benedict's reagent to the test tubes.
3. Using Pasteur pipette add 1 ml of urine to the tubes and mix well.
4. Heat to boil.
5. Cool it in running tap water.
6. Presence of fine yellow, orange or brick-red precipitate indicates a positive reaction.

Observation and Results

Aim

To isolate the pathogenic bacteria from the wound specimens.

Background Information

Wound is an abnormal break in the skin or other tissue, which allows blood to escape. Wounds are of two types—open wound and closed wound. Open wound allows blood to escape from the body. Here the skin is broken. All open wounds are contaminated by germs, which may enter from air, fingers and other parts of the body. Any wound which has not begun to heal properly after 48 hours may be infected. Infection may further spread and cause dangerous illness to human beings.

Staphylococcus aureus is mostly isolated from skin wounds; *Pseudomonas aeruginosa* is associated with infected burns and with hospital-acquired infections. *Escherichia coli* and *Proteus* species are associated with abdominal abscess. *Clostridium perfringens* is found mainly in deep wounds.

Possible Pathogens

Staphylococcus aureus, *Pseudomonas aeruginosa*, *Clostridium* spp., *Escherichia coli*, *Proteus* spp., *Bacteroides*, *Klebsiella* spp., *Streptococcus pyogenes*

Materials Required

Sterile leak-proof container
Sterile cotton swab
Amies transport medium
Blood agar
Neomycin blood agar
Robertson cooked meat medium
MacConkey agar

Specimen Collection

Pus from wound is collected by making an incision in the abscess and draining it or after it ruptures naturally. Collect the pus with a sterile cotton swab and insert it in a leak-proof container of Amies transport medium.

Before inserting into the container, make a smear of the pus on a clean slide and allow it to air-dry.

Label the specimen and send with a request form to reach the microbiology laboratory.

Methodology

Macroscopic and microscopic examination

- Examine the colour and the nature of the specimen.
- Fix the air-dried smear with methanol.
- Stain with gram-staining technique.
- Examine under 40× objective for the pathogens. It may be helpful for media selection.

Culturing

Inoculate the specimen on two blood agars, neomycin blood agar, MacConkey agar and cooked meat medium.

Incubate one blood agar and MacConkey agar aerobically at 35–37°C overnight. Incubate the neomycin blood agar and the other blood agar plate anaerobically at 35–37°C for 48 hours.

Incubate the cooked meat medium at 35–37°C for 72 hours.

Then subject the isolated colony to biochemical tests for species identification.

Observation and Result

Aim

To analyse the aetiology of urinary tract infection.

Background Information

Urine is normally a sterile body fluid. However, unless it is collected properly, it can be contaminated with microbiota from the perineum,prostate, urethra or vagina. The presence of bacteria in urine is called bacteriuria. Significant bacteriuria is usually accompanied by pyuria (pus cells in urine). Infection of the bladder is called cystitis and infection of kidney is called pyelonephritis. *Escherichia coli* is the commonest cause of urinary tract infection.

Possible Pathogens

Staphylococcus saprophyticus, *Pseudomonas aeruginosa*, *Escherichia coli*, *Proteus* sp., *Klebsiella* sp., Haemolytic streptococci, E*nterococcus* sp.

Media

Blood agar, MacConkey agar, cetrimide agar, CLED agar, SS agar, KF streptococcus agar, nutrient agar.

Sample Collection

Mid-stream urine It is collected in sterile, dry, wide necked, leak-proof container. About 20 ml of sample should be collected. Clean catch method is used to collect the mid-stream urine, first voided urine is not collected because it may be contaminated with microbes from lower portion of the urethra. If immediate delivery to the laboratory is not possible, the urine should be refrigerated at 4°C. If a delay of more than 1 hour is anticipated, boric acid should be added to the urine. Specimen containing boric acid need not be refrigerated.

Macroscopy

Appearance, colour and nature of the specimen are noted. Normally freshly passed urine is clear and pale yellow-to-yellow depending on the concentration. Colour and nature of the urine is altered during infection. The appearance of urine during infection is given in Table.

Colour	**Infection**
Cloudy	Bacterial
Red and cloudy	Bacterial and urinary schistosomiasis
Brown and cloudy	Black water fever
Yellow-brown and green-brown	Acute viral hepatitis
Yellow orange	Haemolysis and hepatocellular jaundice
Milky white	Bancroftian filariasis

Microscopic Examination

- 3 loopful of well-mixed fresh urine is placed on a slide and cover with a cover glass.
- This preparation is examined using the 10× and 40× objectives.
- Gram staining is performed using centrifuged urine.

Findings of microscopic examination are bacteria, white cells, pus cells, red cells, yeast cells, epithelial cells, casts (hyaline, waxy, cellular, granular), crystals and parasites.

Culturing

- Approximate number of bacteria per ml of urine can be estimated by using calibrated loop technique (0.002 ml capacity or 1/500 ml or $20 \times 500 = 10{,}000$).
- Urine is diluted up to 10^{-8}.
- Nutrient agar plates are prepared and divided into 8 portions.
- Steak is performed on nutrient agar plate using diluted urine (normal urine has less than 10^4 bacteria).
- Selective medium like blood agar, MacConkey agar, CLED agar, cetrimide agar and SS agar are inoculated using a loopful of urine sample.
- All the plates are incubated under aerobic condition at 37°C for 24 hours.

Observation and Result

Aim

To identify the gastrointestinal pathogens.

Background Information

Normal human intestine harbours more than 500 types of microbes. Among these some of the microbes are considered as pathogens. Pathogens may enter through food and water systems. Most of the intestinal disorders are based on the toxins.

Possible Pathogens

Escherichia coli, *Salmonella enteritidis*, *Shigella* sp., *Campylobacter*, *Vibrio* sp., *Plesiomonas* sp., *Aeromonas* sp., *Yersinia enterocolitica*, *Clostridium perfringens*, *Clostridium difficile*, *Staphylococcus aureus*, *Bacillus cereus*.

Media

GN broth, HE agar, TCBS agar, alkaline peptone water, Yersinia selective medium, biochemical test medium and reagents, cefsulodin irgasan novobiosin agar, cycloserine cefoxin fructose egg yolk agar, campy blood agar, blood agar, blood agar with 10 mg ampicillin, inositol brilliant green bile salt agar.

Specimen

Stool or rectal swab

Specimen Collection

The collection of diarrhoeal stool is not difficult. In cases of diarrhoea, stool specimen should be collected in clean, wide mouthed containers that can be covered with tight fitting lid. In some instances, collection of rectal swab rather than stool specimen may be necessary, particularly in neonates or in severely debilitated adults.

Transport

The specimen should be transported as early as possible. Don't refrigerate the stool specimen because certain species of *Shigella* sp. are susceptible to cooling and drying.

Culturing

- Gram-negative broth and alkaline peptone water are inoculated with few loopful of stool specimen and incubated at 37°C for 4–5 hours and the turbidity is observed. This step is used to enrich the pathogens.
- Hektoen enteric agar is streaked by using gram-negative broth inoculum and TCBS agar with alkaline peptone water inoculum. Plates are observed for the pathogens after 24 hours of incubation at 37°C.
- Rest of the media are streaked by direct method.
- Incubation is performed with either aerobic method or anaerobic method based on the medium and causative agent.
- After the completion of primary plating techniques, bacterial pathogens are identified through various biochemical means.

Observation and Result

Aim

To isolate pathogens from the blood sample.

Background Information

The presence of bacteria in blood is called bacteraemia. Transitory bacteraemia can occur during the course of many infections. Continuous bacteraemia most often suggests an intravascular source of infection. The term septicaemia refers to a severe and often fatal infection of the blood in which bacteria multiply and release toxins in the bloodstream. The symptoms of septicaemia include fever, chills and shock.

Bacteria that can be associated with neonatal septicaemia include *Escherichia coli*, staphylococci, beta haemolytic group B streptococci and other coliforms.

Viridans streptococci is the commonest cause of sub-acute infective endocarditis. In typhoid, *Salmonella typhi* can be detected in the blood of 75–90% patients during the first 10 days of infection.

Possible Pathogens

Staphylococcus aureus, Viridans streptococci, *Streptococcus pneumoniae*, *Streptococcus pyogenes*, *Salmonella typhi*, *Escherichia coli*, *Klebsiella pneumoniae*, *Corynebacterium diptheriae*, *Yersinia pestis*, *Leptospira* sp., *Brucella* sp., Beta haemolytic group B Streptococci, *Proteus* sp., *Haemophilus influenzae*, *Neisseria* sp.

Media

Thioglycolate broth, Tryptone soy diphasic broth, Blood agar, Chocolate agar, MacConkey agar, SS agar, EDTA, materials for WBC and differential count, Giemsa stain.

Sample Collection

Blood collection is performed by needle aspiration procedure. To reduce the contamination during vein puncture, the following method should be followed.

- Wash with green soap.
- Rinse with clean water.
- Apply 1–2% tincture of iodine and allow to dry for 1–2 minutes.
- Remove iodine with 70% alcohol.

Precautions

The following precautions should be taken to ensure 100% microbial isolation. Blood should be collected before antimicrobial treatment and at the time when the patients temperature begins to rise. Blood for culture should be taken by vein puncture, 10–20 ml of blood collected from adult, 1–2.4 ml collected from young infants, 2.4–5 ml collected from old infants, Two-blood culture should be performed from each patient to confirm the causative agent.

Culturing

- Thioglycolate and tryptone soy diphasic medium is prepared and sterilized at 121°C for 15 minutes.
- Blood is collected from the patient by venipuncture procedure.
- The blood sample is divided into three portions. One portion is inoculated into thioglycolate medium and one into tryptone soy diphasic medium.

- Remaining portion is inoculated into a bottle containing EDTA. This blood is used to perform total count, differential count, staining and also streak one loopful of blood into the SS agar.
- All the media are incubated at 37°C.

Examination and Subculturing

- Thioglycolate medium is examined daily for up to 14 days.
- Visible signs of bacterial growth are observed (turbidity above the red cell layer).
- A needle is inserted through rubber liner in the cap and 1ml of broth culture is withdrawn.
- It is inoculated into blood agar, chocolate agar and MacConkey agar.
- Blood agar and chocolate agar are incubated anaerobically for 48 hours and MacConkey agar plate aerobically for overnight.
- Tryptone soy biphasic culture is examined daily for 7 days and twice a week for up to 4 weeks.
- Growth is observed on the agar slope and signs of bacterial growth on broth.
- If growth is present subculture on blood agar, chocolate agar and MacConkey agar.
- Gram-staining is performed to check the preliminary nature of bacterial pathogens.
- If large gram-positive rods resembling C. *perfringens* are seen, subculture on lactose egg yolk milk agar and incubate the plate anaerobically.
- If Brucella is suspected, increased attention should be taken and marked as high risk.

Observation and Result

Introduction

Lower respiratory tract infections involve lung and bronchi. Normally lower respiratory tract is a sterile organ. It is involved in exchange of gases. This organ is most accessible to microorganisms but it is protected with alveolar macrophages and mucus. If microbes grow by breaking the immune response, may lead to severe infections like pneumonia, tuberculosis, etc.

Possible Pathogens

Streptococcus pyogenes, *Klebsiella pneumoniae*, β-haemolytic group B Streptococci, *Pseudomonas aeruginosa*, *Staphylococcus aureus*, *Streptococcus pneumoniae*, *Legionella pneumophila*, *Moraxella catarrhalis*, *Yersinia pestis*, *Bacillus anthracis*, *Mycobacterium tuberculosis*.

Specimen

Early morning sputum is collected from the patients. It contains pooled overnight secretions. More concentrated pathogenic bacteria are present in the early morning sputum. Sterile wide mouthed jar with a tightly firmed screw cap lid is used for sample collection and transported to the laboratory at the earliest.

Media

Blood agar, LJ medium, chocolate agar, buffered charcoal yeast extract agar, MacConkey agar, Bordet-Gengou medium, Baird-Parker agar, cetrimide agar.

Other Materials

Petri plates, blood, pipettes, inoculation loop, conical flask, cotton, etc.

Methodology

Gram Stain

Gram stain can aid in rapid diagnosis and appropriate treatment. It is an efficient method for the assessment of the quality of the sputum based on cellular composition of the specimen. The following cells are observed after staining. They are PMN (polymorphonuclear leucocytes), squamous epithelial cells, ciliated columnar epithelial cells and bacteria. Bartlett grades the sputum sample on the basis of gram stain and the presence or absence of neutrophils and epithelial cells. Table represents the grading of sputum sample.

Number of neutrophils per field	Grade
<10	0
10–25	$+1$
>25	$+2$
Presence of mucus	$+1$
Number of epithelial cells per field	Grade
10–25	-1
>25	-2

- Purulent portion of the sputum sample is smeared on the microscopic slide.
- The smear is fixed by heat and methanol fixing procedure.
- Gram stain technique is adopted to stain the smear.
- The smear is observed under 40× and 100× objectives of compound microscope.

Acid-fast Staining

- Purulent part of the sputum is transferred to clean microscopic slide using a piece of stick.
- Thin smear is prepared (Use a circular movement to spread the specimen).
- The smear is allowed to air dry in a safe place.
- Methanol fixing is done by adding methanol for 2–3 minutes.
- The slide is placed on the water bath.
- Carbol fuschin stain is added over the smear.
- Heat is applied until vapour begins to rise. (Do not over heat).
- Heated stain is allowed to remain on the slide for 5 minutes.
- The smear is washed completely with clean water.
- The smear is decolorized with 3% v/v acid alcohol for 5 minutes.

- With clean tap water smear is washed.
- The smear is stained with malachite green for 1–2 minutes.
- Smear is rinsed again with clean tap water.
- Allowed the smear to air dry and the smear is examined microscopically.

Reporting

More than10 AFB/field	+++
1–10 AFB/field	++
10–100AFB/100field	+
1–9AFB/100field	report the exact number

Culturing

Routine

- Homogenization is performed by adding 5ml of sterile physiological saline.
- Blood agar, chocolate agar and MacConkey agar plates are inoculated with loopful of sputum sample.
- Optochin disc is placed in one corner of chocolate agar plate. This will help to identify *Streptococcus pneumoniae*.
- Blood agar and MacConkey agar plates are incubated aerobically and chocolate agar plate in a CO_2 enriched atmosphere at 35–37°C for up to 48 hours.
- Plates are examined for the growth and colony morphology, and the results are reported.

For *Mycobacterium tuberculosis*

- Sputum is decontaminated by adding equal volumes of sputum and NaOH 40 g/l solution 20 minutes before culturing (Shake at intervals to homogenize the sputum).
- 200 μl of the sputum is inoculated on a slope of Lowenstein-Jensen medium and it is allowed to run down the slope (Slope turn yellow due to alkalinity of the specimen but it will become green again when acid in the medium is neutralized by the NaOH).
- Tubes are incubated at 37°C in rack placed at an angle of about 45° to ensure that the specimen is in contact with the full length of the slope.
- Media is observed after one week of incubation and continued for a further 5–6 weeks (examine twice a week for growth).

For other bacteria

- Mycoplasma is isolated by inoculating the specimen in mycoplasma isolation media (Hi-Media).
- Cetrimide medium is used for isolation and differentiation of *Pseudomonas aeruginosa*.
- Legionella isolation medium is used for the selective cultivation of *Legionella pneumophila*.

Identification

Biochemical tests are adopted along with microscopic and macroscopic examinations to identify pathogenic isolates.

Observation and Result

Aim

To identify the given gram-negative pathogens.

Materials Required

Nutrient agar
MacConkey agar
XLD agar
TCBS agar
SS agar
Peptone water
MR–VP broth
Simmons citrate agar
Triple sugar iron agar
Urea agar base
OF basal medium
Nitrate broth
Decarboxylase medium
Phenylalanine test medium
Esculin agar

Reagents

Gram-staining kit
Capsule-staining kit
Kovac's indole reagent
Methyl red solution
Barrits reagent A and B
Nitrate reagent A and B
Ferric chloride reagent

Preliminary Test

The given organism was streaked on nutrient agar plates to check its purity and incubated at 37°C for 24 hours. After completion of incubation period, observe the plate and note the colony morphology. Single colony was streaked onto a sterile dry nutrient agar slant. This culture was subjected to further analysis.

Colony Morphology

Size	Pinpoint, small, moderate, large.
Colour	Pale yellow, colourless, brown, yellow, black, pink, red, and green.
Form	Punctiform, irregular, circular, filamentous, rhizoid, spindle.
Margin	Entire, undulate, lobate, erose, filamentous, curled.

Elevation	Flat, raised, convex, pulvinate, umbonate, umblicate.
Opacity	Opaque, translucent, transparent.

Gram Staining

Prepare the bacterial smear on a clean transparent glass slide and fix the smear by both methanol fixation and heat-fixing procedure (culture from broth) or heat-fixing procedure (culture from slant). Crystal violet serves as the primary stain which binds to the bacterial cell wall, after treatment with weak solutions of iodine, which serves as the mordant to bind the dye. Some bacterial species because of their chemical nature of cell wall, have the ability to retain the crystal violet even after treatment with an organic decolorizer. These bacteria appears as blue-black when observed under the microscope and are called gram-positive. Certain bacteria lose the crystal violet primary stain when treated with decolorizer, because of the high concentration of lipid in the cell wall. These decolorized bacteria then pick up safranin counterstain and appear pink in colour when observed under the microscope and are called gram-negative.

Use of Selective Medium

Streak the inoculum from the nutrient agar slant on any two selective and differential media. Incubate the culture at 37°C for 24 hours. Observe for colony morphology and nature of colony.

Capsule Staining

If lactose-fermenting colonies are observed on MacConkey agar, perform capsule-staining technique. Capsule can be clearly visualized through Hiss technique. Cover the dried air-fixed smear with crystal violet and heat gently until the steam just begins to rise. Leave the smear for a minute. Wash off the smear and observe under 40× objective for the presence or absence of capsule.

Motility

Bacteria are motile by means of flagella, the number and location of which vary among different species. Motility can be observed directly by placing a drop of culture on a microscopic slide. This technique is called hanging drop technique.

Indole Test

Indole is one of the degradation products obtained from the metabolism of the amino acid tryptophan. Bacteria that possess the enzyme tryptophanase are capable of cleaving tryptophan, thus producing indole, pyruvic acid and ammonia. Indole can be detected in tryptophan test medium by observing a red coloured ring after adding a solution containing-dimethyl aminobenzaldehyde.

Methyl Red Test

Bacteria that follow primarily the mixed acid fermentation route often produce sufficient acid to maintain a pH below 4.4. The MR test provides a valuable characteristic for identifying bacterial species that produce strong acid from glucose. Inoculate MR–VP broth with a pure culture of the test organism. After incubation for 24–48 hours, add 5 drops of methyl red reagent directly to the broth. The development of red colour indicates positive result.

Voges-Proskauer Test

The VP test is used to detect the formation of acetyl methyl carbinol (acetoin). Acetoin is formed as a product of an alternative pathway of glucose fermentation. It is detected by the addition of 5% α-naphthol followed by 40% KOH to the MR–VP medium. Shake the tube gently and allow it to remain undisturbed for 10–15 minutes. Formation of brownish red colour indicates positive result.

Citrate Utilization Test

The principle of the citrate utilization test is to determine the ability of an organism to utilize sodium citrate as the sole source of carbon for metabolism and growth. To perform the test, a single colony from primary isolation medium was streaked on Simmons citrate agar slant. The formation of blue colour in the test medium after the incubation of 24 hrs at 37°C indicates the presence of alkaline products and a positive citrate utilization result. Occasionally, visible growth was detected along the streak line before the conversion of the medium to blue colour. This visible growth also indicates positive result.

Urease Production Test

Microorganisms that possess the enzyme urease that hydrolyses urea, release ammonia and produce a pink-red colour change in the medium. Most commonly, the test is performed using an urea agar slant. The surface of the agar slant was streaked with test organism and allowed to incubate overnight at 35°C. The red colour change on the slant indicates positive result.

Nitrate Reduction Test

Any medium that supports the growth of an organism and contains 0.1% or 1% potassium nitrate or sodium nitrate is suitable for performing the test. Inoculate the medium with the test organism and incubate at 37°C for 18–24 hours. At the end of incubation, add 1 ml of reagent A (alpha naphthalamine) and reagent B (sulphanilic acid) to the test medium. The development of a red colour within 30 seconds indicates the presence of nitrites and represents positive reaction for nitrate reduction. If the colour is not observed, add a pinch of zinc powder. If colour development is observed after the addition of zinc powder the result is ribate negative. If colour is not developed the result is positive. It means nitrate is converted into N_2O or N_2. Red colour is formed only in the presence of nitrite.

Decarboxylation of Lysine, Ornithine and Arginine

Many species of bacteria possess enzymes capable of decarboxylating a specific amino acid in the test medium. This enzyme removes a molecule of carbon dioxide from an amino acid to form alkaline reacting amines. The following are the amino acids most commonly used for decarboxylation process lysine → cadaverine

ornithine → putrosine

arginine → citrulline

The test is performed by inoculating two tubes of Moeller decarboxylase medium, one containing the amino acid to be tested, and the other to be used as a control tube devoid of amino acid. Both tubes are overlaid with sterile mineral oil over the surface. The end point of the reaction is the production of an alkaline pH shift in the medium containing the amino acid and the development of blue colour. The control tube turns yellow as a result of fermentation of glucose and the absence of amino acid decarboxylation.

Phenylalanine Deaminase Test

Certain bacteria have the capacity to degrade phenylalanine with the production of phenylpyruvic acid. Inoculate the test organism on the slant of phenylalanine test medium and incubate overnight at 35°C. The test is positive if a visible green colour develops after the addition of 4 or 5 drops of ferric chloride solution.

Cytochrome Oxidase Activity

Any organism that displaces cytochrome oxidase activity is excluded from the Enterobacteriaceae. Any one of the two methods commonly performs this test:

i. The direct plate technique, in which 2–3 drops of reagent was added directly to the isolated colonies growing on the plate medium.
ii. The indirect paper strip procedure, in which either a few drops of reagent are added to the filter paper and placed on culture, or reagent-impregnated filter paper was rubbed with colonies.

Bacterial colonies having cytochrome oxidase activity develop deep blue colour at the inoculation site within 10 seconds.

Catalase Test

Catalase acts as a catalyst in the breakdown of hydrogen peroxide to oxygen and water. Take the organism with a sterile wooden stick or a glass rod and immerse into the hydrogen peroxide solution. Observe for immediate bubbling and record the result. The culture should not be more than 24 hours old. Nichrome wire should not be used.

TSI Agar Test

This test is used for the differentiation of Enterobacteriaceae members on the basis of sugar fermentation, anaerobic respiration, hydrogen sulphide production and the production of other end products. This test is performed with the help of TSI slants. Inoculate the organism by stab and streak technique. Observe the result after 8 hours up to 48 hours at 37°C.

OF Test (Oxidative–Fermentative)

Inoculate the test organism into two tubes of OF basal medium along with any one of the carbohydrates. Seal one inoculated medium with 1 cm of liquid paraffin to exclude oxygen. Keep the other tube open. Fermentative organisms utilize sugar in both tubes but oxidative organisms utilize sugar in the open tube only. Inoculate the culture using a straight loop and incubate the tubes at 35–37°C for up to 14 days. Phenol red agar base medium may also be used for the detection of acid production from sugars.

Esculin Hydrolysis

Esculin is beta glucose 6,7 dihydroxy coumerin. It is hydrolysed by some bacteria to produce black coloured compound. Inoculate esculin agar with the test organism and incubate at 37°C for 24 hours and observe the result.

Observation

Example

Specification of the culture	Culture number
	Date
	Purity checking date
	Purity of the culture
	Morphology
	Grams nature
	Motility
Growth characters	On XLD agar
	On SS agar
	On MacConkey agar
	On HE agar
	On TCBS agar
	On EMB agar
Biochemical characters	Indole
	Methyl red
	Voges-Proskauer
	Citrate
	Catalase
	Oxidase
	Urease
	Nitrate
	TSI
	Carbohydrate fermentation
	Decarboxylation
	Deaminase
	Esculin hydrolysis

Result

The given test organism was found to be .

Aim

To identify the given gram-positive bacterium.

Materials Required

Nutrient agar
Blood agar
Baird-Parker agar
Mannitol salt agar
Cooked meat medium
Streptococcus selection broth
Saline
Bacitracin
SXT and optochin disc

Growth on Nutrient Agar

The given organism is streaked on nutrient agar plates to check its purity and incubated at 37°C for 24 hours. After completion of incubation period the plates are observed and the colony morphology is noted. Single colony is streaked on sterile dry nutrient agar slant. This culture is subjected to further analysis.

Colony Morphology

Form	Punctiform, irregular, circular, filamentous, rhizoid, spindle.
Margin	Entire, undulate, lobate, erose, filamentous, curled.
Elevation	Flat, raised, convex, pulvinate, umbonate, umblicate.
Size	Pinpoint, small, moderate, large.
Colour	Pale yellow, colourless, brown, yellow, black, pink, red, and green.
Opacity	Opaque, translucent, transparent.

Gram Staining

Prepare a bacterial smear on a clean transparent glass slide and fix the smear by both methanol fixation and heat-fixing procedure (culture from broth) or heat-fixing procedure (culture from slant). Staining was performed by Gram's procedure. Crystal violet serves as the primary stain, binding to the bacterial cell wall after treatment with weak solutions of iodine, which serve as the mordant to bind the dye. Some bacterial species because of their chemical nature of cell wall have the ability to retain the crystal violet even after treatment with an organic decolorizer. These appears blue-black when observed under microscope and these bacteria are called gram-positive. Certain bacteria lose the crystal violet primary stain when treated with decolorizer, because of the high concentration of lipid in the cell wall. These decolorized bacteria then pick up safranin counterstain and appear pink when observed under microscope and are called gram-negative.

Use of Selective Medium

Streak the inoculum from the nutrient agar slant on any two selective and differential media. Incubate the culture at 37°C for 24 hours. Observe for colony morphology and nature of colony.

Capsule Staining

Capsule staining technique is performed when the culture is grown on blood agar. Capsule can be clearly visualized through Hiss technique. Cover the dried air-fixed smear with crystal violet and heat gently until the steam just begins to rise. Leave it for a minute. Wash off the smear and observe under 40× objective for the presence or absence of capsule.

Coagulase Test

Prepare 1 in 10 dilution of the plasma in saline. Transfer 0.5 ml of the diluted plasma into the small test tubes. Inoculate the diluted plasma with 0.1 ml of 18–24-hour-old broth culture. Incubate the tubes at 37°C preferably in a water bath and examine after first, second, third and fourth hours. For performing the coagulase test, the culture should be from blood agar or nutrient agar and not from mannitol salt agar or Baird-Parker agar.

Hippurite Hydrolysis

Inoculate the test culture into media containing hippurite and incubate at 35°C for 20 hours. Centrifuge the culture at 5000 rpm for 5 minutes. Pipette out 0.8 ml of a clear supernatant into a test tube. Add 0.2 ml of ferric chloride reagent and mix. A heavy precipitate that persists for 10 minutes or longer indicates a positive result.

Bile Solubility Test

Grow the organisms in broth for 24 hrs. Prepare 2% bile and normal saline in tubes. Add 2 drops of phenol red indicator to each tube and adjust the pH to 7 by adding a drop of 1/20 N NaOH. Place 0.2 ml of bile and 0.5 ml of saline in separate tubes. Place the broth tubes in a water bath shaker at 37°C. Examine the tubes periodically for 24 hours.

Bacitracin, SXT and Optochin Sensitivity Test

Prepare blood agar plates and swab the given culture over the surface. Place bacitracin, SXT and optochin discs in appropriate places and incubated at 37°C aerobically or anaerobically depending on the nature of culture. Examine the culture after 24 hours.

Salt Tolerance Test

Prepare nutrient agar plates and nutrient broth tubes and add 6% sodium chloride. Sterilize the medium at appropriate temperature. Inoculate the culture and incubate at 37°C for 24 hours. Examine growth and turbidity.

Esculin Hydrolysis

Esculin is beta glucose 6,7 dihydroxy coumerin. It is hydrolysed by some bacteria to produce black coloured compound. Inoculated esculin agar with test organism and incubated at 37°C for 24 hours and observed the result.

CAMP Test

Prepare blood agar plates and inoculate *Staphylococcus aureus* as single line streaks in one edge of the plate. Streak the test organism in another edge in a zig-zag manner. Incubate the plates anaerobically. Observe the plate for growth and junction place of both the cultures.

Gelatin Hydrolysis

Prepare gelatin-containing medium and sterilize. Pour the medium into petri plates and make a single streak with the test organism in the centre place of the plate. Incubate the plate at 37°C for 24 hours. Add 2 ml of mercuric chloride solution. Observe for clear zone in and around the area of the streak.

Starch Hydrolysis

Prepare starch agar plate and streak the test organism on the surface as a single line. Incubate the plate at 37°C for 24 hours. Flood with iodine solution and observe for colour change and clear zone.

Lipid hydrolysis

Inoculate 4 ml of sterile Tween 80 phosphate-buffered substrate with a loopful of test organism. Incubate at 37°C for 18 days. Observe the colour change during incubation.

DNase test

Prepare DNase test agar and divide the plates into 6 equal parts. Inoculate various test organisms in each portion. Incubate at 37°C overnight. Cover the surface of the plate with concentrated hydrochloric acid solution. Clearance around the colonies were observed within 5 minutes of adding acid.

Oxidative-fermentative

Carbohydrate Fermentation (OF test)

Inoculate the test organism into two tubes of OF basal medium along with any one of the carbohydrates. Seal one inoculated medium with 1 cm of liquid paraffin to exclude oxygen. Leave the other tube open. Fermentative organism utilizes sugar in both tubes but oxidative organism utilizes sugar in open tube only. Inoculate the culture using straight loop to the bottom of the tube. Incubate the tubes at 35–37°C for up to 14 days. Phenol red agar base medium is also used for the detection of acid production from sugars.

Observation

Example

Specification of the culture

Culture number

Date

Purity checking date

Purity of the culture

Morphology

Gram staining nature

Growth characters

Biochemical characters

Result

The given test organism was found to be ___________

General Information

The age of fresh faecal specimens is an important factor in the diagnosis of parasitic infections.

Preserve or examine semi formed specimens within one hour of passage.

Preserve or examine formed specimen within the same day.

Preserve or examine liquid or soft stools containing blood or mucus within 30 minutes of passage.

Examine specimen microscopically to determine consistency.

Examine the surface of the faecal specimen for the presence of blood and mucus.

Areas of blood or mucus should be examined for trophozoites of amoebae.

The collection of 3 faecal specimens is usually required to diagnose intestinal parasitic diseases.

Saline and Iodine Wetmount

The microscopic examination of direct smear has several purposes.

1. To assess the worm burden of a patient.
2. To enable quick diagnosis of a heavily infected specimen.
3. To check the motility of an organism.

Materials Required

0.85% NaCl

Lugol's iodine

Procedure

1. Place one drop of 0.85% NaCl on a clean glass slide.
2. Emulsify a small portion of faecal material in a drop of saline.
3. Place a coverslip on each suspension.
4. Systematically scan the suspensions using the 10× objectives. Examine entire coverslip area.
5. If you see something suspicious use the 40× objective for more detailed study.
6. For iodine wet mount, use Lugol's iodine instead of 0.85% NaCl.

Advantages and Disadvantages

Saline mounts have the advantage over fixed and stained smears, in that the motility of trophozoites can be observed. Protozoan cysts also appear more refractile on saline preparation than on iodine preparation.

Definitive identification of either trophozoites or protozoan cysts is difficult in saline mounts because the internal structures are often poorly delineated.

Faecal concentration has became a routine procedure in complete ova and parasite examination and allows the detection of small amount organisms that may be missed by using only a direct wet smear.

SEDIMENTATION

Formalin–Ethyl Acetate

By centrifugation, this concentration procedure leads to the recovery of all protozoa, eggs and larvae present; however, the preparation contains more details than the floatation method. Ethyl acetate is used as an extractor of debris and fat from the faeces and leaves the parasites at the bottom of the suspension. The formalin–ethyl acetate sedimentation concentration is recommended, because it is the easiest to perform, allows recovery of the broad range of organisms, and is least subject to technical error.

Procedure

1. Transfer ½ teaspoon of fresh stool into 10 ml of 10% formalin in a vial or round bottom tube. Mix the stool and formalin thoroughly. Let the mixture stand a minimum of 30 minutes for fixation.
2. Depending upon the amount and viscosity of the specimen, stain a sufficient quantity through wet gauze into a conical 15-ml centrifuge tube to give the desirable amount of sediment.
3. Add 0.85% NaCl almost to the top of the tube and centrifuge for 10 minutes at 2000 rpm. The amount of sediment obtained should be ½ to 1 ml.
4. Decant the supernatant fluid and suspend the sediment in saline. Add saline almost up to the top of the tube and centrifuge again for 10 minutes at 2000 rpm.
5. Decant the supernatant fluid and suspend the sediment at the bottom of the tube in 10% formalin. Fill the tube half full only. If the amount of sediment left in the bottom of the tube is very small or the original specimen contains lot of mucus, do not add ethyl acetate.
6. Add 4–5 ml ethyl acetate. Stopper the tube, shake vigorously for at least 30 seconds.
7. Centrifuge for 10 minutes at 2000 rpm.
8. Four layers should result;
   i. A small amount of sediment in the bottom of the tube.
   ii. A layer of formalin.
   iii. A plug of faecal debris on the top of formalin.
   iv. A layer of ethyl acetate at the top.

9. Free the debris by ringing the plug with an applicator stick; decant all the supernatant fluid. After proper decanting, one or two drops of fluid remaining on the side of the tube may run down into the sediment. Mix this fluid with sediment.
10. If the sediment is still the same, add one or two drops of saline to the sediment mix, add a small amount of material to slides, place a coverslip and examine under 10× objective.
11. The entire coverslip area should be examined.

Result

Protozoan trophozoite and/or cyst eggs and larvae of helminths may be seen and identified. Protozoan trophozoites are less likely to be seen.

FLOATATION METHOD

Aim

To concentrate parasites present in the faecal matter on the basis of specific gravity.

Principle

The floatation procedure permits the separation of protozoan cysts and helminthic eggs from excess debris through the use of liquid with the help of specific gravity. The parasitic elements are recovered in the surface film, and the debris remains at the bottom of the tube. This technique yields a preparation than the sedimentation procedure; however, some helminth eggs do not concentrate well with the floatation method. The specific gravity of the zinc sulphate can be increased, although this usually causes more distortion in the organism present and is not recommended for routine clinical use. To ensure the detection of all possible organisms, both the surface film and sediment must be examined. For most laboratories, this is not a practical approach.

Specific gravity of hookworm egg is 1.055.

Trophozoites deteriorate rapidly in stool specimens, and therefore preservatives like polyvinyl alcohol or the merthiolate-iodine-formaldehyde (MIF) combination are important diagnostic aids.

Trophozoites can be found in diarrhoea. Most infections in formed stool specimens will be detected by examining three specimens passed over a 7–10-day period. Amoebae are difficult to demonstrate in aspirates from extraintestinal abscesses.

Blood parasites can be detected by means of Giemsa and Wright's staining. It is used to differentiate nuclear or cytoplasmic morphology of platelets, RBCs, WBCs and parasites. To observe clear-cut morphology of blood parasites we should prepare blood films. There are two types of blood films, thin blood film and thick blood film. Stained blood films are the most reliable and efficient means for definitive diagnosis of nearly all blood parasites. For some parasites it is recommended to prepare a thin film on one slide, a thick film on another slide and a combination on a third slide.

Specimens

Preparation of Thin Blood Film

Blood from finger puncture or EDTA blood (0.02 g/10 ml) is used.

Place a drop of blood onto the centre of the slide about ½ inch from the end.

Holding a second clean slide at a 40° angle, touch the angled end to the midlength area of the specimen slide. Pull the angled slide back into the blood, and allow the blood to almost fill the end area of the angled slide.

Continue contact with the blood under the lower edge, quickly and steadily moving the angled slide until the blood is used up.

Label the slide appropriately and allow it to air-dry.

Preparation of Thick Blood Film

It is used to differentiate a low parasitemia. For finger puncture, touch a clean glass slide to a large drop of blood standing on the finger until the circle of blood is nearly 1.8–2 cm. For vein puncture blood, place a drop of blood in the centre of the slide. Using either the corner of another slide or an applicator stick, spread the blood into a circle about 1.8–2 cm. Allow the film to air-dry. Do not fix the thick film.

Giemsa Stain

Fix thin blood film in methanol.

Air-dry.

Stain with diluted Giemsa stain for 20 minutes.

Wash by briefly dipping the slide in buffered water.

Air-dry.

Wright's Stain

Prepare thin and thick blood film.

Add wrights stain drop by drop. Count the number of drops needed to cover the surface.

Let it stand for 1–3 minutes.

Add the same drop of buffered water and mix the stain.

After 4–8 minutes, flood the stain from the slide with buffered water.

Wipe the stain.

Air-dry.

Observe under microscope

Plasmodium, *Trypanosoma*, *Leishmania*, *Wuchereria bancrofti* are the parasites detected through Giemsa and Wright's staining procedure.

Aim

1. To determine the effect of antibiotics on microbial growth.
2. To assess the sensitivity pattern of the given microbe.

Background Information

Antibiotics are synthesized by microbial cells that inhibit or arrest the growth of other microorganisms. This process is often called antagonism.

Antibiotics inhibit the growth by such as inhibit protein synthesis, damaging cell wall, interfering with PG layer synthesis, inhibiting nucleic acid synthesis and preventing the formation of cell membrane. Antibiotics are classified into broad spectrum and narrow spectrum antibiotics on the basis of mode of action. Antimicrobials have increased activity and stability, simpler method of administration, better diffusibility into the remote area and greater selective toxicity.

Antibiotic assay is performed by using disc diffusion technique. Two great scientists William Kirby and A.W. Bauer developed it during the year 1966. In this, antibiotics present in the disc diffuse into the remote area. Success depends on, amount of inoculum, nature of disc, moisture content of media and incubation condition. This test is done to determine which antibiotic is effective against the particular pathogen.

Materials Required

Culture

3–4 colonies of pathogen.

Media

Peptone water

Mueller-Hinton agar

This media provides required concentration of Ca^{++} and Mg^{++} ions to growing bacteria. It will avoid false positive and false negative reactions.

Antibiotics

Antibiotic discs for selective pathogens.

General Materials

Petri plates

Beakers

Test tubes

L-rod

Cotton

Antibiotic zone scale

Millimetre scale

pH meter

0.5 McFarland Standard

Solution A

- Barium chloride
- Distilled water 100 ml

Solution B

- Sulphuric acid 1 ml
- Distilled water 100 ml

Mix 0.5 ml of Solution A and 99.5 ml of solution B in a screw cap tube. The mixture is stored in dark place.

Methodology

1. Select three or four colonies of the pathogens of the same species.
2. Inoculate these pathogens into the nutrient broth.
3. Incubate it for few hours at 35°C or until slight turbidity.
4. Prepare Mueller-Hinton agar and pour into the plates up to the uniform thickness of approximately 4 mm.
5. Immerse cotton swab into the bacterial culture-containing tube and then rotate, compress against the wall of the tube so as to express excess fluid.
6. Inoculate the surface of the plate (MHA) with the swab. To ensure uniform and confluent growth, pass the swab three times over the entire surface.
7. Allow the plates to dry for five minutes.
8. Using appropriate technique, place antibiotic discs on the surface of MHA plates. One plate requires at least six discs and distance between each disc should be at least 20 mm.
9. Incubate all plates at 37°C for 18–24 hours.
10. Examine the plates for inhibitory zone and measure the zone in mm and interpret the results as per standard. Mention the results as sensitive, moderate and resistant.

Main Causes of Error

1. Amount of inoculum.
2. Contamination of inoculum.
3. Reduction of disc potency.
4. Error in measuring the zone of inhibition.
5. Poor quality control.

Observation and Result

Aim

To assess the minimal bactericidal concentration (MBC) or minimal inhibitory concentration (MIC) of antibiotics by agar dilution method.

Background Information

Agar dilution is a quantitative method for determining the MIC of the antibiotics against bacteria to be tested. Minimum inhibitory concentration is the lowest or minimum concentration of antimicrobials required to prevent bacterial multiplication under specified condition. This technique is performed when a patient does not respond to adequate treatment through to be adequate, relapses while being treated or when there is immune suppression. Dilution technique measures the MIC. They can also be used to measure the MBC, which is the lowest concentration antimicrobial required to kill bacteria. MIC is reported as the lowest concentration of antimicrobial required to prevent visible growth. It is also measured by subculturing the last tube to show visible growth and all the tubes in which there is no growth. The MBC is lowest concentration required to produce a sterile culture. Dilution technique requires careful standardization and control of inoculum, broth, antimicrobial solutions, incubation time, dilution technique and reading of results. In recent year's semi automated techniques have been developed.

Materials Required

Culture

Pathogenic bacteria

Media

Nutrient broth

Mueller-Hinton broth

Antibiotic Powder

By the recommendation of the physician.

Water bath

0.5 McFarland standard

Sterile test tubes

50 ml conical flasks

Sterilize all the necessary materials appropriately before starting the experiment. Strict aseptic condition should be maintained while handling the cultures.

Preparation of Antibiotics Stock

Weigh 200 mg of antibiotics powder and dissolve in 5 ml of sterile distilled water in sterile screw caped tube (stock 200 mg/5 ml). Mix 0.5 ml of stock solution with 9.5 ml sterile distilled water in sterile screw caped tube (Working solution). Concentration of working solution is 200 μg/ml.

Methodology

1. **Medium Preparation**
   a) Mueller-Hinton agar is prepared in 500 ml conical flasks (180 ml media). Add agar for 200 ml concentration.
   b) The entire medium is equally dispensed to ten 50 ml conical flask (18 ml each) and sterilized at 121°C for 15 minutes in an autoclave.
   c) It is kept ready for adding antibiotic solution. (Maintain this medium in water bath at 55–60°C until use. Avoid solidification.)
2. **Dilution**
   a) Sterilized test tubes are arranged in a test tube rack and labeled as T1 → T10.
   b) 2 ml of sterile saline solution is added to all test tubes under aseptic condition.
   c) One ml of working solution is added to first tube (T1) using sterile pipette.
   d) One ml of solution from T1 is added to T2 and continues the dilution upto final. For each dilution, separate tip or pipette should be used. Maintain equal volume of saline in all tubes after dilution (2 ml).

3. **Preparation of bacterial suspension**
   a) Three to four colonies of the pathogens of the single species are selected.
   b) This is inoculated into peptone water and incubated for 3–4 hours at 35°C.
   c) Turbidity of the suspension is adjusted to match 0.5 McFarland standard and used for MIC assay.
4. **Preparation of agar plate with different concentration of antibiotics**
   a) All the conical flasks (with 18 ml MHA), antibiotic dilutions, petri plates are arranged and label properly in laminar flow chamber.
   b) T1 antibiotic solution is poured into one 50 ml conical flask containing media.
   c) Media and antibiotic solution are mixed properly and poured into appropriately labeled petri plates. Same procedure is followed to remaining solutions T2 to T10. Finally 10 different antibiotics containing plates are prepared and kept ready for inoculation.
5. **Inoculation**
   a) A grid is marked depending on the number of culture available for inoculation.
   b) A loopful of culture is inoculated on the dried surface of the medium, indicated by the rectangular mark below.
   c) Control plate is also inoculated with the test organisms.
   d) All the cultures are allowed to dry and the plates are inverted.
   e) Plates are incubated at 37°C for 18–24 hours.
   f) Plates are observed for visible growth and recorded the results.

Observation

Read the plates for presence or absence of growth. Check the control plate for growth. Control plate must show confluent growth. Read the test plate. The concentration at which growth is completely inhibited is considered as MIC.

Note: In all plates and tubes, indicate dilution/concentration of antibiotic. Express the value as microgram.

Tube No.	Volume of Water	Dilution	Intermediate concentration of antibiotics in μg/ml	Final concentration of antibiotics in medium μg/ml	Observation
T1	2 ml	1 : 01	1000.00	50.00	
T2	2 ml	1 : 02	0500.00	25.00	
T3	2 ml	1 : 03	0250.00	12.50	
T4	2 ml	1 : 04	0125.00	06.25	
T5	2 ml	1 : 05	0067.00	03.38	
T6	2 ml	1 : 06	0033.75	01.69	
T7	2 ml	1 : 07	0016.88	00.84	
T8	2 ml	1 : 08	0008.11	00.12	
T9	2 ml	1 : 09	0004.22	00.21	
T10	2 ml	1 : 10	0002.11	00.10	

Result

Introduction

Histopathology laboratory prepares tissue sections for establishing a histopathogenic diagnosis. The specimen subjected to histopathogenic study mostly comes from the operation theatres.

Materials Required

Microscope
Microtome—microtome knife, knife sharpener
Timer, Bunsen burner
Forceps, scalpels
Dissecting set
General glassware
Equipment for embedding
Slides
Cover glass

Reagents

Phosphate-buffered saline

Sodium chloride	0.8 g
Disodium hydrogen phosphate	0.173 g
Sodium dihydrogen phosphate	0.026 g
Distilled water	100 ml

pH 7.4
50%,70%, 80%, 90% and 100% alcohol solutions
70%, 80%, 90% and 100% xylene solutions
Paraffin wax
DPX mountant

Bouin's fixative

Copper acetate	2.5 g
Picric acid	4 g
Formalin	10 ml
Glacial acetic acid	1.5 ml

Haematoxylin

Haematoxylin	10 ml
100% alcohol	100 ml
Glacial acetic acid	100 ml
Glycerine	100 ml
Distilled water	1000 ml

Mix the solution well and allow to ripen in sunlight for 2 months. After the solution turns deep red, add 2 g of aluminium ammonium sulphate solution.

Eosin

0.5 g of eosin powder was dissolved in 30 ml of distilled water and made up to 100 ml with 70 ml of alcohol.

Physiological Saline

Sodium chloride	0.9 g
Distilled water	100 ml

Procedure

The following procedure was adapted from G.L.Humason,1979, *Animal Tissue Culture Technique*, W.H.Freeman & Co. San Franscisco.

1. Dissect the animal, remove the organs and wash with physiological saline.
2. Cut the organs into small pieces of desired sizes and store in Bouin's fixative solution for 24 hours.
3. Wash the tissues with 70% alcohol.
4. For dehydration, pass the tissues through a series of 70%, 90% and 100% alcohol for a time duration of one hour each.
5. For dealcoholization, keep in 1 : 1 xylene : alcohol solution for 1 hour and in 100% xylene solution till the tissues become translucent.
6. Give the tissues three changes in saturated paraffin in embryo cups or embedding box.
7. Embed the tissues in molten paraffin wax poured into boats. This is also called block making.
8. After the wax solidifies, cut the block into cubes with a single tissue and trim with knife.
9. Attach the blocks to the microtome pegs.
10. Attach the peg with the microtome and take sections of the tissue.
11. Spread the tissues after sectioning on a glass slide by mild heating.
12. Stain and process the slide with the sections.
13. For deparaffinizing, keep the slides in 100% xylene I and 100% xylene II for 10 minutes each.
14. Remove the deparaffinization solution by 100% alcohol I and II solutions for 5 minutes each.
15. Stain the slides by keeping them in haemotoxylin for 5 minutes and wash with water.
16. Destain the slides by dipping in acid alcohol solution and by passing through 50% and 70% alcohol.
17. Counterstain by dipping in eosin solution for 5 minutes.
18. Dehydrate the slides by keeping them in 70%, 90% and 100% alcohol I and II solutions for 5 minutes each.
19. Decolorize and clear the slides by keeping the slides in 100% xylene I and II solutions for 10 minutes each.
20. Dry the slides and make a permanent mount with DPX.

Observation

MEDIA FOR CULTIVATION

1. Gram-negative Broth

Tryptose	20 g
Dextrose	1 g
Mannitol	2 g
Sodium citrate	5 g
Sodium deoxycholate	0.5 g
Dipotassium phosphate	4 g
Monopotassium phosphate	1.5 g
Sodium chloride	5 g
Distilled water	1000 ml
Final pH (at 25°C)	7.0 ± 0.2

2. Nutrient Agar

Peptic digest of animal tissue	5 g
Beef extract	1.5 g
Yeast extract	1.5 g
Sodium chloride	5 g
Agar	15 g
Distilled water	1000 ml
Final pH (at 25°C)	7.4 ± 0.2

3. Hektoen Enteric Agar

Protease peptone	12 g
Yeast extract	3 g
Lactose	12 g
Sucrose	12 g
Salicin	2 g
Bile salt mixture	9 g
Sodium chloride	5 g
Sodium thiosulphate	5 g
Ferric ammonium citrate	1.5 g
Acid fuchsin	0.1 g
Bromothymol blue	0.065 g
Agar	15 g
Distilled water	1000 ml
Final pH (at 25°C)	7.5 ± 0.2

4. Xylose Lysine Decarboxylase

Yeast extract	3 g
L-lysine	5 g
Lactose	7.5 g
Sucrose	7.5 g
Xylose	3.5 g
Sodium chloride	5 g
Sodium deoxycholate	2.5 g
Sodium thiosulphate	6.8 g
Ferric ammonium citrate	0.8 g
Phenol red	0.08 g
Distilled water	1000 ml
Final pH (at 25°C)	7.4 ± 0.2

5. Salmonella-Shigella Agar

Peptic digest of animal tissue	5 g
Protease peptone	5 g
Beef extract	5 g
Lactose	10 g
Bile salt mixture	8.5 g
Sodium citrate	10 g
Sodium thiosulphate	8.5 g
Ferric citrate	1 g
Brilliant green	0.00033
Neutral red	0.025
Agar	15 g
Distilled water	1000 ml
Final pH (at 25°C)	7.0 ± 0.2

6. Bismuth Sulphite Agar

Peptic digest of animal tissue	10 g
Beef extract	5 g
Dextrose	5 g
Disodium phosphate	4 g
Ferrous sulphite	0.3 g
Bismuth sulphite	8 g
Brilliant green	0.025 g
Distilled water	1000 ml
Agar	20 g
Final pH (at 25°C)	7.7 ± 0.2

7. MacConkey Agar

Bactone peptone	17 g
Peptone	3 g
Lactose	10 g
Bile salt mixture	1.5 g
Sodium chloride	5 g
Agar	13.5 g
Neutral red	0.03 g
Crystal violet	0.001 g
Distilled water	1000 ml
pH	7.0

8. Nutrient Broth

Peptic digest of animal tissue	5 g
Beef extract	1.5 g
Yeast extract	1.5 g
Sodium chloride	5 g
Distilled water	1000 ml
Final pH (at 25°C)	7.4 ± 0.2

9. Mueller-Hinton Agar

Beef infusion	300 g
Casein acid hydrolysate	17 g
Starch	1.5 g
Agar	17 g
Distilled water	1000 ml
Final pH (at 25°C)	7.3 ± 0.2

10. Bile Esculin Agar Medium

Peptone	5 g
Beef extract	3 g
Oxgall (bile)	40 g
Esculin	1 g
Ferric citrate	0.5 g
Agar	15 g
pH	7

11. Alkaline Peptone Water

Peptone	5 g
Sodium chloride	5 g
Distilled water	100 ml
pH	8.6–9

12. Amies Transport Medium

Charcoal pharmaceutical neutral	10 g
Sodium chloride	3 g
Sodium hydrogen phosphate	1.15 g
Potassium dihydrogen phosphate	02 g
Potassium chloride	0.2 g
Sodium thioglycolate	1 g
Calcium chloride	0.1 g
Magnesium chloride	0.1 g
Agar	4 g
pH	7.1–7.3

13. Blood Agar

Peptone	5 g
Yeast extract	2 g
Sodium chloride	5 g
Agar	15 g
Blood	5 ml/100 ml
pH	7.2–7.6

14. Crystal Violet Blood Agar

Add 1ml of 0.02% aqueous solution of crystal violet to every 100 ml of sterile blood agar.

15. Neomycin Blood Agar

Prepare stock solution containing 70,000 μg/ml of neomycin by dissolving 0.5 g neomycin sulphate in 5 ml sterile water. From this prepare a working solution of 17,500 μg/ml by mixing 2 ml of stock solution with 6 ml of sterile water. Add 1 ml of working solution to 250 ml of blood agar to give a concentration of 70 μg/ml.

16. Kanamycin Blood Agar

Prepare 1.5 g kanamycin in 100 ml of water. Add 1 ml of this solution to every 100 ml of blood agar.

17. Chocolate Agar

Prepare blood agar base and sterilize. Then add defibrinated blood, heat the medium in a 70°C water bath until it becomes brown in colour. This takes about 10–15 minutes.

18. Cetrimide Agar

Peptone	20 g
Potassium sulphate	10 g
Magnesium chloride	1.4 g
Cetyl methylammonium bromide	0.3 g

Agar	15 g
Distilled water	1000 ml
pH	7–7.4

19. Mannitol Salt Agar

Peptone	10 g
Mannitol	10 g
Sodium chloride	75 g
Phenol red	0.025 g
Agar	15 g
Distilled water	1000 ml
pH	7.3–7.7

20. Sabouraud Dextrose Agar

Mycological peptone	10 g
Dextrose	100 g
Agar	15 g
Distilled water	1000 ml
pH	5.4 – 5.8

21. Dorset Egg Medium

Nutrient broth	20 ml
Whole fresh egg	80 ml
pH	7.2 – 7.4

22. Loeffler Serum Medium

Tryptose	10 g
Dextrose	5 g
Sodium chloride	5 g
Horse serum	30 ml
Distilled water	1000 ml
pH	7–7.4

23. Lowenstein-Jensen Acid Medium

Homogenized whole egg	275 ml
Hydrochloric acid1 mol/l (1N)	8 ml
Salt glycerol solution*	153 ml
Malachite green (20 g/l)	2.75 ml
Penicillin G	25000 IU
*Salt glycerol solution	
Potassium dihydrogen phosphate	6.3 g
Magnesium sulphate	0.3 g
Glycerol	12 ml

Distilled water	600 ml
pH	6.4–6.8

24. Modified Tinsdale Medium

Proteose peptone	2 g
Sodium chloride	0.5 g
Serum	10 ml
Sodium hydroxide 0.1 mol/l	6 ml
L-cystine (4 g/l)	6 ml
Potassium tellurite 10 g/l	3 ml
Sodium thiosulphate 25 g/l	1.7 ml
Distilled water	1000 ml
Agar agar	2 g
Make up to 100 ml with distilled water	
pH	7.5

MEDIA AND REAGENTS FOR BIOCHEMICAL TESTS

25. Indole Test

Peptone Water

Peptone	1 g
Sodium chloride	0.5 g
Distilled water	100 ml
pH	7.4

Kovac's Indole Reagent

Paradimethylamino benzaldehyde	2 g
Iso amyl alcohol	30 ml
Concentrated HCl	10 ml

26. MR-VP Broth (or) Glucose Phosphate Peptone Water

Peptone	0.5 g
Glucose	0.5 g
Dipotassium hydrogen phosphate	0.5 g
Distilled water	100 ml
pH	7.4–7.6

Methyl Red Solution

Methyl red	0.05 g
Ethanol	28 ml
Distilled water	22 ml

27. Voges-Proskauer Test

MR-VP broth (or)	
Glucose Phosphate Peptone Water	
Naphthol	
α-naphthol	5 g
Ethyl alcohol	100 ml
Potassium Hydroxide	
Potassium hydroxide	40 g
Distilled water	100 ml

28. Citrate Utilization Test

Simmons citrate agar	
Magnesium sulphate	0.2 g
Ammonium dihydrogen phosphate	1 g
Dipotassium phosphate	1 g
Sodium citrate	2 g
Sodium chloride	5 g
Bromothymol blue	0.08 g
Agar agar	15 g
Distilled water	1000 ml
pH	6.8–7.0

29. Catalase Test

3% hydrogen peroxide

30. Triple Sugar Iron Test

Triple sugar iron agar	
Peptic digest of animal tissue	10 g
Casein enzyme hydrolysate	10 g
Yeast extract	3 g
Beef extract	3 g
Lactose	10 g
Saccharose	10 g
Dextrose	1 g
Ferrous sulphate	0.2 g
Sodium chloride	5 g
Sodium thiosulphate	0.3 g
Phenol red	0.024 g
Agar agar	15 g
Distilled water	1000 ml
pH	7.4

31. Nitrate Reduction Test

Nitrate Broth	
Peptone	5 g
Beef extract	3 g
Potassium nitrate	1 g
Distilled water	1000 ml
pH	6.8–7.2
Sulphanilic Acid	
Sulphanilic acid	0.16 g
Glacial acetic acid	5.7 ml
Distilled water	14.3 ml
Alpha Naphthalamine Reagent	
Alpha naphthalamine	0.1 g
Glacial acetic acid	5.7 ml
Distilled water	14.3 ml
Zinc powder	a pinch

32. Urease Test

Urea Agar Base	
Peptic digest of animal tissue	1 g
Dextrose	1 g
Sodium chloride	5 g
Monopotassium phosphate	2 g
Urea	20 g
Phenol red	0.012
Distilled water	1000 ml
pH	6.8–7

33. Oxidative–Fermentative Test of Basal Medium

Casein	2 g
Sodium chloride	5 g
Dipotassium phosphate	0.3 g
Bromothymol blue	0.08 g
Agar agar	15 g
Distilled water	1000 ml
pH	6.8–7
Sterile Liquid Paraffin	
Glucose	
Maltose	
Sucrose	
Lactose	
Xylose	

Mannose
Trehalose

34. Carbohydrate Fermentation Medium

Peptone	10 g
Sodium chloride	5 g
**D-glucose	5 g
Bromocresol purple	0.03 g
Agar agar	13 g
Distilled water	1000 ml
pH	7.1

Carbohydrates are sterilized through filtration.
**it is replaced by other sugars.

35. Oxidase Test

Oxidase Reagent

Tetra methyl *p*-phenylene diamine hydrochloride	0.1 g
Distilled water	10 ml

36. Amino Acid Deaminase Test

Yeast extract	3 g
Amino acid	2 g
Sodium chloride	5 g
Disodium hydrogen phosphate	1 g
Agar agar	15 g
Distilled water	1000 ml
pH	6.8–7.2
Phenylalanine	1%

Note: Should not use meat extract or protein because of their varying natural content of phenylalanine. Yeast extract serves as the carbon and nitrogen source.

Ferric Chloride Reagent

Ferric chloride	5 g
Conc. HCl	25 ml
Distilled water	10 ml

37. Decarboxylase Test

Moeller Decarboxylase Medium

Peptone	5 g
Meat extract	5 g
Glucose	0.5 g
Pyridoxol	5 mg
Bromocresol purple (1 in 500 ml)	5 ml
Cresol red (1 in 500 ml)	2.5 ml

Distilled water	1000 ml
pH	6

L-Lysine, L-Ornithine, L-Arginine

These aminoacids are added in 1% concentration.

38. Phosphate Buffer

Stock solution A

Sodium dihydrogen phosphate	27.6 g
Distilled water	1000 ml

Stock solution B

Disodium hydrogen phosphate	28.39 g
Distilled water	1000 ml

For pH 7 buffer

Solution A	39 ml
Solution B	61 ml

39. Litmus Milk Reduction Test

Skimmed milk powder	2 g
Distilled water	20 ml
Litmus	trace amounts
pH	6.8

40. Lipid Hydrolysis

Tween 80 phosphate-buffered substrate medium

Phosphate buffer pH 7	40 ml
Tween 80	0.2 ml
Neutral red (1 g/l)	0.8 ml
pH	6.8–7.2

41. Starch Hydrolysis

Starch Agar

Peptone	5 g
Beef extract	3 g
Soluble starch	2 g
Agar agar	15 g
Distilled water	1000 ml
pH	7

Iodine

Potassium iodide	20 g
Iodine	10 g
Distilled water	1000 ml

42. Gelatin Hydrolysis

Nutrient gelatin	
Peptone	5 g
Beef extract	3 g
Gelatin	0.4%
Agar agar	15 g
Distilled water	1000 ml
pH	6.8 – 7.2
Gelatin precipitin reagent	
Conc. HCl	20 ml
Distilled water	80 ml
Mercuric chloride	15 g

43. Casein Hydrolysis

Skim agar medium	
Skimmed milk powder	100 g
Peptone	5 g
Agar agar	20 g
Distilled water	1000 ml
pH	7.2

44. Esculin Hydrolysis

Brain heart infusion agar	40 g
Esculin	1 g
Ferric chloride	0.5 g
Distilled water	1000 ml
pH	6.8–7.2

45. Melanoate Utilization Medium

Yeast extract	1 g
Ammonium sulphate	2 g
Dipotassium hydrogen phosphate	0.6 g
Potassium dihydrogen phosphate	0.4 g
Sodium chloride	2 g
Sodium malonate	3 g
Bromothymol blue	0.025 g
Distilled water	1000 ml
pH	7.4

46. Salt Tolerance Test Medium

6.5% Sodium Chloride Broth

Heart infusion broth	25 g
Sodium chloride	60 g
*Indicator	1 ml
(Bromocresol purple in 100 ml of 95% ethanol)	
Glucose	1 g
Distilled water	1000 ml

*Indicator may be omitted

SPOTTERS

To understand practical aspects of medical microbiology, one should know few identification characters of bacteria, fungus, protozoa, parasites and media used for cultivation. This will be helpful to students to score best marks during practical examinations.

- E.*coli* is a common member of the normal flora of the large intestine. As long as these bacteria do not acquire genetic elements encoding for virulence factors, they remain benign commensals.
- It is the head of the large bacterial family, Enterobacteriaceae, the enteric bacteria, which are facultative anaerobic gram-negative rods. Its nutritional requirement is very simple, grows very fastly and its genome is a well known one.
- It can respond to environmental signals such as chemicals, pH, temperature, osmolarity, etc.
- Virulence determinants of pathogenic E. *coli* are Adhesins, CFAI/CFAII fimbriae, intimin (non-fimbrial adhesin,Invasins, haemolysins, flagella, LT toxin, ST toxin, Shiga-like toxin, cytotoxins, and endotoxin (LPS).
- It causes gastroenteritis in humans. It is an important cause of acute watery diarrhoea in adults and children.

Types of *E.coli*

Five classes of E. *coli* that cause diarrhoeal diseases are now recognized. They are

1. Enterotoxigenic *E. coli* (ETEC),
2. Enteroinvasive *E. coli* (EIEC),
3. Enterohaemorrhagic *E. coli* (EHEC),
4. Enteropathogenic *E. coli* (EPEC), and
5. Enteroaggregative *E. coli* (EAggEC).

- It is grown on EMB, XLD, MacConkey and Hektoen enteric agar.

- It is a gram-negative facultative anaerobe.
- *Shigella* is a non-motile organism.
- Shigellae are the principal agents of bacillary dysentery.

- Dysentery is characterized by the daily loss of 200 to 300 ml of serum protein in the faeces.
- The genus *Shigella* is differentiated into four species:
  1. *S. dysenteriae* (serogroup A, consisting of 12 serotypes)
  2. *S. flexneri* (serogroup B, consisting of 6 serotypes)
  3. *S. boydii* (serogroup C, consisting of 18 serotypes) and
  4. *S. sonnei* (serogroup D, consisting of a single serotype)
- Serogroups A, B, and C are very similar physiologically, while S. *sonnei* can be differentiated from the other serogroups by positive β-D-galactosidase and ornithine decarboxylase biochemical reactions. The identification of shigellae by species in the clinical laboratory is usually accomplished by slide agglutination using commercially available, absorbed rabbit antisera.
- In DNA hybridization studies, E*scherichia coli* and S*higella* species cannot be differentiated on the polynucleotide level.
- GN broth enriches S*higella* from faeces and Hektoen enteric agar is used for isolation.

- It is a gram-negative rod-shaped bacteria.
- Motility brought about by peritrichous flagella.
- They are facultative anaerobes.
- Metabolism is of respiratory and fermentative type.
- The members of this genus cause varieties of infections in human beings.
- *Salmonella typhi* causes typhoid.
- *Salmonella enteritis* causes salmonellosis.

Cultural Characters

1. On MacConkey agar, it produces small, circular, translucent, NLF colonies.
2. In Wilson-Blair bismuth sulphite medium, the colonies are jet black with metallic sheen colonies.
3. On salmonella-shigella agar, black centred colonies are observed.
4. On XLD agar slight pink colonies with black centres are observed.
5. On deoxycholate citrate agar, black colonies are developed.

- *Vibrio cholerae* is the causative agent of cholera.
- It is a gram-negative, comma-shaped bacillus with single polar flagellum.
- It was first isolated in pure culture by Robert Koch in 1883.

- Most vibrios have relatively simple growth factor requirements and will grow in synthetic media with glucose as the sole source of carbon and energy. However, since vibrios are typically marine organisms, most species require large amount of NaCl or seawater base for optimal growth.
- Vibrios vary in their nutritional versatility, but some species will grow on more than 150 different organic compounds such as carbon and energy sources, occupying the same level of metabolic versatility as *Pseudomonas*.
- In liquid media, vibrios are motile by polar flagella.
- Generation time for *Vibrio cholerae* is less than 30 minutes.
- In humans it produces an infection called cholera with a symptom called diarrhoea (rice-water stool).
- It grows well on TCBS agar and produces yellow-coloured colonies.

- *Haemophilus influenzae* is a small, non-motile, gram-negative bacterium belonging to the family Pasteurellaceae. Encapsulated strains of *Haemophilus influenzae* isolated from cerebrospinal fluid are coccobacilli.
- The organism may appear gram-positive unless the gram stain procedure is very carefully carried out.
- *H. influenzae* is highly adapted to its human host.
- It is present in the nasopharynx of approximately 75 per cent of healthy children and adults.
- Haemophilus "loves haeme", more specifically it requires a precursor of haeme.
- Nutritionally, *Haemophilus influenzae* prefers a complex medium and requires preformed growth factors that are present in blood, specifically **X factor** (i.e., haemin) and **V factor** (NAD or NADP).
- In the laboratory it is usually grown on chocolate agar, which is prepared by adding blood to an agar base at 80°C. The heat releases X and V factors from the RBCs and turns the medium a chocolate brown colour.
- The bacterium grows best at 35–37°C and has an optimal pH of 7.6.
- *Haemophilus influenzae* is generally grown in the laboratory under aerobic conditions or under slight CO_2 tension (5% CO_2).

- *Bordetella pertussis* was first isolated in pure culture in 1906 by Bordet and Gengou from the sputum of children.
- Formerly it was called *Haemophilus pertussis*.
- It is a small ovoid coccobacillus, non-motile and non-sporing bacteria.

- It is capsulated, but tends to lose the capsule on repeated cultivation.
- Its bipolar metachromatic granules were demonstrated by toluidine blue staining.
- The bacteria are nutritionally fastidious and are usually cultivated on rich media supplemented with blood. On blood agar the organism grows slowly and requires 3–6 days to form pinpoint colonies.
- The organisms are strict aerobes and grow best at 35–36°C.
- It does not ferment sugars, produce indole, reduce nitrates, utilize citrate, hydrolyse urea and catalase, and is oxidase-positive.
- It is a delicate organism being killed readily by heating, drying and disinfectants.
- Freshly isolated strains have fimbriae.
- The organism is sensitive to unsaturated fatty acids,sulphides and peroxides.

- It is also called *Meningococcus*.
- It is a gram-negative coccus, usually seen in pairs.
- *Neisseria meningitidis* strains are grouped on the basis of their capsular polysaccharide into 12 serogroups.
- *Neisseria meningitidis* is usually cultivated in a peptone blood-based medium in a moist chamber containing 5–10 % of CO_2.
- New York city agar medium and modified Thayer-Martin agar act as selective media.

- Gonorrhoea is caused by *Neisseria gonorrhoeae*.
- It is also called *Gonococcus*.
- Neisser first described it in gonorrhoeal pus in 1879.
- Bumm in 1885 cultured the coccus and proved its pathogenicity.
- It is a gram-negative coccus.
- It is a fastidious organism, 20% of strains require glutamine for primary isolation.
- Growth occurs best at pH 7.2–7.6 and at the temperature of 35–36°C.
- Some strains require 5–10% carbon dioxide.
- A popular selective medium is the Thayer-Martin medium which contains vancomycin, colistin, nystatin and trimethoprim lactate.
    - Vancomycin inhibits gram-positive bacteria.
    - Colistin inhibits gram-negative bacteria except *Proteus*.
    - Nystatin inhibits yeast cells.
    - Trimethoprim lactate inhibits *Proteus*.
- Gonococci ferment only glucose and not maltose.

- *Staphylo* means "grape-like clusters", which is due to three-planar cell division.
- Staphylococci are gram-positive spherical bacteria.
- Von Reckling Heusen first observed it in human pyogenic lesions in 1871.
- Sir Alexander Ogston demonstrated the causative role of *Staphylococcus* in 1880.
- The most significant pathogen of this genus is S. *aureus*.
- Staphylococci are among the most resistant of non-sporing bacteria.
- They are non-motile and facultative anaerobes.
- Fermentation of glucose produces lactic acid.
- They are catalase-positive, coagulase-positive and ferment mannitol.
- They produce golden yellow colonies on nutrient agar.
- They produce beta haemolysis on blood agar.
- They utilize potassium tellurite and reduce it to telluramine and produce black colonies.
- S.*epidermidis* produces white colonies on nutrient agar and non-β-haemolytic colony on blood agar.

- Streptococci comprises a group of gram-positive cocci that grow in pairs and chains.
- Billroth first described it in 1874.
- In 1879, Pasteur isolated similar type of organisms in the blood of patients with puerperal sepsis.
- Rosenbach (1884) isolated the cocci from human suppurative lesions and gave the name S. *pyogenes*.
- They are facultative anaerobes with fermentative metabolism and are catalase-negative.

- It causes epidemic or even pandemic forms of pneumonia.
- It remains a leading cause of morbidity and mortality in humans of all ages.
- Pneumonia is the sixth leading cause of death in the United States.
- *Streptococcus* was first isolated by Pasteur and Sternbery independently during the year 1881 from human saliva.

Characters

- Facultative gram-positive coccus.
- Also called *Diplococcus* and often called *Pneumococcus*.
- Capsulated.

- Non-motile.
- Alpha-haemolytic.
- Sometimes lancet-shaped.
- Inulin fermenter.
- Bile solubilizer.

- Corynebacteria are gram-positive, aerobic, non-motile and rod-shaped bacteria.
- They form characteristic irregular shaped, club-shaped or V-shaped arrangements in normal growth.
- They undergo snapping movements just after cell division, which gives them the characteristic arrangements resembling Chinese letters.
- Three strains of *Corynebacterium diphtheriae* were recognized by McLeod on the basis of growth and other characters: gravis, intermedius and mitis.
- The gravis strain has a generation time (*in vitro*) of 60 minutes; the intermedius strain has a generation time of about 100 minutes; and the mitis strain has a generation time of about 180 minutes. The faster growing strains typically produce a larger colony on most growth media.
- It produces a disease called diphtheria.

- They are usually shorter gram-positive rods.
- Capsules may be observed by direct observation of smear from wounds but are not uniformly demonstrable in culture.
- Spores are central or subterminal but are rarely seen in artificial cultures.
- It is a non-motile, aerotolerant anaerobe.
- In BAP, it produces double zone of haemolysis.
- It produces invasive enzymes like proteases, DNases, collagenases, etc.
- It produces alpha toxin (most importantly lecithinase and phospholipase C) and is phage-encoded. Lecithin is the major lipid in cell membranes. Lecithinase causes leakage, oedema and myonecrosis in capillary endothelial cells.

$$\text{Lecithin} \xrightarrow{\text{lecithinase}} \text{Diglyceride} + \text{Phosphorylcholine}$$

- It is quite long and thin organism.
- It is usually gram-positive and produces terminal spores, so it appears like a drumstick.
- Motility is by peritrichous flagella.

- It is an obligate anaerobe.
- Optimum temperature is 37°C and pH is 7.4.
- Moderately fastidious, it requires vitamins and amino acids for growth.
- It is non-capsulated.
- It grows in cooked meat medium with turbidity and some gas formation within 48 hours, there is no digestion of meat.
- It does not ferment any carbohydrate.
- In horse blood agar, alpha hemolysis is observed first, then beta haemolysis is observed due to prolonged incubation and production of haemolysins.
- It is indole-positive, methyl-red and Voges-Proskauer negative.
- It is motile and has different O antigens.
- Ten types of organisms were identified on the basis of flagellar antigens and had single type of somatic antigen. Type IV is a non-flagellar organism.
- Organism produces 2 distinct toxins
  i. Tetanolysin
    - It disturbs the activity of RBC and WBC.
    - It is oxygen-labile and heat-labile.

  ii. Tetanospasmin
    - It is responsible for the symptoms of tetanus.
    - Toxins are released during autolysis of cell.

- It is an acid-fast bacillus. It is due to mycolic acid content of the cell wall.
- It is a fairly large, non-motile, rod-shaped bacterium.
- It is a non-spore former, and is non-capsulated.
- In tissues these are thin straight rods, occurring singly, in pairs or in clumps.
- They are weakly gram-positive, and obligate aerobes.
- They grow slowly, the generation time being 6–12 hours.
- Optimum temperature is 37°C and pH is 6.4–7.
- They grow only in media containing egg, asparagine, potatoes and serum.
- *Factors responsible for pathogenesis* *Mycobacterium* has large quantities of lipids. It includes mycolic acids (long-chain fatty acid C78-C90), waxes and phosphatides. Muramyl dipeptide complexed with mycolic acid can cause granuloma formation. Phosphatide induces tubercle formation. Lipids can cause accumulation of macrophages and neutrophils.
- Virulent strains of tubercle bacilli form microscopic serpentine cords in which bacilli are arranged in parallel chains.

- M. *leprae* causes leprosy. It was the first bacillus isolated from humans.
- It is one of the least understood bacteria.
- Hanson, a Norwegian physician first observed it.
- It is a straight or slightly curved rod showing considerable morphological variations.
- Organisms are found singly or in large masses termed globi. Large numbers of bacilli may be packed in the cells in an arrangement that suggests packets of cigars.
- It is one of the acid-fast bacilli.
- The presence of a phenolase in M. *leprae* obtained from lepromatous skin nodules provides the simplest test for separating M. *leprae* from other mycobacteria.
- It grows well on footpads of ninebanded armadillo at 30°C. This temperature is obtained by controlling air temperatures at 20–25°C.

- Syphilis is caused by the spirochaete *Treponema pallidum*.
- *Treponema* is a gram-negative, thin, motile, spiral-shaped bacterium in the order Spirochaetales.
- The name spirochaete is derived from the Greek words for "coiled hair."
- The spirochaetes are able to swim in viscous environments (e.g. oral cavity, intestinal tract).
- T. *pallidum* requires the presence of tissue culture cells, a microaerobic environment and serum components for growth.
- Its fastidious nature may account for its obligate parasitism.
- This microorganism is responsible for natural disease in humans only, although rabbits can be infected experimentally.
- T. *pallidum* cannot be grown in cell-free cultures.
- Limited growth has been achieved in cultured rabbit epithelial cells, but replication is slow (doubling time, 30 hours) and can be maintained for only a few generations.

- It is a non-enveloped, icosahedral, linear dsDNA virus.
- It causes pharyngitis, pneumonia, diarrhoea. Some strains cause sarcomas.
- Disease is transmitted through droplets; virus infects respiratory epithelium and damages the cells, which cause various symptoms.
- Virus causes cytopatheic effect in cell culture.
- It is identified by fluorescent antibody or complement fixation test.
- Live vaccine is used for prevention.

- It causes warts in human.
- It is non-enveloped, icosahedral, circular, double-stranded DNA virus.
- Direct contact of skin or genital lesions transmits the virus.
- After entry, viral genes E6 and E7, inhibits activities of tumour suppressor genes and cause warts.
- DNA hybridization tests are used for diagnosis.
- Alpha interferon is used for treatment.

- It causes mumps.
- It is an enveloped virus with helical symmetry.
- Genome is single-stranded negative-sense RNA.
- Respiratory droplet is a source of infection.
- From upper respiratory tract, the virus spreads to local lymph nodes and then via the bloodstream to other organs, especially parotid glands, testes, ovaries and meninges.
- Virus can be isolated by cell culture.
- Haemadsorption test is also useful.
- No antiviral therapy.

- *Penicillium* sp. is ubiquitous and omnipresent throughout the world. They are found in soil and decaying vegetation.
- Pulmonary infection, onychomycosis, cutaneous lesions, bladder infection are due to *Penicillium* species.
- *Penicillium* is one of the most common laboratory contaminants.
- Penicillin is produced by *Penicillium chrysogenum*.
- Colony is flat granular and typically blue-green.
- Modified SDA with chloramphenicol is used for cultivation.
- Phialides have blunt tips.
- Chains of conidia from the phialides.
- Phialides may be arranged in whorls.

- The pulmonary infection caused by the organism is sometimes called Valley fever.
- It is a dimorphic fungus.

- It is the most virulent of all the agents of human mycosis.
- Modified SDA, BHIA are used for cultivation.
- Optimum temperature for growth is 25°C to 30°C.
- Rapidly growing colonies with early appearance of white cottony aerial mycelium and areas of adherent surface hyphae.
- Septate hyaline hyphae of varying widths are seen microscopically. Fertile arthroconidiating hyphae are wider than vegetative hyphae.
- The arthroconidia are usually single-celled, barrel-shaped or rectangular.

- It is found worldwide and is endemic in North America.
- The pulmonary infections caused by this species are sometimes called Darlings disease.
- Nickname of histoplasmosis (a chronic granulomatous infection) is Spelunkers's disease.
- It is a dimorphic fungus.
- Blood agar plate and SDA are used for cultivation.
- At 37°C, it resembles the cells of *Candida*.
- Macroconidia are hyaline, unicellular, and relatively large, with spherical or pyriform shape. As the macroconidia age, they become tuberculate, that is, they form finger-like projections on the thick wall of the conidium.
- Mature macroconidia are sometimes described as resembling sunflowers in bloom.
- It forms waxy, wrinkled light brown colony of yeast at 37°C.
- At 25°C they may appear white to brown at first, later they become wooly, as aerial hyphae develop.

- It causes endemic infections.
- Disease caused by blastomyces is called blastomycosis.
- Blood agar plates, BHIA or SDA are used for cultivation.
- Optimum temperature for growth is 25–30°C.
- Globose to pyriform "lollipops" are formed on conidiophores of varying length at 25–30°C.
- It forms waxy, wrinkled, light brown colony of yeast cells at 37°C.
- Small round or pear shaped conidia grow directly on the septate hyphae.

- It is a working yeast.
- Various strains are used in industry to make bread, beer, wine and industrial alcohol.
- PDA is used for the isolation of *Saccharomyces cerevisiae*.

- On Rose Bengal agar, milky white colonies are formed.
- It is sensitive to cycloheximide and produces blastoconidia, but neither germ tube nor chlamydospores are formed. Pseudohyphae may be formed.
- It produces single multilateral budding yeast.
- It is a facultative anaerobe.

- It is found worldwide.
- The yeast is able to survive in pigeon's gut.
- It causes cryptococcosis in humans.
- It produces capsule, capsule production being enhanced by inoculating a plate of chocolate agar and incubating it at 37°C.
- It is a thin-walled, globose or oval-shaped yeast.
- It forms smooth mucoid colonies on Sabouraud dextrose agar.

- C.*albicans* is recognized as the most frequently encountered fungal opportunist and is now regarded as the most common cause of serious fungal disease.
- The clinical diseases caused by candida are collectively called candidiasis.
- It is found worldwide on fruits and vegetables.
- For culture, BAP, (Blood agar plate) and modified SDA (Sabouraud dextrose agar) with cycloheximide are used.
- Candida species develop as entire, white, pasty, convex colonies that initially resemble staphylococci. The colonies may produce pseudohyphal fringes around the periphery.
- In a wet preparation C.*albicans* demonstrates blastoconidia on pseudohyphae.
- Forms germ tubes within 3 hours at 35°C.

- *Aspergillus* is a group of fungi that is frequently isolated from human beings.
- These organisms cause opportunistic infections.
- Some species are used in the industry for the production of some enzymes.
- About 167 species are identified, of these 16 species are the aetiological agents of aspergillosis.
- The organism is found worldwide in soil, air, on mouldy storage grains and on decaying vegetables.
- Important species are, A. *niger,* A. *flavus,* A. *fumigatus and* A. *terreus.*

Identifying features of various agents are,

A. niger

- Coarse black granules against the creamy colony.
- Globose vesicles with biseriate phialides.
- Large echinulate jet black conidia in chains.

A. flavus

- Yellow to yellow-green colonies.
- Globose to subglobose vesicles with uniseriate or biseriate phialides.
- Conidial heads that radiate are loosely formed.

A. fumigatus

- Bluish green to grey colonies.
- Flask-shaped vesicles.
- Uniseriate arrangement of phialides.
- Compact columnar arrangement of conidia.

A. terreus

- Smallest aspergilli.
- Cinnamon to buff brown colonies.
- Dome-shaped vesicles with biseriate phialides.
- Long and compact conidial heads.
- Submerged hyphae that may form globose conidia.

- On modified SDA at 25°C after 2–4 days, colonies of Mucor are wooly and rapidly fill the entire petri plate with an abundant matted mycelium.
- The colony is white at first and becomes grey or yellow.
- Broad irregular hyphae that are aseptate, septate or sparsely septate.
- Branching sporangiophores with columellae support the sporangia filled with sporangiospores.
- Absence of rhizoids.

- The organism is found worldwide as a commensal on smooth skin of humans.
- It causes Tinea versicolor.

- Skin scrapings are submitted for diagnosis.
- Modified SDA supplemented with olive oil and antibiotics is used for cultivation.
- It is a lipophilic yeast.
- Shiny or pasty white to cream-coloured colonies are formed after 1–2 weeks.
- The phialoconidia are thick-walled, round or oval in shape. They typically occur in clusters.
- Individual cells look like yeast.

- The organism is a plant parasite.
- Human infection caused by this parasite is called black piedra.
- Black piedra is a nodular infection of the hair shaft.
- Colour of the colony on SDA is greenish black or black.
- Asci and ascospores are rarely seen in culture.
- Hair is the specimen for diagnosis.

- It is the only human pathogen in this genus.
- Organism finds worldwide distribution.
- It is a highly contagious fungus.
- Skin scraping is a specimen.
- It produces white, downy somewhat scanty colony.
- Centre of the colony may be folded.
- Hyphae are thin hyaline, septate and branched.
- Microconidia are absent.
- Abundant, characteristic snow shoe or paddle-shaped macroconidia, with thin smooth walls.

- It is found worldwide as a zoophilic pathogen.
- It is highly contagious and easily transmitted from animal to human.
- It causes *Tinea corporis* and *Tinea capitis*.
- Modified SDA, DTM, PDA and modified corn meal agar are used for cultivation.
- Positive to hair perforation test.
- Rare microconidia.
- Characteristic spindle-shaped macroconidia with beaked tips.
- Hyphae are hyaline, septate and branched.

- Vegetative hyphal structures are racquet hyphae chlamydoconidia.
- It produces yellow pigmented conidial colony.

- It is found in soil as a saprophyte.
- It causes mycotic keratitis.
- SDA with chloramphenicol is used for cultivation.
- Optimum temperature for growth is 25–30°C.
- Dematiaceous hyphae and dark unbranched conidiophores.
- Darkly pigmented clavated proconidia produced laterally through pores along the conidiophores.

- Organisms found throughout the world.
- This fungus can infect nail and skin of both children and adults, but the hair is not invaded.
- SDA and DTM are used for cultivation.
- Colonies are white downy to fluffy or yellow or tinged with red.
- Hyaline, septate hyphae and branched.
- Hair perforation test negative.

- Organisms found throughout the world.
- It causes ringworm infection in cattle.
- It causes highly inflammatory infection in human.
- Best growth at 37°C.
- Hyphae are hyaline, septate and distorted.
- SDA, DTM are used for cultivation.

- Organisms found throught out the world.
- It causes sporotrichosis in humans.
- SDA,BHIA are used for cultivation.
- Best growth at 25–37°C.
- At 25–37°C
    - Leathery black or mottled black colonies are formed.
    - Thin septate hyphae.
    - Tapering sympodial conidiophores.

- At 35–37°C
    - Cream to tan yeast like colony.
    - Hyaline unicellular budding yeast forms, with a variety of forms.
    - Delicate connection between mother and daughter cells.

- The organism is found throughout the world.
- It is the most common agent of dermatophyte infection.
- It is a highly contagious fungus.
- It is associated with *Tinea cruris*, *Tinea corporis*, *Tinea pedis*, *Tinea capitis*, *Tinea barbae* and *Tinea unguium*.
- DTM, SDA and CMA are used for cultivation.
- Flat, granular, creamy yellow to tan or reddish brown with a buff, yellow-brown or reddish brown reverse.
- Hair perforation test positive.
- Cigar-shaped macroconidia with thin, smooth walls.

- It is seen as a long tube-like projection extending from the yeast.
- It is a characteristic feature of *Candida albicans*.
- It is produced by inoculation of culture on human serum and incubated at 37°C for 2–4 hours.
- A drop of suspension is examined under the microscope and observe for pseudohyphae.
- There is no septum in germ tube.
- Demonstration of germ tube is known as Reynolds-Braude phenomenon.

- It was first seen by Leeuwenhoek in 1681 while examining his own stool.
- It is worldwide in distribution.
- It is present in the duodenum and upper part of ileum.
- The *Giardia* life cycle involves two stages: the trophozoite and the cyst.
- The G. *lamblia* trophozoite is easily recognized under a microscope.
- It is about 12 to 15 μm long, shaped like a tennis racket.
- The dorsal surface is convex and the ventral surface is concave with a sucking disc, and has two nuclei that resemble eyes, structures called median bodies that resemble a mouth, and four pairs of flagella that look like hair; these combine to give the stained trophozoite the eerie appearance of a face.

- The flagella help these organisms to migrate to a given area of the small intestine, where they attach by means of an adhesive disc to epithelial cells and thus maintain their position despite peristalsis.
- It is bilaterally symmetrical. Anterior end is broad and the posterior end tapers to a sharp point.
- The *Giardia* cyst, the form usually seen in the faeces, is ovoid, 6 to 12 μm long, and can often be seen to contain two to four nuclei at one end and prominent diagonal fibrils. Flagella and sucking disc are seen inside the cytoplasm.

- It causes Ascariasis.
- It is an intestinal nematode.
- It is transmitted by food contaminated with soil containing eggs.
- Heavy worm burden in intestine can cause intestinal obstruction or malnutrtion.
- Eggs are visible in faeces, and this is a diagnostic tool.
- Mebandazole is used for treatment.
- Prevention is by proper disposal of human waste.
- Human is the only host.

- It causes Schistosomiasis.
- It is a bisexual blood fluke.
- Important species are
  - S. *mansoni* with a large lateral spine
  - S.*japanicum* with a small lateral spine
  - S. *hematobium* with a terminal spine
- Transmission is by penetration of skin by cercariae.
- Humans are the definite host.
- Snail is the intermediate host.
- Eggs in tissue induce inflammation, granuloma, fibrosis and obstruction.
- Eggs are visible in faeces.
- Prazigrantel is used for treatment.
- Proper disposal of waste is a way of prevention.

- It causes filariasis.
- It is a tissue nematode.

- Bites of female mosquito deposits infective larvae which penetrate wounds and produce microfilaria. These circulate in the blood and block lymphatic vessels and cause elephantiasis.
- Blood examination reveals microfilaria.
- Diethylcarbamazine is used for treatment.
- Mosquito control prevents infection.

- Blood agar is used for the differentiation of *Staphylococcus*, *Streptococcus* and others, based on haemolysis.
- 5% defibrinated sheep blood is recommended for the preparation of blood agar.
- Three haemolytic patterns are described in blood agar.
  1. Alpha haemolysis (partial haemolysis)
  2. Beta haemolysis (complete haemolysis)
  3. Gamma haemolysis (no haemolysis)
- Coagulase-positive *Staphylococcus aureus* produce beta haemolysis.
- *Streptococcus pyogenes* produce beta haemolysis when it is incubated anaerobically.
- Based on the type of supplements added to the blood agar its name will vary. Some of these are
  1. Neomycin blood agar
  2. Ampicillin blood agar
  3. Tellurite ceffixime blood agar
  4. Kanamycin blood agar
- Supplements are used to select particular species from the human body.
- Satellism pattern is described when H. *influenzae* is cross-inoculated with *Staphylococcus aureus*.

- It is a selective cum differential medium used for the isolation of Gram-negative enteric organisms.
- Eosin and methylene blue are used as indicators in EMB medium.
- Gram-positive organisms are inhibited by the action of eosin and methylene blue mixture.
- Those organisms that ferment one or two sugars appear as black or dark-centred colonies, e.g. Enterobacter.
- E.*coli* produces metallic sheen colonies because it ferments all sugars available in the medium and reduces the pH of the medium.
- Colonies that do not ferment lactose or sucrose appear transparent.
- Adding 5% agar inhibits swarming motility of proteus.
- *Pseudomonas aeruginosa* produces umbonate colonies with filamentous margin, usually violet in appearance.

- HE agar is recommended for the isolation of *Salmonella* and *Shigella* from faecal specimens.
- It contains lactose, sucrose and salicin.
- Bromothymol blue and acid fuchsin act as indicators.
- *Salmonella* and *Shigella* do not ferment any of the sugars available in the medium and produce blue or green coloured colonies.
- In the presence of acid, bromothymol blue produces yellow colour, and the acid fuchsin produces red colour. The combination of yellow and red produces orange, pink or salmon colour.
- Expected colonies are

  Salmonella—blue-green with black centres.

  Shigella—small greenish colonies.

  E.*coli*, *Enterobacter*, *Klebsiella*—salmon-coloured colonies.

- It is a selective cum differential media used for the isolation of enteric pathogens.
- Selectivity nature is increased through the addition of deoxycholate.
- Lysine is incorporated into the medium to offset the acid production.
- If the organism ferments only xylose, the colour of the colony is pink, depletion of xylose leads to lysine decarboxylation which causes reversion of pH.
- Phenol red acts as an indicator.
- Lactose and sucrose fermentors produce yellow colonies.
- Sodium thiosulphate is a sulphur source.
- Ferric ions act as indicator for hydrogen sulphide production.
- **Colony morphology**

  E. *coli*, *Enterobacter*, *Klebsiella*, *Serratia* and *Citrobacter*—yellow colonies.

  Citrobacter freundii, *Proteus*—yellow colonies with black centre.

  Shigella,*Pseudomonas*—red colonies.

  Salmonella and Edwardsiella—red colonies with black centre.

- It is used for the differentiation of enteric organisms.
- Lactose is the main carbohydrate, which acts as a differential source based on lactose utilization.
- Lactose-utilizing microorganisms produce pink coloured colonies , which are referred to as LF colonies.

- Non-lactose-fermenting colonies produce yellowish colonies and are called NLF colonies.
- Gram-positive organisms are inhibited by bile salts in the medium.
- Colony morphology

  Escherichia coli—pink coloured rough colonies.

  Klebsiella—pink coloured mucoid colonies.

  Salmonella , *Shigella*, *Pseudomonas* produce NLF colonies.

- It is sometimes called CHOC agar.
- It is used for the cultivation of fastidious microorganisms like *H.influenzae*, *N. meningitidis* and *N. gonorrhoeae*.
- It provides X and V factor for the growth of fastidious microorganisms.
- It is prepared by heating blood agar at 50°C for 10 minutes in a water bath.
- The name was given because of its colour after heating.
- Additional enrichment and inhibitory substances are added to the medium to enhance the growth of specific microorganisms.

- It is an acronym of Cystine lactose electrolyte deficiency medium.
- CLED agar with bromothymol blue is recommended for the isolation, enumeration and identification of urinary pathogens on the basis of lactose fermentation.
- Restricted availability of electrolytes control swarming motility of proteus.
- Colony morphology

  Escherichia coli—yellow colonies.

  Proteus vulgaris—large bluish colonies.

  Salmonella typhi—small bluish colonies.

- It was originally developed by Robertson for the cultivation of anaerobes from wound.
- It is also called Robertson cooked meat medium.
- It contains beef infusion and muscle protein.
- Two types of reactions are observed in this medium.
- Proteolytic activity is indicated by the formation of black colour.
- Saccharolytic reaction is indicated by the formation of reddish colour.
- *Clostridium tetani* produces proteolytic reaction.
- *Clostridium perfringens* produces saccharolytic reaction.

- It is a selective medium used for the isolation of *Pseudomonas aeruginosa* from pus, sputum, drains, etc.
- It is also used for determining the ability of an organism to produce fluorescein and pycocyanin.
- Cetrimide (Cetyltrimethyl ammonium bromide) is incorporated in the medium to inhibit microbes other than *P. aeruginosa*.
- *P. aeruginosa* colonies may appear pigmented blue or blue-green.
- Cetrimide acts as a quaternary ammonium compound, which causes release of N_2, phosphorus from bacterial cells other than *P. aeruginosa*.

- It is recommended for the isolation and cultivation of *Vibrio*.
- Oxgall and sodium citrate inhibit gram-positive bacteria.
- Bromothymol blue and thymol blue act as indicators.
- Alkaline pH improves the recovery of *Vibrio cholerae*.
- Colony morphology

  Vibrio cholerae—yellow-coloured colonies due to sucrose fermentation.

  V. *paramaemolyticus*—bluish green colonies.

  Pseudomonas and *Aeromonas*—blue-green colonies.

- It is used for the isolation of M. *tuberculosis*.
- Malachite green, penicillin and nalidixic acid inhibit the growth of a majority of contaminants.
- Egg and potato starch enhance the growth of M. *tuberculosis* producing rough, warty, dry, granular greenish coloured colonies.
- Medium is greenish blue in color. The RNA acts as stimulant and helps to the increase isolation rate of *Mycobacterium*.

- It is also called Rajhans medium.
- It is recommended for identification and differentiation of Salmonella species .
- Salmonella is differentiated on the basis of propylene glycol fermentation.
- Sodium deoxycholate inhibits the growth of gram-positive bacteria.
- Colony morphology

  E. coli—blue colonies

  S. enteritidis—red colonies

  Shigella flexneri—colourless colonies

- It is a selective medium for the isolation of coagulase-positive *Staphylococcus aureus*.
- It was developed by Baird-Parker.
- Lithium chloride and potassium tellurite are incorporated into the medium as selective agent.
- Glycine is added to the medium to enhance the growth of *Staphylococcus*.
- Egg yolk emulsion is used to identify the proteolytic character of the isolate.
- During growth, S. *aureus* utilizes potassium tellurite, converts it into telluramine and produces black colonies.

Some other spotters for which information can be obtained from the chapters are listed below.

1. *Entamoeba histolytica*
2. Syphilis
3. Leprosy
4. Tetanus
5. Gas Gangrene
6. Anthrax
7. Rocky Mountain Spotted Fever
8. Chickenpox
9. Measles
10. Poliomyelitis
11. Rabies
12. Rubella (3-day measles or German measles)
13. Histoplasmosis
14. Coccidioidomycosis (Valley fever)
15. Cryptococcosis
16. Blastomycosis
17. Paracoccidioidomycosis

APPENDIX I

BRIEF SUMMARY OF VARIOUS INFECTIONS OF THE HUMAN BODY

Disease	Causative agent	Symptoms	Pathogenesis	Lab diagnosis	Epidemiology	Prevention and treatment
Syphilis	*Treponema pallidum*	Chancre, fever, rash, stroke, nervous system deterioration.	Primary lesion (chancre) appears at site of inoculation, heals after 2–6 weeks, *T. pallidum* invades vascular system and is carried throughout the body, causing fever, rash, mucous membrane lesions, damage to brain.	Dark-field illumination, direct fluorescent antibody staining techniques, serological tests, Treponemal and non-treponemal tests.	Sexual contact with infected partner.	Education, use of condoms, treatment of sexual contacts. Penicillin is the drug of choice.
Pneumonia	*Streptococcus pneumoniae*	Cough, fever, rust-coloured sputum. Shortness of breath, chest pain.	Colonization of the alveoli incites inflammatory response, inflammatory cells fill the alveoli.	Stained smears of sputum show pus and gram-positive diplococci, culture of sputum and blood show alpha-haemolytic colonies of gram-positive diplococci.	Inhalation of infected droplets.	Vaccine is available and it contains 23 capsular antigens. Penicillin and erythromycin are the drugs used for treatment.

(Contd.)

BRIEF SUMMARY OF VARIOUS INFECTIONS OF THE HUMAN BODY (Continued)

Disease	Causative agent	Symptoms	Pathogenesis	Lab diagnosis	Epidemiology	Prevention and treatment
Pneumonia	*Klebsiella pneumoniae*	Chills, fever, cough, chest pain and bloody, mucoid sputum.	Colonization of the alveoli incites inflammatory response, inflammatory cells fill the alveoli; destruction of lung tissue and abscess formation are common, infection spreads by blood to other body tissues.	Stained smears of sputum show pus and large, gram-negative, rod-shaped bacteria; cultures of sputum and blood show mucoid colonies. Inhalation of infected droplets, mild infections are common.	Inhalation of infected droplets, aspiration of mucous.	No vaccine is available. Cephalosporin with an amyloglycoside is the drug of choice.
Pneumonia	*Mycoplasma pneumoniae*	Cough, fever, sputum production, headache, fatigue and muscle ache.	Cells attach to specific receptors on the respiratory epithelium; inhibition of ciliary motion and destruction of cells occur.	About two-thirds of cases develop cold agglutinins, CF test using *M.pneumoniae* antigen reveals a rise in antibody titre in about 60% of cases, culture often takes a week or more.	Inhalation of infected droplets, mild infections are common.	No vaccine is currently available. Tetracycline or erythromycin is used for treatment.
Pertussis (Whooping cough)	*Bordetella pertussis*	Nasal congestion and mild cough followed by spasms of violent cough, vomiting and convulsions.	Colonization of the surfaces of the upper respiratory tract and tracheobronchial system, ciliary action is slowed; toxin released when organisms disintegrate and causes death of epithelial cells and a rise in the number of lymphocytes in the bloodstream.	Smears of nasopharyngeal secretion stained by fluorescent antibody, ELISA test for toxin in respiratory secretion.	Inhalation of infected droplets.	Whole cell and acellular vaccines are available for immunization of children. No cure. Erythromycin is given to limit spread.

Tuberculosis	*Mycobacterium tuberculosis*	Fever, weight loss, cough and sputum production.	Colonization of the alveoli incites inflammatory response, ingestion by macrophages follows; organisms survive ingestion and are carried to lymph nodes, lungs and other body tissues, tubercle bacilli multiply and form granuloma.	Sputum treated with 40% KOH is inoculated in LJ medium. Smears examined for acid-fast bacilli. Biochemical testing, cord formation and gene probe analysis. Mantoux test.	Inhalation of infected droplets.	BCG vaccination. Isoniazid, ethambutol, rifampin and streptomycin are used for treatment.
Cholera	*Vibrio cholerae*	Abrupt onset of massive, painless diarrhoea, generally followed by vomiting without nausea, muscle cramps, severe dehydration and shock.	Heat-labile enterotoxin causes excessive secretion of water and electrolytes by the intestinal epithelium.	Recovery of *V.cholerae* from cultures of faeces or vomits on special media like TCBS.	Ingestion of faecally contaminated food or water and some natural sources.	Careful hand washing, vaccination gives partial protection. Electrolytes and glucose given intravenously. Tetracycline is given orally.

(Contd.)

BRIEF SUMMARY OF VARIOUS INFECTIONS OF THE HUMAN BODY (Continued)

Disease	Causative agent	Symptoms	Pathogenesis	Lab diagnosis	Epidemiology	Prevention and treatment
Gastritis or peptic ulcer	*Helicobacter pylori*	Pain in the stomach.	Organism secretes urease, which produces ammonia from urea that damage gastric mucosa.	Gram-staining, culture of campy blood agar with antibiotics like vancomycin, trimethoprim, polymyxin and amphotericin B. Urease test is performed. *Helicobacter* is urease-positive.	By ingestion.	Amoxicillin, metranidazole peptobismol are drugs used. No vaccine is available.
Shigellosis	Various species of *Shigella*	Fever, diarrhoea,vomiting, pus and blood in faeces. Convulsions and painful joints.	Invasion and multiplication within intestinal epithelial cells, death of the cells, followed by inflammation and ulceration.	Isolation of organisms from faeces by using GN broth as enrichment medium and Hektoen enteric agar as selective cum differential medium.	Infected person via faecal-oral route, ingestion of contaminated food and water.	Sanitation. Ampicillin, trimethoprim, chlorotetracycline are drugs of choice.

Gonorrhoea	*Neisseria gonorrhoeae*	In men, inflammation in urethra, pain on urination, sterility, arthritis. In women, infection of fallopian tube leads to ectopic pregnancy, arthritis and sterility.	Organism attaches to epithelial cells by pili, which also interfere with phagocytosis, escape from immune mechanisms.	Gram-stained smears of purulent discharge; smears often reveals gram-negative diplococci. Growth on New York city agar medium and modified Thayer-Martin agar.	By sexual contact.	Education, use of condoms, ceftriaxone, penicillin and tetracycline.
Leprosy	*Mycobacterium leprae*	Progressive nerve damage, chronic skin lesions, ulcerative lesions of mucous membranes, deformed face, loss of fingers or toes.	Invasion of small nerves of skin, multiplication in macrophages; course of disease depends on immune response of host, immune macrophages limits the growth of organisms, attack of immune cells against infected nerve cells produces nerve damage leading to deformity.	Microscopic inspection of skin scraping or biopsies of infected tissues are examined for AFB. Lepromin test is performed.	Direct contact with organisms from lesions of mucous membrane or skin.	Dapsone and rifampin are used for treatment.

(Contd.)

BRIEF SUMMARY OF VARIOUS INFECTIONS OF THE HUMAN BODY (Continued)

Disease	Causative agent	Symptoms	Pathogenesis	Lab diagnosis	Epidemiology	Prevention and treatment
Meningo-coccal meningitis	*Neisseria meningitidis*	Cold, headache, fever, pain, stiffness of neck and back, purplish spot on skin.	Meningococci adhere by pili and colonize upper respiratory tract, enters bloodstream and are carried to the meninges and spinal fluid. Inflammatory response obstructs normal outflow of fluid, increased pressure caused by obstructed flow and oedema impairs brain function. Damage to motor nerves produces paralysis. Endotoxin released from bacteria causes shock.	Smears and cultures of centrifuged CSF.	Close contact.	Polysaccharide vaccine is available. Penicillin, ceftriaxone and chloramphenicol are drugs of choice.
Tetanus	*Clostridium tetani*	Restlessness, irritability, headache, muscle pain, spasm in jaw, abdomen or back	Tetanus results from the action of exotoxin: toxin enters the circulation and is carried to the brain: toxin acts by attaching to nerve cells and preventing them from releasing a chemical that normally inhibits contraction of opposing muscle, uncontrolled nerve impulse in spasms of muscles.	Wound tissues may be cultured for organisms.	Contamination of wound, creation of anaerobic environment	Immunization at ages of 2 months, 4, 6,18 months, booster dose at 5th year and at 10-year intervals. Tetanus immunoglobulin, cleaning of wound and penicillin.

Typhoid fever	*Salmonella typhi*	Fever, headache, abdominal pain, intestinal rupture and shock.	Organisms enter into the intestine and multiply enormously and cause inflammation, release of organism in blood leads to shock and fever. Organisms lodge in spleen and other organs.	Growth of selective cum differential medium like SS agar. Blood, faeces and urine act as specimen.	Ingestion of contaminated foods.	Oral attenuated vaccines, are available. Ampicillin, chloramphenicol and trimethoprim are drugs of choice.
Diphtheria	*Corynebacterium diphtheriae*	Sore throat, fever, fatigue, malaise, pseudo-membrane formation in throat, heart and kidney failure.	Infection in upper respiratory tract; exotoxin released and absorbed by bloodstream, toxin kills cells by interfering with protein synthesis and damage kidney, heart and nerves.	Recovery of microorganisms is by using selective medium. Schick's test is also useful.	Inhalation of infected droplets, direct contact with patient.	Immunization with diphtheria toxoid for 2 months, 4th, 6th, 18th months, booster dose at 5th year and at 10-year intervals. Erythromycin is the drug used.
Brucellosis	*Brucella* spp.	Influenza, undulating fever, fatigue, enlarged lymph node	Organisms localize in reticuloendothelial cells. Endotoxin induces symptoms. No exotoxins.	Gram stain, culture on blood agar and serology.	Transmission via unpasteurized milk.	Pasteurization. Vaccination of cattle. Tetracycline is used for treatment.

(Contd.)

BRIEF SUMMARY OF VARIOUS INFECTIONS OF THE HUMAN BODY (Continued)

Disease	Causative agent	Symptoms	Pathogenesis	Lab diagnosis	Epidemiology	Prevention and treatment
Plague (Black death)	*Yersinia pestis*	Pain and swelling of lymph nodes, high fever, myalgia, septic shock, pneumonia	After inoculation, organisms spread to the regional lymph nodes, which become swollen and tender called as bubonic plague. Organisms reach blood and are disseminated throughout the body. Exotoxin-related symptoms arise in the body. Virulent factors are envelope antigen, endotoxin, exotoxin, V antigen and W antigen.	Giemsa staining, Wayson staining, fluorescent antibody staining.	Endemic in wild rodents, especially rats.	Controlling of rat population Tetracycline and streptomycin are used for treatment.
Leptospirosis	*Leptospira interrogans*	Fever, chills, headache, meningitis, liver damage and kidney failure.	After ingestion, organisms pass through mucous membrane or skin. They circulate in the blood and multiply in various organs and produce various symptoms.	Dark field microscopic examination, agglutination reactions, culturing of blood and urine.	Transmission via animal urine.	Penicillin G, Vaccination of domestic livestock.

Common cold	Rhinoviruses	Scratchy throat, nasal discharge, malaise, headache and cough.	Viruses lodge on respiratory epithelium, starting infection that spreads to adjacent cells, ciliary action ceases and cells slough; mucous secretion increases and inflammatory reaction occurs, infection stopped by interferon production and antibody production.	Nasal washings, are inoculated in human diploid fibroblast cells. Recovery between 1–7 days. The viruses are identified by cytopatheic effect. Differentiation is done by neutralizing sera.	Inhalation of infected droplets.	Hand washing, avoiding contact with people having cold.
Influenza	Influenza virus	Fever, muscle ache, lack of energy, head ache, sore throat, cough	Infection of upper respiratory epithelium, cells destroyed and virus released to infect other cells.	Recovery of viruses from nasal secretions using cell culture, electron microscopy.	Inhalation of infected droplets	Amantidine and rimantadine are drugs used. Vaccine is available.
Poliomyelitis	Polioviruses 1,2,3	Headache, fever, stiff neck, nausea, pain, paralysis.	Virus infects the throat and intestine, circulate via bloodstream, and enters some motor nerve cells of the brain and spinal cord. Infected nerve cells lyses upon release of mature virus.	Culture of poliovirus from throat and stool.	Faecal-oral route, contaminated water.	Oral polio vaccine, artificial ventilation for respiratory paralysis.

(Contd.)

BRIEF SUMMARY OF VARIOUS INFECTIONS OF THE HUMAN BODY (Continued)

Disease	Causative agent	Symptoms	Pathogenesis	Lab diagnosis	Epidemiology	Prevention and treatment
Rabies	Rabies virus	Fever, headache, nausea, vomiting, sore throat, spasm of the muscles of mouth and throat, coma and death.	During incubation period virus multiplies at the site of bite, then travels via nerves to the CNS where it multiplies and spreads outward via multiple nerves to infect heart and other organs.	Fluorescent antibody-stained smear from conjunctiva, fatty tissue of neck or brain show negri inclusion bodies.	Bite of rabid animal, usually dog.	Avoiding suspected animals, immunization of pets, immunization of wildlife.
Chickenpox	Varicella-zoster virus	Small, itchy bumps and blisters over skin and mucous membranes and fever.	Virus multiplies in the upper respiratory tract and spread via blood to the skin, cytopatheic effect of virus is seen in epidermis, vesicles formed contain serum, polymorphonuclear leucocytes and intranuclear inclusion bodies are found.	Fluorescent antibody technique to detect viral antigens, virus isolation in tissue culture.	Transmission by respiratory route, highly infectious.	Passive immunization with Zoster immunoglobulin (ZIG), non-asprin medicine to relieve itching. Acyclovir used for severe cases.
Rubella (3-day measles or German measles)	Rubella virus	Malaise, headache, fever, rash beginning on forehead and face, enlarged lymph nodes behind the ear.	Following replication, virus spreads to all parts of the body and crosses the placenta.	Inoculation of tissue culture with throat washing.	Human is the only source.	Live attenuated rubella vaccine administered to children at 15 months of age.

Measles	Rubeola virus	Rash, malaise, fever, conjunctivitis, cough and nasal discharge.	Virus multiplies in respiratory tract, carried by blood to various parts of the body, notably, skin, lungs and brain, and creates damage to respiratory epithelium, leads to secondary infection.	Complement fixation test, haemagglutination inhibition test and cultivation of virus in tissue culture.	Acquired by respiratory route highly contagious.	Live attenuated rubella vaccine administered to children at 15 months of age.
Infectious mononucleosis	Epstein – Barr virus	Fatigue, fever, sore throat, enlargement of lymph nodes and spleen, meningitis.	Productive infection of epithelial cells of throat and salivary duct, latent infection of B lymphocytes, activation of B and T lymphocytes.	Viral isolation, nucleic acid hybridization, indirect immuno-fluorescence, ELISA.	Spread by saliva	Acyclovir reduces EB virus shedding. There is no vaccine.
Encephalitis	*Arbovirus*	Headache, chillness, fever, nausea, vomiting, generalized pain, within 24–48 hours, mental confusion, tremors, convulsions and coma.	The virus enters into the human body through mosquito or tick bite. After entry, the virus multiplies in non-neural tissue and is present in the blood 3 days before first sign of involvement of the CNS. Then the virus multiplies in the brain cells which are injured and destroyed, and encephalitis becomes apparent.	Neutralizing haemagglutination inhibiting antibodies are detected within few days, CF antibodies appear later.	Mosquito or tick-borne.	No specific treatment is available.

(Contd.)

BRIEF SUMMARY OF VARIOUS INFECTIONS OF THE HUMAN BODY (Continued)

Disease	Causative agent	Symptoms	Pathogenesis	Lab diagnosis	Epidemiology	Prevention and treatment
Histo-plasmosis	*Histoplasma capsulatum*	Mild respiratory symptoms, malaise, fever, chest pain and cough.	Spores undergo metamorphosis to tissue phase; granuloma formation	Stained smears of ulcer exudates, bone marrow, sputum	Inhalation of airborne spores from soil	Amphotericin B, ketaconazole and Itraconazole are drugs of choice.
Coccidioido mycosis (Valley fever)	*Coccidioides immitis*	Fever, cough, chest pain, loss of appetite and weight, pain in joints, skin	After lodging in lung, arthrospores develop into spherules that mature and discharge their endospores, each of which can then develop into another spherule.	Microscopic examination of sputum or pus for spherules, blood test for antibodies.	Inhalation of spores from soil	Control of dust. Amphotericin B and ketaconazole are drugs of choice.
Crypto-coccosis	*Cryptococcus neoformans*	Headache, vomiting, confusion, weight loss, fever, paralysis, coma and death.	Infection starts in lungs; organisms multiply and enters bloodstream and carried to various parts of the body. Phagocytosis inhibited and opsonins neutralized; meninges and adjacent tissues of brain are infected.	CSF examination with India ink.	Inhalation of dust containing pigeon droppings.	Amphotericin B with flucystosine or itraconazole.

Malaria	Four species of *Plasmodium*	Violent chills and fever alternating with feeling healthy.	Cell rupture and release protozoa that causes fever.	Blood examination and serology	Transmitted by bite of mosquito Anopheles.	Chloroquine and primaquine are drugs used. Vaccine is under development.
African sleeping sickness	*Trypanosoma brucei*	Fever, sleeplessness, rash, enlargement of lymph nodes and headache.	Organisms multiply at the site and then enter blood and lymphatic circulation, as new cycles of parasites are released.	Microscopic examination and serology.	Bite of tsetse fly.	Protective clothing, use of insecticides. Suramin is the drug of choice.

APPENDIX II

Model Questions

1. Why are some microorganisms called normal flora?
2. How does a bacterial capsule contribute to the virulence of the bacterium?
3. What is the mechanism of bile esculin test?
4. What are the characteristics of pandemic disease?
5. What are the attributes of an ideal antimicrobial agent?
6. What are the similarities between exotoxins and endotoxins?
7. Why are some mycotic diseases of human called opportunistic mycoses?
8. What are dermatomycoses?
9. State the clinical manifestations of infections due to *Corynebacterium diphtheriae*.
10. List the characteristics of *Bordetella pertussis*.
11. State the characteristics of disseminated gonococcal infections.
12. List the intracellular bacterial pathogens.
13. For what infections are the following tests done, Shick's test, Dick's test, Mantoux test and Frie test?
14. State four differences between *Salmonella typhi* and *Salmonella paratyphi A*.
15. State the characteristics of oocysts of malarial parasite.
16. List the organisms that cause dental caries.
17. State the uses of protein A.
18. Give the significance of Kirby–Bauer test in chemotherapy.
19. Give examples of preventable and non-preventable nosocomial infections.
20. Define nosocomial infection.
21. List four differences between mycoplasma and virus.
22. Name different kinds of interferons and their cells of origin.
23. List properties of OPA proteins of *Neisseria gonorrhoeae*.
24. State four differences between *Shigella dysenteriae* and S*almonella typhi*.
25. State the differences between male and female gametocytes of *Plasmodium vivax*.
26. List the differences between *Epidermophyton* and *Microsporum*.

27. State four features of interferon.
28. Define antigenic shift and antigenic drift. Give examples.
29. List four important properties of pilins of *N. gonorrhoeae*.
30. State four morphological differences between *Microsporum gypseum* and *M. canis*.
31. State two tests for the demonstration of spirochaetes in primary syphilis cases.
32. Define tuberculin units (TU).
33. Name the factors involved in the pathogenesis of pertussis.
34. List the differences between the trophozoites of *E. histolytica* and *E. coli*.
35. State two characteristic features of rings of *P. falciparum*.
36. List four RNA viruses that cause meningitis in man.
37. Define a street virus with example.
38. State the differences between *Corynebacterium diphtheriae* and *L. acidophilus*.
39. Give the differences between *Trichophyton mentagrophytes* and *T. rubrum*.
40. How is *Streptococcus pneumoniae* typed? Explain.
41. List four selective media used for the isolation of *V. cholerae*.
42. List four dimorphic moulds pathogenic to man.
43. List four haemagglutinating viruses.
44. List four methods to assay virus infectivity.
45. List four bacteria that cause meningitis in man.
46. State the characteristics of negri bodies.
47. What are the factors that predispose one to pneumococcal infection?
48. List the selective media used for the isolation of *Salmonella*.
49. Enumerate the technique to concentrate *Entamoeba* cysts.
50. State four differences between *Candida albicans and Cryptococcus neoformans*.
51. List eight organisms of normal flora of vagina of prepubescent females.
52. State the pathogenic attributes of *Streptococcus pneumoniae*.
53. Give a list of four bacterial pathogens that grow intracellularly.
54. List the characters of *Ureaplasma urealyticum*.
55. Name four bacterial zoonotic diseases.
56. Give four characteristics of *P. vivax* with diagram.
57. List four fungal pathogens that cause subcutaneous mycoses in man.
58. Give a brief note on SSPE.
59. State the principle of pulse polio vaccination.
60. How do you cultivate influenza virus.
61. List eight organisms of normal nasal flora.
62. State the pathogenic attributes of *S. aureus*.

63. List the differences between *N. meningitidis and H. influenzae*.
64. Enumerate the agents causing ophthalmia neonatorum.
65. Give the properties of *Shigella dysentriae* exotoxin.
66. Describe the structure and characteristics of influenza virus neuraminidase.
67. List the differences between polio and rhinoviruses.
68. State the properties of fusion proteins of paramyxovirus.
69. Name eumycotic agents producing black granules.
70. List four characteristics of trophozoites of *P. malariae* with diagrams.
71. List the selective media used for the cultivation of *Legionella*.
72. List the selective and differential media for *Staphylococcus*.
73. Define comedo and acne vulgaris.
74. Explain about ectopic pregnancy.
75. What is PID and explain?
76. Which bacterium is called Koch's weak bacillus?
77. What is a reservoir host?
78. Differentiate saprophyte from parasite.
79. Which sample of urine is collected from a patient for bacterial culture?
80. What is a catheter?
81. In how many ways viruses can be transmitted directly from one human being to another?
82. Name the diseases transmitted by *Culex* mosquitoes.
83. What is a wetmount?
84. What is coagulase test?
85. How many different stages are there for *Entamoeba histolytica*?
86. Name two important filarial parasites.
87. What is reinfection?
88. What is septicaemia?
89. What is pyaemia?
90. Explain gram stain technique.
91. Define epidemic.
92. Write a note on rice-water stool.
93. Write short note on *Aedes aegypti?*
94. What is parotitis?
95. Define nosocomial infection?
96. Write a note on iatrogenic infection.
97. What are antifols?

98. Why don't ionidazoles affect human cells?
99. *Vibrio cholerae* does not satisfy Koch's postulates. Why?
100. Comment on quelluing reaction.
101. Comment on *Risus sardonicus*.
102. How does *N. gonorrhoeae* establish itself in the hostile mucosa?
103. Comment on Dane particle.
104. Comment on Kopliks spot.
105. What is cytostome?
106. What is quartan malaria?
107. What are the major differences between infectious and non-infectious diseases and between communicable and non-communicable infectious diseases?
108. How do drug-resistant hospital infections arise and how can they be treated and prevented?
109. How can epidemiologists recognize an infectious disease in a population?
110. Compare tetanospasmin with botulinum toxin. Add a note on how toxins contribute to the pathology of each disease.
111. What are the different causative agents of hepatitis and how do they differ from one another? How can one avoid hepatitis?
112. Describe the antigenic structure of the influenza A virus and explain why this virus is capable of causing pandemics.
113. List the reasons why malaria is still one of the most serious of all infectious diseases that affect humans.
114. How are fungal diseases categorized?
115. Describe lepra reactions.
116. Describe the types, chemical nature and the characteristics of pneumococcal capsule.
117. Describe the prophylaxis of tetanus.
118. Give an account of the laboratory diagnosis of food poisoning due to *Clostridium botulinum*.
119. Describe the attributes of bacteria that make them pathogenic to man.
120. Describe the cultivation methods of *Entamoeba histolytica* and malarial parasite.
121. List the agents causing actinomycotic mycetoma. Give their differentiating features.
122. Describe the agents causing chromoblastomycosis. Add a note on the morphology of any two of them.
123. Describe in detail the replication of double-stranded DNA virus.
124. Describe the replication of paramyxoviral nucleic acid.

125. Describe the properties of *Shigella dysenteriae* exotoxins.
126. Describe the structure and mode of action of toxin of *Clostridium tetani*.
127. List the difference between paramyxo and orthomyxoviruses.
128. Describe three non-treponemal antibody tests for syphilis.
129. List different typing methods of *Salmonella typhi* for epidemiological purpose.
130. List the morphological features of agents of chromoblastomycosis.
131. Write short notes on Wood's Lamp examination for the diagnosis of dermatophytosis.
132. List the important properties of picornaviruses.
133. List the pathogenesis and pathology of coxsackie viruses.
134. List the differences between *M. tuberculosis and M. leprae.*
135. State the difference between *Salmonella* and *Shigella.*
136. Give the lab diagnosis of extraintestinal amoebiasis.
137. Describe the toxigenicity test for *C. diphtheriae*.
138. Write a brief note on hepatitis E virus.
139. Differentiate ectothrix and endothrix.
140. Give an account of the structure and functions of haemagglutinins of influenza virus.
141. Give an account of rabies immunization schedule.
142. Explain the morphology and identification characteristics of *Candida albicans*.
143. Explain the morphology and identification characteristics of *Cryptococcus neoformans*.
144. Describe about disseminated gonococcal infections.
145. Write a note a rhinosporidiasis.
146. Write a note on sporotrichosis.
147. Classify and describe the characteristics of pili of *E. coli*. State their functions.
148. How will you demonstrate *M.tuberculosis* or its components in the specimens?
149. Describe the action of botulinum toxin and its detection methods.
150. Describe the pathogenesis and lab diagnosis of diphtheria. Add a note on toxigenicity testing.
151. Describe the differences between *C. immitis* and *Paracoccidioides brasiliensis.*
152. Describe the life cycle of *F. hepatica* and its lab diagnosis.
153. What are the normal microflora of the skin and eyes?
154. What are non-specific lines of defence provided by the skin?
155. How can a fungus and parasite be detected in a clinical specimen?
156. Name and describe some biochemical tests used to identify an unknown bacterium from the patient's specimen.

157. Which body site can be affected by gonorrhoea. Add a note on its pathogenicity?
158. Why do HIV infections lead to opportunistic infections and development of cancer?
159. How does *E. histolytica* cause the disease?
160. Describe the epidemiology, pathogenesis and lab diagnosis of malarial infection.
161. Compare a common source epidemic to a host-to-host epidemic. Site at least one example of each.
162. Why is *Mycobacterium tuberculosis* such a widespread respiratory pathogen. Add a note on the factors that aid in controlling the spread of tuberculosis, its diagnosis and treatment.
163. List the differences between exotoxins and endotoxins.
164. Explain the serological diagnosis of typhoid.
165. Give an account of the toxins and enzymes of *Staphylococcus.*
166. Explain the treatment measures of staphylococcal infections.
167. Write an account of pathogenesis, diagnostic tests and treatment of cholera.
168. Write an account of pathogenesis, diagnostic tests and treatment of tetanus.
169. Write an account of pathogenesis, diagnostic tests and treatment of influenza.
170. Write an account of pathogenesis, diagnostic tests and treatment of post-natal rubella.
171. Write an account of pathogenesis, diagnostic tests and treatment of leprosy.
172. How will you demonstrate *T. pallidum* or its components in the specimen/ tissues?
173. Give two examples each of preventable and non-preventable nosocomial infections. State the reasons.
174. List the differences between bacillary and amoebic dysentery.
175. Describe the characteristic difference between septic and aseptic meningitis.
176. Describe hair penetration test and its uses.
177. Describe the morphological characteristics of agents of chromoblastomycosis.
178. Describe the difference between orthomyxo and paramyxoviruses.
179. Describe the multiplication of myxoviral nucleic acid and its importance.
180. Describe various concentraton techniques used for the detection of cysts of *E. histolytica.*
181. With a neat diagram, explain the morphology and characteristics of *W. bancrofti* microfilariae. How is it differentiated from other microfilariae.
182. Describe the factors that determine the outcome of host–parasite relationship.
183. Discuss the specialized strategies used by pathogens for survival and multiplication.

184. Describe the types of streptococci based on their action on RBCs in blood agar medium. Add a note on antigenic structure and classification of streptococci.
185. Discuss serodiagnosis of typhoid fever and paratyphoids.
186. Explain meningitis. Add a note on aetiology with examples.
187. Explain the mode of action of tetanus toxin.
188. Describe briefly about antiviral agents.
189. Which is the widely used method for screening HIV antibodies? Explain. How will you confirm it?
190. With suitable examples differentiate protozoan parasites from helminthes.
191. Discuss the sources, route of spread and control of hospital infections.
192. Define pathogenicity. Describe the nature of pathogenicity.
193. Explain various non-specific resistance factors.
194. How will you collect specimens for microbiological investigation?
195. Write an account on serodiagnosis.
196. Give a resume on the epidemiology, pathogenesis and treatment of tuberculosis.
197. Discuss the different types of staphylococcal infections.
198. Draw and label a HIV particle.
199. Describe a hepatitis virus type B.
200. What is zoonosis? Give five examples.
201. Mention the epidemiology and pathogenesis of Ascariasis.
202. Discuss the process of host–parasite interactions with reference to opportunistic pathogens.
203. Give an account on macrolides.
204. Sketch a brief note on tuberculosis.
205. Write short notes on enteric fever.
206. Sketch a brief note on zoonotic diseases.
207. Describe about syphilis.
208. Write short note on measles.
209. Give brief notes on food-borne hepatitis.
210. Give an account on mycoplasmal infections.
211. Discuss the pathogenesis of filariasis and add a note on control measures.
212. What are the methods and principles of epidemiological studies?
213. How is immunofluorescence used to identify *Treponema pallidum*? Why are serologic methods critical for the identification of pathogens?
214. Describe the pathogenesis and pathology of each of the following and add a note on identification of *Shigella dysenteriae* and *Mycobacterium leprae.*

215. Explain the pathogenesis, identification and control of the following:
    i. Mycoplasmal infections
    ii. Bacterial food poisoning
216. Describe the pathogenesis, pathology and diagnosis of rabies and measles.
217. How do infections caused by *Entamoeba histolytica* occur? Add a note on pathogenesis, identification and prevention.
218. With suitable example, describe diseases due to bacterial toxins. Add a note on laboratory diagnosis of any one of them.
219. Describe the procedures used for the direct demonstration *of T. pallidum* from clinical specimens.
220. Describe the opportunistic fungal infections. Add a note on the methods of direct demonstration of fungal components in blood.
221. Classify the viruses causing hepatitis. Describe the pathogenesis and prevention of hepatitis A virus infection.
222. Describe in detail the laboratory diagnosis of tuberculosis. Add a note on antimicrobial sensitivity of *Mycobacterium tuberculosis.*
223. Describe in detail laboratory diagnosis of leprosy. Add a note on the current status of leprosy vaccines.
224. Discuss antiviral agents and their mode of action.
225. Describe the characteristics of *Mycoplasma*. Name the diseases caused by them and discuss their lab diagnosis.
226. Classify *Entamoeba histolytica*. Discuss intestinal amoebiasis and its laboratory diagnosis.
227. Discuss the morphological differentiation of *P. vivax and P. falciparum*.
228. Describe viral encephalitis.
229. Describe *H. influenzae* meningitis.
230. Describe the cultivation of *Clostridium tetani.*
231. List the agents of systemic mycosis. Add a note on histoplasmosis.
232. Describe the lab diagnosis of gonococcal infections.
233. Discuss the role of aspergilli in human infections.
234. Discuss the etio-pathogenesis of anthrax and its lab diagnosis.
235. Discuss the etio-pathogenesis of plague and its lab diagnosis.
236. Give the prophylaxis of hepatitis infections.
237. Explain the post-exposure prophylaxis of rabies.
238. Discuss various methods used to trace the source of hospital-acquired infections with suitable examples.
239. Discuss in detail about hospital waste disposal. Add a note on the government guidelines on infectious waste disposal.

240. Write an account on host–microbe interaction and add a note on the virulence factors that enhance the pathogenicity.
241. Describe the specialized methods and precautions necessary for successful isolation of anaerobic pathogens.
242. Give an account on staphylococcal skin infections.
243. List the viral respiratory diseases, symptoms and pathogenicity.
244. What are the three categories of fungal infections? Distinguish superficial infections from systemic mycoses.
245. Give a schematic description of collection and processing of urine.
246. Enumerate the virulent factors of microbes and its significance.
247. Give a detailed account on pathogenesis, lab diagnosis, treatment and prevention of *Streptococcus pneumoniae* infection.
248. Give a detailed account on pathogenesis, lab diagnosis, treatment and prevention of hepatitis.
249. Give a detailed account on pathogenesis, lab diagnosis, treatment and prevention of syphilis.
250. Give a detailed account on pathogenesis, lab diagnosis, treatment and prevention of amoebiasis.
251. Describe the agents causing gastroenteritis. Describe the laboratory diagnosis of cholera.
252. Describe the procedure and interpretation of treponemal tests for syphilis.
253. List the dimorphic fungi causing systemic infections. Discuss the lab diagnosis of histoplasmosis.
254. Describe the characteristics of atypical pneumonia. Add a note on the lab diagnosis of *M. pneumoniae* infections.
255. Describe the characteristics of methicillin-resistant *S. aureus* (MRSA). What are the strategies adopted to control hospital-borne infections due to MRSA?
256. Discuss the source of hospital infections. How are they detected and controlled?
257. Give an account of the normal flora of human system.
258. Discuss the role of non-specific factors in disease resistance.
259. Describe the different types of staining and microscopic methods in identification of bacteria.
260. Write about the proper collection of clinical specimens, storage and transportation to the laboratory.
261. Describe the morphology, pathogenesis and laboratory diagnosis of *Mycobacterium leprae*.
262. Describe about measles.

263. What is superficial mycosis? Classify them. Explain the diseases with examples.
264. Describe the development of malarial plasmodia. Add a note on the treatment and prophylactic measures.
265. Write an essay on host–microbe interactions.
266. Give a detailed outline of the general methods and identification of pathogenic bacteria.
267. Explain the epidemiology and laboratory treatment of syphilis and gonorrhoea.
268. Briefly describe the epidemiology and treatment of leprosy and cholera.
269. How will you diagnose rabies and encephalitis? Add a note on their prevention?
270. Write an essay on the diagnosis and treatment of rubella and herpes simplex infections.
271. Describe the life cycle of malarial infections.
272. What are hospital-acquired infections?
273. Give an account of the normal flora of human body and typing of bacterial isolates.
274. Give an account of the bacillary dysentery and mumps.
275. Give a detailed account on antifungal agents.
276. Describe pneumonia.
277. Describe poliomyelitis.
278. Give an account on hospital-borne infections
279. What are the responsibilities of an epidemiologist?
280. Give an account on viral encephalitis.

GLOSSARY

AB toxins Activity and structure of many exotoxins are based on the AB model. B portion of the toxin is responsible for toxin binding to a cell but does not directly harm it. The A portion enters the cell and disrupts its function.

Acid-fast Bacteria like the mycobacteria that cannot be easily decolorized with acid alcohol after being stained with dyes such as basic fuchsin.

Acid-fast staining A staining procedure that differentiates bacteria based on their ability to retain a dye when washed with an acid-alcohol solution.

Acquired enamel pellicle A membranous layer on the tooth enamel surface formed by selectively adsorbing glycoproteins (mucins) from saliva. This pellicle confers a net negative charge to the tooth surface.

Acquired immune deficiency syndrome (AIDS) An infectious disease syndrome caused by the human immunodeficiency virus and is characterized by the loss of a normal immune response, followed by increased susceptibility to opportunistic infections and an increased risk of some cancers.

Acquired immunity A type of specific immunity that develops after exposure to a suitable antigen or is produced after antibodies are transferred from one individual to another.

Active carrier An individual who has an overt clinical case of a disease and who can transmit the infection to others.

Active immunization The induction of active immunity by natural exposure to a pathogen or by vaccination.

Acute infections Virus infections with a fairly rapid onset that last for a relatively short time.

Acute viral gastroenteritis An inflammation of the stomach and intestines, normally caused by Norwalk and Norwalk-like viruses, other caliciviruses, rotaviruses, and astroviruses.

Acyclovir A synthetic purine nucleoside derivative with antiviral activity against herpes simplex virus.

Adenosine 5-triphosphate (ATP) The triphosphate of the nucleoside adenosine, which is a high-energy molecule or has high phosphate group transfer potential and serves as the cell's major form of energy currency.

Adhesin A molecular component on the surface of a microorganism that is involved in adhesion to a substratum or cell. Adhesion to a specific host tissue usually is a preliminary stage in pathogenesis, and adhesins are important virulent factors.

Adjuvant Material added to an antigen to increase its immunogenicity. Common examples are alum, killed *Bordetella pertussis*, and an oil emulsion of the antigen, either alone (Freund's incomplete adjuvant) or with killed mycobacteria (Freund's complete adjuvant).

Agglutination reaction The formation of an insoluble immune complex by the cross-linking of cells or particles.

AIDS-related complex (ARC) A collection of symptoms such as lymphadenopathy (swollen lymph glands), fever, malaise, fatigue, loss of appetite, and weight loss. It results from an HIV infection and may progress to frank AIDS.

Airborne transmission The type of transmission of infectious organism in which the pathogen is truly suspended in the air and travels over a metre or more from the source to the host.

Alpha haemolysis A greenish zone of partial clearing around a bacterial colony growing on blood agar.

Alpha-proteobacteria One of the five subgroups of proteobacteria, each with distinctive 16S rRNA sequences. This group contains most of the oligotrophic proteobacteria; some have unusual metabolic modes such as methylotrophy, chemolithotrophy, and nitrogen fixing ability. Many have distinctive morphological features.

Alternative complement pathway An antibody-independent pathway of complement activation that includes the C3–C9 components of the classical pathway and several other serum protein factors (e.g. factor B and properdin).

Alveolar macrophage A vigorously phagocytic macrophage located on the epithelial surface of the lung alveoli where it ingests inhaled particulate matter and microorganisms.

Amantadine An antiviral compound used to prevent type A influenza infections.

Ames test A test that uses a special *Salmonella* strain to test chemicals for mutagenicity and potential carcinogenicity.

Aminoglycoside antibiotics A group of antibiotics synthesized by *Streptomyces* and *Micromonospora*, which contain a cyclohexane ring and amino sugars; all aminoglycoside antibiotics bind to the small ribosomal subunit and inhibit protein synthesis.

Amoebiasis (amoebic dysentery) An infection caused by amoebae, often resulting in dysentery; usually it refers to an infection by *Entamoeba histolytica.*

Amoeboid movement Moving by means of cytoplasmic flow and by the formation of pseudopodia (temporary cytoplasmic protrusions of the cytoplasm).

Amphotericin B An antibiotic from a strain of *Streptomyces nodosus* that is used to treat systemic fungal infections; it is also used topically to treat candidiasis.

Anorexia Loss of appetite.

Anthrax An infectious disease of animals caused by ingesting *Bacillus anthracis* spores. It can also occur in humans and is sometimes called woolsorter's disease.

Antigenic drift A small change in the antigenic character of an organism that allows it to avoid attack by the immune system.

Antigenic shift A major change in the antigenic character of an organism that alters it to an antigenic strain unrecognized by host immune mechanisms.

ASO Antistreptolysin O (titre).

Aspergillosis A fungal disease caused by a species of *Aspergillus*.

Autogenous infection An infection that results from a patient's own microbiota, regardless of whether the infecting organism became part of the patient's microbiota subsequent to admission to a clinical care facility.

Autolysins Enzymes that partially digest peptidoglycan in growing bacteria so that the peptidoglycan can be enlarged.

Bacteraemia The presence of viable bacteria in the blood.

Bacterial vaginosis A sexually transmitted disease caused by *Gardnerella vaginalis*, *Mobiluncus* spp., *Mycoplasma hominis*, and various anaerobic bacteria. Although a mild disease, it is a risk factor for obstetric infections and pelvic inflammatory disease.

Bacteriophage (phage) typing A technique in which strains of bacteria are identified based on their susceptibility to a variety of bacteriophages.

Balanitis Inflammation of the glans penis usually associated with *Candida*. It is a sexually transmitted disease.

Beta-haemolysis A zone of complete clearing around a bacterial colony growing on blood agar. The zone does not change significantly in colour.

Beta-proteobacteria One of the five subgroups of proteobacteria, each with distinctive 16S rRNA sequences. Members of this subgroup are similar to the alpha-proteobacteria metabolically, but tend to use substances that diffuse from organic matter decomposition in anaerobic zones.

Biopsy Removal of some tissue from the body for examination to establish a diagnosis.

Black piedra A fungal infection caused by *Piedraia hortae* that forms hard black nodules on the hairs of the scalp.

Botulism A form of food poisoning caused by a neurotoxin (botulin) produced by *Clostridium botulinum* serotypes A–G. It is sometimes found in improperly canned or preserved food.

Bubo A tender, inflamed, enlarged lymph node that results from a variety of infections.

Buboe The infected lymph nodes associated with the bubonic plague.

Candidiasis An infection of *Candida*, a dimorphic fungi, commonly involving the skin.

Carbuncle Skin lesion typically caused by *Staphylococcus aureus*, and resembling a boil with multiple heads.

Caries Suppuration and subsequent decay of hard tissue.

Carrier An infected individual who is a potential source of infection for others and plays an important role in the epidemiology of a disease.

Caseous lesion A lesion resembling cheese or curd; cheesy. Most caseous lesions are caused by *M. tuberculosis*.

Casual carrier An individual who harbours an infectious organism for only a short period.

Catheter A tubular surgical instrument for withdrawing fluids from a cavity of the body, especially one for introduction into the bladder through the urethra for the withdrawal of urine.

Cellulitis A diffuse spreading infection of subcutaneous skin tissue caused by streptococci, staphylococci, or other organisms. The tissue is inflamed with oedema, redness, pain, and interference with function.

Cerebrospinal fluid (CSF) Fluid bathing the ventricles of the brain and circulating around the spaces surrounding the brain and spinal cord.

Cervical lymphadenopathy Swollen glands in the neck.

Chancre The primary lesion of syphilis, occurring at the site of entry of the infection.

Chancroid A sexually transmitted disease caused by the gram-negative bacterium *Haemophilus ducreyi*. Worldwide, chancroid is an important cofactor in the transmission of the AIDS virus. Also known as genital ulcer disease due to the painful circumscribed ulcers that form on the penis or entrance to the vagina.

Chemotherapeutic agents Compounds used in the treatment of disease that destroy pathogens or inhibit their growth at concentrations low enough to avoid an undesirable damage to the host.

Chickenpox A highly contagious skin disease, usually affecting 2 to 7-year-old children; it is caused by the Varicella–zoster virus, which is acquired by droplet inhalation into the respiratory system.

Cholecystectomy Surgical removal of the gall bladder.

Choleragen The cholera toxin; an extremely potent protein molecule elaborated by strains of *Vibrio cholerae* in the small intestine after ingestion of faeces-contaminated water or food. It acts on epithelial cells to cause hypersecretion of chloride and bicarbonate and an outpouring of large quantities of fluid from the mucosal surface.

Chromoblastomycosis A chronic fungal infection of the skin, producing wart-like nodules that may ulcerate. It is caused by the black moulds *Phialophora verrucosa* or *Fonsecaea pedrosoi*.

Clostridial myonecrosis Death of individual muscle cells caused by clostridia. Also called gas gangrene.

Coagulase An enzyme that induces blood clotting; it is characteristically produced by pathogenic staphylococci.

Coccidioidomycosis A fungal disease caused by *Coccidioides immitis* that exists in dry, highly alkaline soils, and is also known as valley fever, San Joaquin fever, or desert rheumatism.

Cold sore A lesion caused by the herpes simplex virus; usually occurs on the border of the lips or nares. Also known as a fever blister or herpes labialis.

Colicin A plasmid-encoded protein that is produced by enteric bacteria and binds to specific receptors on the cell envelope of sensitive target bacteria, where it may cause lysis or attack of specific intracellular sites such as ribosomes.

Coliform A gram-negative, non-sporing, facultative rod that ferments lactose with gas formation within 48 hours at 35°C.

Comedo A plug of dried sebum in an excretory duct of the skin.

Commensal Living on or within another organism without injuring or benefiting the other organism.

Common cold An acute, self-limiting, and highly contagious virus infection of the upper respiratory tract that produces inflammation, profuse discharge, and other symptoms.

Common-source epidemic An epidemic that is characterized by a sharp rise to a peak and then a rapid, but not as pronounced, decline in the number of individuals infected; it usually involves a single contaminated source from which individuals are infected.

Common vehicle transmission The transmission of a pathogen to a host by means of an inanimate medium or vehicle.

Communicable disease A disease associated with a pathogen that can be transmitted from one host to another.

Concatemer A long DNA molecule consisting of several genomes linked together in a row.

Congenital (neonatal) herpes An infection of a newborn caused by transmission of the herpesvirus during vaginal delivery.

Congenital rubella syndrome A wide array of congenital defects affecting the heart, eyes, and ears of a foetus during the first trimester of pregnancy, and caused by the rubella virus.

Congenital syphilis Syphilis that is acquired *in utero* from the mother.

Convalescent carrier An individual who has recovered from an infectious disease but continues to harbour large numbers of the pathogen.

Cryptococcosis An infection caused by the basidiomycete, *Cryptococcus neoformans*, which may involve the skin, lungs, brain, or meninges.

Dane particle A 42-nm spherical particle, one of the three infectious particles that are seen in hepatitis B virus infections. The Dane particle is the complete virion.

Delta-proteobacteria One of the five subgroups of proteobacteria, each with distinctive 16S rRNA sequences. They are chemoorganotrophic bacteria that usually are either predators on other bacteria or anaerobes that generate sulphide from sulphate and sulphite.

Dental plaque A thin film on the surface of teeth consisting of bacteria embedded in a matrix of bacterial polysaccharides, salivary glycoproteins, and other substances.

Dermatomycosis A fungal infection of the skin. It is a general term that comprises the various forms of tinea, and it is sometimes used to specifically refer to athlete's foot (tinea pedis).

Diphtheria An acute, highly contagious childhood disease that generally affects the membranes of the throat and less frequently the nose. It is caused by *Corynebacterium diphtheriae*.

Dipicolinic acid A substance present at high concentrations in the bacterial endospore. It is thought to contribute to the endospore's heat resistance.

DPT (diphtheria–pertussis–tetanus) vaccine A vaccine containing three antigens that is used to immunize people against diphtheria, pertussis or whooping cough, and tetanus.

Droplet nuclei Small particles (1 to 4 mm in diameter) that represent what is left from the evaporation of larger particles (10 mm or more in diameter) called droplets.

Dysphagia Difficulty in swallowing due to pain.

Dyspnoea Difficulty with breathing.

Dysuria Difficulty or pain on passing urine.

Ebola virus haemorrhagic fever An acute infection caused by a virus that produces varying degrees of haemorrhage, shock, and sometimes death.

Ectoparasite A parasite that lives on the surface of its host.

Ectopic pregnancy Pregnancy in which implantation occurs outside the uterus. Development of ectopic pregnancy is a life-threatening condition for the mother.

Ehrlichiosis A tick-borne (*Dermacentor andersoni*, *Amblyomma americanum*) rickettsial disease caused by *Ehrlichia chaffeensis*. Once inside leucocytes, a non-specific illness develops that resembles rocky mountain spotted fever.

Elementary body A small, dormant body that serves as the agent of transmission between host cells in the chlamydial life cycle.

Encephalopathy Cerebral inflammation. This most often manifests as disorientation, mental confusion, excitability or abnormal behaviour.

Encystations The formation of a cyst.

Endemic (murine) typhus A form of typhus fever caused by the rickettsia *Rickettsia typhi* that occurs sporadically in individuals who come into contact with rats and their fleas.

Endemic disease A disease that is commonly or constantly present in a population, usually at a relatively steady low frequency.

Endogenous infection An infection caused by an individual's own normal body microbiota.

Endogenous pyrogen A substance such as the lymphokine interleukin-1, which is produced by host cells and induces a fever response in the host. It is also called simply a pyrogen.

Endoparasite A parasite that lives inside the body of its host.

Endotoxin The heat-stable lipopolysaccharide in the outer membrane of the cell wall of gram-negative bacteria that is released when the bacterium lyses, or sometimes during growth, and is toxic to the host.

Enterotoxin A toxin specifically affecting the cells of the intestinal mucosa, causing vomiting and diarrhoea.

Enzyme-linked immunosorbent assay (ELISA) A technique used for detecting and quantifying specific antibodies and antigens.

Epidemic A disease that suddenly increases in occurrence above the normal level in a given population.

Epidemic (louse-borne) typhus A disease caused by *Rickettsia prowazekii* that is transmitted from person to person by the body louse.

Epidemiologist A person who specializes in epidemiology.

Epidemiology The study of the factors determining and influencing the frequency and distribution of disease, injury, and other health-related events and their causes in defined human populations.

Epsilon-proteobacteria One of the five subgroups of proteobacteria, each with distinctive 16S rRNA sequences. They are slender gram-negative rods, some of which are medically important (*Campylobacter* and *Helicobacter*).

Erysipelas An acute inflammation of the dermal layer of the skin, occurring primarily in infants and persons over 30 years of age with a history of streptococcal sore throat.

Erythema infectiosum A disease in children caused by the parvovirus B19. This disease is common in children between 4 and 11 years of age and is sometimes called fifth disease, since it was the fifth of six erythematous rash diseases in children in an older classification.

Erythromycin An intermediate spectrum macrolide antibiotic produced by *Streptomyces erythraeus*.

Eumycota A division of fungi in some classification systems. These are the true fungi consisting of the zygomycetes, ascomycetes, basidiomycetes, and chytridiomycetes.

Exfoliative toxin or exfoliatin An exotoxin produced by *Staphylococcus aureus* that causes the separation of epidermal layers and the loss of skin surface layers. It produces the symptoms of the scaled skin syndrome.

Exotoxin A heat-labile, toxic protein produced by a bacterium as a result of its normal metabolism or because of the acquisition of a plasmid or prophage that redirects its metabolism. It is usually released into the bacterium's surroundings.

Extracutaneous sporotrichosis An infection by the fungus *Sporothrix schenckii* that spreads throughout the body.

Fastidious Demanding in requirements for nutrients, atmospheric conditions and/or temperature of incubation.

Fever A complex physiological response to disease mediated by pyrogenic cytokines and characterized by a rise in core body temperature and activation of the immune system.

Fungaemia The presence of fungi in the bloodstream.

Furuncle Infection of a single hair follicle, typically caused by *Staphylococcus aureus*—a boil.

Gamma-proteobacteria One of the five subgroups of proteobacteria, each with distinctive 16S rRNA sequences. This is the largest subgroup and is very diverse physiologically; many important genera are facultatively anaerobic chemoorganotrophs.

Gas gangrene A type of gangrene that arises from dirty, lacerated wounds infected by anaerobic bacteria, especially species of *Clostridium*. As the bacteria grow, they release toxins and ferment carbohydrates to produce carbon dioxide and hydrogen gas.

Gastritis Inflammation of the stomach.

Gastroenteritis An acute inflammation of the lining of the stomach and intestines, characterized by anorexia, nausea, diarrhoea, abdominal pain, and weakness. It has various causes including food poisoning due to such organisms as *E. coli*, *S. aureus*, *Campylobacter* (campylobacteriosis), and *Salmonella* species; consumption of irritating food or drink; or psychological factors such as anger, stress, and fear. It is also called enterogastritis.

Genital herpes A sexually transmitted disease caused by the herpes simplex virus type 2.

Ghon complex The initial focus of parenchymal infection in primary pulmonary tuberculosis.

Giardiasis A common intestinal disease caused by the parasitic protozoan *Giardia lamblia*.

Gingivitis Inflammation of the gingival tissue.

Gingivostomatitis Inflammation of the gingiva and other oral mucous membranes.

Glomerulonephritis An acute inflammatory disease of the renal glomeruli.

Gnotobiotic Animals that are germ-free (microorganism-free) or live in association with one or more known microorganisms.

Gonococci Bacteria of the species *Neisseria gonorrhoeae*—the organism causing gonorrhoea.

Gonorrhoea An acute infectious sexually transmitted disease of the mucous membranes of the genitourinary tract, eye, rectum, and throat. It is caused by *Neisseria gonorrhoeae*.

Guillain–Barr syndrome A relatively rare disease affecting the peripheral nervous system, especially the spinal nerves, but also the cranial nerves. The cause is unknown, but it most often occurs after an influenza infection or flu vaccination. Also called French polio.

Gumma A soft, gummy tumour occurring in tertiary syphilis.

Haematuria The presence of blood in the urine.

Haemolysis The disruption of red blood cells and release of their haemoglobin. There are several types of haemolysis when bacteria such as streptococci and staphylococci grow on blood agar. In α-haemolysis, a narrow greenish zone of incomplete haemolysis forms around the colony. During β-haemolysis, a clear zone of complete haemolysis without any obvious colour change is formed.

Haemolytic uraemic syndrome A kidney disease characterized by blood in the urine and often results in kidney failure. It is caused by enterohaemorrhagic strains of *Escherichia coli* O157:H7 that produce a Shiga-like toxin, which attacks the kidneys.

Haemoptysis Coughing up blood. Because it is aerated in the lungs, the blood is often bright red, and if force is used to expel the sputum, it may also be frothy.

Harborage transmission The mode of transmission in which an infectious organism does not undergo morphological or physiological changes within the vector.

Hay fever Allergic rhinitis; a type of atopic allergy involving the upper respiratory tract.

Healthy carrier An individual who harbours a pathogen, but is not ill.

Hemiplegia Paralysis of one side of the body.

Haemorrhagic fever A fever usually caused by a specific virus that may lead to haemorrhage, shock, and sometimes death.

Hepatitis Any infection that results in inflammation of the liver. Also refers to liver inflammation as such.

Hyperaemic Gorged with blood. This will give a very red appearance to the tissue.

Hypertension Raised blood pressure.

Impetigo A superficial cutaneous disease, most commonly seen in children, characterized by crusty lesions, usually located on the face; the lesions typically have vesicles surrounded by a red border. It is the most frequently diagnosed skin infection caused by *S. pyogenes* (also by *S. aureus*).

Inclusion bodies Granules of organic or inorganic material lying in the cytoplasmic matrix of bacteria.

Incubatory carrier An individual who is incubating a pathogen in large numbers but is not yet ill.

Index case The first disease case in an epidemic within a given population.

Infection The invasion of a host by a microorganism with subsequent establishment and multiplication of the agent. An infection may or may not lead to overt disease.

Infectious disease Any change in the state of health in which a part or all of the host's body cannot carry on its normal functions because of the presence of an infectious agent or its products.

Infectious disease cycle (chain of infection) The chain or cycle of events that describes how an infectious organism grows, reproduces, and is disseminated.

Infectious dose 50 (ID50) Refers to the dose or number of organisms that will infect 50% of an experimental group of hosts within a specified time period.

Infectious mononucleosis An acute, self-limited infectious disease of the lymphatic system caused by the Epstein–Barr virus and characterized by fever, sore throat, lymph node and spleen swelling, and the proliferation of monocytes and abnormal lymphocytes.

Infectivity Infectiousness; the state or quality of being infectious or communicable.

Inflammation A localized protective response to tissue injury or destruction. Acute inflammation is characterized by pain, heat, swelling, and redness in the injured area.

Influenza or flu An acute viral infection of the respiratory tract, occurring in isolated cases, epidemics, and pandemics. Influenza is caused by three strains of influenza virus, labelled types A, B, and C, based on the antigens of their protein coats.

Interferon (IFN) A glycoprotein that has non-specific antiviral activity by stimulating cells to produce antiviral proteins which inhibit the synthesis of viral RNA and proteins. Interferons also regulate the growth, differentiation, and/or function of a variety of immune system cells. Their production may be stimulated by virus infections, intracellular pathogens (chlamydiae and rickettsias), protozoan parasites, endotoxins, and other agents.

Intermediate host The host that serves as a temporary but essential environment for the development of a parasite and to complete its life cycle.

Invasiveness The ability of a microorganism to enter a host, grow and reproduce within the host, and spread throughout its body.

Ionizing radiation Radiation of very short wavelength or high energy that causes atoms to lose electrons or ionize.

Ischaemia Deprivation of blood supply.

Jarisch–Herxheimer reaction The sudden worsening of a patient's condition following penicillin therapy for syphilis. It is caused by the sudden release of toxins from the dying and dead treponemes.

Kirby-Bauer method A disc diffusion test to determine the susceptibility of a micro-organism to chemotherapeutic agents.

Koch's postulates A set of rules for proving that a microorganism causes a particular disease.

Koplik's spots Lesions of the oral cavity caused by the measles (rubeola) virus that are characterized by a bluish white speck in the centre of each.

Laceration Wound with jagged edges.

Lancefield system One of the serologically distinguishable groups (as group A, group B) into which streptococci can be divided.

Laproscopy Surgical investigation of the abdominal cavity.

Legionnaires' disease A pulmonary form of legionellosis, resulting from infection with *Legionella pneumophila*.

Leishmanias Zooflagellates, members of the genus *Leishmania*, that cause the disease leishmaniasis.

Leishmaniasis The disease caused by the protozoa called leishmanias.

Lepromatous (progressive) leprosy A relentless, progressive form of leprosy in which large numbers of *Mycobacterium leprae* develop in skin cells, killing the skin cells and resulting in the loss of features. Disfiguring nodules form all over the body.

Leprosy or Hansen's disease A severe disfiguring skin disease caused by *Mycobacterium leprae*.

Lethal dose 50 (LD50) Refers to the dose or number of organisms that will kill 50% of an experimental group of hosts within a specified time period.

Leucocytosis Increased number of white blood cells in the blood.

Lumbar puncture Procedure in which a needle is introduced into the lumbar region of the spine to withdraw a sample of cerebrospinal fluid.

Lymphadenopathy Swollen glands.

Maduromycosis A subcutaneous fungal infection caused by *Madurella mycetoma*; also termed an eumycotic mycetoma.

Malaise Discomfort, feeling awful.

Malaria A serious infectious illness caused by the parasitic protozoan *Plasmodium*. Malaria is characterized by bouts of high chills and fever that occur at regular intervals.

Marburg viral haemorrhagic fever An acute infection caused by a virus that produces varying degrees of haemorrhage, shock, and sometimes death.

Measles A highly contagious skin disease that is endemic throughout the world. It is caused by a morbilli virus in the family Paramyxoviridae, which enters the body through the respiratory tract or through the conjunctiva.

Membrane attack complex (MAC) The complex complement components (C5b–C9) that create a pore in the plasma membrane of a target cell, leading to cell lysis. C9 probably forms most of the actual pore.

Membrane-disrupting exotoxin A type of exotoxin that lyses host cells by disrupting the integrity of the plasma membrane.

Meningitis A condition that refers to inflammation of the brain or spinal cord meninges (membranes). The disease can be divided into bacterial or septic meningitis (caused by bacteria) and aseptic meningitis syndrome (caused by non-bacterial sources).

Metachromatic granules Inclusion bodies in bacterial cells that alter the colour of particular stains.

Micturition Passing urine.

Miliary tuberculosis An acute form of tuberculosis in which small tubercles are formed in a number of organs of the body because of dissemination of *M. tuberculosis* throughout the body by the bloodstream. It is also known as reactivation tuberculosis.

Minimal inhibitory concentration (MIC) The lowest concentration of a drug that will prevent the growth of a particular microorganism.

Minimal lethal concentration (MLC) The lowest concentration of a drug that will kill a particular microorganism.

Morbidity rate A measure of the number of individuals who become ill as a result of a particular disease within a susceptible population during a specific time period.

Mortality rate The ratio of the number of deaths from a given disease to the total number of cases of the disease.

Mould A large group of fungi that cause mould or mouldiness and that exist as multicellular filamentous colonies; also the deposit or growth caused by such fungi. Moulds typically do not produce macroscopic fruiting bodies.

MSU Mid-stream specimen of urine.

Mucociliary escalator Mucus lines the respiratory tract, and cilia beat to remove the mucus out of the lungs. This is a continual process so any particle that penetrates into the lungs is trapped in the mucus and swept out up to the epiglottis where it can be swallowed.

Mumps An acute generalized disease that occurs primarily in young children and is caused by a paramyxovirus that is transmitted in saliva and respiratory droplets. The principal manifestation is swelling of the parotid salivary glands.

Myalgia Muscle pain.

Mycoplasma Bacteria that are members of the class Mollicutes and order Mycoplasmatales: they lack cell walls and cannot synthesize peptidoglycan precursors. Most require sterols for growth. They are the smallest organisms capable of independent reproduction.

Mycoplasmal pneumonia A type of pneumonia caused by *Mycoplasma pneumoniae*. Spread involves airborne droplets and close contact.

Mycosis Fungal infection.

Myeloma Tumour of the bone marrow.

Myringotomy Surgical incision of the tympanic membrane.

Necrosis Death of tissue.

Necrotizing enterocolitis Destruction of both small and large intestines.

Necrotizing fasciitis Infection that leads to the destruction of the musculature underlying skin.

Neurotoxin A toxin that is poisonous to or destroys nerve tissue; especially the toxins secreted by *C. tetani*, *Corynebacterium diphtheriae*, and *Shigella dysenteriae*.

Nongonococcal urethritis (NGU) Any inflammation of the urethra not caused by *Neisseria gonorrhoeae*.

Nosocomial infection An infection that develops within a hospital (or other type of clinical care facility) and is produced by an infectious organism acquired during the stay of the patient.

Nosocomial Literally, disease associated with, and often restricted to, hospital; hospital-acquired.

Oedema Collection of fluid in a tissue causing swelling which, if indented, only slowly regains its former shape.

Ophthalmia neonatorum A gonorrhoeal eye infection in a newborn, which may lead to blindness. It is also called conjunctivitis of the newborn.

Opportunistic microorganism or pathogen A microorganism that is usually free-living or a part of the host's normal microbiota, but may become pathogenic under certain circumstances, such as when the immune system is compromised.

Outbreak The sudden, unexpected occurrence of a disease in a given population.

Pandemic An increase in the occurrence of a disease within a large and geographically widespread population (often refers to a worldwide epidemic).

Panzootic The wide dissemination of a disease in an animal population.

Papular Pimple-like.

Parasitaemia The presence of parasites in the bloodstream.

Parasite An organism that lives on or within another organism (the host) and benefits from the association by harming its host. Often the parasite obtains nutrients from the host.

Parasitism A type of symbiosis in which one organism adversely affects the other (the host), but cannot live without it.

Paroxysm Fit, sudden attack.

Pathogen Any agent that causes disease. (e.g. Bacteria, virus, fungi, etc.)

Pathogenic potential The degree to which a pathogen causes morbid signs and symptoms.

Pathogenicity The condition or quality of being pathogenic, or the ability to cause disease.

Pathogenicity island A large segment of DNA in some pathogens that contains the genes responsible for virulence; often it codes for the type III secretion system that allows the pathogen to secrete virulence proteins and damage host cells. A pathogen may have more than one pathogenicity island.

PCR The polymerase chain reaction, used to amplify DNA lying between two target sequences.

Peptic ulcer disease A gastritis caused by *Helicobacter pylori*.

Peripheral neuropathy Inflammation of the peripheral nerves with pain, loss of function, altered sensation, etc.

Petechia Small spot caused by leakage of blood under the skin.

Photophobia Light intolerance.

Piedra A fungal disease of the hair in which white or black nodules of fungi form on the shafts.

Plague An acute febrile, infectious disease, caused by the bacillus *Yersinia pestis*, which has a high mortality rate; the two major types are bubonic plague and pneumonic plague.

Plasmodium A stage in the life cycle of myxomycetes (plasmodial slime moulds); a multinucleate mass of protoplasm surrounded by a membrane. Also, a parasite of the genus *Plasmodium*.

Pleuritic pain Pain in the chest or side associated with inflammation of the pleural membranes surrounding the lungs.

Polymorphs Polymorphonuclear leucocytes, phagocytic cells.

Prevalence rate Refers to the total number of individuals infected at any one time in a given population regardless of when the disease began.

Prion Putative infectious protein that is associated with spongiform encephalopathies such as scrapie, BSE and Creutzfeldt-Jakob disease.

Propagated epidemic An epidemic that is characterized by a relatively slow and prolonged rise and then a gradual decline in the number of individuals infected. It usually results from the introduction of an infected individual into a susceptible population, and the pathogen is transmitted from person to person.

Prostration Extreme exhaustion and collapse.

Psittacosis A disease due to a strain of *Chlamydia psittaci*, first seen in parrots and later found in other birds and domestic fowl (in which it is called ornithosis). It is transmissible to humans.

Puerperal fever An acute, febrile condition following childbirth, characterized by infection of the uterus and/or adjacent regions and is caused by streptococci.

Pulmonary anthrax A form of anthrax involving the lungs; also known as Woolsorter's disease.

PUO Pyrexia of unknown origin. This term is now more commonly referred to as FUO (Fever of unknown origin.)

Purulent Resembling or containing pus.

Pyogenic cocci *Staphylococcus aureus* and the haemolytic streptococci associated with the formation of pus (*puon* is the Greek for 'pus').

Pyrexia Fever, elevated body temperature.

Quellung reaction The increase in visibility or the swelling of the capsule of a microorganism in the presence of antibodies against capsular antigens.

Rabies An acute infectious disease of the central nervous system, which affects all warm-blooded animals (including humans). It is caused by an ssRNA virus belonging to the genus *Lyssavirus* of the family Rhabdoviridae.

Retroviruses A group of viruses with RNA genomes that carry the enzyme reverse transcriptase and form a DNA copy of their genome during their reproductive cycle.

Reverse transcriptase (RT) An RNA-dependent DNA polymerase that uses a viral RNA genome as a template to form a DNA copy. This is a reverse of the normal flow of genetic information, which proceeds from DNA to RNA.

Reye's syndrome An acute, potentially fatal disease of childhood that is characterized by severe oedema of the brain and increased intracranial pressure, vomiting, hypoglycaemia, and liver dysfunction. The cause is unknown but is almost always associated with a previous viral infection (e.g. influenza or Varicella-zoster virus infections).

Rheumatic fever An autoimmune disease characterized by inflammatory lesions involving the heart valves, joints, subcutaneous tissues, and central nervous system. The disease is associated with haemolytic streptococci in the body. It is called rheumatic fever because two common symptoms are fever and pain in the joints similar to that of rheumatism.

Ringworm The common name for a fungal infection of the skin, even though it is not caused by a worm and is not always ring-shaped in appearance.

Rocky Mountain spotted fever A disease caused by *Rickettsia rickettsii*.

Rubella A moderately contagious skin disease that occurs primarily in children 5 to 9 years of age that is caused by the rubella virus, which is acquired by droplet inhalation into the respiratory system. It is commonly called German measles.

Scarlet fever A disease that results from infection with a strain of *Streptococcus pyogenes* that carries a lysogenic phage with the gene for erythrogenic (rash-inducing) toxin. The toxin causes shedding of the skin. This is a communicable disease spread by respiratory droplets.

Septic shock Sepsis associated with severe hypotension despite adequate fluid resuscitation, along with the presence of perfusion abnormalities that may include, but are not limited to, lactic acidosis, oliguria, or an acute alteration in mental status. Gram-positive bacteria, fungi, and endotoxin-containing gram-negative bacteria can initiate the pathogenic cascade of sepsis leading to septic shock.

Septicaemia A disease associated with the presence of pathogens or bacterial toxins in the blood.

Serotyping A technique or serological procedure that is used to differentiate between strains (serovars or serotypes) of microorganisms that have differences in the antigenic composition of a structure or product.

Sign An objective change in a diseased body that can be directly observed (e.g. a fever or rash).

Smallpox Once a highly contagious, often fatal disease caused by a poxvirus. Its most noticeable symptom was the appearance of blisters and pustules on the skin. Vaccination has eradicated smallpox throughout the world.

Snapping division A distinctive type of binary fission resulting in an angular or a palisade arrangement of cells, which is characteristic of the genera *Arthrobacter* and *Corynebacterium*.

Sporadic disease A disease that occurs occasionally and at random intervals in a population.

Sporotrichosis A subcutaneous fungal infection caused by the dimorphic fungus *Sporothrix schenckii*.

Sputum The mucous secretion from the lungs, bronchi, and trachea that is ejected through the mouth.

Staphylococcal scalded skin syndrome (SSSS) A disease caused by staphylococci that produces an exfoliative toxin. The skin becomes red (erythema) and sheets of epidermis may separate from the underlying tissue.

Streptolysin-O (SLO) A specific haemolysin produced by *Streptococcus pyogenes* that is inactivated by oxygen (hence the "O" in its name). SLO causes beta-haemolysis of blood cells on agar plates incubated anaerobically.

Streptolysin-S (SLS) A product produced by *Streptococcus pyogenes* that is bound to the bacterial cell but may sometimes be released. SLS causes beta haemolysis on aerobically incubated blood-agar plates and can act as a leucocidin by killing leucocytes that phagocytose the bacterial cell to which it is bound.

Superantigen Bacterial proteins that stimulate the immune system much more extensively than do normal antigens. They stimulate T cells to proliferate non-specifically through simultaneous interaction with class II MHC proteins on antigen-presenting cells and variable regions on the β-chain of the T-cell receptor complex. Examples include streptococcal scarlet fever toxins, staphylococcal toxic shock syndrome toxin-1, and streptococcal M protein.

Suppuration Formation of pus.

Suture Stitch(es) used to close a wound.

Symptom A change during a disease that a person subjectively experiences (e.g. pain, bodily discomfort, fatigue, or loss of appetite). Sometimes the term symptom is used more broadly to include any observed signs.

Synergy The condition where the combined action of two antimicrobials is greater than the sum of their effects when used alone.

Systematic epidemiology The field of epidemiology that focuses on the ecological and social factors that influence the development of emerging and re-emerging infectious diseases.

Tachycardia Rapid heart beat and consequent pulse rate.

Tachypnoea Rapid breathing.

TB skin test Tuberculin hypersensitivity test for a previous or current infection with *Mycobacterium tuberculosis*.

Tetanolysin A haemolysin that aids in tissue destruction and is produced by *Clostridium tetani*.

Tetanospasmin The neurotoxic component of the tetanus toxin, which causes the muscle spasms of tetanus. Tetanospasmin production is under the control of a plasmid gene.

Tetanus An often fatal disease caused by the anaerobic, spore-forming bacillus *Clostridium tetani*, and characterized by muscle spasms and convulsions.

Tinea A name applied to many different kinds of superficial fungal infections of the skin, nails, and hair, the specific type (depending on characteristic appearance, etiologic agent, and site) usually designated by a modifying term.

Toxaemia Poisoning of the blood caused by toxins.

Toxic shock syndrome A staphylococcal disease that most commonly affects females who use certain types of tampons during menstruation. It is associated with the production of toxic shock syndrome toxin (TSST-1 toxin) by certain strains of *Staphylococcus aureus*.

Toxic shock-like syndrome (TSLS) A disease caused by an invasive group A streptococcal infection that is characterized by a rapid drop in blood pressure, failure of many organs, and a very high fever. It probably results from the release of one or more streptococcal pyrogenic exotoxins.

Toxigenicity The capacity of an organism to produce a toxin.

Toxin A microbial product or component that can injure another cell or organism at low concentrations. Often the term refers to a poisonous protein, but toxins may be lipids and other substances.

Toxoplasmosis A disease of animals and humans caused by the parasitic protozoan, *Toxoplasma gondii*.

TPPA test *Treponema pallidum* particle agglutination test. This is a test that has replaced the TPHA test—*Treponema pallidum* haemagglutination test. In the TPPA test, stained gelatin particles substitute for avian red blood cells used in the TPHA test.

Tracheostomy Making an opening through the throat into the trachea.

Trachoma A chronic infectious disease of the conjunctiva and cornea, producing pain, inflammation and sometimes blindness. It is caused by *Chlamydia trachomatis* serotypes A–C.

Transurethral resection Removal of part of an enlarged prostate gland via the urethra.

Traveller's diarrhoea A type of diarrhoea resulting from ingestion of certain viruses, bacteria, or protozoa normally absent from the traveller's environment. One of the major pathogens is enterotoxigenic *Escherichia coli*.

Trichomoniasis A sexually transmitted disease caused by the parasitic protozoan *Trichomonas vaginalis*.

Trophozoite The active, motile, feeding stage of a protozoan organism. In the malarial parasite, it is the stage of schizogony between the ring stage and the schizont.

Trypanosomiasis An infection by trypanosomes that live in the blood and lymph of the infected host.

Tubercle A small, rounded nodular lesion produced by *Mycobacterium tuberculosis*.

Tuberculoid (neural) leprosy A mild, non-progressive form of leprosy that is associated with delayed-type hypersensitivity to antigens on the surface of *Mycobacterium leprae*. It is characterized by early nerve damage and regions of the skin that have lost sensation and are surrounded by a border of nodules.

Tuberculosis An infectious disease of humans and other animals resulting from an infection by a species of *Mycobacterium* and characterized by the formation of tubercles and tissue necrosis, primarily as a result of host hypersensitivity and inflammation. Infection is usually by inhalation, and the disease commonly affects the lungs (pulmonary tuberculosis), although it may occur in any part of the body.

Tuberculous cavity An air-filled cavity that results from a tubercle lesion caused by *M. tuberculosis*.

Tularemia A plague-like disease of animals caused by the bacterium *Francisella tularensis* ssp. *tularensis* (Jellison type A), which may be transmitted to humans.

Tumour A growth of tissue resulting from abnormal new cell growth and reproduction (neoplasia).

Typhoid fever A bacterial infection transmitted by contaminated food, water, milk, or shellfish. The causative organism is *Salmonella typhi*, which is present in human faeces.

Vaccine A preparation of either killed microorganisms; living, weakened (attenuated) microorganisms; or inactivated bacterial toxins (toxoids). It is administered to induce development of the immune response and protect the individual against a pathogen or a toxin.

Vector-borne transmission The transmission of an infectious pathogen between hosts by means of a vector.

Vehicle An inanimate substance or medium involved in the transmission of a pathogen.

Venereal syphilis A contagious, sexually transmitted disease caused by the spirochaete *Treponema pallidum*.

Viraemia The presence of viruses in the bloodstream.

Virulence The degree or intensity of pathogenicity of an organism as indicated by case fatality rates and/or ability to invade host tissues and cause disease.

Virulence factor A bacterial product, usually a protein or carbohydrate, that contributes to virulence or pathogenicity.

Weil–Felix reaction A test for the diagnosis of typhus and certain other rickettsial diseases. In this test, the blood serum of a patient with suspected rickettsial disease is tested against certain strains of *Proteus vulgaris* (OX-2, OX-19, OX-K). The agglutination reactions, based on antigens common to both organisms, determine the presence and type of rickettsial infection.

White piedra A fungal infection caused by the yeast *Trichosporon beigelii* that forms light-coloured nodules on the beard and mustache.

Whitlow Infection of the nail bed, typically caused by *Staphylococcus aureus*, but also associated with herpes simplex virus.

Widal test A test involving agglutination of typhoid bacilli when they are mixed with serum containing typhoid antibodies from an individual having typhoid fever. It is used to detect the presence of *Salmonella typhi* and *S. paratyphi*.

Wort The filtrate of malted grains used as the substrate for the production of beer and ale by fermentation.

Yellow fever An acute infectious disease caused by a flavivirus, which is transmitted to humans by mosquitoes. The liver is affected and the skin turns yellow in this disease.

Zoonosis A disease of animals that can be transmitted to humans.

INDEX

E

F

G

H

I

J

K

L

Printed in Dunstable, United Kingdom

87838133R00387